D0081439

Historical Seismograms and Earthquakes of the World

Edited by

W. H. K. Lee
U. S. Geological Survey
Menlo Park, California

H. Meyers
World Data Center A
National Oceanic and Atmospheric Administration
Boulder, Colorado

K. Shimazaki
Earthquake Research Institute
University of Tokyo
Tokyo, Japan

ACADEMIC PRESS, INC.
Harcourt Brace Jovanovich, Publishers
San Diego New York Berkeley Boston
London Sydney Tokyo Toronto

Academic Press Rapid Manuscript Reproduction

ACADEMIC PRESS, INC.
1250 Sixth Avenue
San Diego, California 92101

United Kingdom Edition published by
ACADEMIC PRESS INC. (LONDON) LTD.
24-28 Oval Road, London NW1 7DX

Library of Congress Cataloging-in-Publication Data

Historical seismograms and earthquakes of the world / edited by W.H.K. Lee, H. Meyers, K. Shimazaki.
p. cm.
Papers presented at the Symposium on Historical Seismograms and Earthquakes, held Aug. 27-28, 1985 in Tokyo, Japan.
Includes index.
ISBN 0-12-440870-2 (alk. paper)
1. Seismology—Congresses. 2. Earthquakes—Congresses. I. Lee, William Hung Kan, Date. II. Meyers, H. (Herbert), Date. III. Shimazaki, K. (Kunihiko), Date. IV. Symposium on Historical Seismograms and Earthquakes (1985 : Tokyo, Japan)
QE531.H57 1987
551.2′2—dc19 87-28993
CIP

PRINTED IN THE UNITED STATES OF AMERICA
88 89 90 91 9 8 7 6 5 4 3 2 1

To

James F. Lander

for his leadership in national and international seismological programs

Contents

IV. Individual Historical Earthquakes

V. Earthquake History

VI. Seismicity and Tectonics

VII. Filming and Processing of Historical Seismograms

VIII. Historical Seismograms from Various Observatories

Preface

A symposium on Historical Seismograms and Earthquakes was held August 27–28, 1985, during the General Assembly of the International Association of Seismology and Physics of the Earth's Interior (IASPEI) in Tokyo, Japan. Sixty papers were presented, of which 51 papers are included in this Proceedings volume. The papers are grouped in eight sections according to the subject matter: I. Introduction, II. Analysis of Historical Seismograms, III. Earthquake Catalogs and Databases, IV. Individual Historical Earthquakes, V. Earthquake History, VI. Seismicity and Tectonics, VII. Filming and Processing of Historical Seismograms, and VIII. Historical Seismograms from Various Observatories. However, such grouping is rather subjective, and some papers could have been grouped differently.

Ever since civilization began, earthquakes have caused much suffering to mankind. In the past two decades, more than half a million people were killed by earthquakes. In order to mitigate earthquake hazards, it is important to learn from past earthquakes. Because seismograms comprise the basic observational data for earthquake study, a summary of available seismograms and completed analyses is essential. Because post-1963 seismograms are more uniform and accessible, we concentrate on pre-1963 "historical" seismograms. The purpose of this volume is to provide a summary of historical seismograms and earthquakes of the world. It is intended for seismologists, geologists, geophysicists, and engineers who wish to learn more about past earthquakes, how to study them, and what materials are available.

This volume is published under the auspices of the IASPEI/UNESCO Working Group on Historical Seismograms and Earthquakes. The Working Group is indebted to the following agencies for financial support: IASPEI, Pan American Institute for Geography and History (PAIGH), Tokyo Kaijo Kagami Foundation, U.S. Geological Survey (USGS), and U.S. National Geophysical Data Center of NOAA. We thank R. D. Adams (IASPEI), E. R. Engdahl (USGS), A. Giesecke (PAIGH), M. Hashizume (UNESCO), R. M. Hamilton (USGS), K. Kitazawa (UNESCO), and J. F. Lander (NOAA) for their encouragement and support.

This volume is dedicated to James F. Lander, who recently retired from the U.S. Federal Government. He was instrumental in formulating the original recommendations for preserving historical seismograms. At the onset, Lander and Lee prepared a plan to establish an international library of significant seismograms. The plan was circulated among seismologists and then presented to the IASPEI General Assembly in August, 1977. The plan was so well received by that Assembly that the Sub-Commission on Data Exchange established the first working group for copying historical seismograms. As Deputy Director of the U.S. National Geophysical Data Center and Director of World Data Center A for Solid Earth Geophysics, Mr. Lander provided much-needed support and encouragement to the IASPEI/UNESCO Working Group on Historical Seismograms and Earthquakes. It is because of these efforts and his many years of leadership in national and international seismological programs that this volume is dedicated to him on the occasion of his retirement.

All manuscripts in this volume were reviewed by two or more reviewers, according to usual scientific practice. Authors were encouraged to revise their manuscripts according to the reviewers' comments, and galley proofs were sent to the authors for proofreading. However, because of difficulties in communication, a few papers have had to be presented without authors' revisions or proofreading. In these cases, we did our best in editing and proofreading.

We thank the following colleagues for reviewing one or more of the manuscripts: K. Abe, R. D. Adams, A. Boissonnade, G. Brady, S. Billington, D. Brillinger, H. Bungum, P. W. Burton, D. Denham, G. A. Eiby, E. R. Engdahl, K. Fujita, J. R. Goodstein, R. B. Herrmann, H. Kanamori, C. Kisslinger, L. Knopoff, O. Kulhánek, J. Lander, J. J. Lienkaemper, A. Lopez Arroyo, R. Masse, S. Miyamura, J. J. Mori, E. A. Okal, M. Ohtake, W. Person, C. R. Real, W. A. Rinehart, P. Rodriquez-Tome, K. Satake, H. Sato, H. N. Srivastava, S. K. Singh, S. Stein, C. Stephens, S. S. Su, T. N. Taggart, F. Tajima, M. Takeo, T. Toppozada, A. Udias, R. A. Uhrhammer, T. Usami, T. Utsu, C. von Hake, F. T. Wu, and R. Yerkes.

In addition, we wish to thank our editorial assistants: Jerry Coffman, Janice Ellefson, and Jim Thordsen for their dedicated efforts in rendering the manuscripts to camera ready form with uniformity in style. Our thanks also to Dee Simpson of Academic Press, who provided many valuable suggestions in the course of preparation.

The entire manuscript was typeset using the PC TeX software by Personal TeX, Inc. on an IBM PC/AT personal computer with a Cordata laser printer. We are indebted to Mr. Lance Carnes, President of Personel TeX, Inc., for providing us the software and the laser printer at a greatly reduced price. Our special thanks to Jim Thordsen for typesetting and arranging the final manuscript.

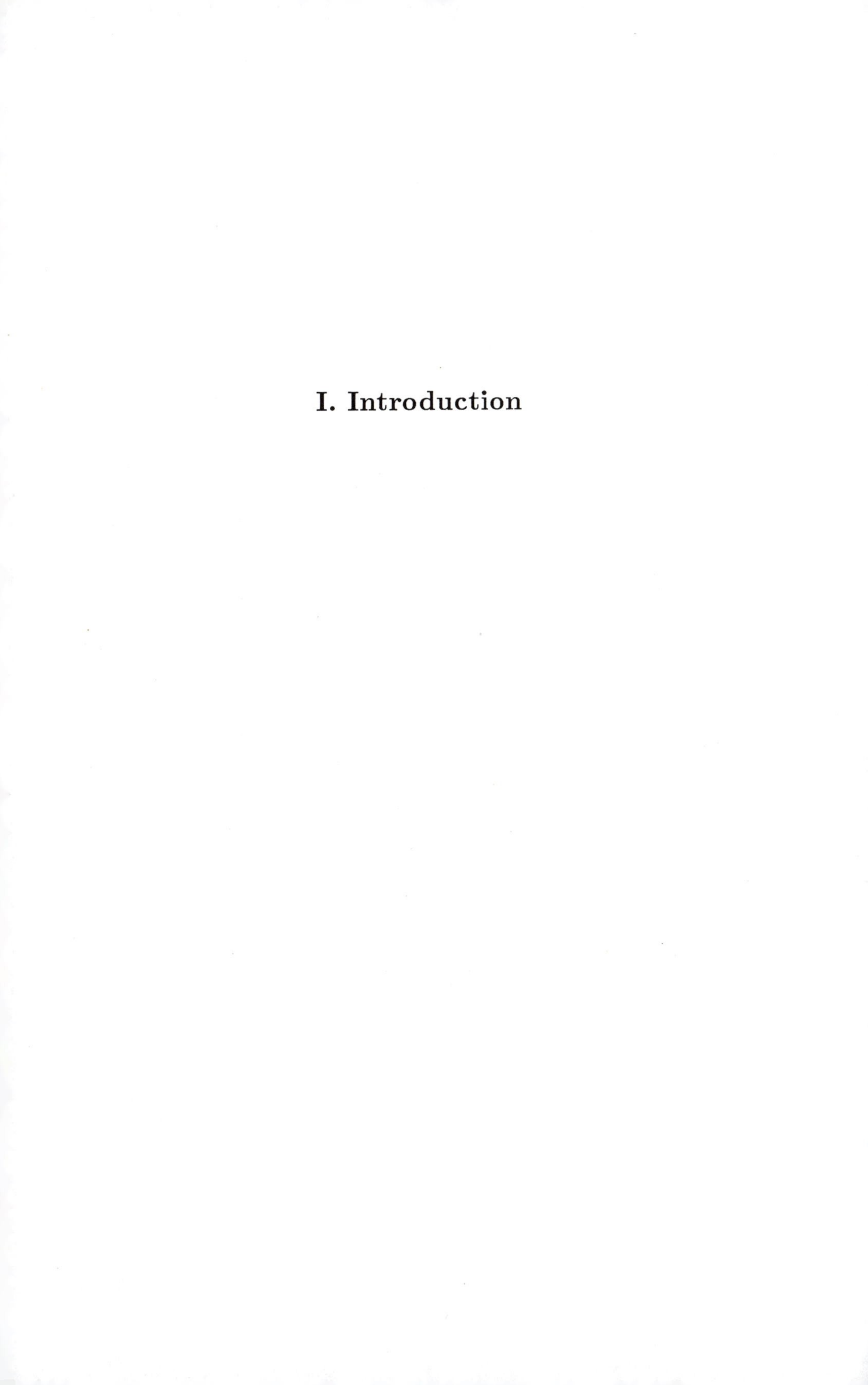

I. Introduction

Introduction to the Symposium on Historical Seismograms and Earthquakes

W. H. K. Lee
MS 977, U. S. Geological Survey
345 Middlefield Road, Menlo Park, CA 94025, USA

H. Meyers
World Data Center A, NOAA
325 Broadway, Boulder, CO 80303, USA

K. Shimazaki
Earthquake Research Institute
University of Tokyo, Tokyo 113, Japan

ABSTRACT

This paper serves as an introduction to the Symposium on Historical Seismograms and Earthquakes, which was held during the 23rd General Assembly of the International Association of Seismology and Physics of the Earth's Interior (IASPEI), August 27-28, 1985, in Tokyo, Japan. The proceedings of this Symposium consist of 36 papers presented in 6 sessions, and 15 written contributions. The topics can be roughly divided into: (1) earthquake data analysis, (2) historical earthquakes and catalogs, and (3) historical seismograms. The Symposium was organized by the IASPEI/Unesco Working Group on Historical Seismograms and Earthquakes. By historical seismograms, we mean seismograms recorded before the establishment of the World-Wide Standardized Seismograph Network (WWSSN), i.e., before 1963. In this paper, we briefly describe the origin of this Working Group and summarize its accomplishments so far.

1. Introduction

Seismograms are the basic observational records to study earthquakes and the earth's interior. They are recorded at observatories all over the world and are usually stored locally. Because of their size and fine resolution, seismograms are not easily reproducible. Consequently, seismologists spend large amounts of time and effort collecting significant sets of seismograms for their studies, and progress can be slow.

The importance of uniform instrumentation and centralized distribution of seismograms was recognized long ago. In the early 1960's, a global seismic network called WWSSN was set up by the United States with the cooperation of many countries to cover a large part of the earth (Oliver and Murphy, 1971). Because of easy access to WWSSN seismograms, great advances in seismology were accomplished, especially in developing the concept of plate tectonics, which revolutionized the earth sciences (see, e.g., Sykes, 1967; Isacks, Oliver, and Sykes, 1968).

Instrumental seismology is nearly 100 years old, while WWSSN seismograms cover only the last 23 years. Naturally, it is very desirable to consolidate the seismograms recorded before the WWSSN era. The importance of historical seismograms for geophysical and seismological research is discussed in several papers of this Symposium (see, e.g., Kanamori, 1987; Singh, 1987; Stein *et al.*, 1987).

The idea of filming seismograms for preservation and distribution has occurred to many seismologists. A few observatories actually practice it on a limited scale. However, the Historical Seismogram Filming Project is intended to create a film library of significant seismograms prior to 1963 that will rival the WWSSN seismogram collection. This Project is the major mission of the IASPEI/Unesco Working Group on Historical Seismograms and Earthquakes.

2. Working Group on Historical Seismograms and Earthquakes

An idea to film pre-WWSSN seismograms was formed during a discussion between Igor Nersesov and Willie Lee in late 1976. They speculated how nice it would be if seismograms of "old" earthquakes were easily available, and sent a joint letter to some 30 seismologists around the world for comments. They were encouraged by favorable responses and proceeded to obtain official blessing. In early 1977, Willie Lee and Jim Lander prepared a report, "A plan for establishing an international library of significant seismograms", and asked IASPEI to consider endorsing such a project. Lee and Lander's proposal was well received at the IASPEI General Assembly in August, 1977. Subsequently, IASPEI passed the following resolution:

"Noting that seismograms recorded at observatories around the world are basic for research on earthquakes and the structure of the Earth and that many of the early seismograms have been lost in war, through natural hazards and deterioration and, therefore, it is essential that seismograms of significant earthquakes be systematically collected and preserved by making photographic copies at observatory sites, and be made available through the World Data Centres. IASPEI urges that seismological observatories around the world cooperate with a copying programme by providing access to historical seismograms to be photographed on site and by preparing supporting observatory data to accompany the copies."

Following up on this resolution, the IASPEI Sub-Commission on Data Exchange established a working group for copying historical seismograms with Dr. Jorgen Hjelme as Chairman. In 1978, Willie Lee obtained funding from the U. S. Geological Survey to begin the Historical Seismogram Filming Project in collaboration with the World Data Center A. The early efforts are summarized in Meyers and Lee (1979).

In July 22-24, 1981, the United Nations Educational, Scientific and Cultural Organization (Unesco) sponsored a meeting of experts on historical seismograms during the IASPEI General Assembly in London, Ontario, Canada. A joint IASPEI/Unesco Working Group on Historical Seismograms was established with the following membership: S. Alsinawi, D. Denham, E. Kausel, R. M. Kebeasy, W. H. K. Lee (Chairman), H. Meyers, A. R. Ritsema, K. Shimazaki, O. E. Starovoit, and J. Yamamoto. This meeting generated considerable interest in many nations to participate in the Historical Seismogram Filming Project. The proceedings of this meeting were summarized in a Unesco report released in September, 1981 (see Appendix 1).

In December 20-22, 1982, the Working Group convened a regional workshop at the Earthquake Research Institute, University of Tokyo, Tokyo, Japan. The primary purposes of this Workshop were (1) to gain interest and cooperation from Asian seismological observatories to participate in the Historical Seismogram Filming Project, and (2) to evaluate the existing seismograms recorded by the Asian observatories. Six technical sessions were held with over 50 participants. The proceedings of this meeting were summarized in a report to Unesco in March, 1983 (see Appendix 2).

In August 18-19, 1983, the Working Group convened a workshop in conjunction with the IASPEI General Assembly in Hamburg, Federal Republic of Germany. The Hamburg Workshop was specially organized to discuss the status of historical seismic data for Latin America and Europe. It was divided into six sessions with a total of 29 presentations from representatives of 19 countries and 4 international organizations. The proceedings of this meeting were summarized in a report to Unesco in October, 1983 (see Appendix 3).

It is clear from the proceedings of 1981, 1982, and 1983 meetings that in addition to filming pre-WWSSN seismograms, the Working Group is actively engaged in organizing auxiliary earthquake information (such as station bulletins), and promoting research in studying instrumental and pre-instrumental earthquakes. Consequently, the name of the Working Group has been recently changed to "Historical Seismograms and Earthquakes".

3. Historical Seismogram Filming Project

The Working Group established in 1981 consisted of members who were willing to serve as regional coordinators for World Data Center A to carry out the Historical Seismogram Filming Project. The actual work of preparing and filming historical seismograms and studying earthquakes, however, has been conducted by many individuals and institutions. We are greatly indebted to them for their labors, and many of them are presenting their work in this Symposium.

The current status of the Historical Seismogram Filming Project is presented by Glover and Meyers (1987) in this Symposium and appears in more detail in Glover *et al.* (1985). In brief, over 500,000 seismograms have been filmed so far. Station bulletins from 450 stations around the world have also been microfilmed as part of the Project. At present, many countries are participating, including China, Egypt, Germany, India, Japan, Philippines, Peru, USA, and USSR.

4. Discussion

Papers in this Proceedings cover a wide range of topics. We are pleased with the world-wide interest in making historical seismograms more accessible by inventorying and filming, in studying instrumental and pre-instrumental earthquakes, and in applying modern analysis on historical seismograms to gain valuable insight into several seismological problems.

REFERENCES

Glover, D. P. and H. Meyers (1987). Historical Seismogram Filming Project: current status, *this volume*, p. 373-379.

Glover, D. P., H. Meyers, R. B. Herrmann, and M. Whittington (1985). Inventory of filmed historical seismograms and station bulletins at World Data Center A, *World Data Center A Report SE-37*, 218 pp.

Isacks, B. L., J. Oliver, and L. R. Sykes (1968). Seismology and the new global tectonics, *J. Geophys. Res.*, **73**, 5855-5899.

Kanamori, H. (1987). Importance of historical seismograms for geophysical research, *this volume*, p. 16-33.

Meyers, H., and W. H. K. Lee (1979). Historical Seismogram Filming Project: first progress report, *World Data Center A Report SE-22*, 68 pp.

Oliver, J., and L. Murphy (1971). WWNSS: seismology's global network of observing stations, *Science*, **174**, 254-261.

Singh, S. K. (1987). Analysis of historical seismograms of large Mexican earthquakes ($M_S \geq 7.0$): summary of important results, *this volume*, p. 70-84.

Stein, S., E. A. Okal, and D. A. Wiens (1987). Application of modern techniques to analysis of historical earthquakes, *this volume*, p. 85-104.

Sykes, L. R. (1967). Mechanism of earthquakes and nature of faulting on the mid-oceanic ridges, *J. Geophys. Res.*, **72**, 2131-2153.

APPENDIX 1. Summary of Final Report of the Meeting of Experts on Historical Seismograms, London, Ontario, Canada, July 22-24, 1981 by Unesco and IASPEI, September, 1981.

1. Introduction

The meeting was opened on behalf of the Director-General of Unesco by Dr. Sibrava, the Director of the Division of Earth Sciences. He introduced the draft agenda and pointed out that the meeting was jointly organized by Unesco and IASPEI. He emphasized that Unesco recognized the importance of collecting and preserving seismograms of major historical earthquakes, for research and education purpose as well as archiving old records. Dr. Adams, the Secretary-General of IASPEI, introduced the participants (see Section 9) and expressed the Association's thanks to Unesco for its support. He pointed out that such a joint venture is a new approach for the seismological community. He believed that the results of the meeting would benefit both organizations.

2. Purpose of Project

During discussions in meetings convened by Unesco on the creation of a Global Seismic Data Bank in 1978 and 1980, an ongoing U.S. project of filming historical seismograms was reported. Since 1977 World Data Center A for Solid Earth Geophysics (WDC-A) and U.S. Geological Survey (USGS) have undertaken a programme of copying historical seismograms of significant earthquakes. Dr. Lee briefly introduced the historical background of the WDC-USGS project on filming historical seismograms.

The effectiveness of microfilmed records was discussed and many members recognized their tremendous merits. The limited availability of microfilm readers particularly in developing countries was raised. Full size copies are still regarded as being the most useful for training and digitizing. Additionally, Group members from the United States commented on the economic reasons to support microfilmed copies: low cost of reproduction, rapid development of inexpensive and convenient microfilm readers, saving of storing space, ease and economy of mailing, etc.

The Group then concluded that the main task of this meeting is to provide Unesco with the necessary elements, scientific background and specialists' needs, to enable Unesco to initiate a project on historical seismogram copying on an international scale.

3. Seismogram Library

According to his experience as a lecturer at the International Institute for Seismology and Earthquake Engineering (IISEE) in Tokyo, Dr. Shimazaki mentioned that many young scientists from the developing world, trained at IISEE, are having difficulties in obtaining raw seismic data, including seismograms, when they return to their home countries. For this reason, some of them are compelled to give up their research in seismology. Dr. Lee then explained his ideas for an international library of seismograms which stores a complete set of the records and serves individual scientists by lending microfilm records or hard copies. As an alternative idea, the Group suggested establishing several regional libraries instead of a single international one. These libraries may be situated at the regional seismological centers and will encourage regional seismological studies. The Group requested Unesco to search for a funding source to realize this idea.

4. Work Already Undertaken or Planned

Mr. Meyers explained the history and present situation of the project carried out by the WDC-A, with major funding from USGS. Since 1977 WDC-A and USGS have jointly operated the Historical Seismogram Filming Project which was originated essentially to film the U.S. records. It was, however, considered desirable to expand the project onto a global scale. A series of progress reports have been published by WDC-A: (1) Report SE-22, Nov. 1979, (2) Report SE-24, Aug. 1980, and (3) Report SE-28, July 1981. He stated that the procedures described in the three reports could be used for similar programs planned in other countries. He said that most of the continuous station filming has been done on a rotary camera because of its ease and speed of operation. The early smoked-paper records, however, have to be filmed on a 16-mm planetary camera because of their fragile nature and because they require both direct and transmitted light. The most serious problem to date has been that of missing records. For some stations and intervals, as many as two-thirds of the event records have been lost or misplaced. In the past, access to these records was not rigidly controlled, and many were probably misfiled or not returned.

Related to these difficulties, Dr. Kausel stated that serious problems lay with records in most South American countries. Storage space is often limited and records are generally not kept in chronological order. Another difficulty is that most of the older records are folded for storing and tend to fall apart with frequent use. He also pointed out problems in knowing time corrections. Dr. Lee proposed that all station operators be urged to announce in advance the destruction of old records so that action might be taken to preserve them on microfilm. Dr. Shimazaki reported that the International Latitude Observatory at Mizusawa, Japan, had just initiated a three-year project to microfilm its station records from 1902. After several experiments, the Observatory decided to use 70 mm film, and to film only significant earthquakes larger than magnitude 7 which occurred adjacent to Japan.

5. Major Historical Earthquakes

The traditional definition of a major earthquake is that its magnitude be greater than or equal to 7. However, it was pointed out that this magnitude criterion precluded important earthquakes of smaller magnitude. WDC-A has resolved this problem by considering two sets of major historical events: (a) those of magnitude greater than or equal to 7 and (b) events located in and around the United States of smaller magnitude, but important for other criteria (see WDC-A Report SE-22). This policy could be adopted for selection of events whose records are to be copied in other countries. A useful guide in identifying significant earthquakes has been published by WDC-A: "Catalogue of Significant Earthquakes 2000 BC-1979: Including Quantitative Casualties and Damage" (Report SE-27, July 1981).

6. Available Seismological Records

The President of the European Seismological Commission (ESC), Dr. Ritsema, reported recent activities related to this project. He mentioned that, upon request by WDC-A, ESC has identified seismological stations where historical seismograms are stored in Europe. For this purpose the Commission had sent questionnaires to selected members for their suggestions of five to ten stations in western Europe that would be willing to participate in the U.S.-organized project. Some stations posed financial conditions in participation, but if these difficulties could be solved they would hope to join the project actively. He had received positive replies from several observatories to participate. They are stations at Toledo (records available since 1909), Stuttgart (1910), Göttingen (1903), Uppsala (1904), Vienna (1906), Uccle (1906), De Bilt (1908). In addition, Dr. Lopez Arroyo stated that the following stations in Spain may offer access to records for WDC-USGS project. They possess records from the early 1900's, for instance, San Fernando station has data from 1899, Cartuja from 1903, Toledo from 1909, etc. He warned that these old seismograms are fragile, not stored in good condition, and urged copying quickly before destruction.

Dr. Nikolaev expressed the possible participation from observatories in the eastern European countries. They are: Alma Ata which has records from 1922, Baku from 1940, Vladivostok from 1929, Irukutsk from 1901, Tbilisi from 1940, etc. In addition, several stations from the regions such as German Democratic Republic and Bulgaria, are now in a position to join in the activities. He stressed that copying work may be carried out by local scientists if appropriate sets of copying equipment are provided. It is preferable to work with mobile equipment at individual stations instead of gathering records at a central point. Prof. Miyamura commented that some stations stored old records from other stations in the past, such as Göttingen and Leningrad. This fact should be taken into account in selecting stations whose records may be copied. Dr. Denham explained the situation in Australia. Some stations, such as Riverview, started observing from early in this century, and he expects its records would be in good condition. It was, however, very unfortunate that floods in 1970 destroyed many of the old records from the Brisbane station. The main problem in his country would probably not be the lack of funds, but to obtain technical staff to carry out the copying. However, with a reasonable lead time, the available records could be annotated properly for copying. It was also stressed that it is important to copy not only seismograms but also station bulletins as well as station logs and operational notes to help in providing additional information on timing and instrumental constants.

7. Selection of Appropriate Records

Dr. Lee estimated that there were about 350 seismographic stations in operation around the world prior to the establishment of the WWSSN in 1963. Since each station on average produced about 1,000 seismograms per year, there are approximately 25 million seismograms from the 1880's to 1962. Thus the cost of filming every existing record prior to 1963 is clearly beyond the anticipated financial support for this project; consequently, seismograms to be recorded must be carefully selected. Practically only a few percent of the existing seismograms could be filmed. The cost of filming a seismogram depends on many factors. The unit cost for filming a chronological file of seismograms of a given station is a factor of two or more cheaper than for filming seismograms of a given event from many individual stations.

From a scientific viewpoint, larger earthquakes are much preferable than smaller ones, and high-quality stations than poorer ones. For seismic hazard evaluation, it is important to have a complete record of earthquakes down to certain thresholds of magnitude. One approach is to film two types of seismograms: (1) several critical stations will be chosen for chronological filming of all seismograms, and (2) about 2,000 events will be selected on the basis of magnitude ≥ 7 or of scientific interest to be filmed at about 30 selected stations around the world. Because of varying time systems used in labeling the earlier records, it is also important to film the day before and the day after a given selected event. This will ensure that the selected event will be filmed and afford an opportunity to include some foreshocks and aftershocks.

The group requested Unesco to consider the possibility of organizing a series of regional meetings to study and select the stations in each region whose seismograms were most appropriate for filming and adding to the Historical Seismogram Film Library.

8. Recommendations for Future Activities by Unesco and Other Organizations

(1) Noting that steady progress has been made by the World Data Centre (WDC) - U.S. Geological Survey (USGS) project in filming historical seismograms of selected U.S. stations, the present Group, aware of the continuing deterioration of existing seismograms, urges that continuing support be given to the WDC/USGS project to complete a significant library of historical seismograms,

(2) Noting that several nations have been planning and undertaking filming of their historical seismograms, the present Group urges that WDC's coordinate and provide guidance to assure that the resulting film libraries of historical seismograms contain the relevant information in a useful form,

(3) Noting that the cost of filming historical seismograms may be beyond the financial capabilities of many nations, the present Group urges that Unesco and other UN agencies seek ways to provide the necessary support (such as cameras, films, local logistics, etc.) to film historical seismograms,

(4) Noting that an IASPEI Working Group on Copying Historical Seismograms already exists under the Commission of Practice, the present Group urges that Unesco and other interested organizations receive technical guidance and necessary information for preparing project proposals from IASPEI and other appropriate groups,

(5) Noting that difficulties may arise in selecting the appropriate seismograms and in gaining logistic support in many developing nations, the Group recommends Unesco convene regional workshops (e.g. Asia, South America, etc.) to gain national interest and cooperation. The Group further urges that support be provided to evaluate the existing seismograms for selection and filming purpose,

(6) Noting the importance of educating seismologists from developing nations, the present Group urges Unesco and other UN agencies to provide support for convening training courses, and for updating a technical manual for studying seismograms, and for preparing a handbook summarizing the content of the Historical Seismogram Library. The Group further recommends depositing a set of historical seismograms at World Data Centers and at regional centers, and supporting scientists to use these records (e.g. by purchase of film viewers, copying selected seismograms, travel assistance, etc.),

(7) Noting that the proposed project to film historical seismograms will only preserve less than 5% of the existing records, the Group urges national organizations to make efforts to preserve their collections and to inform IASPEI if any sets of seismograms, station bulletins, logs and other information are in danger of destruction,

(8) Recommends that a short guide be prepared by experts to help in selecting and preparing seismograms for filming,

(9) Noting the importance of historical seismograms in many fields of seismological research, the Group recommends that any national or international project that needs historical seismograms should include filming seismograms in its project.

9. List of Participants in the Meeting of Experts on Historical Seismograms

Experts: A. A. Hughes (UK), E. Kausel (Chile), J. F. Lander (USA), W. H. K. Lee (USA), A. Lopez Arroyo (Spain), H. Meyers (USA), V. Nikolaev (USSR), K. Shimazaki (Japan), J. M. Van Gils (Belgium), and J. Yamamoto (Mexico).

Officials: R. D. Adams (IASPEI), V. Sibrava (Unesco), and K. Kitazawa (Unesco).

Observers: D. Denham (Australia), E. Hurtig (German Democratic Republic), R. M. Kebeasy (Egypt), N. V. Kondorskaya (USSR), O. Kulhánek (Sweden), M. Maamoun (Egypt), S. Miyamura (Japan), A. R. Ritsema (Netherlands), and A. A. Solovjev (USSR).

APPENDIX 2. Summary of Report on the Regional Workshop on Historical Seismograms held at the University of Tokyo, Tokyo, Japan, December 20-22, 1982 by IASPEI/Unesco Working Group on Historical Seismograms, March, 1983.

1. Introduction

As recommended by the Unesco Committee of Experts on historical seismograms during the IASPEI General Assembly in London, Ontario, Canada, 1981, a Regional Workshop on Historical Seismograms was held from December 20-22, 1982, at the Earthquake Research Institute, University of Tokyo, Tokyo, Japan. This Regional Workshop emphasized seismograms from Asia, Australia, and Oceania area.

The Workshop participants were welcomed by Prof. T. Rikitake, Chairman of the Japanese National Committee for Geodesy and Geophysics, and by Prof. D. Shimozuru, Director of the Earthquake Research Institute, University of Tokyo. Prof. Z. Suzuki, Vice-President of IASPEI, welcomed the participants on behalf of IASPEI. Dr. M. Hashizume explained the reasons for the Workshop on behalf of Unesco.

2. Technical Sessions

Six technical sessions were held during the Workshop. In the first session, Dr. W. H. K. Lee described the Historical Seismogram Filming Project and progress towards an international earthquake data bank. He noted that since the project started in 1977, about 400,000 seismograms had been filmed, and the goal of a library of about 2 million seismograms might be reached in a decade if enough countries participate. The keynote lecture of the Workshop was given by Prof. H. Kanamori. He used many examples to illustrate the importance of historical seismograms for geophysical research. Finally, Dr. J. R. Goodstein described the filming of seismograms and related materials currently in progress at the California Institute of Technology, Pasadena, CA. She noted that considerable efforts had been spent in sorting and annotating records.

In the second session, Prof. T. Utsu reported a catalog of large ($M \geq 6$) and damaging earthquakes in Japan for the years 1885-1925. In preparing this catalog, Prof. Utsu used many old records for revision and corrections. It was then followed by a visit to the Earthquake Prediction Data Center and the Seismogram Archive, Earthquake Research Institute, University of Tokyo. At the Earthquake Prediction Data Center, Drs. M. Mizoue and K. Tsumura explained their processing and analyses of earthquake data for earthquake prediction purposes. At the Seismogram Archive, Dr. K. Shimazaki displayed many examples of old seismograms dating back to the beginning of instrumental seismology in the late 19th century.

The third session was devoted to the use of historical seismograms and other historical records in studying the earthquake history of various regions. Prof. T. Usami reported on his study of historical earthquakes in Japan. Dr. J. Taggart described the U.S. National Earthquake Catalog Project. It was then followed by Dr. K. Abe on the determination of magnitudes towards uniform earthquake catalogs, and Dr. R. J. Geller on source studies of pre-WWSSN earthquakes. Finally, Mr. R. A. White reported on his studies of historical earthquakes in Central America, and Prof. Y. S. Xie on his studies of Chinese historical earthquakes.

The fourth session was a visit to the Japan Meteorological Agency. Dr. N. Yamakawa gave an introductory lecture on the various seismological activities of the Japan Meteorological Agency (JMA). It was then followed by an extensive tour of the JMA facilities for processing and analysis of seismic data. The tour was conducted by Dr. M. Yamamoto, and the Workshop participants were welcomed by Dr. S. Suyehiro.

The fifth session began with Dr. R. D. Adams describing the activities of the International Seismological Centre, and Mr. H. Meyers reporting the role of the World Data Center in the Historical Seismogram Filming Project. It was then followed by eight papers on the status of seismograms in various regions: Dr. Y. Umeda (Abuyama Observatory, Japan), Rev. S. S. Su (the Philippines), Mr. H. K. Lam (Hong Kong), Dr. D. Denham (Australia), Drs. K. Hosoyama and M. Ooe (Mizusawa Observatory, Japan), Dr. Y. B. Tsai (Taiwan), Mr. R. Soetardjo (Indonesia), and Dr. H. N. Srivastava (India). In addition, Dr. R. Inoue described the digitization and processing of the JMA strong motion records from several great earthquakes.

The sixth session began with three additional papers on the status of seismograms: China (Mr. K. Qu), the Philippines (Mr. R. Valenzuela), and the Oceanic area (abstract by Mr. G. A. Eiby and comments by Dr. R. D. Adams). It was then followed by a general discussion on a plan to film seismograms in Asia, Australia, and the Oceania area. The participants then passed eight resolutions given below.

3. Resolutions by the Participants

(1) Realizing the great value to practical seismological research (particularly to assessment of earthquake hazards) of those historical seismograms still existing throughout the world and the danger of their further deterioration or loss, urges all appropriate agencies, both national and international, to undertake a program to microfilm these seismograms and associated records and bulletins and to adequately preserve the originals whenever practical.

(2) Noting with satisfaction that the United States of America has already made considerable progress in filming its historical seismograms and associated materials and in providing copies to the World Data Center A, urges that this national program be completed as soon as possible.

(3) Noting that many developing countries do not have the technical equipment or experience to carry out the microfilming of their seismological records and aware of the experience of the World Data Center A in this field, urges Unesco or other appropriate agencies to provide the necessary support to World Data Centers to assist countries in participating in the Historical Seismogram Filming Project.

(4) Noting that the major part of any program for microfilming seismological records is the adequate sorting, preparation, and annotation of materials to be copied, recommends that funds be made available to national agencies to enable them to undertake this preparatory work, as well as to complete the filming.

(5) Realizing the importance of encouraging national and regional seismological research in countries participating in the microfilming program, recommends that at least two copies are made when seismological records are filmed, one of which shall remain at a suitable agency in the country of origin, and one to be deposited at a World Data Center.

(6) Desiring to achieve the best available geographic distribution of stations selected for the microfilming of seismograms for the entire instrumental period up to the 1960's, recommends as a minimum the complete microfilming of seismograms and associated records at the following key stations, subject to the approval and cooperation of local authorities: (a) A selection to be made from Abuyama, Mizusawa, and stations in Tokyo operated by the University of Tokyo, (b) One station to be selected from China mainland (In a letter to Dr. K. Shimazaki on Feb. 1, 1983, Mr. K. Qu suggested that seismograms from Sheshan (Zikawei) (1904-1962), and Lanzhou (1954-1962) be filmed), (c) Taipei, (d) Baguio, (e) Jakarta (Batavia), (f) Riverview, (g) Perth, (h) One station from the Indian peninsula, (i) Wellington, and (j) Apia. In addition, all the stations in the region are encouraged to film seismograms of selected earthquake events.

(7) Realizing that it is essential to make full use of those seismograms that are copied, recommends that researchers publish papers fully describing their techniques of analysis.

(8) Aware of the success of the present Regional Workshop on Historical Seismograms, recommends that additional Regional Workshops be held and, in particular, that a Workshop on seismograms from the Latin American and European regions be held in conjunction with the General Assembly of the International Union of Geodesy and Geophysics in Hamburg, Germany, August 1983.

APPENDIX 3. Summary of Report on the Workshop on Historical Seismograms Held at the International Union of Geodesy and Geophysics XVIII General Assembly, Hamburg, Federal Republic of Germany, August 18-19, 1983 by IASPEI/Unesco Working Group on Historical Seismograms, October 1983.

1. Introduction

The Hamburg Workshop was specifically organized to discuss the status of historical seismic data for Latin America and Europe. Since it is unlikely that an additional workshop will be held on this subject, reports for other regions were included as well. The workshop was divided into six sessions as follows: Introduction and Welcome, Europe, Latin America, Other Regions, General, and Recommendations. There was a total of 29 presentations from representatives of 19 countries and 4 international organizations.

2. Technical Sessions

In the first session, H. Meyers, serving as Acting Chairman in place of ailing W. H. K. Lee, described the purpose of the workshop and previous activities of the IASPEI/Unesco Working Group on Historical Seismograms. E. R. Engdahl, Vice Chairman of the IASPEI Commission on Practice, welcomed the participants on behalf of IASPEI. He noted that thus far more than 500,000 seismograms have been filmed as part of the Historical Seismogram Filming Project, and emphasized the importance of the activities that were to be covered during the workshop. M. Hashizume, representing Unesco, described the importance of historical seismic data, and the Unesco interests in having these data available for the analysis of seismic risks, particularly in areas where the recurrence rate of significant earthquakes is very low and for regions where much data do not exist. He mentioned that both these conditions occur frequently in developing nations.

The second, third, and fourth sessions provided an opportunity for the participants to describe the status of historical seismograms in their country. Generally, all countries indicated that there is activity relating to the preservation of historical seismograms. The level of activity varies from country to country. The importance in preserving original seismograms, whether filmed or not, was stressed by several participants. The first task which appears to have been completed by most countries is the identification of the older seismic stations and the instrumentation. In some countries, work has not yet been completed in identifying whether the records still exist or where they are located. For many countries, particularly in Europe, the records have generally been located and some of them are well organized. In some cases filming has been done, or at least tests in various filming procedures have been started. Almost all of the reports indicated two serious conditions which prevail: (1) Many of the oldest records, particularly those which are smoked paper, are deteriorating and are in fragile condition, and (2) Many records for significant events are missing. In past years these records were lent to seismologists in other countries and probably never returned. An appeal was made for the return of seismograms to the station of origin.

During the fifth session, the speakers reflected on some of the experiences gained thus far in the Historical Seismogram Filming Project. The status of a related project, including microfilm collection of historical station bulletins, was described. Most of the participants of the Sessions prepared abstracts in advance which provided details on the status of historical seismogram projects. The possibility of publishing full papers or extended abstracts for the Tokyo and Hamburg meeting is being considered.

The last session was devoted to recommendations. The workshop participants made ten recommendations which were eventually written into four substantial resolutions, incorporating all of the aspects of the recommendations. The critical aspects of these four resolutions were reconstructed into one major resolution which was submitted to IASPEI and eventually passed as an official IASPEI resolution.

3. Resolutions Adopted by the Working Group

(1) The Working Group notes and approves the progress in copying and preserving historical seismograms and other seismological documents of continuing importance to research, re-affirms the resolutions of its meeting in Tokyo in 1982, Dec. 20-22, and urges IASPEI and Unesco to continue their valued support, and to consider the following additional resolutions:

(2) Realizing that continued delay in collecting and copying historical records will result in further deterioration and loss of irreplaceable data, The Working Group recommends: (a) That World Data Centers and Regional Commissions compile lists of earthquakes believed to have been large, destructive, or to have occurred in some unusual location, or to possess some other special rarity; (b) That Centro Regional de Sismologia para America del Sur and Pan American Institute of Geography and History ascertain what records from stations in South and Central America still exist, where they are stored, and what their physical condition is; and recommend at least two stations in the region for immediate filming

of all records; (c) That the USSR be asked to extend its existing programme of copying to include copying of all records from at least one selected station; (d) That the European Seismological Commission select at least two stations in Western and Central Europe whose records should be completely copied, and report their selection to their General Assembly in 1984; (e) That developing countries with long-established stations such as those at Helwan (Egypt) and Tacubaya (Mexico) be given special encouragement and assistance to copy their records.

(3) Realizing that for some studies even the best copies are inferior to originals, and that documents and archival material other than instrumental records contains data of importance for seismological research, urges: (a) that IASPEI and Unesco broaden the scope of the Working Group to consider other historical material, such as catalogues, station bulletins, notebooks containing unpublished materials, instrumental constants, time corrections and the like, and documents describing pre-instrumental earthquakes, and to assess the desirability and methods available for preserving them, (b) that World Data Centers A and B should publish a comprehensive list of seismograms that may still be consulted, and should complete the microfilming of its global collection of station bulletins, (c) that all observatories retain their original records, and store them in the manner best calculated to preserve them and extend their useful life, and (d) that participating nations make their holding of historical material available to World Data Centers for copying.

(4) Concerned at the number of records of important earthquakes missing from surviving files of historical seismograms, urges that observatories and individual seismologists holding records borrowed from other stations, or copies of missing originals, endeavor to return them to the most appropriate organization in their country of origin without delay.

Importance of Historical Seismograms for Geophysical Research

Hiroo Kanamori
Seismological Laboratory, California Institute of Technology
Pasadena, California 91125

ABSTRACT

Among the most important data in geophysics are the seismograms which have produced most of the basic and quantitative information concerning the seismic source and the Earth's interior. Modern seismographic instruments provide high quality seismograms with a wide dynamic range and frequency band. The analysis of these seismograms, together with the recent developments in theory and methodology, has resulted in an order of magnitude increase, both in quantity and quality, of our knowledge of the Earth's interior and physics of earthquakes. However, the earthquake cycle is a long-term process so that study over a long period of time is essential for a thorough understanding of the earthquake phenomenon. Furthermore, since earthquakes may not repeat in exactly the same way, detailed analyses of earthquakes in the past are important.

For more than a century, a large number of seismograms have been recorded at many stations in the world, but many of them have not been fully utilized mainly because theories and methods had not been developed well enough to fully interpret the seismograms at the time when the earthquakes occurred. In many cases, methods developed in later years enabled seismologists to investigate existing records of earlier events, searching for further information. In this paper, we illustrate the importance of historical seismograms for various geophysical studies.

1. Global Seismicity

During the past 80 years, about 10 earthquakes larger than surface-wave magnitude 7 ($M_S \geq 7$) occurred every year in the world. Table 1 lists the earthquakes with $M_S \geq 8$ (from Abe, 1981). Recently Abe and Noguchi (1983) revised M_S for the events for the period 1897 to 1912. The revised values are given in the parentheses in Table 1. The magnitude-frequency relation is shown in Table 2 and Figure 1. In terms of the energy released in seismic waves, this level of activity corresponds to about 4.5×10^{24} ergs/year. The spatial distribution of these great earthquakes is shown in Figure 2. The temporal variation of earthquake energy release (Figure 3) shows a pronounced peak during the period from 1952 to 1965. The data for the 19th century shown in Figure 3 are obtained from tsunami data (Abe, 1979) and are somewhat incomplete; however, it is probably true that during the period from 1835 to 1900 there is no peak comparable to that for the period 1952 to 1965. It is clear that the global seismic activity is very non-uniform in time, at least on a time scale of 100 years or so. This is one of the reasons why seismicity data over a long period of time are essential.

Table 1a. Large Earthquakes with $M_S \geq 8.0$ for the Period 1904 to 1985

Date	Time	Region	Lat.	Long.	M_S	M_W
1904 06 25	21 00.5	Kamchatka	52	159	8.0 (7.4)	
1905 04 04	00 50.0	E. Kashmir	33	76	8.1 (7.5)	
1905 07 09	09 40.4	Mongolia	49	99	8.4 (7.6)	8.4
1905 07 23	02 46.2	Mongolia	49	98	8.4 (7.7)	8.4
1906 01 31	15 36.0	Ecuador	1	–81.5	8.7 (8.2)	8.8
1906 04 18	13 12.0	California	38	–123	8.3 (7.8)	7.9
1906 08 17	00 10.7	Aleutian Is.	51	179	8.2 (7.8)	
1906 08 17	00 40.0	Chile	–33	–72	8.4 (8.1)	8.2
1906 09 14	16 04.3	New Britain	–7	149	8.1 (7.5)	
1907 04 15	06 08.1	Mexico	17	–100	8.0 (7.7)	
1911 01 03	23 25.8	Turkestan	43.5	77.5	8.4 (7.8)	
1912 05 23	02 24.1	Burma	21	97	8.0 (7.7)	
1914 05 26	14 22.7	W. New Guinea	–2	137	8.0	
1915 05 01	05 00.0	Kurile Is.	47	155	8.0	
1917 06 26	05 49.7	Samoa Is.	–15.5	–173	8.4	
1918 08 15	12 18.2	Mindanao Is.	5.5	123	8.0	
1918 09 07	17 16.2	Kurile Is.	45.5	151.5	8.2	
1919 04 30	07 17.1	Tonga Is.	–19	–172.5	8.2	
1920 06 05	04 21.5	Taiwan	23.5	122	8.0	
1920 12 16	12 05.8	Kansu, China	36	105	8.6	
1922 11 11	04 32.6	Chile	–28.5	–70	8.3	8
1923 02 03	16 01 41	Kamchatka	54	161	8.3	8.5
1923 09 01	02 58 36	Kanto	35.25	139.5	8.2	7.9
1924 04 14	16 20 23	Mindanao	6.5	126.5	8.3	
1928 12 01	04 06 10	Chile	–35	–72	8.0	
1932 05 14	13 11 00	Molucca Passage	0.5	126	8.0	
1932 06 03	10 36 50	Mexico	19.5	–104.25	8.2	
1933 03 02	17 30 54	Sanriku	39.25	144.5	8.5	8.4
1934 01 15	08 43 18	Nepal/India	26.5	86.5	8.3	
1934 07 18	19 40 15	Santa Cruz Is.	–11.75	166.5	8.1	
1938 02 01	19 04 18	Banda Sea	–5.25	130.5	8.2	8.5
1938 11 10	20 18 43	Alaska	55.5	–158.0	8.3	8.2
1939 04 30	02 55 30	Solomon Is.	–10.5	158.5	8.0	
1941 11 25	18 03 55	N. Atlantic	37.5	–18.5	8.2	
1942 08 24	22 50 27	Peru	–15.0	–76.0	8.2	
1944 12 07	04 35 42	Tonanki	33.75	136.0	8.0	8.1
1945 11 27	21 56 50	W. Pakistan	24.5	63.0	8.0	
1946 08 04	17 51 05	Dominican Rep.	19.25	–69.0	8.0	
1946 12 20	19 19 05	Nankaido	32.5	134.5	8.2	8.1
1949 08 22	04 01 11	Queen Char. Is.	53.75	–133.25	8.1	8.1
1950 08 15	14 09 30	Assam	28.5	96.5	8.6	8.6
1951 11 18	09 35 47	Tibet	30.5	91.0	8.0	7.5
1952 03 04	01 22 43	Tokachi-Oki	42.5	143.0	8.3	8.1
1952 11 04	16 58 26	Kamchatka	52.75	159.5	8.2	9.0
1957 03 09	14 22 28	Aleutian Is.	51.3	–175.8	8.1	9.1
1957 12 04	03 37 48	Mongolia	45.2	99.2	8.0	8.1
1958 11 06	22 58 06	Kurile Is.	44.4	148.6	8.1	8.3
1960 05 22	19 11 14	Chile	–38.2	–72.6	8.5	9.5
1963 10 13	05 17 51	Kurile Is.	44.9	149.6	8.1	8.5
1964 03 28	03 36 14	Alaska	61.1	–147.5	8.4	9.2
1965 02 04	05 01 22	Aleutian Is.	51.3	178.6	8.2	8.7
1968 05 16	00 48 57	Tokachi-Oki	40.9	143.4	8.1	8.2
1977 08 19	06 08 55	Sumbawa	–11.2	118.4	8.1	8.3
1985 09 19	13 17 38	Mexico	18.2	–102.6	8.1	8.0

Table 1b. Some Large Earthquakes With M_W Around 8

Date	Time	Region	Lat.	Long.	M_S	M_W
1958 07 10	06 15 56	Alaska	58.3	–136.5	7.9	7.7
1966 10 17	21 41 57	Peru	–10.7	–78.6	7.8	8.1
1969 08 11	21 27 36	Kurile Is.	43.4	147.8	7.8	8.2
1970 05 31	20 23 28	Peru	–9.2	–78.8	7.6	7.9
1974 10 03	14 21 29	Peru	–12.2	–77.6	7.6	8.1
1975 05 26	09 11 52	Azores	36.0	–17.6	7.8	7.7
1976 08 16	16 11 05	Mindanao Is.	6.2	124.1	7.8	8.1
1978 11 29	19 52 49	Mexico	16.1	–96.6	7.6	7.6
1979 12 12	07 59 03	Colombia	1.6	–79.4	7.6	8.2
1980 07 17	19 42 23	Santa Cruz Is.	–12.5	165.9	7.7	7.9
1985 03 03	22 47 07	Chile	–33.1	–71.9	7.8	8.0

Table 2. Frequency-Magnitude Relation, 1904-1980

Interval				Cumulative		
$\leq M_S <$		n1	n2	$M_S \geq$	N1	N2
7.0	7.2	252	252	7.0	720	720
7.2	7.4	162	162	7.2	468	468
7.4	7.6	104	104	7.4	306	306
7.6	7.8	97	97	7.6	202	202
7.8	8.0	52	52	7.8	105	105
8.0	8.2	25	22	8.0	53	53
8.2	8.4	17	13	8.2	28	31
8.4	8.6	8	10	8.4	11	18
8.6	8.8	3	3	8.6	3	8
8.8	9.0		1	8.8		5
9.0	9.2		2	9.0		4
9.2	9.4		1	9.2		2
9.4	9.6		1	9.4		1

Note:

(1) n1 is the number of events within the magnitude range of .2 for the period 1904 to 1980.
(2) n2 is the same as n1 in (1) with M_S replaced by M_W for the ten largest events with $M_W \geq 8.5$ listed below.
(3) N1 is the total number of events larger than the given M_S for the period 1904 to 1980.
(4) N2 is the same as N1 in (3) with M_S replaced by M_W for the ten largest events with $M_W \geq 8.5$ listed below: 1960 Chile ($M_S = 8.5; M_W = 9.5$), 1964 Alaska (8.4; 9.2), 1957 Aleutian Is. (8.1; 9.1), 1952 Kamchatka (8.2; 9.0), 1906 Colombia (8.6; 8.8), 1965 Aleutian Is. (8.2; 8.7), 1950 Assam (8.6; 8.6), 1938 Banda Sea (8.2; 8.5), 1963 Kurile Is. (8.1;8.5), 1922 Chile (8.3; 8.5).

For most events which occurred after 1960, a relatively complete set of seismograms is available for detailed studies, but for the events prior to 1960, the data are usually incomplete. In many cases, only a few records at most were used for the determination of the source parameters. There are still many unresolved problems. The solution of these problems requires more complete sets of seismograms.

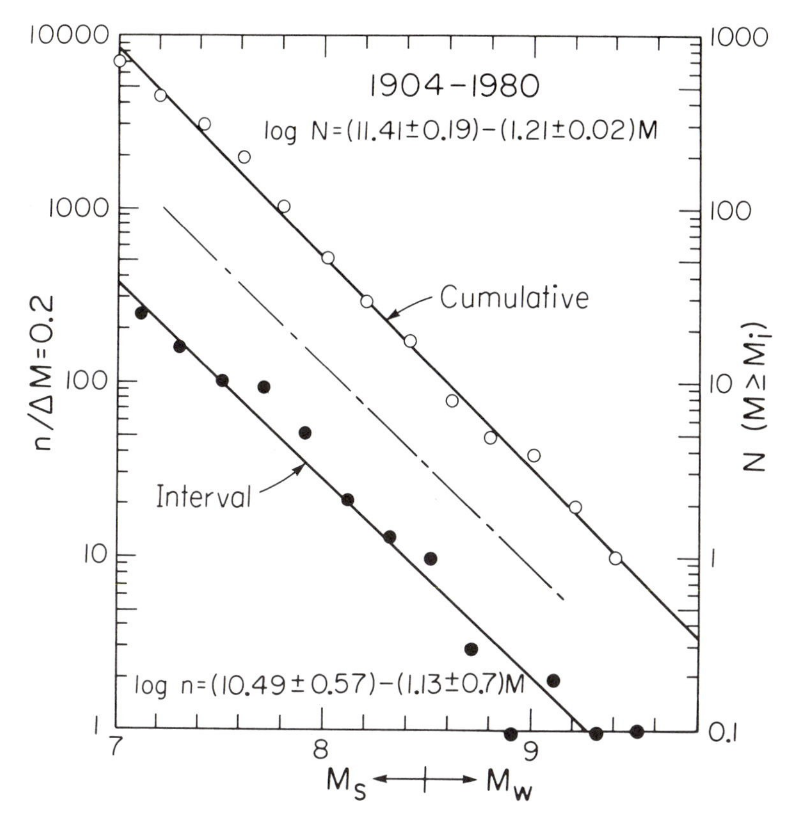

Figure 1. The magnitude–frequency relation for the earthquakes in the world during the period from 1904 to 1980. The surface-wave magnitude M_S listed in Abe (1981) is used for all the events except those with $M_W \geq 8.5$ for which M_W is used. The ordinate on the left is for the interval frequency (n2 in Table 2, the number of events per 0.2 magnitude unit for the period 1904 to 1980), and that on the right is for the cumulative frequency (N2 in Table 2). The straight lines are the least–squares fit for the entire range.

2. Normal Fault vs. Thrust Events at Subduction Zones.

Most large earthquakes at subduction zones have thrust mechanism. However, there are some very large normal-fault events, notable examples being the 1933 Sanriku earthquake ($M_W = 8.4$, Kanamori, 1971a) and the 1977 Sumbawa earthquake ($M_W = 8.3$, Given and Kanamori, 1980; Silver and Jordan, 1982). The thrust events represent slip between the plate boundaries, while the normal-fault events represent failure within the oceanic plate. For the purpose of evaluating earthquake potential of seismic gaps (e.g. McCann, *et al.*, 1979), distinction between thrust events and normal-fault events is essential. For some old events, (e.g. 1917 and 1919 Tonga Islands earthquakes, see Figure 2), the mechanism is still uncertain. Since the polarity of P-wave first motions is often ambiguous on old seismograms, the direct determination of the mechanism is not always possible. However, normal-fault

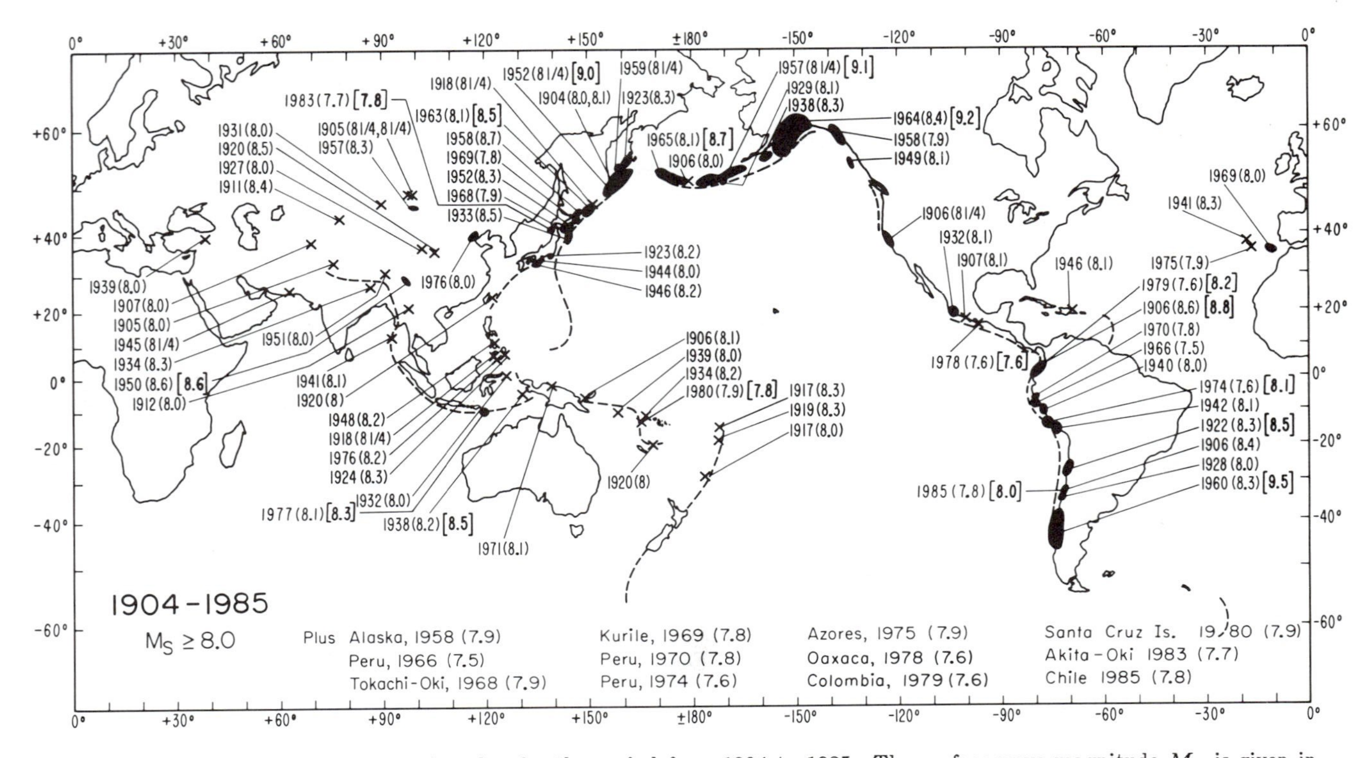

Figure 2. Great and large earthquakes for the period from 1904 to 1985. The surface-wave magnitude M_S is given in the parentheses and M_W is given in the brackets for some earthquakes, including the ten largest earthquakes. Major rupture zones are indicated by dark zones. This figure is modified from that in Kanamori (1978), and the magnitude values differ slightly from those listed in Table 1. (Kanamori, 1982)

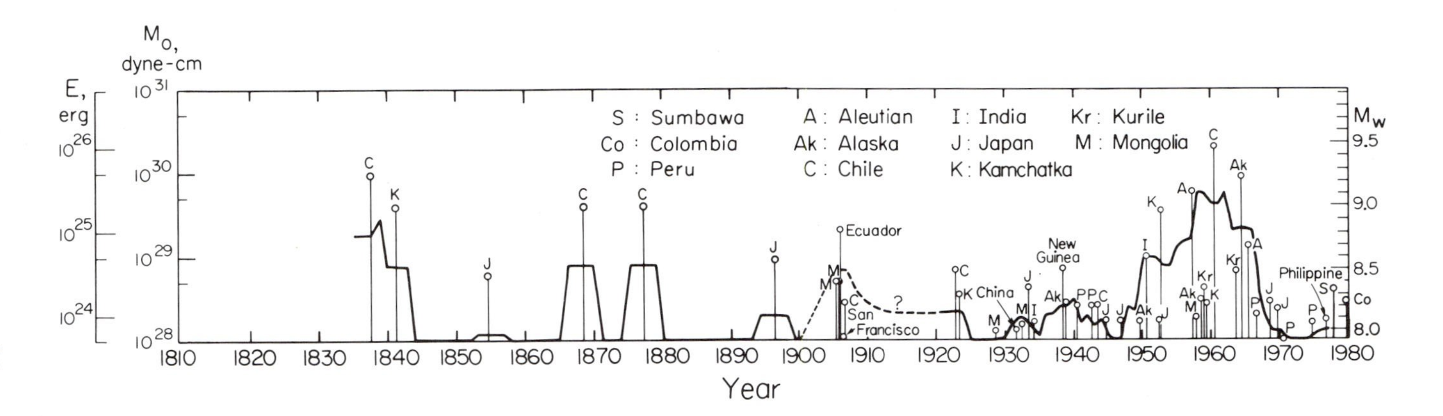

Figure 3. The seismic moment M_o, seismic energy E, and the magnitude M_W of great and large earthquakes as a function of time. The solid curve shows unlagged 5-year running average (in ergs per year) taken at the center of the interval (modified from Kanamori, 1977). For the period from 1835 to 1900 tsunami magnitude M_t (Abe, 1979) is used to estimate M_o, $\sim E$, and M_W.

events often have a very distinct first motion while thrust events have an emergent beginning. This difference can be used to distinguish normal-fault mechanism from thrust mechanism. Recently, Kanamori and McNally (1982) applied this method to the records from the 1906 Ecuador-Colombia earthquake ($M_W = 8.8$) and concluded that it is a thrust event.

Singh *et al.* (1985a) found that the great Oaxaca, Mexico, earthquake of 1931, which had been generally considered a large interplate thrust event, is a lithospheric normal-fault earthquake in the subducted Cocos plate (Figure 4). This result has an important bearing on the evaluation of seismic potential along the Mexican subduction zone.

3. Rupture Process of Large Earthquakes.

The rupture process of large earthquakes has been one of the important research subjects since the beginning of instrumental seismology. Imamura (1932, p.267) examined the seismograms of the great Kanto earthquake of 1923 and identified several discrete events during the main shock sequence. This type of multiple shock has been studied by various investigators, particularly for several large earthquakes such as the 1960 Chilean earthquake (Nagamune, 1971), and the 1964 Alaskan earthquake (Wyss and Brune, 1967).

Many recent studies have demonstrated that mechanical heterogeneities on the fault plane have important effects on generation of strong motions, precursory seismicity patterns, and fault mechanics. In this context, Hartzell and Heaton (1985) examined Benioff long-period seismograms of large subduction-zone earthquakes to study the regional variation of the nature of subduction zones. Since the occurrence of great earthquakes is infrequent, the use of historical seismograms is crucial to this type of study.

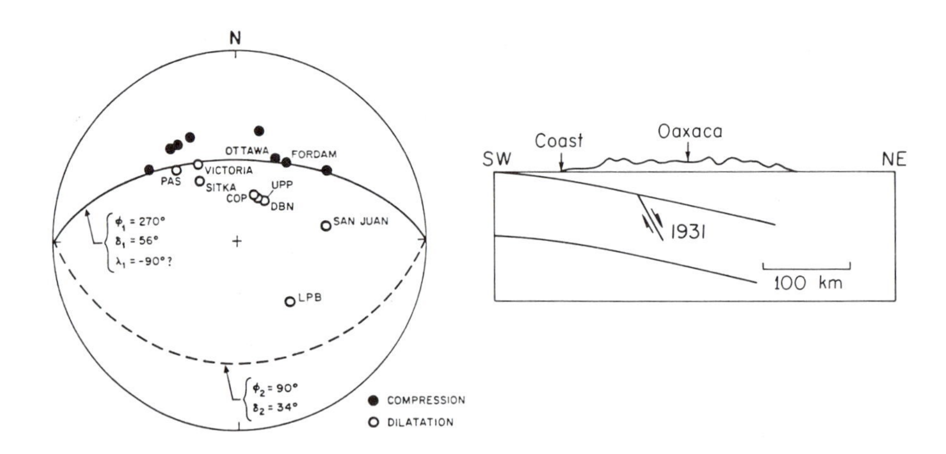

Figure 4. First-motion data of 1931 Oaxaca earthquake on the equal area projection of the lower focal hemisphere (left). The SW-NE cross section perpendicular to the coast line near Oaxaca. The mechanism of the 1931 event is schematically shown (right). (After Singh *et al.*, 1985a)

Boore (1977) investigated the rupture process of the 1906 San Francisco earthquake using old seismograms recorded by Ewing seismographs. From the arrival times of P- and S-waves and their amplitudes at Mt. Hamilton, within 35 km of the San Andreas fault, he could locate the epicenter and infer the distribution of slip along the fault.

4. Study of Seismic Gaps.

In order to evaluate the seismic potential of a seismic gap, it is important to study the nature of earthquakes which have occurred in the gap. For most seismic gaps, however, the repeat time is much longer than the length of available instrumental data so that quantitative studies are very difficult.

For the Parkfield, California, earthquake sequence, the repeat time is relatively short, about 22 years. Earthquakes with $M = 6$ are known to have occurred there in 1857, 1881, 1901, 1922, 1934, and 1966. Bakun and McEvilly (1984) examined seismograms of the 1922, 1934, and 1966 earthquakes recorded in Europe, North America, and South America, and concluded that these events are very similar in size, and can be considered as characteristic events occurring in Parkfield (Figure 5). This finding is the basis of the Parkfield earthquake prediction experiment currently conducted in California.

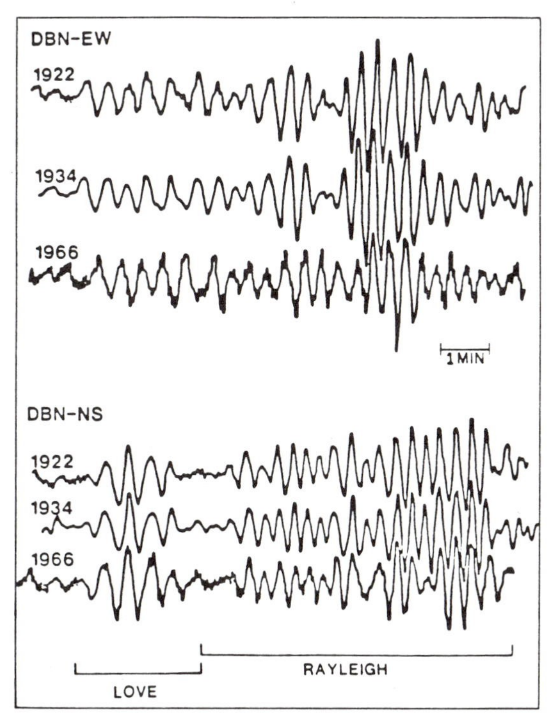

Figure 5. Seismograms of the Parkfield earthquakes of 1922, 1934, and 1966 recorded by Galitzin seismographs at Debilt (After Bakun and McEvilly, 1984).

Thatcher *et al.* (1975) analyzed several seismograms of large earthquakes which occurred on the San Jacinto fault, California, to determine the spatial-temporal pattern of seismic moment release pattern along the San Jacinto fault. On the basis of this analysis, Thatcher *et al.* identified a seismic slip gap where the amount of seismic slip for the previous few decades is significantly smaller than in the adjacent segments. This seismic slip gap is located near Anza, southern California, and is extensively monitored by various seismic instruments.

Recently several investigators examined the possibility of large subduction-type earthquakes along the plate boundary between the Juan de Fuca plate and the North American plate (along the Oregon-Washington coast) (Figure 6). Although no large earthquakes are known to have occurred here at least for the past 150 years, comparison of this subduction zone with other subduction zones in the world suggests that this subduction zone may be capable of generating large thrust earthquakes (Heaton and Kanamori, 1984).

One important feature of this subduction zone is the very young (about 10 My) subducting Juan de Fuca plate. Can a subduction zone with such a young subducting plate generate a large earthquake? To answer this question, it is important to examine seismicity of subduction zones with similar characteristics.

A recent study by Singh *et al.* (1985b) has an important bearing on this problem. A large ($M_S = 8.2$) earthquake occurred in Jalisco, Mexico, in 1932. The epicenter of this earthquake is very close to the triple junction between the Cocos, the North American, and the Rivera plates. However, the exact location and the nature of this event is not well known until recently. Singh *et al.* (1984, 1985b) examined old seismograms from several Mexican stations to determine the aftershock area of this earthquake, and concluded that the 1932 Jalisco earthquake occurred on the boundary between the Rivera and the North American plates (Figure 6). The geometry and age of this plate boundary are very similar to those of the Juan de Fuca plate boundary. It is important to note that a large earthquake did occur at a subduction zone very similar to the Juan de Fuca boundary.

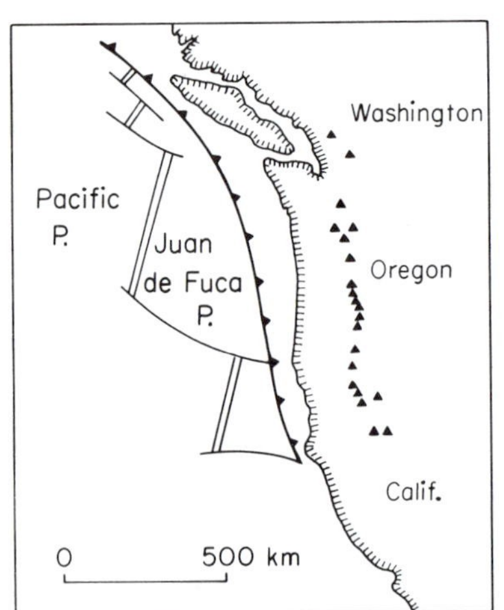

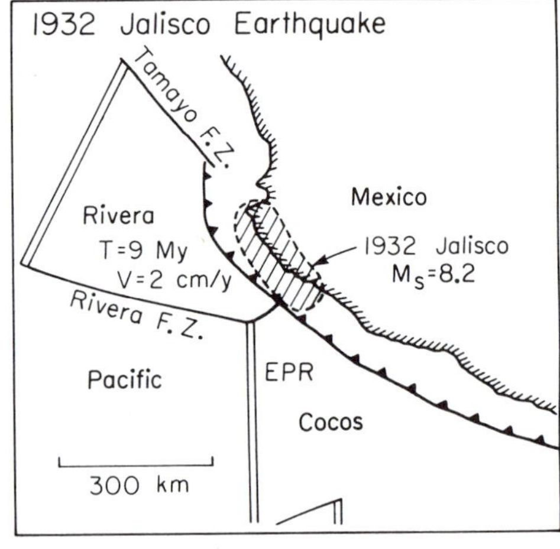

Figure 6. Comparison of the geometry of subduction zone along the coast of Oregon and Washington (left), and Jalisco (right). The rupture zone of the 1932 Jalisco earthquake is indicated (After Singh *et al.*, 1985b).

5. Regional Seismotectonics.

Historical seismograms have been effectively used to study seismotectonics of various regions. Chen and Molnar (1977) studied many large earthquakes in central Asia, and determined the spatial and temporal deformation patterns in the area.

Stein and Okal (1978) studied several large events which occurred in the Indian Ocean, and determined the nature of internal deformation of the Indian plate.

Stein *et al.* (1979) analyzed the seismograms of the 1933 Baffin Bay ($M_S = 7.3$) earthquake, and suggested that it is associated with stresses induced by removal of Pleistocene glacial loads.

Doser (1985) analyzed seismograms of the 1959 Hebgen Lake, Montana, earthquake which is the largest recorded earthquake in the inter-mountain region of the United States.

These studies clearly demonstrate that in order to investigate the pattern of current tectonic activity or to evaluate long term seismic risk in a certain region, we need seismicity data for a relatively long period of time. Unfortunately, very often high quality regional data are available only for very recent years, and the records for old events are too incomplete to be analyzed by conventional methods such as the P-wave first-motion method.

However, recent developments in the computational technique of synthetic seismograms have made possible the determination of the fault mechanism, seismic moment and other fault parameters from a relatively small number of records. In particular, if the medium response in the region can be calibrated by using more recent events for which complete data are available, a very detailed analysis can be made on the old records to retrieve important source parameters.

Various methods are now available to analyze seismograms recorded at various distances (see e.g. Helmberger, 1983). Since these methods are applicable to events as small as $M_S = 5.5$, there is a good possibility that we can drastically expand our data base on seismicity and mechanism for various regions if good collections of historical seismograms become available.

6. Moment Determination of Old Events.

For most large events that occurred after 1960, relatively accurate (a factor of 4 in the worst case) determinations of seismic moments have been made. However, for many events prior to 1960, accurate moment values are not available because of incomplete data.

Several investigators (e.g. Brune and King, 1967; Brune and Engen, 1969; Kanamori and Miyamura, 1970; Kanamori, 1971a, 1971b, 1972a, 1972b, 1976; Okal, 1976, 1977; Abe, 1976; Ben-Menahem, 1977; Fukao and Furumoto, 1979) determined the magnitudes and seismic moments of old events by analyzing historical seismograms. Although these determinations are not as accurate as for more recent events, they are important for a better understanding of global seismicity.

Kanamori (1977) lists many large events for which no direct measurements were available. Since then, Wang (1981) determined seismic moments of many of these events using the historical seismograms collected from various stations in the world.

Singh *et al.* (1984) systematically analyzed P-waveforms of several large shallow earthquakes along the Mexican subduction zone, and clarified the pattern of seismic energy release along the subduction zone.

If relatively complete data sets of old seismograms for major events are available, this type of study can be made more easily and thoroughly. There are still many more important events which have not been examined thoroughly. Examples are the 1957 Fox Island (Aleutian Islands) earthquake, the 1923 Kamchatka earthquake and the 1906 Valparaiso (Chile) earthquake.

7. Strong-motion Seismology.

Since near-field strong motion records of very large earthquakes are seldom available, strong motion records from historical earthquakes are very important.

In an attempt to estimate the amplitude of strong ground motions generated by the 1906 San Francisco earthquake, Jennings and Kanamori (1979) examined old seismograms recorded by Ewing duplex pendulum seismographs (Figure 7). Although the quality of these seismograms is rather poor, Jennings and Kanamori calibrated one of the old seismographs which recorded the San Francisco earthquake, and estimated its local magnitude to be in the range $6\frac{3}{4}$ to 7. Despite the large uncertainty involved in this determination, it is the only M_L value for earthquakes with $M_S = 8$.

Some long-period strong motion seismographs developed in the early days of seismology recorded near field strong motions of several Japanese earthquakes (e.g. 1943 Tottori earthquake; 1930 Izu earthquake; 1948 Fukui earthquake). These records have been used to estimate the gross rise time of fault motion (Kanamori, 1972a, 1973; Abe, 1974a, 1974b, 1978). Although these results are subject to large uncertainties because of the very limited number of records, they provide useful information on the mechanics of faulting.

8. Tsunami Earthquakes and Other Unusual Events

The 1896 Sanriku and the 1946 Aleutian Islands earthquakes generated very large tsunamis despite their relatively small earthquake magnitude, about 7.5 to 8 (Kanamori, 1972c). Although several mechanisms have been proposed (slow slip, shallow steep fault, landslide, etc.), the problem is still unresolved.

The 1929 Grand Banks, Canada, earthquake is not only the largest historical earthquake in Atlantic Canada, but also a very unique event in that it cut Trans-Atlantic cables in 28 places, and generated large tsunamis. Heezen and Ewing (1952) suggest that a large submarine landslide caused by this earthquake transformed into a turbidity current which ruptured the Trans-Atlantic cables. The details of the seismological nature of this event, however, are still uncertain, and more detailed studies of this unusual event using historical seismograms are desirable.

Ben-Menahem (1975) analyzed a few seismograms of the 1908 Siberian explosion, and concluded that this explosion consists of an air explosion at a height of 8.5 km and a ballistic wave. This is probably the only event of this nature that was recorded by seismic instruments.

The Spanish deep focus earthquake of 1954 (m_b = 7.1; depth = 630 km) is a very unique event, because it is the only large deep focus earthquake that is not associated with a well defined Benioff zone. In view of its uniqueness, Chung and Kanamori (1976) made a detailed study of this event using historical seismograms recorded mainly in Europe. They found that the Spanish earthquake is a very complex event which is atypical of deep focus earthquakes. The complexity and the long duration of the rupture process may be responsible for the considerable damage in Granada and Malaga, Spain.

Further analyses of these events, using more complete sets of historical seismograms, and investigations into other unusual events are important.

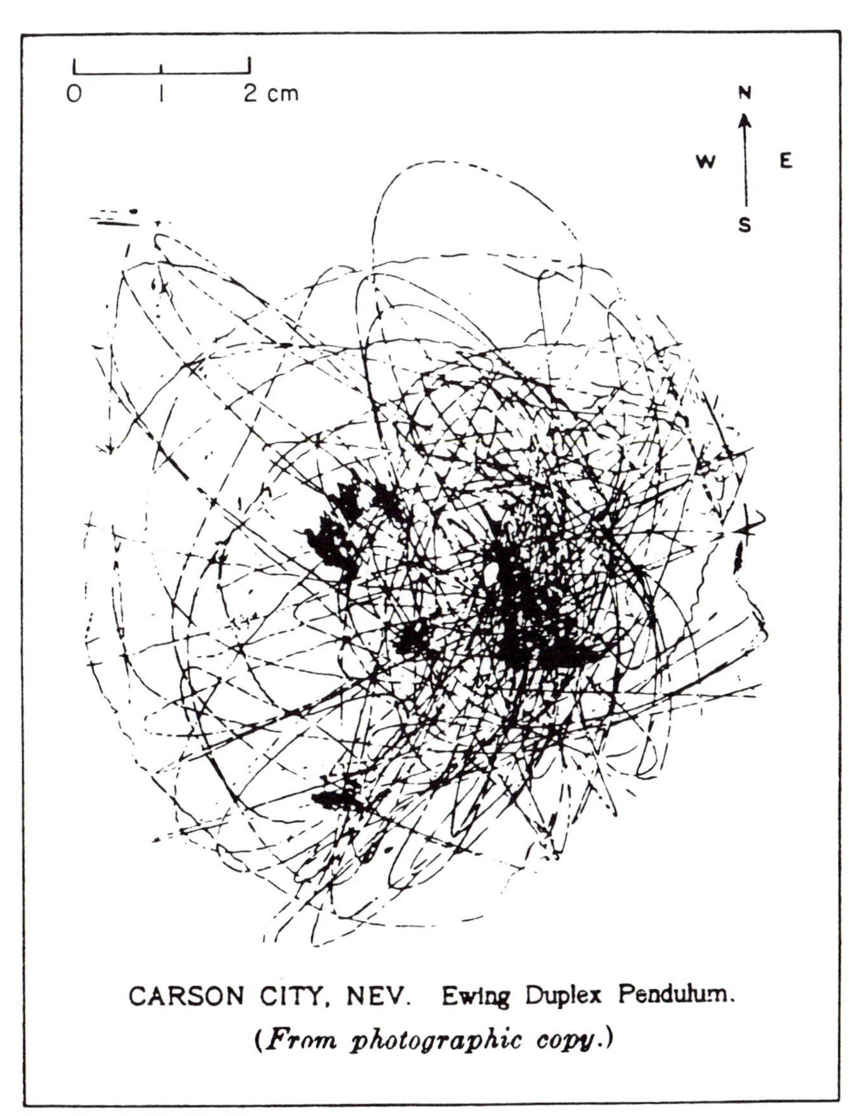

Figure 7. Seismogram of the 1906 San Francisco earthquake recorded by the Ewing duplex pendulum seismograph at Carson City, Nevada. The local magnitude M_L is determined from this seismogram.

9. Difficulties in the Use of Historical Seismograms.

The quality of old seismograms varies greatly from station to station as a function of time, and difficulties are encountered frequently in analyzing historical seismograms.

Most common problems include: (1) unknown instrument constants, (2) missing or uncalibrated time marks; (3) large solid friction between the stylus and paper, (4) no damping device attached to the seismograph; (5) cross coupling between the N-S and E-W components, and (6) unknown polarity.

Because of these difficulties, there is no standard method for the analysis of old records. Sometimes, the signal duration provides useful information on the size of the earthquake even if the instrumental constants are unknown. Even when the time marks are missing, some distinct phases on the record serve as "time marks". The problem of the solid friction becomes particularly serious when an attempt to use long-period (longer than 100 sec) waves for the seismic moment determination is made. When damping is not provided or is very weak, extreme caution must be exercised for the use of the amplitude data. For seismographs in which one pendulum is used to record vectorial horizontal motions, (e.g. Wiechert seismograph), cross coupling between the two horizontal components sometimes occurs. In this case, the conventional method of rotating seismograms into transverse and radial components cannot be applied.

10. Conclusion

Although the quality of historical seismograms is non-uniform and it is more difficult to use them than standardized modern seismograms, old seismograms, if properly interpreted, provide invaluable information on earthquakes in the past, and every effort should be made to save them, regardless of their quality, from possible loss and to make copies in an easily readable form.

REFERENCES

Abe, K. (1974a). Seismic displacement and ground motion near a fault: The Saitama earthquake of September 21, 1931, *J. Geophys. Res.*, **79**, 4393-4399.

Abe, K. (1974b). Fault parameters determined by near- and far-field data: The Wakasa bay earthquake of March 26, 1963, *Bull. Seism. Soc. Am.*, **64**, 1369-1382.

Abe, K. (1977). Mechanisms of the 1938 Shioya-Oki earthquakes and their tectonic implications, *Tectonophys.*, **41**, 269-289.

Abe, K. (1978). Dislocations, source dimensions and stresses associated with earthquakes in the Izu peninsula, Japan, *J. Phys. Earth*, **26**, 253-274.

Abe, K. (1979). Size of great earthquakes of 1873-1974 inferred from tsunami data, *J. Geophys. Res.*, **84**, 1561-1568.

Abe, K. (1981). Magnitudes of large shallow earthquakes from 1904 to 1980, *Phys. Earth Planet. Int.*, **27**, 72-92.

Abe, K., and S. Noguchi (1983). Revision of magnitudes of large shallow earthquakes, 1897-1912, *Phys. Earth Planet. Int.*, **33**, 1-11.

Bakun, W. H., and T. V. McEvilly (1984). Recurrence models and Parkfield, California, earthquakes, *J. Geophys. Res.*, **89**, 3051-3058.

Ben-Menahem, A. (1975). Source parameters of the Siberian explosion of June 30, 1908, from analysis and synthesis of seismic signals at four stations, *Phys. Earth Planet. Int.*, **11**, 1-35.

Ben-Menahem, A. (1977). Renormalization of the magnitude scale, *Phys. Earth Planet. Int.*, **15**, 315-340.

Ben-Menahem, A., M. Rosenman, and D. G. Harkrider (1970). Fast evaluation of source parameters from isolated surface-wave signals, *Bull. Seism. Soc. Am.*, **60**, 1337-1387.
Boore, D. M. (1977). Strong-motion recordings of the California earthquake of April 18, 1906, *Bull. Seism. Soc. Am.*, **67**, 561-577.
Brune, J. N., and C. Y. King (1967). Excitation of mantle Rayleigh waves of period 100 seconds as a function of magnitude, *Bull. Seism. Soc. Am.*, **57**, 1355-1365.
Brune, J. N., and G. R. Engen (1969). Excitation of mantle Love waves and definition of mantle wave magnitude, *Bull. Seism. Soc. Am.*, **59**, 923-933.
Chen, W. P., and P. Molnar (1977). Seismic moments of major earthquakes and the average rate of slip in Central Asia, *J. Geophys. Res.*, **82**, 2945-2969.
Chung, W. Y., and H. Kanamori (1976). Source process and tectonic implications of the Spanish deep-focus earthquake of March 29, 1954, *Phys. Earth Planet. Int.*, **13**, 85-96.
Doser, D. (1985). Source parameters and faulting processes of the 1959 Hebgen Lake, Montana earthquake sequence, *J. Geophys. Res.*, **90**, in press.
Fukao, Y., and M. Furumoto (1979). Stress drops, wave spectra and recurrence intervals of great earthquakes; implications of the Etorofu earthquake of 1958 November 6, *Geophys. J. Roy. Astr. Soc.*, **57**, 23-40.
Given, J. W., and H. Kanamori (1980). The depth extent of the 1977 Sumbawa Indonesia earthquake (abstract), *EOS*, **61**, 1044.
Hartzell, S. H., and T. H. Heaton (1985) Teleseismic time functions for large shallow subduction zone earthquakes, *Bull. Seism. Soc. Am.*, **75**, in press.
Heaton, T., and H. Kanamori (1984). Seismic potential associated with subduction in the northwestern United States, *Bull. Seism. Soc. Am.*, **74**, 933-941.
Heezen, B. C., and M. Ewing (1952). Turbidity currents and submarine slumps, and the 1929 Grand Banks earthquake, *Am. J. of Sci.*, **250**, 849-873.
Helmberger, D. V. (1983). Theory and application of synthetic seismograms, in *Earthquakes: Observation, Theory and Interpretation*, H. Kanamori and E. Boschi, Editors, North-Holland, Amsterdam, 174-222.
Imamura, A. (1937). *Theoretical and Applied Seismology*, Maruzen, Tokyo, 358 pp.
Jennings, P. C., and H. Kanamori (1979). Determination of local magnitude, M_L, from seismoscope records, *Bull. Seism. Soc. Am.*, **69**, 1267-1288.
Kanamori, H. (1971a). Seismological evidence for a lithospheric normal faulting - the Sanriku earthquake of 1933, *Phys. Earth Planet. Int.*, **4**, 289-300.
Kanamori, H. (1971b). Faulting of the great Kanto earthquake of 1923 as revealed by seismological data, *Bull. Earthq. Res. Inst.* Tokyo Univ., **49**, 13-18.
Kanamori, H. (1972a). Determination of effective tectonic stress associated with earthquake faulting. The Tottori earthquake of 1943, *Phys. Earth Planet. Int.*, **5**, 426-434.
Kanamori, H. (1972b). Tectonic implications of the 1944 Tonankai and the 1946 Nankaido earthquakes, *Phys. Earth Planet. Int.*, **5**, 129-139.
Kanamori, H. (1972c). Mechanism of Tsunami earthquakes, *Phys. Earth Planet. Int.*, **6**, 346-359.
Kanamori, H. (1973). Mode of strain release associated with major earthquakes in Japan, *Ann. Rev. Earth Planet. Sci.*, **1**, 213-239.
Kanamori, H. (1976). Re-examination of the Earth's free oscillations excited by the Kamchatka earthquake of November 4, 1952, *Phys. Earth Planet. Int.*, **11**, 216-226.
Kanamori, H. (1977). The energy release in great earthquakes, *J. Geophys. Res.*, **82**, 2981-2987.
Kanamori, H. (1983). Global Seismicity, in *Earthquakes: Observation, Theory and Interpretation*, H. Kanamori and E. Boschi, Editors, North-Holland, Amsterdam, 596-608.
Kanamori, H., and S. Miyamura (1970). Seismometrical re-evaluation of the Great Kanto earthquake of September 1, 1923, *Bull. Earthq. Res. Inst.*, **48**, 115-125.
Kanamori, H., and K. C. McNally (1982). Variable rupture mode of the subduction zone along the Ecuador - Colombia coast, *Bull. Seism. Soc. Am.*, **72**, 1241-1253.
Macelwane, J. B., and F. W. Sohon (1932). *Theoretical Seismology*, Part 2 by F. W. Sohon, Wiley, New York.
McCann, W. R., S. P. Nishenko, L. R. Sykes, and J. Krause (1979). Seismic gaps and plate tectonics: Seismic potential for major boundaries, *Pageoph*, **117**, 1087-1147.

Nagamune, T. (1971). Source regions of great earthquakes, *Geophys. Mag.*, **35**, 333-399.
Okal, E. A. (1976). A surface-wave investigation of the rupture mechanism of the Gobi-Altai (Dec. 4, 1957) earthquake, *Phys. Earth Planet. Int.*, **12**, 319-328.
Okal, E. A. (1977). The July 9 and 23, 1905, Mongolian earthquakes: A surface wave investigation, *Earth Planet. Sci. Lett.*, **34**, 326-331.
Richter, C. F. (1958). *Elementary Seismology*, W. H. Freeman and Company, San Francisco and London, 768 pp.
Silver, P. G., and T. H. Jordan (1982). Optimal estimation of scalar seismic moment, *Geophys. J. Roy. Astr. Soc.*, **70**, 755-788.
Singh, S. K., G. Suarez, and T. Dominguez (1985a). The great Oaxaca, Mexico, earthquake of 15 January 1931: Lithospheric normal faulting in subducted Cocos plate, *Nature*, **317**, 56-58.
Singh, S. K., L. Ponce, and S. P. Nishenko (1985b). The great Jalisco, Mexico earthquake of 1932 and the Rivera subduction zone, *Bull. Seism. Soc. Am.*, **75**, 1301-1314.
Singh, S. K., T. Dominguez, R. Castro, and M. Rodriguez (1984). P waveform of large, shallow earthquakes along the Mexican subduction zone, *Bull. Seism. Soc. Am.*, **74**, 2135-2156.
Stein, S., and E. A. Okal (1978). Seismicity and tectonics of the Ninetyeast ridge area: Evidence for internal deformation of the Indian plate, *J. Geophys. Res.*, **83**, 2233-2245.
Stein, S., N. H. Sleep, R. Geller, S. C. Wang, and G. Kroeger (1979). Earthquakes along the passive margin of eastern Canada, *Geophys. Res. Lett.*, **6**, 537-540.
Thatcher, W., J. A. Hileman, and T. C. Hanks (1975). Seismic slip distribution along the San Jacinto fault zone, southern California and its implications, *Geol. Soc. Am. Bull.*, **86**, 1140-1146.
Wang, S. C. (1981). Tectonic Implications of Global Seismicity Studies, *Ph.D thesis*, Stanford University, 70 pp.
Wyss, M., and J. N. Brune (1967). The Alaska earthquake of 28 March 1964: A complex multiple rupture, *Bull. Seism. Soc. Am.*, **57**, 1017-1023.

APPENDIX

In order to evaluate the usefulness of historical seismograms for various seismological research, the recording thresholds for various seismographs have been computed and compared with that of the WWSSN seismograph. For this purpose, we choose eight representative seismographs listed in Table 3, and compute the amplitude response curves which are shown in Figure 8.

For the mechanical seismographs, the instrument response can be given by 3 constants, pendulum period, τ, damping ratio, ϵ, (or damping constant h), and the static magnification V (see, Richter, 1958, p. 219). The response of the Galitzin seismograph is often given by 4 constants, l, T, A_1, and k (see Macelwane and Sohon, 1932, Vol. 2, p. 84). T is the period of the pendulum and the galvanometer (for the Galitzin seismograph, the pendulum and the galvanometer always have the same period.), and k is the transfer factor. Then the peak magnification is given by $V_m = 0.32\, k\, A_1\, T/\pi l$.

In Figure 8, the response of a representative IDA (International Deployment of Accelerographs) instrument is also shown for comparison.

At a period range from 1 to 5 sec, the responses of the Milne-Shaw, Galitzin-1, Wiechert-2 and Mainka seismographs are more or less comparable to the WWSSN LP instrument. Therefore, for body-wave studies the records obtained by these seismographs should be almost as useful as those obtained by the WWSSN (LP) seismograph. In order to compare the recording thresholds for long-period surface waves, we compute the peak-to-peak amplitude of the ground displacement which

Table 3. Instrumental Constants

Mechanical Seismograph

Instrument	Pendulum Period, τ, (sec)	Damping Ratio, ϵ	Damping Constant, h^*	Static Magnification, V
Wiechert 1	5.0	4	0.404	80
Bosch-Omori 1	12.0	4	0.404	40
Bosch-Omori 2	30.0	5.3	0.469	10
Mainka	10.0	2	0.215	120
Wiechert 2	12.6	3.4	0.363	180
Milne-Shaw	12.0	20	0.690	250

* h and ϵ are related by $\epsilon = \exp[\pi h / \sqrt{1 - h^2}]$.

Electro–Magnetic Seismograph*

Instrument	Pendulum Period, (sec)	Galvanometer Period, (sec)	Maximum Gain, V_m
Galitzin 1	12.0	12.0	580
Galitzin 2	25.0	25.0	310
WWSSN LP	15.0	100.0	1500

* Damping constants of the pendulum and the galvanometer are assumed to be 1.0, and the coupling constant is assumed to be 0.02.

would give rise to a trace amplitude of 5 mm (peak-to-peak). We consider this trace amplitude to be a reasonable lower bound of the amplitude of usable records. These threshold amplitudes of ground motions are shown by heavy solid curves in Figure 9.

Next, we compute the amplitude (peak-to-peak) of ground motions resulting from earthquakes with various magnitudes. For this purpose, we use a vertical strike-slip earthquake at a depth of 10 km, and compute the amplitude of the vertical component of Rayleigh waves at a distance of 90° in the azimuth of the maximum radiation pattern 45° away from the fault strike). The amplitude of ground motions depends upon the mechanism and the wave type, but the amplitude computed here should give an approximate magnitude of the amplitude of seismic surface waves to be observed. The results are shown by thin solid curves in Figure 9. For very large events ($M_W > 8$) the amplitude of observed surface waves at short periods would be considerably smaller than predicted by these curves because of the finiteness of the source.

Figure 9 shows that for events larger than $M_W = 8$, the Galitzin-1, Galitzin-2, Wiechert-2 and Milne-Shaw seismographs would provide usable data at periods 50 to 100 sec. For events larger than $M_W = 8.5$, long-period data up to 200 sec may be retrieved from the records obtained by these instruments. Records from Galitzin-2 seismograph may be used to study events with $M_W = 7$ to 7.5. Therefore combined use of records obtained by these different types of seismographs would provide valuable information on the long-period source characteristics.

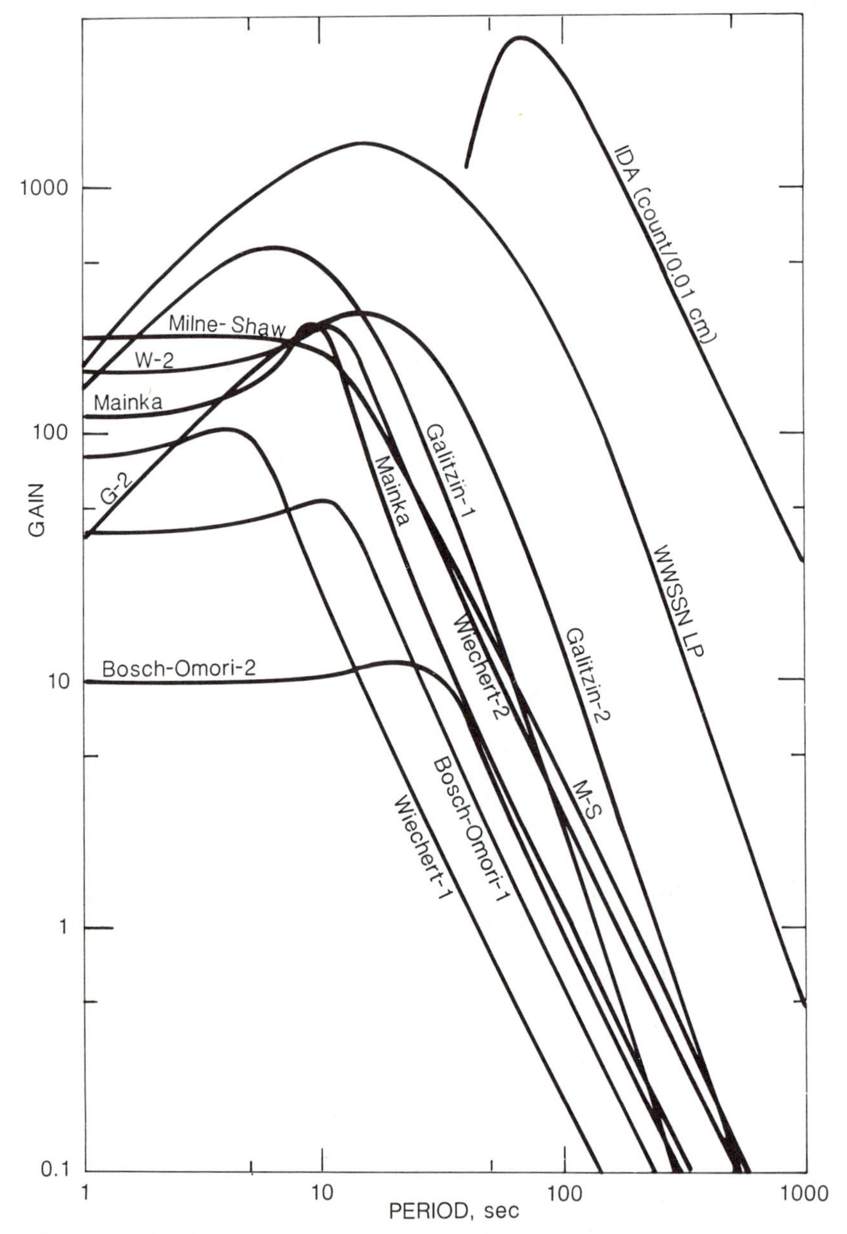

Figure 8. Amplitude response curves for various seismographs. For IDA, the gain refers to the number of counts for ground displacement with the amplitude of 0.01 cm. For the constants of the seismographs, see Table 3.

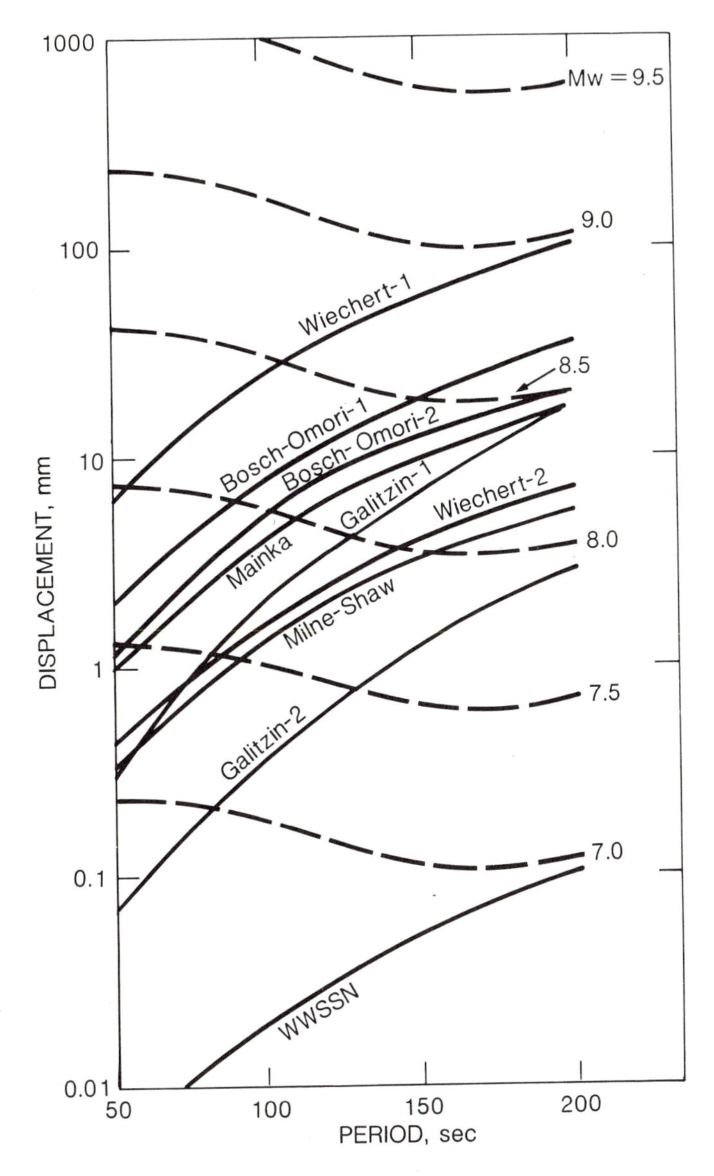

Figure 9. Recording thresholds for various seismographs. The solid curves indicate the amplitudes of ground motions (peak-to-peak) which would give rise to a trace amplitude of 5 mm (peak-to-peak) on the seismogram recorded by various seismographs. For the constants of the seismographs, see Table 3. The dashed curves indicate calculated amplitudes (peak-to-peak) of surface waves with periods 50 to 200 sec at a distance of 90° from reference earthquakes with magnitude M_W. The reference earthquake is a vertical strike-slip event at a depth of 10 km. The displacement is computed for the azimuth of the maximum radiation pattern. The calculation is based on equation (52) of Ben-Menahem *et al.* (1970).

II. Analysis of Historical Seismograms

Magnitudes and Origin Times from Milne Seismograph Data: Earthquakes in China and California, 1898–1912

Katsuyuki Abe
Earthquake Research Institute, University of Tokyo
Bunkyo-ku, Tokyo, Japan 113

ABSTRACT

Milne seismographs of similar design were distributed all over the world around the turn of this century. Voluminous reports on times and amplitudes based on records of the Milne network were systematically published for the period from 1897 through 1912. Nevertheless, the Milne data were scarcely used, mainly owing to the lack of a damping device. Recently, Abe and Noguchi (1983) calibrated the instrument response by using various sets of known magnitude data, and successfully obtained the simple expression which gives fairly accurate estimates of surface-wave magnitude M_S from the Milne data. Actually, M_S values from their method are on average only 0.02 lower than M_S values derived from damped seismograms for 49 shallow earthquakes which occurred in the world during 1910-1912. Using their method and the Milne data, we determine M_S values of shallow earthquakes which occurred in China and California for 1898-1912. Station corrections to be applied on account of variability of instrument and ground at various stations are calculated to make the M_S determination more reliable. Arrival time data of maximum phases in the Milne reports are useful for determining instrumental origin times.

1. Introduction

Undamped seismographs were gradually superseded by more advanced, damped instruments around the turn of this century. During this transition period of instrumental seismology, John Milne succeeded in distributing his seismographs of similar design to many different parts of the world; by 1912 they were in operation at more than 30 stations. Data of times and maximum amplitudes were systematically collected from the Milne stations and published by British Association for the Advancement of Science. These reports covering period from 1897 to 1912 are very important for a systematic investigation of the seismic activity in the dawn of instrumental seismology.

The Milne seismographs having natural period of 17.5 sec or so were almost standardized (Kanamori and Abe, 1979). However, the principal defect is the absence of a damping device. Owing to its defect, the Milne data were scarcely used (Gutenberg, 1956; Kanamori and Abe, 1979). In a previous study, Abe and Noguchi (1983) succeeded in calibrating the instrument response through comparison of amplitudes measured by the Milne instruments with those by various other instruments, so that surface-wave magnitudes for shallow earthquakes of early date can now be estimated on a certain basis.

In this study, the surface-wave magnitudes are determined by using the Milne data for earthquakes that occurred from 1898 to 1912 in China and California. In addition to this, a technique is presented to estimate the origin times from arrival time data of surface waves. The present study is an extension of the previous work on the quantification of large earthquakes of early date, and small earthquakes as well as large ones are treated here.

2. Materials and Methods

2.1. Materials Used

In this study, reported amplitudes and times based on records of undamped Milne seismographs are used. These data are published in *Report* (BAAS, 1899) and *Circulars* (BAAS, 1900-1912). The former covers readings for 1897-1899, and the latter for 1899-1912.

The maximum trace amplitudes on the Milne seismograms are reported in units of mm. The minima of the reported amplitudes are 0.1 mm at most stations, and often below 0.1 mm at Guildford, Victoria and Toronto. In *Report* and *Circulars*, arrival times are regularly reported in units of min or the tenth of min. The nature of the time data is discussed later.

2.2. Magnitudes

Adhering to the original definition of Gutenberg (1945), Abe and Noguchi (1983) calculated surface-wave magnitudes, M_S, by

$$M_S = \log(A_t/G) + 1.656 \log \Delta + 1.818 + s \tag{1}$$

where A_t is the maximum trace amplitude in μm on the single component of Milne seismograms, G is the effective magnification of Milne instruments, Δ is the epicentral distance in degrees, and s is the station correction. Assuming that $s = 0$, they adjusted G to make the calculated magnitude equal to M_S derived from amplitude data based on various types of seismographs and experimental data based on a newly-built Milne seismograph. They concluded that $G = 15.5$ is most satisfactory over a wide range of magnitude. We apply this result to the present study.

Gutenberg and Richter (1954) calculated magnitudes (denoted here by M_{GR}) for 49 world earthquakes of 1910-1912. For comparison, the values of M_S have been recalculated by using the original method of Gutenberg (1945) and the amplitude data given in the Gutenberg and Richter's worksheets (for details, Abe, 1981); here we call these magnitudes M_S(GR). Figure 1 shows a comparison of M_S(GR) with M_S from the Milne data for 49 shocks. It is seen that M_S from the Milne data is essentially equal to M_S(GR) over a wide range; M_S is only 0.02 ± 0.19 lower on average than M_S(GR). Figure 2 shows a comparison between M_S derived from the Milne data and M_{GR} taken directly from Gutenberg and Richter (1954). It is evident that M_{GR} deviates from M_S, particularly for smaller and deeper shocks. This deviation originates from a somewhat ill definition of the M_{GR} scale (Abe, 1981). It should be emphasized here that M_S used in this study is different from M_{GR}.

In the course of the previous study and the later work, it has been observed that the calculated magnitudes from the Milne data are above average at certain stations, below at others. This observation is considered to be related chiefly to

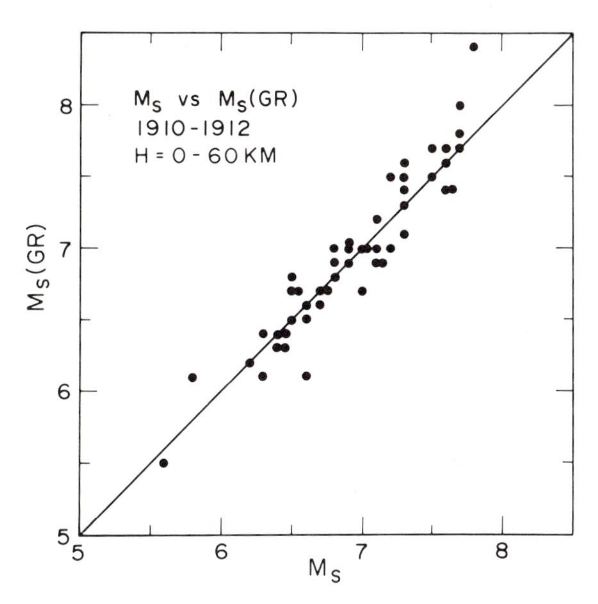

Figure 1. Comparison between M_S (GR) and M_S for shallow earthquakes of 1910-1912. M_S (GR) is surface-wave magnitude calculated from unpublished worksheets of Gutenberg and Richter (1954). M_S is surface-wave magnitude determined from Milne data. It is seen that M_S is essentially equivalent to M_S(GR) over a wide range.

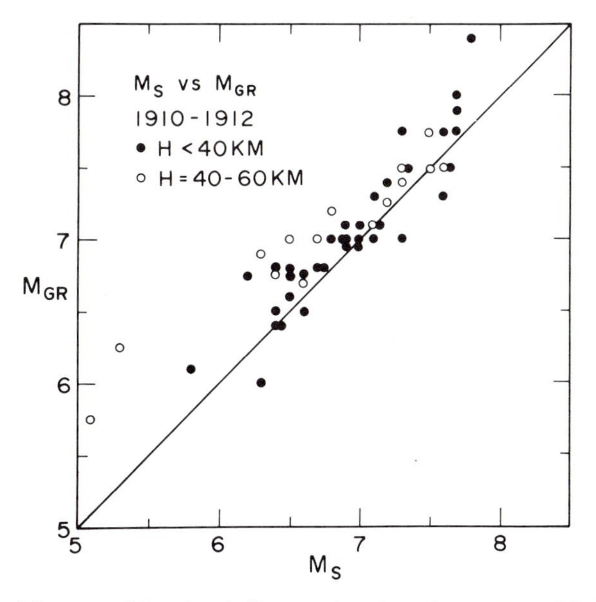

Figure 2. M_S versus M_{GR} for shallow earthquakes of 1910-1912. M_S is surface-wave magnitude determined from Milne data, and M_{GR} is magnitude given by Gutenberg and Richter (1954). Note that M_{GR} deviates from M_S.

variability of instruments and ground at various stations. The station correction at a particular station is defined by

$$s = \sum_i (M_i - m_i) / n \tag{2}$$

where M_i is the average value of M_S of a particular earthquake, estimated from a number of Milne stations, m_i is the magnitude of the same event at a particular station, and n is the number of earthquakes for which estimates of M_i and m_i are available. By using world earthquakes (1904-1912), the residuals for each station have been calculated, and combined to obtain station corrections. Data for great shocks have not been used, because of a saturation problem (Kanamori and Abe, 1979). It has been assumed in the calculation that the correction remained constant at individual stations during the whole period. Fifty earthquakes with more than ten individual observations have been selected for the calculation of residuals. The station correction, the number of observations and the standard deviations are summarized in Table 1, together with coordinates of the stations. The values of s are given to two decimals in order to avoid the accumulation of errors in rounding off. The standard deviation is about 0.2 or 0.3 units of M_S. There are blank entries for six out of 47 stations, for which the correction could not be calculated, because of only few observations available. The design of particular instruments at Shide and San Fernando, called Type A in the reports, are different from that of wide use, and those data are not used in this study. The result in Table 1 is fairly consistent with that of Gutenberg (1956), but the total number of earthquakes used here is much larger. The corrections have little effect on magnitude estimates except when the number of stations is small.

2.3. Origin Times

In *Report* and *Circulars*, arrival times of "Commencement" of the first preliminary tremors and arrival times of "Maximum Waves" are regularly reported, and times of beginning of "Large Waves" are often supplemented at some stations. The data for four shocks are exhibited in Figs. 3-6, where open circles show the data of "Commencement", cross symbols "Large Waves" and closed circles "Maximum Waves". Figs. 3 and 4 show the data for two Chinese earthquakes, and Figs. 5 and 6 show the data for California earthquakes, respectively. Travel time curves of P-wave, S-wave and surface-wave are drawn for reference. Characteristic features immediately show up in these figures: (1) the reported time of "Commencement" often refers to S-phase or even a later phase, particularly for small shocks; (2) the time of "Large Waves" often refers to the beginning of surface-waves or even S-phase; and (3) the data points of "Maximum Waves" cluster about a line indicating a constant velocity. The detection of the first arrival of P-waves and surface-waves at distant stations may have been difficult with low gain of the Milne instruments. Consequently, the timing accuracy of the initial-phase data is obviously too poor to determine hypocenters by ordinary inversion techniques.

Table 1. Station Corrections to be Added

Station	Lat	Long	*s*	*s.d.*	*n*
Adelaide	-34.93	138.58	0.07	0.24	18
Ascension Island	-7.95	-14.35	0.02	0.20	6
Azores (Ponta Delgada)	37.73	-25.68	0.27	0.29	15
Baltimore	39.28	-76.62	0.01	0.35	11
Batavia	-6.18	106.78	0.14	0.33	16
Beirut	33.90	35.47	0.12	0.20	23
Bidston (Liverpool)	53.40	-3.07	-0.11	0.27	41
Bombay (Colaba)	18.88	72.80	0.07	0.29	36
Cairo (Helwan)	29.87	31.33	0.04	0.22	30
Calcutta (Alipore)	22.53	88.33	-0.07	0.28	30
Cape of Good Hope	-33.93	18.47	0.01	0.34	32
Chacarita (Buenos Aires)	-34.58	-58.47			
Christchurch	-43.52	172.62	0.13	0.28	14
Cocos (Keeling Is.)	-12.20	96.90			
Colombo	6.93	79.83	0.16	0.24	21
Cordoba (Pilar)	-31.67	-63.85	0.21	0.26	8
Cork	51.88	-8.47			
Edinburgh	55.92	-3.18	0.04	0.21	42
Eskdalemuir	55.32	-3.20	-0.04	0.27	23
Fernando Noronha	-3.83	-32.42	0.02	0.26	10
Guildford	51.25	-0.58	0.07	0.30	18
Haslemere	51.08	-0.72	-0.25	0.23	33
Honolulu	21.32	-158.05	-0.02	0.34	35
Irkutsk	52.27	104.32	0.04	0.28	12
Kew	51.47	-0.32	-0.11	0.20	41
Kodaikanal (Madras)	10.23	77.45	0.18	0.28	36
Lima	-12.05	-77.05	0.19	0.27	7
Malta (Valetta)	35.90	14.52	-0.10	0.16	22
Mauritius	-20.08	57.88	0.03	0.28	22
Paisley	55.83	-4.43	0.04	0.23	39
Perth	-31.95	115.83	0.11	0.23	21
Rio Tinto (Huelva)	37.77	-6.63	-0.15	0.33	13
San Fernando	36.45	-6.20	-0.34	0.34	38
Seychelles (Mahe)	-4.08	55.08	-0.02	0.22	5
Shide	50.70	-1.32	-0.30	0.26	36
Stonyhurst	53.83	-2.47	-0.28	0.24	26
St. Helena	-15.92	-5.73			
St. Vincent (Cape Verde)	16.50	-24.00	-0.13	0.33	6
Strassburg	48.58	7.77			
Sydney	-33.87	151.20	0.24	0.24	25
Tiflis	41.72	44.78			
Tokyo	35.70	139.75	0.06	0.30	13
Toronto	43.65	-79.38	0.20	0.34	37
Trinidad	10.67	-61.50	0.25	0.22	15
Victoria	48.38	-123.32	0.31	0.27	32
Wellington	-41.27	174.80	0.11	0.16	8
West Bromwich	52.53	-2.00	-0.43	0.27	15

s = station correction.
s.d. = standard deviation.
n = number of observations.

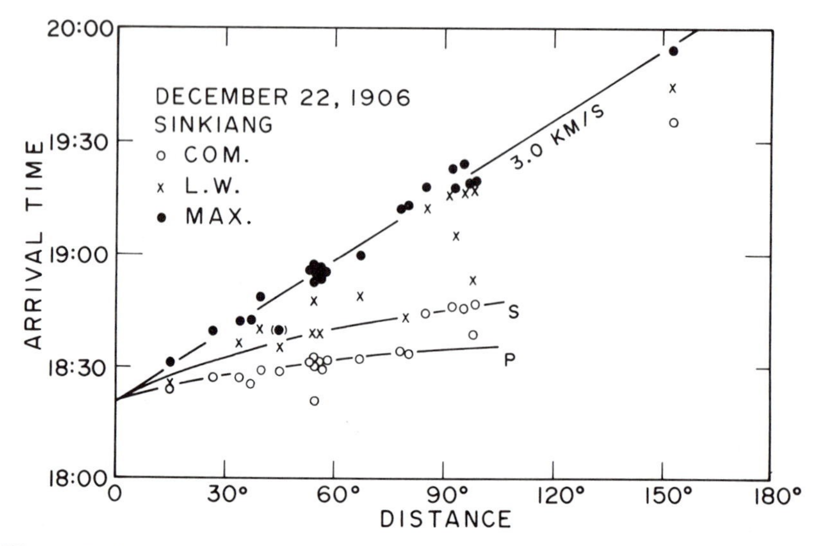

Figure 3. Arrival time data from Milne stations for the Sinkiang earthquake of 1906 ($M_S = 7.3$).

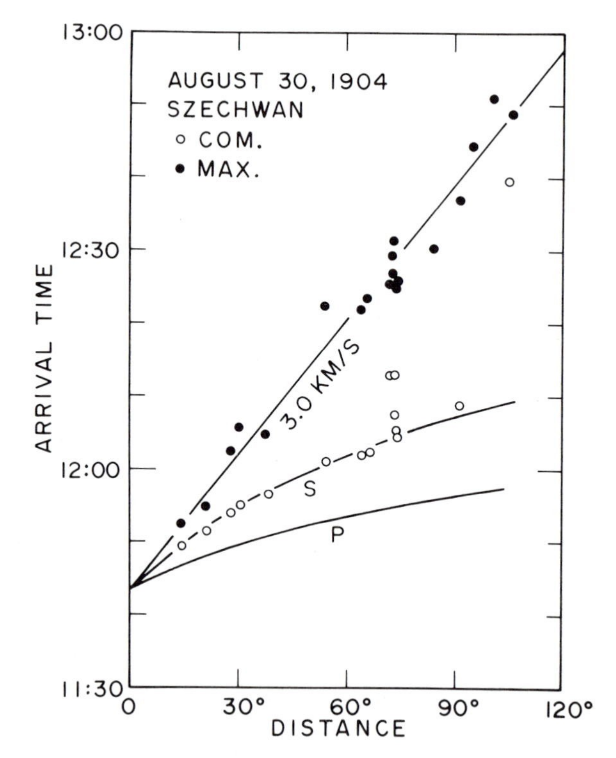

Figure 4. Arrival time data from Milne stations for the Szechwan earthquake of 1904 ($M_S = 6.9$).

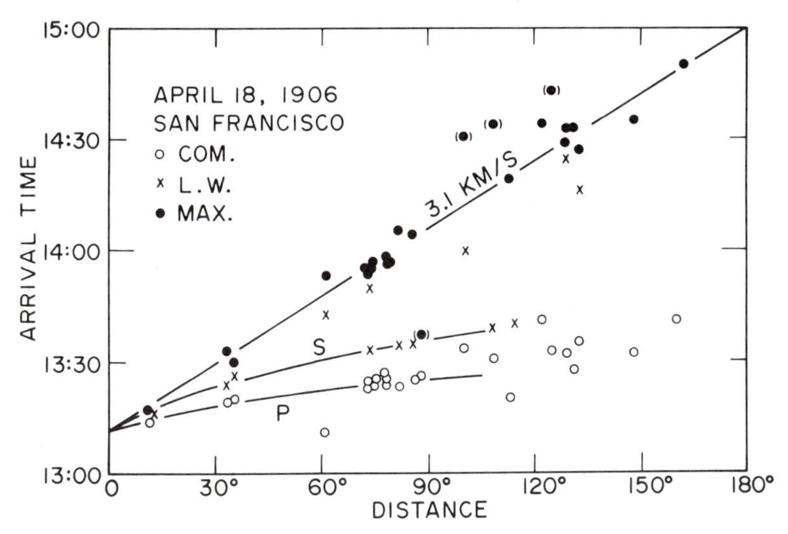

Figure 5. Arrival time data from Milne stations for the San Francisco earthquake of 1906 ($M_S = 7.8$).

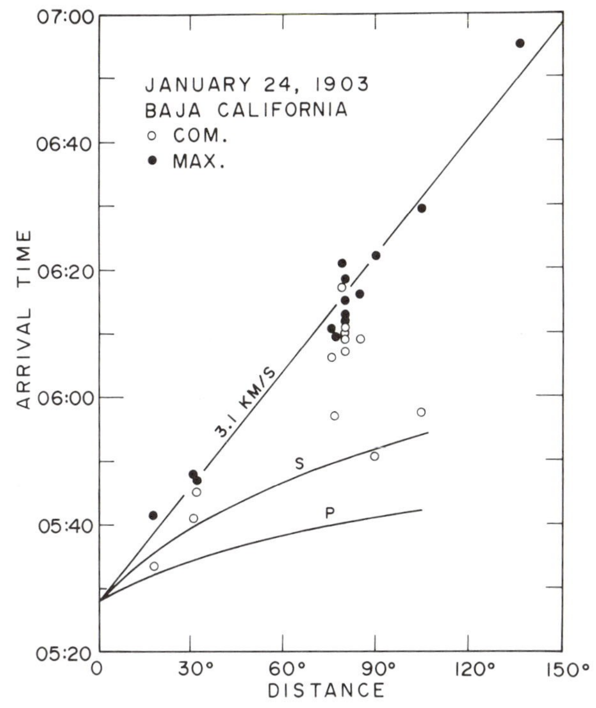

Figure 6. Arrival time data from Milne stations for the Lower California earthquake of 1903 ($M_S = 6.6$).

In the present study, the "Maximum Waves" data alone are used for estimates of instrumental origin times. The origin time is estimated by

$$t_c = \sum_i (t_i - \Delta_i / U) / n \tag{3}$$

where t_c is the origin time of an earthquake, t_i is the reported time of "Maximum Waves" at a particular station, Δ_i is the epicentral distance, U is the group velocity and n is the number of Milne stations used. Assuming that the group velocity is constant for individual regions, we have determined the velocity from the data given in Figs. 3-6 by the least-squares fit. Excluding few points of apparently large deviation (bracketed in Figs. 3 and 5), we obtain $U = 3.0$ km/s for Chinese earthquakes from Figs. 3 and 4, and $U = 3.1$ km/s for California earthquakes from Figs. 5 and 6. Using these values of U, single-station origin times are calculated and averaged. Then, extreme values which are different by more than one standard deviation are excluded, and the remainder values are re-averaged to obtain the origin time of an earthquake. All the time data in this paper are given in GMT.

From the above method, the origin time for the 1906 Sinkiang earthquake has been estimated to be 18:21.0 (Figure 3); the data used are given later in Table 3. This time is equal to the time given by Gutenberg and Richter (1954). Similar good agreement is obtained for the 1906 San Francisco earthquake. Its origin time has been estimated to be 13:12.5 from the Milne data (Figure 5). This time is close to the time, 13:12.0, obtained by Gutenberg and Richter (1954), and 13:12:21, determined by Bolt (1968). In later sections, results are compared further with only a few cases where origin times were formerly determined on an instrumental basis. It is considered from the comparisons that the estimates for well-observed shocks are accurate within one minute. For shocks of earlier date and small shocks of later date, the data are not sufficient and results are probably less accurate.

3. Chinese Earthquakes

The primary sources of earthquake data are catalogs of Gu (1983) and Lee *et al.* (1978). Earthquake catalog of Utsu (1982) is useful for supplementing some earthquakes near Taiwan. All the shallow shocks of 1898-1912 in these catalogs have been surveyed to assign magnitudes and origin times from the Milne data. The results for 32 shocks are listed in Table 2. The earthquakes for which no Milne data were available are omitted from the table.

No reports from any Milne station have been found for many shocks of magnitude 6 or less in the Gu's catalog. In the catalog of Lee *et al.* (1978), 28 shallow shocks of magnitude 6 or greater are listed for the period 1900-1912. For three out of the 28 shocks, no data have been found. It is suggested that the smallest earthquakes recognizable from the Milne data alone are considered to be $M_S = 6$ or so.

Following are some remarks and additional information concerning some of the earthquakes listed.

A series of Taiwan earthquakes from CH12 to CH17 in 1906, except CH13, started with the large earthquake of CH11; the reported loss of life is 1258 (Richter, 1958). The origin time of CH12 is given 02h by Lee *et al.* (1978), but the correct time has been proved to be 03h in GMT. The origin time for CH14 has not been estimated owing to the lack of the time data of "Maximum Waves". Irkutsk ($\Delta \sim 31°$) station reports several earthquakes during the period from March to April; the smallest

Table 2. Magnitudes of Shallow Earthquakes in China

EVENT	YEAR	MO	DY	HR:MN	° N	° E	M†	M_S #
CH01	1901	2	15	08:00.3	26.0	100.1	6	6.3
CH02	1902	8	22	03:03.2	39.9	76.2	8.25	7.7
CH03	1902	8	30	21:49.2	40	77		6.9
CH04	1902	11	4	11:33.4	36	96	7	6.4
CH05	1902	11	21	07:06.2	23	121	7.25	6.8
CH06	1903	6	7	09:03.6	25	122		6.4
CH07	1903	9	7	07:16.3	22.7	121.1	6	6.3
CH08	1904	4	24	06:41.0	23.5	120.5	6	6.4
CH09	1904	8	30	11:43.2	31.2	100.9	6	6.9
CH10	1904	11	5	20:24.8	23.5	120.3	6.25	6.1
CH11	1906	3	16	22:43.1	23.6	120.5	6.75	6.8
CH12	1906	3	26	03:25.8	23.7	120.5	6	6.0
CH13	1906	3	27	22:55.0	24.5	118.5	6.5	6.0
CH14	1906	4	6	16:53*	23.4	120.4	5.5	5.9
CH15	1906	4	7	22:37.0	23.4	120.4	6.25	6.1
CH16	1906	4	13	19:15.6	23.4	120.4	6.5	6.7
CH17	1906	4	13	23:51.3	23.4	120.4	6.25	6.4
CH18	1906	12	22	18:21.0	43.9	85.6	8	7.3
CH19	1908	1	11	03:36.0	23.7	121.4	6.5	6.8
CH20	1908	2	9	18:14.4	26	100	7.3	6.7
CH21	1908	7	1	07:25.9	24.0	122.0		6.2
CH22	1908	8	20	09:53.4	32	89	7.0	6.7
CH23	1909	11	21	07:38.0	24.4	121.8	7.3	6.7
CH24	1910	1	8	14:47.5	35	122	6.75	6.5
CH25	1910	6	17	05:30.2	21	121	6.5	6.0
CH26	1910	9	1	00:45.1	22.7	121.7	6.5	7.0
CH27	1910	9	1	14:23.8	24.1	122.4	6.5	6.8
CH28	1910	11	14	07:38.0	24.5	122	6.25	6.8
CH29	1911	3	24	03:15.1	24	122	6	5.9
CH30	1911	10	14	23:24.0	31	80.5	6.75	6.5
CH31	1912	11	3	06:07.9	23.5	122	5.5	6.1
CH32	1912	12	24	18:09.7	23.8	121.8	6.25	6.5

* = Origin time from Gu (1983).
† = Magnitude given in Lee *et al.* (1978) and Gu (1983).
# = Surface-wave magnitude from Milne data.

shocks have been estimated to be of $M_S = 5.6$ or so. Additional earthquakes for which reliable estimates have been possible from the Milne data are the shock of $M_S = 6.0$ at 20:13.1 on March 28 and the shock of $M_S = 5.9$ at 04:56.0 on April 7, where the epicenter (23.4° N, 120.4° E) has been assumed.

The magnitude of CH18 (Sinkiang earthquake of December 22, 1906) is assigned 8 in the published catalogs. This value is close to $M_{GR} = 7.9$ given by Gutenberg and Richter (1954), who obtained it from readings at Göttingen, Jena, Osaka and Zikaway. It has been pointed out by Abe and Noguchi (1983) and Abe (1984) that magnitudes in Gutenberg and Richter (1954) for earthquakes of early date are systematically overestimated. The present study has reduced the magnitude to $M_S = 7.3$. Isoseismal data given in Gu (1983) indicate that the area of strong intensity for CH18 is considerably smaller than that for the large shock (CH02, $M_S = 7.7$) of August 22, 1902. The Milne data used are shown in Table 3, which contains six entries: station names, epicentral distances, reported times of

Table 3. Sinkiang Earthquake of December 22, 1906

Station	Distance (deg)	Arrival time (hr:mn)	Origin time (hr:mn)	A (mm)	M_S
Irkutsk	15.0	18:31.2	18:21.9	>17.0	(6.85)
Calcutta	21.4			14.0	6.91
Bombay	27.2	18:39.1	18:22.3	9.0	7.03
Kodaikanal	34.3	18:42.2	18:21.0	5.0	7.05
Colombo	37.2	18:42.3	18:19.3	11.5	7.45
Beirut	39.7	18:48.5	18:24.0	2.5	6.79
Helwan	44.9	18:39.5	(18:11.8)	2.2	6.75
Batavia	53.4	18:55.6	18:22.6	2.9	7.09
Edinburgh	54.5	18:52.5	18:18.8	6.5	7.36
Kew	55.2	18:55.0	18:20.9	11.4	7.46
Paisley	55.2	18:56.7	18:22.6	14.0	7.70
Haslemere	55.6	18:54.3	18:20.0	14.3	7.42
Bidston	55.7	18:56.8	18:22.4		
Shide	56.1	18:53.5	18:18.8	>15.0	(7.40)
San Fernando	67.0	18:59.3	18:17.9	8.9	7.26
Azores	77.7	19:11.9	18:23.9	1.4	7.18
Perth	80.4	19:12.8	18:23.1	1.0	6.89
Victoria	84.7	19:17.7	(18:25.4)	4.0	7.73
Toronto	91.8	19:23.0	(18:26.3)	3.2	7.58
Honolulu	92.8	19:18.2	18:20.9	6.5	7.68
Baltimore	95.7	19:24.3	(18:25.2)	6.5	7.73
Sydney	97.7	19:18.2	(18:17.8)	0.65	6.98
Cape of Good Hope	98.6	19:19.0	18:18.1	3.8	7.52
Cordoba	153.1	19:54.6	28:20.0	0.9	7.41
Average			18:21.0		7.28

Values in the parentheses are not used in the average.

"Maximum Waves", origin time calculated from (3), maximum trace amplitudes and values of M_S. The station correction has been incorporated in the M_S estimates.

Ten shocks in Table 2 are listed by Duda (1965), and their values of M_S are on average 0.5 smaller than the Duda's magnitudes. The largest difference is 1.0. A discussion about the scale used by Duda has been made elsewhere (Abe, 1981, 1984). Duda lists the earthquake of March 2, 1906. Its epicenter (43° N, 80° E) is located marginally outside China. The origin time and M_S from the 17 Milne data have been estimated to be 06:15.1 $\pm$ 1.7 and 6.36 $\pm$ 0.18, respectively. This value of M_S is considerably smaller than 7.3 given by Duda.

Data of instrumentally determined origin times are very few for early shocks. The time for CH18 has been noted before. The origin times for CH24 and CH30 are given by Gutenberg and Richter (1954). The time for CH24 has been estimated to be 14:47.5 $\pm$ 1.0 from 19 Milne data, while Gutenberg and Richter (1954) calculated it to be 14:49.5 from 6 stations. There is no difference for CH30.

The earthquake of CH28 with $M_S = 6.8$ is not listed in the previous catalog of Abe and Noguchi (1983) which lists world earthquakes with $M_S = 6.8$ or above. This magnitude has been increased from 6.7 owing to the inclusion of station corrections. For the same reason, values of M_S for CH03, CH09 and CH18 are different by 0.1 unit from the previous catalog.

4. California Earthquakes

Earthquakes listed in Toppozada *et al.* (1978), Richter (1958) and Coffman and von Hake (1973) have been surveyed. The results for 21 earthquakes are summarized in Table 4.

Most of the Milne stations are distributed far from the California region, and only five stations are at distances shorter than 70°: Victoria ($\Delta \sim 15°$), Toronto ($\Delta \sim 30°$), Baltimore ($\Delta \sim 30°$), Honolulu ($\Delta \sim 35°$), and Trinidad ($\Delta \sim 60°$). The reports from Toronto started in September, 1897, and those from Victoria started in October, 1898. The data from these two stations are complete for 1899 and later years, and are very useful. The reports from other three stations started in 1901, and their reportings are so discontinuous or infrequent that the data are only partly useful, particularly for large shocks. Owing to such a biased distribution, the data for small shocks are insufficient. Magnitudes with a dagger symbol in Table 4 indicate that they have been determined from only one or two stations.

Richter (1958) lists three large shocks with magnitude 6 or $6\frac{1}{2}$: September 20, 1907, November 4, 1908 and May 15, 1910. If these magnitudes are correct, the shocks are large enough to be recorded at many stations. However, no Milne data have been reported from any stations, not even Victoria and Toronto. In addition, no data have been found for five shocks of magnitude 5.5 listed in Toppozada *et al.* (1978). Probably these shocks may have smaller magnitudes on the instrumental basis.

Table 4. Magnitudes of Earthquakes in California Region

EVENT	YEAR	MO	DY	HR:MN	° N	° W	M^*	M_S #
CA01	1898	4	15	07:06.7	39	124		6.7†
CA02	1899	7	6	20:14.5	37	121.5		4.7†
CA03	1899	7	22	20:31.1	34.5	117.5		5.6†
CA04	1899	12	25	12:24.9	33.5	116.5		6.4
CA05	1901	3	3	07:42.3	36	120.5	5.5	6.4
CA06	1902	7	28	06:57*	34.6	120.4		5.0†
CA07	1903	1	24	05:27.8	31.5	115.5	7	6.6
CA08	1903	6	11	13:13.9	37.6	121.8	5.5	5.4†
CA09	1903	8	3	06:50.7	37.3	121.8	5.5	5.3†
CA10	1906	3	3	20:23.8	33	117	4.5	5.3†
CA11	1906	4	18	13:12.5	37.7	122.5	8.3	7.8
CA12	1906	4	19	00:27.7	32.5	115.5	6.0	6.2
CA13	1906	4	23	09:10.3	41	124		6.4
CA14	1907	8	11	12:19.3	40.8	124.2		5.0†
CA15	1908	8	18	10:59*	40.8	124	5.0	4.9†
CA16	1909	5	18	01:19*	41	124		4.9†
CA17	1909	10	29	06:48.5	40.5	124.2	6.0	5.8
CA18	1910	3	11	06:54.2	36.9	121.8	5.5	5.8
CA19	1910	3	19	00:09.4	40	125	6.0	6.0
CA20	1910	8	5	01:31.2	40.8	124.2	6.8	6.6
CA21	1911	7	1	21:59.7	37.3	121.8	6.6	6.5

* = Richter (1958) and Toppozada *et al.* (1978).
\# = Surface-wave magnitude from Milne data.
† = Determination from only one or two station data.

Sieh (1978) and others list the earthquake originating at Parkfield on March 21, 23:45 (PST), 1901. No Milne data have been found for this shock. Ample evidence indicates that the actual date is March 2 in PST or March 3 in GMT (McAdie, 1907; BAAS, 1911; Townley and Allen, 1939). The Milne data for this shock (CA05) are shown in Table 5. Similarly, the actual date for the earthquake of October 28, 1909, in Richter (1958), is October 29. The earthquake of CA07 is not listed by Toppozada *et al.* (1978), because it originated in northern Baja California. Richter (1958) assigned magnitude 7+. The Milne data give $M_S = 6.6$, as shown in Table 6.

The San Francisco earthquake of April 18, 1906 (CA11), was formerly assigned $M_{GR} = 8\frac{1}{4}$ by Gutenberg and Richter (1954), and later revised to be $M_S = 7.8$ from the Milne data by Abe and Noguchi (1983). The station corrections have been incorporated in the present estimate, and the previous value of $M_S = 7.8$ has been unchanged to a tenth of a unit.

Table 5. Parkfield Earthquake of March 3, 1901

Station	Distance (deg)	Arrival time (hr:mn)	Origin time (hr:mn)	A (mm)	M_S
Victoria	12.5	07:42.5	07:45.0	8.0	(6.66)
Toronto	32.2	08:06.4	(07:47.2)	0.6	6.10
Edinburgh	74.1	08:26.6	07:42.3	1.0	6.77
Edinburgh	74.1	08:28.5	07:44.2	0.75	6.64
Bidston	75.8	08:26.2	07:40.9	0.3	6.11
Kew	78.4	08:22.0	(07:35.1)	0.5	6.35
Shide	78.5	08:26.2	07:39.3	0.5	6.17
Average			07:42.3		6.36

Values in the parentheses are not used in the average.

Table 6. Baja California Earthquake of January 24, 1903

Station	Distance (deg)	Arrival time (hr:mn)	Origin time (hr:mn)	A (mm)	M_S
Victoria	17.9	05:41.5	05:30.8	6.1	6.80
Toronto	30.9	05:48.1	05:29.6	2.0	6.60
Baltimore	32.1	05:47	05:27.8		
Edinburgh	75.7	06:10.5	05:25.2	1.0	6.78
Bidston	77.2	06:09.5	(05:23.3)	1.0	6.65
Cordoba	79.3	06:20.9	(05:33.5)		
Shide	79.7	06:15.1	05:27.5	1.0	6.48
Shide	79.7	06:13.2	05:25.6	1.2	6.56
Shide	79.7	06:12.2	(05:24.6)	0.7	6.32
Kew	79.7	06:18.5	05:30.9	1.0	6.67
San Fernando	85.4	06:16.0	05:24.9	0.9	6.44
Irkutsk	89.6	06:22.2	05:28.6	0.5	6.60
Tiflis	104.8	06:29.6	05:26.9		
Perth	136.7	06:55	(05:33.3)	0.25	6.67
Average			05:27.8		6.60

Values in the parentheses are not used in the average.

The origin time for CA13 is assigned 08:48 by Toppozada *et al.* (1978), but 09:10 by Coffman and von Hake (1973). The time data from 10 Milne stations indicate 09:10.3 ± 1.1. Data of instrumentally determined origin times are available only for three California shocks of 1904-1912. The time for CA11 has been noted before. Gutenberg and Richter (1954) give the times for CA20 and CA21. The difference between their times and the present results is 0.3 and 0.4 min, respectively.

5. Conclusion

Using the amplitude and time data of surface-wave maxima measured by undamped Milne seismograms, we have determined the surface-wave magnitudes M_S and origin times of shallow earthquakes in China and California for 1898-1912. The station corrections have been calculated from a number of the Milne data, and incorporated to make the magnitude estimate more reliable. A comparison between the M_S values and the catalog magnitudes is shown in Figure 7, plotted from Tables 2 and 4. The scatter is considered to be mainly due to an inhomogeneity of magnitude scales used in the previous catalogs. The estimates of origin times from the Milne data are considered to be accurate within one minute for well-observed shocks. The use of the Milne data permits the fairly accurate determination of M_S and origin times in shallow earthquakes where the Milne data are available, and extends the period of instrumental seismology.

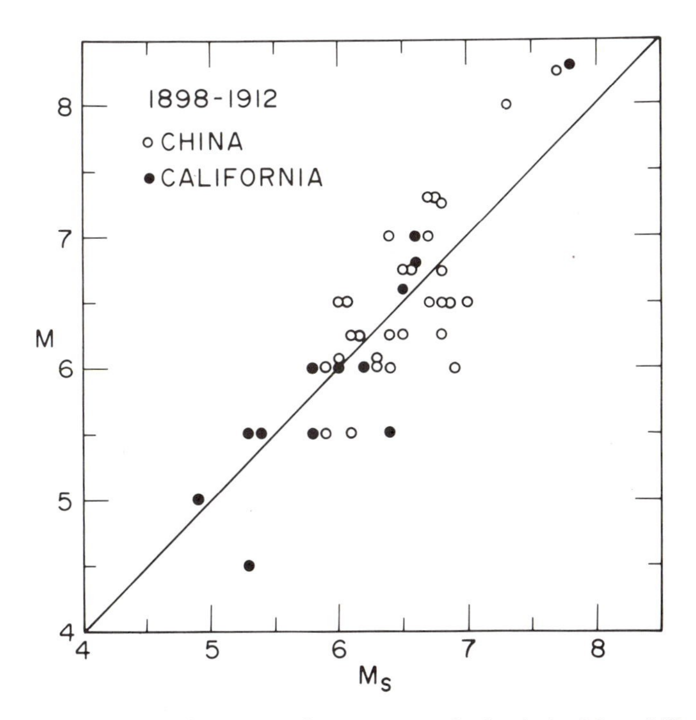

Figure 7. Comparison between surface-wave magnitudes derived from Milne data and magnitudes given in previous catalogs. The data are taken from Tables 2 and 4.

REFERENCES

Abe, K. (1981). Magnitudes of large shallow earthquakes from 1904 to 1980, *Phys. Earth Planet. Interiors*, **27**, 72-92.

Abe, K. (1984). Complements to "Magnitudes of large shallow earthquakes from 1904 to 1980", *Phys. Earth Planet. Interiors*, **34**, 17-23.

Abe, K., and S. Noguchi (1983). Revision of magnitudes of large shallow earthquakes, 1897-1912, *Phys. Earth Planet. Interiors*, **33**, 1-11.

BAAS (1899). *Fourth Report on Seismological Investigations*, British Association for the Advancement of Science, London, 161-238.

BAAS (1900-1912). *Circulars of the Seismological Committee*, British Association for the Advancement of Science, London, Nos. 1-27.

BAAS (1911). *Sixteenth Report on Seismological Investigations*, British Association for the Advancement of Science, London, 26-36.

Bolt, B.A. (1968). The focus of the 1906 California earthquake, *Bull. Seismol. Soc. Am.*, **50**, 457-471.

Coffman, J. L. and C. A. von Hake (1973). *Earthquake History of the United States*, Publ. 41-1, U.S. Dept. Commerce, 208 pp.

Duda, S. J. (1965). Secular seismic energy release in the circum-Pacific belt, *Tectonophysics*, **2**, 409-452.

Gu, Gongxu (1983). *Catalogue of Chinese Earthquakes (1831 B.C.-1969 A.D.)*, Science Press, Peking, 894 pp. (In Chinese).

Gutenberg, B. (1945). Amplitudes of surface waves and magnitudes of shallow earthquakes, *Bull. Seismol. Soc. Am.*, **35**, 3-12.

Gutenberg, B. (1956). Great earthquakes 1896-1903, *Trans. Am. Geophys. Union*, **37**, 608-614.

Gutenberg, B. and C. F. Richter (1954). *Seismicity of the Earth and Associated Phenomena*, 2nd Edition., Princeton Univ. Press, Princeton, NJ, 310 pp.

Kanamori, H. and K. Abe (1979). Reevaluation of the turn-of-the-century seismicity peak, *J. Geophys. Res.*, **84**, 6131-6139.

Lee, W. H. K., F. T. Wu, and S. C. Wang (1978). A catalog of instrumentally determined earthquakes in China (Magnitudes ≥ 6) compiled from various sources, *Bull. Seismol. Soc. Am.*, **68**, 383-398.

McAdie, A. G. (1907). Catalogue of Earthquakes on the Pacific Coast, Smithsonian Misc. Collections, Smithsonian Inst. City of Washington, No. 1721, 64 pp.

Richter, C. F. (1958). *Elementary Seismology*, Freeman, San Francisco, CA, 768 pp.

Sieh, K. E. (1978). Central California foreshocks of the great 1857 earthquake, *Bull. Seismol. Soc. Am.*, **68**, 1731-1749.

Toppozada, T. R., D. L. Parke, and C. T. Higgins (1978). Seismicity of California 1900-1931, *Calif. Div. Mines Geol. Spec. Rep. 135*, Sacramento, CA, 39 pp.

Townley, M. D. and M. W. Allen (1939). Descriptive catalog of earthquakes of the Pacific coast of the United States, *Bull. Seismol. Soc. Am.*, **29**, 1-297.

Utsu, T. (1982). Catalog of large earthquakes in the region of Japan from 1885 through 1980, *Bull. Earthquake Res. Inst., Tokyo Univ.*, **57**, 401-463, (In Japanese).

Analysis of Historical Seismograms from the Potsdam Station (GDR) in the Period 1902-1910

E. Hurtig and G. Kowalle
Central Institute Physics of the Earth, Academy of Science of GDR
DDR-1500 Potsdam, Telegrafenberg, German Democratic Republic

1. German Seismology at the Beginning of the 20th Century

The former German Empire had a long seismological tradition. In 1904 Gerland published proposals for a station network, distinguishing between the central station (in Strasbourg), main regional stations, and stations of 2nd order. This paper outlines the history of four stations within what is now the territory of the German Democratic Republic during the years 1902 to 1910. These stations are:

Potsdam: main station for the province of Brandenburg.
Leipzig: main regional station for the Kingdom of Saxony.
Plauen: auxiliary station of Leipzig.
Jena: main regional station for the Thuringian dukedoms.

Figure 1 shows the location of these stations, and Table 1 their instrumentation when routine observations began. An excellent international review of the state of development of seismometers and recording equipment at the end of the last century, Ehlert (1898), shows that all four stations were well equipped by the standards of those times. They produced excellent seismograms of both teleseismic and local earthquakes. Up to the First World War, all these stations were well maintained and their records are well interpreted and documented. Figure 2 shows the data obtained from the first records of the Leipzig station (Etzold, 1902). Similar detailed bulletins were published for all the stations, but unfortunately, many of the original seismograms were lost during the Second World War.

2. The Potsdam Station from 1902 to 1910

2.1 Instrumentation

The seismological observations at Potsdam were carried out in the former Geodetic Institute by Hecker from 1902 to 1908, and by Meissner in 1909 and 1910. A special earthquake recording house had been built with elaborate precautions to keep the temperature variations as small as possible. Routine observations began in April 1902, continuing the earlier work of Rebeur-Paschwitz and Hecker (Hurtig, 1981).

The following instruments were used in the period 1902-1910: (1) A horizontal pendulum with two components. This instrument was in operation from April 1, 1902 to June 7, 1909. Its characteristics are given in Table 2. (2) A Wiechert astatic seismometer. This instrument was in operation from October 13, 1903 to the mid 1970s, and its characteristics are listed in Table 3. The natural period was determined in November 1905, and again on September 26, 1910 by Schweydar. He found large changes (E-component: 12.5 sec; N-component: 19.8 sec) which suggest that the period was very unstable, and that the data given in Table 3 should not be accepted without further investigation.

Figure 1. Seismological stations operating early in the 20th Century, and now in the territory of the GDR.

Table 1. Seismological Stations in the Territory of the GDR at the Beginning of the 20th Century

Station	Start date	Instrumentation
Potsdam	April 1, 1902	v. Rebeur-Paschwitz (horizontal pendulum)
	October 13, 1903	Wiechert 1100 kg, EW, NS
Jena	May 1900-1902	Reubeur-Ehlert (horizontal pendulum)
	Interruption	
	April 1905	Wiechert 1200 kg NS, EW
		Straubel (vertical seismometer)
Leipzig	March 1902	Wiechert 1100 kg, NS, EW
Plauen	July 27, 1905	Wiechert 200 kg, NS

Tabellarische Übersicht über die von Wiecherts astatischem Pendelseismometer vom 28. März bis 15. Juli 1902 in Leipzig gelieferten Seismogramme von Fernbeben.

No.	Datum	Beginn in MEZ der 1. Vorphase	Beginn in MEZ der 2. Vorphase	Beginn in MEZ der Hauptphase	Ende der Aufzeichnung	Geradliniges Maß der größten Amplitude in mm während der 1. Vorphase NS-Componente	1. Vorphase OW-Componente	2. Vorphase NS-Componente	2. Vorphase OW-Componente	Hauptphase NS-Componente	Hauptphase OW-Componente	Epizentrales Gebiet und dessen Entfernung von Leipzig in km
1.	28. März	$15^{h} 58^{m} 20^{s}$	$16^{h} 10^{m} 30^{s}$	$16^{h} 33^{m}$ —^{s}	$17^{h} 29^{m}$ —^{s}	1	3	2	2,75	3	4	Molukken, ca. 11 500 km
2.	2 April	ca. 6^{h}	—	—	ca. $7^{h} 20^{m}$ —^{s}	—	—	—	—	0,75	1	unbekannt
3.	5. April	—	—	ca. $20^{h} 41^{m}$ —^{s}	ca. $20^{h} 52^{m}$ —^{s}	—	—	—	—	0,5	0,75	unbekannt
4.	12. April	—	—	$1^{h} 13^{m} 6^{s}$	$2^{h} 24^{m} 6^{s}$	—	—	—	—	1,25	1,5	Irkutsk-Baikalsee, 6000 km
5.	19. April	$3^{h} 32^{m} 35^{s}$	$3^{h} 42^{m} 56^{s}$	$4^{h} 1^{m} 30^{s}$	$6^{h} 8^{m} 30^{s}$	2	7	8	56	35	71	Guatemala, 9500 km
6.	6. Mai	$3^{h} 58^{m} 11^{s}$	—	$4^{h} 1^{m} 49^{s}$	ca. $4^{h} 5^{m}$ —^{s}	—	—	—	—	2,50	1,5	Südwestfrankreich, 1200 km
7.	8. Mai	ca. $4^{h} 2^{m} 14^{s}$	—	—	ca. $4^{h} 33^{m} 38^{s}$	—	—	—	—	6	1,5	unbekannt
8.	25. Mai	$17^{h} 58^{m} 22^{s}$	—	$18^{h} 25^{m} 57^{s}$	$18^{h} 55^{m}$ —^{s}	0,75	0,75	—	—	1,5	0,5	unbekannt
9.	26. Mai	$5^{h} 17^{m} 50^{s}$	$5^{h} 21^{m} 25^{s}$	$5^{h} 23^{m} 45^{s}$	$5^{h} 43^{m}$ —^{s}	3,5	2,5	—	—	5	1,5	unbekannt
10.	11. Juni	$7^{h} 32^{m} 16^{s}$	—	$7^{h} 45^{m} 46^{s}$	ca. $8^{h} 30^{m}$ —^{s}	—	—	—	—	8	8	unbekannt
11.	19. Juni	$10^{h} 23^{m} 39^{s}$	—	$10^{h} 25^{m} 2^{s}$	$10^{h} 27^{m} 30^{s}$	—	—	—	—	2,5	3	Hall-Innsbruck, 460 km
12.	5. Juli	$15^{h} 59^{m} 24^{s}$	$16^{h} 1^{m} 58^{s}$	$16^{h} 3^{m} 24^{s}$	ca. $16^{h} 30^{m}$ —^{s}	1,5	1,25	—	—	27	92	Macedonien, ca. 1400 km
13.	6. Juli	$14^{h} 22^{m} 30^{s}$	—	ca. $15^{h} 12^{m}$ —^{s}	ca. $16^{h} 24^{m}$ —^{s}	1,3	0,75	—	—	1	0,50	unbekannt
14.	9. Juli	—	—	ca. 5^{h}	ca. $5^{h} 30^{m}$ —^{s}	—	–	—	—	1,5	0,75	Südpersien, ca. 4600 km

Figure 2. First data from the Leipzig station (Etzold, 1902). [List of distant earthquakes from March 28 to July 15, 1902 recorded by the Wiechert astatic pendulum seismometer at Leipzig].

Table 2. Characteristics of the Horizontal Pendulum in Operation at Potsdam (April 1, 1902 to June 7, 1909)

Mass	85g	
Length of the arm	17.1 cm	
Damping coefficient	1.5:1	Apr 1, 1902 - Aug 1906
	4.1:1	August - October 1906
	7.5:1 (E-comp)	since October 1906
	5.0:1 (N-comp)	since October 1906
Natural period	18 sec	
Distance pendulum mirror-recording drum	3.08 m	
Magnification	36	
Recording speed	36 cm/hr	
Components	NW-SE	Apr 1, 1902 - Aug 20, 1903
	NE-SW	Apr 1, 1902 - Aug 20, 1903
	N-S	since Oct 13, 1903
	E-W	since Oct 13, 1903

Table 3. Characteristics of the Astatic Wiechert Seismometer at Potsdam

In operation since	Oct 13, 1903
Mass	1100 kg
Natural period	about 14 sec
Damping coefficient	about 5:1
Recording speed	about 64 cm/hr
Magnification	E: 180 – Oct 13, 1903 - November 1905
	N: 205 – Oct 13, 1903 - November 1905
	E: 130 – since November 1905
	N: 133 – since November 1905

2.2 Records of Large Earthquakes

Since historical seismograms and seismological readings of strong and destructive earthquakes may be of special interest for re-analysis, the catalogs of Abe (1981), Ganse and Nelson (1981), and Richter (1958) were examined to determine which earthquakes with $M \geq 7.0$ were recorded at the Potsdam station. We found that all shallow, intermediate, and deep focus earthquakes above magnitude 7 between 1902 and 1910 listed in these catalogs were recorded at the Potsdam station (Table 4).

Many earthquakes well-recorded at the Potsdam station were not found in the catalogs mentioned above. For example, of the 30 earthquakes in 1906 recorded with large amplitudes, only 10 could be identified. Four more shocks appear in Kárník's (1968) Catalog of Earthquakes (1901-1955) for the European area and the Catalogue Régional des Tremblements de Terre 1906 (Scheu, 1911) and 14 events are labelled by Scheu (1911) as "obs. micros".

Table 4. Strong Earthquakes ($M_S \geq 7.0$) Recorded at the Potsdam Station (1902-1910)

Year	Date	Arrival time (G.C.T.)	Latitude	Longitude	M
1902	Apr 19	02:36:12	14 N	91 W	8.3
	Aug 22	~ 03	40 N	77 E	8.6
	Sep 22	02:01:09	18 N	146 E	8.1
	Sep 23	20:32:28	16 N	93 W	8.4
1903	Jan 04	05:26	20 N	175 W	8.0
	Jan 14	(1)	15 N	98 W	8.3
	Feb 01	09:44:04	48 N	98 E	7.8
	Feb 27	01:08:03	48 N	98 E	7.8
	May 13	06:54:43	17 S	168 E	7.9
	Jun 02	13:28:18	57 N	156 W	8.3
	Aug 11	04:36:22	36 N	23 E	8.3
	Dec 28	03:10:24	7 N	127 E	7.8
1904	Jan 20	15:04:38	7 N	79 W	7.7
	Apr 04	10:05:46	41.8 N	23.2 E	7.3
	Jun 07	08:38:29	40 N	134 E	7.9
	Jun 25	14:57:24	52 N	159 E	7.9
	Jun 25	21:12:24	52 N	159 E	8.0
	Jun 27	00:20:33	52 N	159 E	7.9
	Jul 24	10:56:06	52 N	159 E	7.5
	Aug 24	21:12:25	30 N	130 E	7.7
	Aug 24	22:06:43	64 N	151 W	7.7
	Aug 30	12:02:43	31 N	101 E	7.5
	Oct 03	03:14:31	12 N	58 E	7.1
	Dec 20	05:57:10	8.5 N	83 W	7.6
1905	Jan 22	02:57:05	1 N	123 E	8.4
	Feb 14	08:58:20	53 N	178 W	7.9
	Apr 04	00:58:45	33 N	76 E	8.1
	Jun 02	05:51:36	39 N	132 E	7.9
	Jul 06	16:33:08	39.5 N	142.5 E	7.8
	Jul 09	09:49:38	49 N	99 E	8.4

(*continued*)

Table 4. *Continued*

Year	Date	Arrival time (G.C.T.)	Latitude	Longitude	M
1905	Sep 08	01:46:27	38.8 N	16.1 E	7.3
	Sep 15	06:14:06	55 N	165 E	7.6
	Nov 08	22:09:46	40 N	24 E	7.5
1906	Jan 21	14:01:14	34 N	138 E	8.4
	Jan 31	15:49:16	1 N	81.5 E	8.7
	Mar 16	23:05	23.6 N	120.5 E	7.0
	Apr 18	13:24:50	38 N	123 W	8.3
	Aug 17	00:22:41	51 N	179 E	8.2
	Aug 17	01:39	33 S	72 W	8.4
	Sep 14	16:24:30	7 S	149 E	8.1
	Sep 28	15:37:42	2 S	79 W	7.9
	Nov 19	07:37:31	22 S	109 E	7.5
	Dec 22	18:31:35	43.5 N	85 E	7.9
1907	Jan 04	05:32	2 N	94.5 E	7.6
	Apr 15	06:21:19	17 N	100 W	8.0
	Apr 18	~ 21:00	14 N	123 E	7.6
	Apr 19	00:05:48	13.5 N	123 E	7.5
	May 25	14:12:32	51.5 N	147 E	7.9
	Jun 25	18:12:46	1 N	127 E	7.9
	Sep 02	16:13:15	52 N	173 E	7.8
	Oct 16	14:10:18	28 N	112.5 W	7.5
	Oct 21	04:33:02	38 N	69 E	7.7
1908	Mar 26	23:16:18	18 N	99 W	8.1
	May 05	06:36	3 N	123 E	7.3
	Nov 06	07:21:06	45 N	150 E	7.1
	Dec 12	13:05:42	26.5 N	97 E	7.6
	Dec 28	04:23:58	38 N	15.5 E	7.2
1909	Jan 23	02:54:56	33 N	53 E	7.4
	Feb 22	09:41	18 S	179 W	7.9
	Mar 13	14:41:11	31.5 N	142.5 E	8.3
	Jun 03	18:53:51	2 S	101 E	7.7
	Jun 08	06:05	26.5 S	70.5 W	7.6
	Jul 07	21:48:24	36.5 N	70.5 E	8.1
	Jul 30	11:04:48	17 N	100.5 W	7.4
	Aug 14	06:43	35.5 N	136 E	7.0
	Oct 20	23:49:45	30 N	68 E	7.1
	Nov 10	06:25:22	32 N	131 E	7.9
	Dec 09	23:46:12	12 N	144.5 E	7.1
1910	Jan 01	11:13:06	16.5 N	84 W	7.0
	Jan 22	08:53:10	67.5 N	17 W	7.0
	Apr 12	00:34:09	25.5 N	122.5 E	8.3
	Jun 29	11:06:06	32 S	176 W	7.1
	Sep 09	01:25:12	51.5 N	176 W	7.0
	Nov 09	06:23	16 S	166 E	7.9
	Nov 26	05:00:30	14 S	167 E	7.3
	Dec 10	09:48:22	11 S	162.5 E	7.4
	Dec 13	11:47:42	8 S	31 E	7.4
	Dec 16	15:03:00	4.5 N	126.5 E	7.6

(1) instrumental problems, time not exactly determined.

REFERENCES

Abe, K. (1981). Magnitudes of large shallow earthquakes from 1904-1980, *Phys. Earth Planet. Interiors*, **27**, 72-93.

Ehlert, R. (1898). Zusammenstellung, Erläuterung und kritische Beurtheilung der wichtigsten Seismometer mit besonderer Berücksichtigung ihrer praktischen Verwendbarkeit, *Beitr. Geophys.*, **3**, 350-475.

Etzold, F. (1902). Das Wiechertsche astatische Pendelseismometer der Erdbebenstation Leipzig und die von ihm gelieferten Seismogramme von Fernbeben, *Ber. Kgl. Säch. Gesell. Wiss. Leipzig, math.-phys. Kl. Leipzig*, 283-326.

Ganse, R. A. and J. B. Nelson (1981). Catalog of significant earthquakes, 2000 B.C. - 1979, including quantitative casualties and damage, *World Data Center A Report SE27*, 145 pp.

Hurtig, E. (1981). Ernst August v. Rebeur-Paschwitz und seine Bedeutung für die Entwicklung der Seismologie, Veröff, *Zentralinst. Phys. d. Erde*, **64**, 54-59.

Kárník, V. (1968). *Seismicity of the European Area: Part 1*, Academia Praha.

Richter, C. F. (1958). *Elementary Seismology*, W. H. Freeman & Co., San Francisco, 768 pp.

Scheu, E. (1911). *Catalogue Régional des Tremblements de Terre ressentis pendant l'annee 1906*, Strasbourg.

Historical Seismograms and Interpretation of Strong Earthquakes

N. V. Kondorskaya and Yu. F. Kopnichev
Institute of the Physics of the Earth
USSR Academy of Sciences, Moscow, USSR

ABSTRACT

Discussed are difficulties involved in utilization of historical seismograms and some aspects of the interpretation of the world's strongest earthquakes.

1. Introduction

The importance of historical seismograms of the world's strongest earthquakes for geophysical studies is obvious, therefore, great emphasis is placed on the project of microfilming historical seismograms, which was supported by Unesco and was carried out from 1977 within the purview of the special Working Group of IASPEI headed by Dr. W. Lee.

At present great headway has been achieved in setting up a bank of microfilms, and evidently the time is ripe for developing a program to prepare a unified catalog of world earthquakes. Here use will be made, certainly, of the experience gained in compiling existing catalogs and new interpretation methods.

The present paper discusses some difficulties encountered in the analysis of microfilms and also some possibilities of interpreting records of the strongest earthquakes in regard to specifics of previous years (inadequacy of instrumentation, records of short-period instruments, etc.).

2. Difficulties in Microfilm Analysis

We analyzed microfilms of numerous stations – Tucson, Bermuda, Bozeman, Columbia, Vieques, Lincoln, Rapid City, Denver, Salt Lake, Sitka, Ukiah, Pasadena – kindly supplied to us by Dr. H. Meyers of WDC-A for selective earthquakes, and microfilms of the oldest Soviet stations – Pulkovo, Tashkent, Tiflis, and Irkutsk.

It should be noted, first of all, that the work with microfilms at the present stage is very laborious since the microfilm reels of all stations abroad include records of all seismograms. Consequently, finding the seismograms of a specific earthquake may require much time, although some reels contain records of only one earthquake. The quality of many records is at times inadequate; the records do not provide the information necessary for further processing – they do not contain minute marks, time corrections, and other details of presentations; specifications of instrumentation are not always found. We believe that at the present stage the need is to undertake a thorough review of microfilms to select records of earthquakes and to supplement the library of microfilms with descriptions of the quality of these records using a unified format.

It will be useful, in our opinion, to establish archives of microfilms not only by stations, but also for each earthquake. Such preparatory work will help to attract seismologists to handling microfilms.

3. Interpretation of Seismograms

One of the simplest and highly important elements in the interpretation of old records is determination of dynamic parameters in groups of body (P and S) and surface (L) seismic waves and applying corresponding calibrating functions. In preparing a New Catalog of the USSR territory from ancient times to 1977, Kondorskaya and Shebalin (1982) did not have access to world-wide seismograms, thus they analyzed mainly seismological bulletins. In Russia, the Bulletin of Permanent Central Seismological Commission (BPCSC), issued in 1902-1908 under the editorship of G.V. Levitsky, is a very useful publication. Acknowledging the imperfection of seismic instrumentation used at that time, those who compiled the BPCSC tried their utmost to preserve for the future the principal characteristics of these records – maximum amplitude and oscillation period. Without delving into the history of the development of the seismological observations in the USSR (see Kirnos *et al.*, 1957), we will draw attention only to the basic factor which impeded work on the new catalog, i.e., while substituting the old instruments of Bosch, Milne, Elert, and Tselner with more up-to-date seismographs of B.B. Galitzin, the formal approach to observation data took the upper hand. Inability to accurately evaluate the actual ground displacement was assumed enough to remove from the bulletins of 1911 all dynamic characteristics of seismic waves. This tendency is also apparent in well known global publications, such as the ISS and the BCIS bulletins. In 1923, the publication of amplitude data in the ISS was terminated and for a long time afterward no evaluations of earthquake size appeared except for indirect indications – the number of stations recording earthquakes. For measurements of $(A/T)_{\max}$ in a group of body and surface waves or of magnitudes, we advise the use of the recommendations of the Commission on Practice (Recommendations of the Commission on Practice, 1967), and the calibrating functions of Vanek *et al.* (1962) and Vanek *et al.* (1980).

Evaluations of the duration τ of earthquake waves on historical seismograms can probably be used to estimate magnitudes. The following equations were obtained by Kondorskaya and Shebalin (1982):

$$M_\tau = 2.4 \ \log_{10} \bar{\tau} + 1.6 \qquad \text{(for shallow earthquakes)}$$

$$M_\tau = 2.4 \ \log_{10} \bar{\tau} + 2.2 \qquad \text{(for deep earthquakes),}$$

where $\bar{\tau}$ is the derived duration in minutes.

The above duration $\bar{\tau}$ is calculated with regard to the effective magnification of instruments; correlations between $\bar{\tau}$ and τ for various types of devices is specified experimentally. M_τ is based on surface waves recorded on horizontal instruments for shallow earthquakes, while for deep earthquakes it is a conventional characteristic the nature of which is still to be verified.

Records of early years were in most cases obtained from short-period instruments. We believe that these records of historical seismograms can be used to estimate the spatial-temporal parameters of the sources from the characteristics of short-period radiation of strong earthquakes.

Kopnichev (1985) shows that relatively short-period ($T < 3$ sec) seismic waves carry important information on incoherent radiation of strong earthquakes and on absorbing and disseminating characteristics of medium. Incoherent radiation originates in the unevenness of rupture propagation and friction at a rupture and, possibly, at various secondary ruptures, according to Kostrov (1967).

Because short-period waves are easily scattered by small-scale inhomogeneities of the Earth, short-period source radiation is essentially complicated with dispersed waves. Therefore, it is worthwhile to undertake bandpass filtering of oscillations as well as statistical screening of dispersed waves in order to specify, in pure form, the effects related to the source. An important conclusion from this investigation is that the source radiates practically nothing in the time intervals when the envelope of the strong earthquake P-wave attenuates in a manner similar to the coda of P-waves for weak earthquakes. This important conclusion enables us to determine easily the instant of stopping of the rupture propagation using the envelope of the P-wave of a strong earthquake, which includes the moment of source impulse termination. Figure 1 shows the envelope of P-waves and P-codas in the Aleutian earthquake.

The envelope is plotted on the basis of the local maximum amplitudes on records of the TCHISS instruments with the one hertz frequency. It is seen that for weak earthquakes (M = 5.0) maximum amplitudes are reached in one to two seconds with their subsequent monotonous attenuation corresponding to dispersed waves – P-coda. At the same time for stronger earthquakes (M = 7.2) oscillations build up for nearly 12 seconds, after which the source practically comes to a still since the envelope goes similar to the P-coda of a weak signal. For the earthquake with a magnitude M = 8.2 the record is more complicated – several large impulses of incoherent radiation may be found there. Similarity to the envelope of P-codas of a weak earthquake is not observed until 180 sec after the P-wave arrival. This value may be assumed a typical duration of incoherent radiation for great earthquakes. Thus, the analysis of short-period records of historical seismograms of the strongest earthquakes help to obtain new information on the source mechanism, the depth, and the length of the radiation pulse, which are essential in particular for specifying the tsunami-generating capacity of earthquakes employing a relatively simple method.

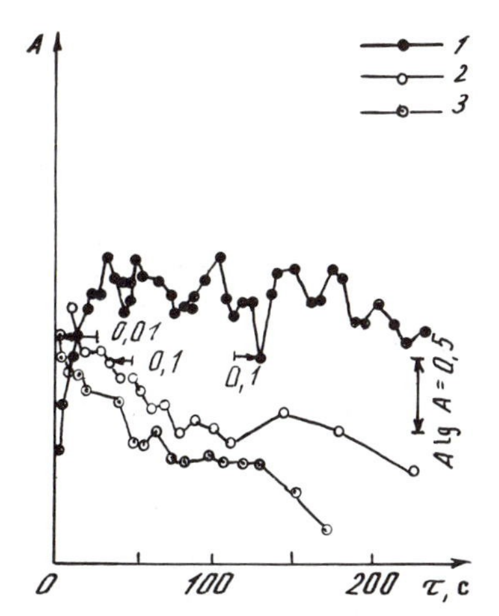

Figure 1. Envelopes of P-waves and P-codas of the Aleutian earthquake. (1,2) – 04.02.65, (1) – $M = 8.2$, (2) – $M = 7.3$; (3) – 15.02.65, $M = 5.0$. Talgar (TCHISS stations 1.0-1.6 Hz); arrows – levels in mm/s.

To use the proposed method, it is necessary to carry out spectral-temporal analysis of P-wave records on short period instruments. This involves digital sampling of the records at intervals not exceeding 0.2 sec and follow-up fast Fourier transform (FFT) of the time series; the amplitude Fourier spectrum is multiplied by the characteristic of the respective TCHISS filters (bandpassed between 1.0 to 1.6 Hz at the 0.7 level) (Zapolsky, 1971). Then the reverse Fourier transform is made. As a result, we get a record equivalent to the record of the TCHISS channel. The envelope of the record is plotted and compared to average standard curves to reveal the rise time and also to determine the total duration of oscillations (τ) through comparison of the codas of a weak earthquake from the same focal region with the envelope of the strong earthquake under study. Studies of specific characteristics of short-period radiation enabled us to identify a strong relationship between the amplitude of the velocity increment in the first P-wave arrivals at teleseismic distances and source mechanism (Figure 2).

For shallow earthquakes ($h < 100$ km) amplitudes build up slower for normal faults, quicker for strike-slip faults and overthrusts, and very quick for reverse faults. This relationship is not valid for deep earthquakes.

Thus, we may trace the distribution of oceanic earthquakes by two parameters: by the inclination of the P-wave envelope at the beginning of amplitude growth and by the time τ_m of reaching the maximum amplitude in the train of P-waves. The first parameter provides information on the mechanism and depth of the earthquake source. The second parameter provides information on the duration of one or several great pulses making up short-period radiation (for most "single pulse" sources contributing to strong earthquakes, the time τ_m practically coincides with τ_o – the duration of the propagation process in the source). Figure 3 shows the envelopes of some oceanic earthquakes (they all gave rise to intensive tsunami waves). It is seen that the envelope in stretches of amplitude growth are close to average relationships obtained through analysis of combined continental and oceanic earthquakes with various displacements in the source. The distribution of τ_m for tsunamigenic and non-tsunamigenic earthquakes is given in Figure 4. A sharp distinction between tsunamigenic and non-tsunamigenic events is seen and threshold values of τ_m differ for earthquakes with different source mechanisms.

It is necessary to emphasize that the data from one station are quite enough to provide information on the tsunamigenic capacity of earthquakes and the displacements in the source, since the above parameters are relatively stable according to data of various stations. In this context we undertook a preliminary interpretation of the records of a strong Japan earthquake of December 20, 1946 made at the Tucson station with short-period Benioff instruments (Figure 5). Plotting of the envelope and estimation of τ_m equal to 139 sec indicate that this earthquake was tsunamigenic and its source mechanism was close to an overthrust, which is in accord with other determinations (Abe, 1980).

Very recently we received a record of this earthquake, made with a short-period device at the Pasadena station, which agrees well with the results of the Tucson station.

Short-period records of historical earthquakes may be used in the evaluation of source parameters employing the method suggested by Koyama and Zhang (1985), where the smoothed envelope of a wave packet is plotted and the short-period seismic moment is calculated.

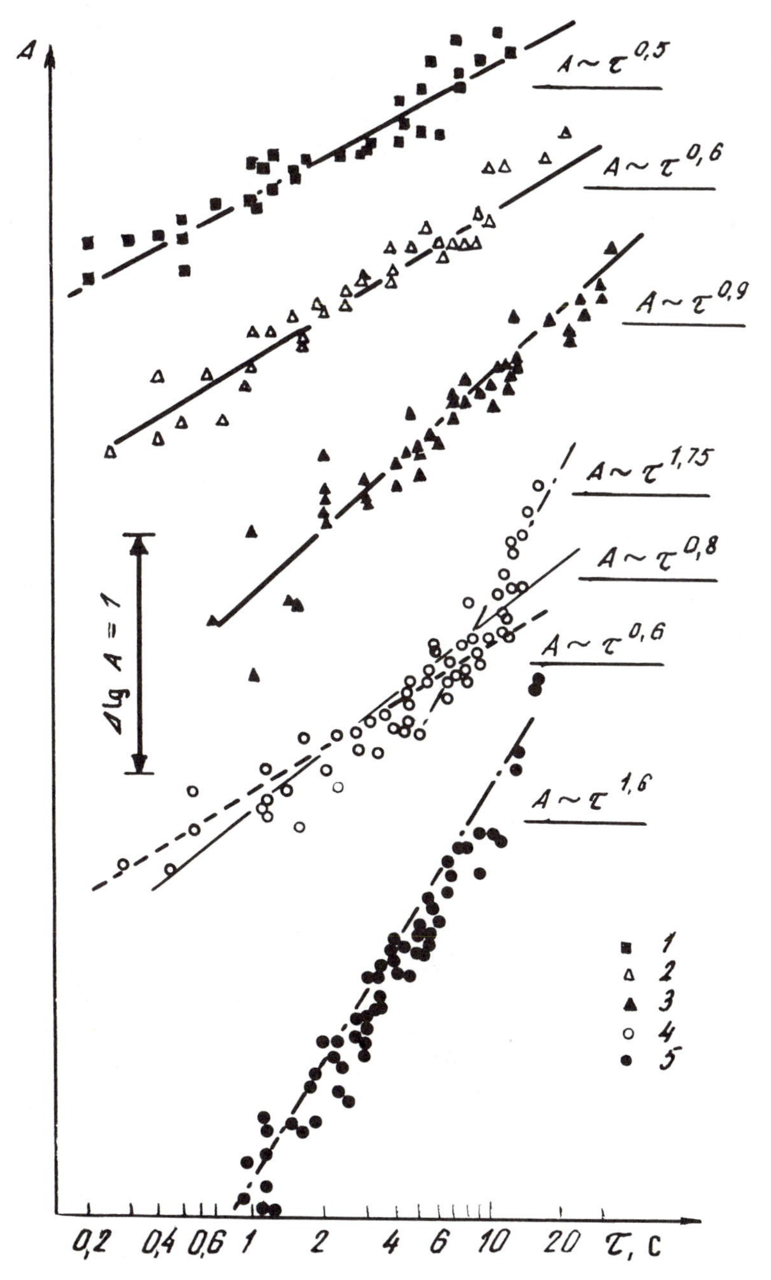

Figure 2. Graphs of P-wave envelopes at the beginning of amplitude growth. 1 – deep earthquakes ($h > 100$ km); 2-5 – shallow earthquakes ($h \leq 100$ km); 2 – normal faults and strike-slip faults; 3 – strike-slip faults; 4 – overthrusts and diagonal-thrusts; 5 – reverse faults and diagonal reverse faults. Talgar, Garm, Chusal, TCHISS stations (1.0-1.6 Hz).

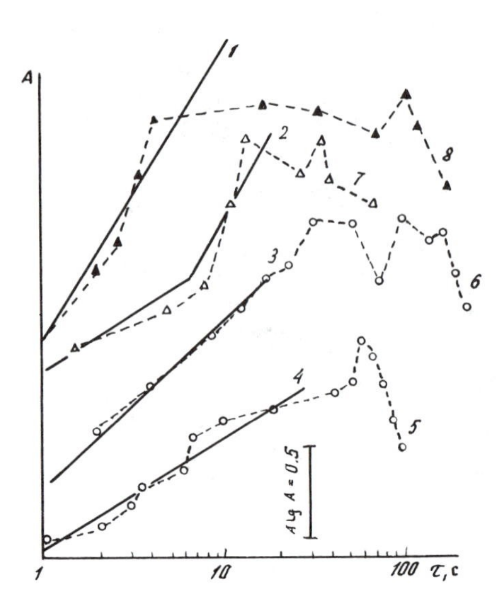

Figure 3. P-wave envelopes of tsunamigenic earthquakes: 1-4 – average relationships $A(\tau)$ for various mechanisms: 1 – reverse faults; 2 – overthrusts; 3 – strike-slip faults; 4 – normal faults; 5 – Hawaii, 29.10.75, $M = 7.2$; 6 – Aleutians, 04.02.65, $M = 8.2$; 7 – Shikotan, 11.08.69, $M = 8.1$; 8 – Niigata, 16.06.64, $M = 7.5$. Talgar, Garm, TCHISS stations (1.0-1.6 Hz).

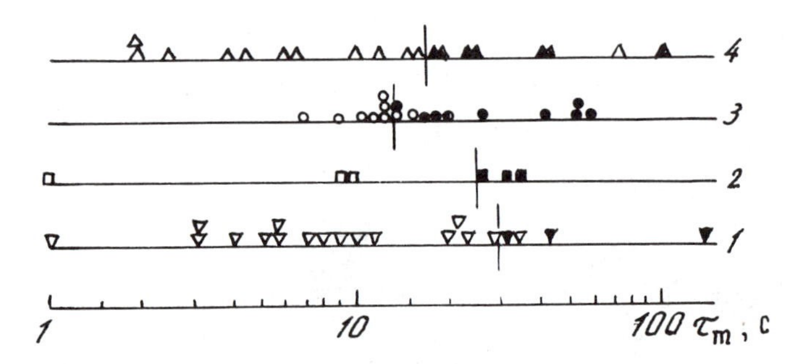

Figure 4. Values of τ_m for tsunamigenic (filled marks) and non-tsunamigenic earthquakes. 1 – normal faults; 2 – strike-slip faults; 3 – overthrusts; 4 – reverse faults.

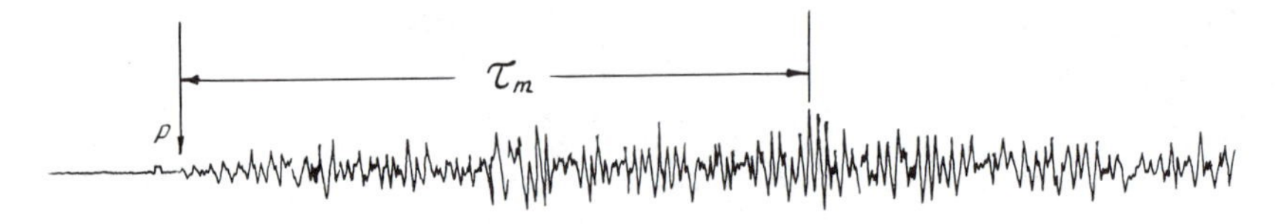

Figure 5. The Japan earthquake of December 20, 1946 (Tucson seismic station).

REFERENCES

Abe, K. (1975). Determination of static and dynamic fault parameters: the Saitama earthquake of July 1, 1960, *Tectonophysics*, **27**, 223-238.

Kirnos, D. P., D. A. Kharin, and N. V. Shebalin (1956). Progress in seismic observations in the USSR. In *Earthquakes in the USSR*, 20-45.

Kondorskaya, N.V. and N.V. Shebalin, editors (1982), *New Catalog of Strong Earthquakes in the USSR from Ancient Times through 1977.* (English translation of original 1980 publication in Russian), Report SE-31, WDC-A for Solid Earth Geophysics, 608 + XI.

Kopnichev, Yu.F. (1985). *Short-period seismic wavefields*, Moscow, Nauka.

Kostrov, B.V. (1975). *Source mechanisms of a tectonic earthquake*, Moscow, Nauka, 176 pp.

Koyama, J. and Si-Hua Zhang (1985). Excitation of short-period body waves by great earthquakes, *Phys. Earth Planet. Interiors*, **37**, 108-123.

Recommendations of the Commission on Practice of IASPEI (1967).

Vanek, J., et al. (1962). Standardization of magnitude scales, *Izv. Fizika Zemli AN SSSR.*

Vanek, J., N. V. Kondorskaya, and L. Christoskov (1980). Magnitude in the seismic practice of PV and PV_s waves, BAN, Sofia.

Zapolsky, K. K. (1971). Frequency-selective TCHISS stations. In *Experimental Seismology*, Moscow, Nauka, 20-36.

The Status, Importance, and Use of Historical Seismograms in Sweden

Ota Kulhánek
Seismological Department
Uppsala University, Uppsala, Sweden

The birth of observational seismology in Sweden dates back to the beginning of the 20th century, to be more specific, to October 1904, when a Wiechert-type instrument (1000 kg, horizontal pendulum) was installed in the Observatory Park in central Uppsala. This seismograph, the first in the country, has been in continuous operation since that time, with practically unchanged characteristics during the entire period of more than eight decades. Amplitude characteristics of the instrument's two horizontal components are shown in Figure 1. At this writing, almost 60,000 Wiechert records have been produced, all stored in archives at the Seismological Department, in Uppsala. Uppsala Wiechert seismograms were systematically analyzed from October 1904 to December 1955 (except for the time from June 1905 to June 1906). All collected relevant information is listed in annual Seismological Bulletins from Uppsala. The Bulletins provide the date, phase identification and corresponding arrival times. For large events, periods and amplitudes, epicentral distances, focal regions and often even brief commentaries are also given. Only exceptionally (more recent events), information on focal depth and first-motion polarity is included. Uppsala Wiechert records constitute one of the most extensive, complete, homogeneous and well documented seismogram series available, this in respect to the long operation time, practically unchanged instrumental constants and regular daily record readings. This and similar series are invaluable for studying earthquake history of various regions and for comparative investigations of large historical earthquakes.

Several other seismograph stations operated temporarily in Sweden at the beginning of this century. In 1906, an 80 kg horizontal Wiechert seismograph was deployed at Vassijaure in northern Sweden. In 1915, the apparatus was moved to Abisko where its operation ceased on July 12, 1951. The Abisko station was also equipped with a three-component Galitzin system having a photographic recording via mirror galvanometers. During the period 1917-1953, a Wiechert seismograph, similar to that installed at Uppsala, was in operation in Lund in southern Sweden. Station name, seismometer used, recording type and time of operation of historical Swedish seismograph stations are given in Table 1. The station locations are shown in Figure 2. As follows from the table, during the period 1917-1951, there were three seismograph stations operating in Sweden, covering the country relatively well in the N-S direction. All records are now stored in Uppsala. However, whereas the Uppsala records were continuously analyzed, seismograms from Vassijaure/Abisko and Lund were not.

Considerable efforts had been spent to establish a convenient seismogram storage system in Uppsala. After many years of improvisation, all records are now deposited in the Seismological Department's building in a permanent archive facility; they are accessible within minutes for inspection and/or further processing.

In spite of the vast and continuous instrumental modernization taking place at the Uppsala station since the early 1950's, the veteran Wiechert pendulum is still

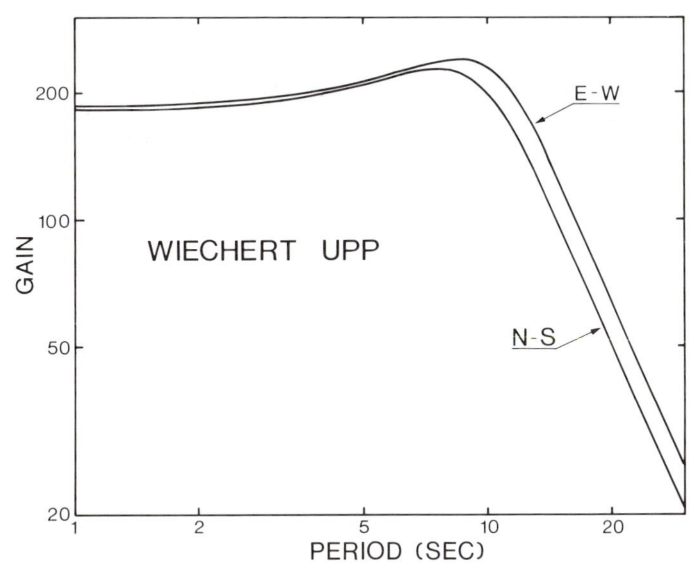

Figure 1. Amplitude response curves for the Uppsala Wiechert horizontal instrument.

Table 1. Historical Seismograph Stations in Sweden

Station	Lat. °N	Long. °E	Seismometer	Recording	Time of Operation
Uppsala	59.86	17.63	Wiechert E,N	smoked paper	1904-
Vassijaure	68.42	18.18	Wiechert Z	smoked paper	1906-1915
Abisko	68.34	18.82	Wiechert Z	smoked paper	1915-1951
			Galitzin E,N,Z	photogr. paper	1919-1943
Lund	55.70	13.19	Wiechert E,N	smoked paper	1917-1953

operating without interruptions. The available historical, as well as more recent records, are invaluable in revising the existing and in compiling new earthquake catalogues, in making re-evaluations and/or determinations of magnitudes and seismic moments and in illuminating source processes, for example. When determining magnitudes for larger data sets (in time and space) essentially two requirements are crucial. First, the same magnitude has to be applied throughout the event series. Second, the way in which magnitudes were determined should be explicitly specified to make calculations of other source parameters possible. For example, when compiling the earthquake catalogue for Turkey for the period 1913-1970 (Alsan *et al.*, 1975), surface-wave magnitudes, M_S, deduced from Uppsala Wiechert records were introduced. There were only a few Turkish earthquakes with clearly recorded body phases in Uppsala which a priori disqualifies the use of body-wave magnitudes, m_b. Besides, M_S-values often are more stable when compared with corresponding m_b estimates. Many earthquakes in Turkey occur at shallow depths, and so focal-depth influence does not create any additional problem. Uppsala Wiechert records are available for the entire time interval considered, which in turn, guarantees magnitudes of high homogeneity, an essential requirement for any statistical analysis.

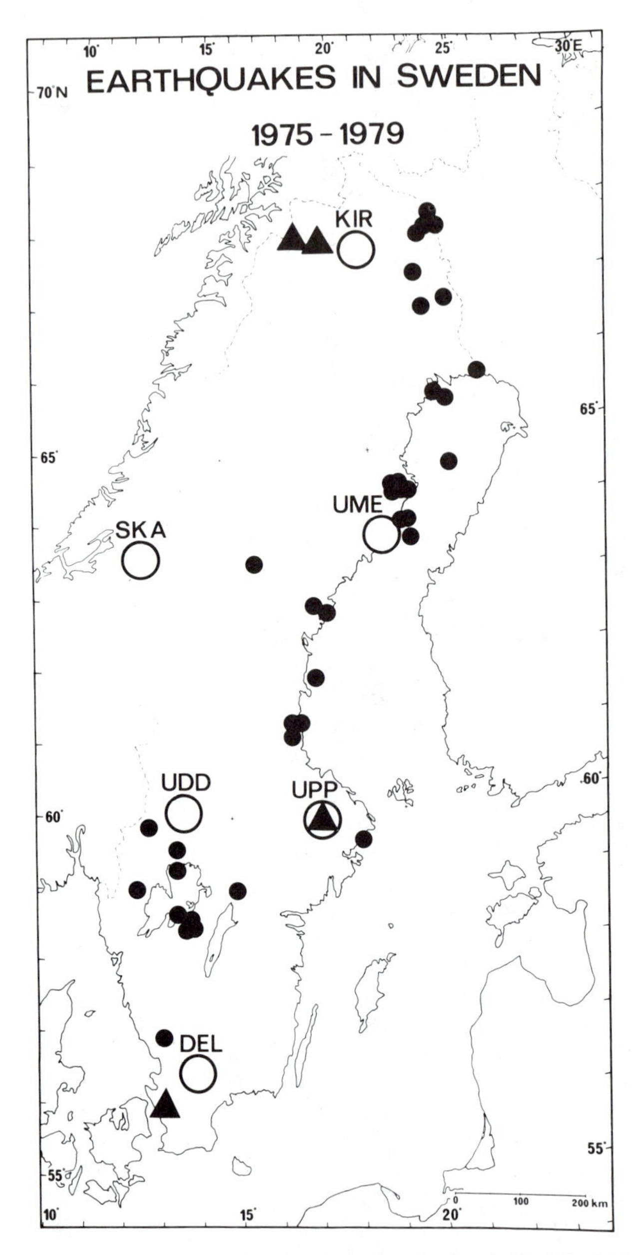

Figure 2. Distribution of Swedish stations during the first half of the 20th century (triangles); current Swedish seismograph station network (open circles); epicenters of Swedish earthquakes during the period 1975-1979 (solid circles).

More recently, Singh *et al.*, (1984) used Uppsala Wiechert seismograms and station bulletins to estimate surface-wave magnitudes, M_S. Amplitude readings, A_E and A_N, of 20-sec surface waves were taken from the Uppsala bulletin and whenever possible also checked on Wiechert seismogram copies. It has been shown that Uppsala amplitude data are complete for $M_S \geq 6.6$ from 1906 to 1952 and for $M_S \geq 6.3$ from 1953 to 1966. The increase in the magnitude sensitivity is most likely due to an increase in the number of instruments in Sweden, in the 1950's and 1960's, as well as to improved communication among seismological observatories.

Successful attempts have lately been made to compare seismograms of historical and more recent earthquakes from the same area, recorded at the same seismograph station, i.e. by the same instrument, in order to extract information on focal processes. For example Bakun and McEvilly (Clark, 1984) compared De Bilt records from three Parkfield earthquakes which took place in 1922, 1934 and 1966. They quickly revealed that the records remarkably resemble each other, indicating that the "same earthquake" is taking place over and over. Following the above procedure, we compared in a similar way, two major shocks that occurred in northern Algeria in 1954 and 1980. Corresponding source parameters are listed in Table 2. The following steps were performed on the data: (1) The original (i.e. no enlargements have been introduced) smoked-paper seismograms from both events were digitized at non-equal time intervals. (2) Correction for the offset of the arm pivot was introduced. (3) The signal was corrected for the finite arm length. (4) Data were interpolated to produce equally spaced (0.5 sec) points and smoothed by a moving averaging over three successive points. In Figure 3, the four steps are demonstrated on real Wiechert records. Comparison of corrected records from both events, amplified to the same maximum amplitude level is made in Figure 4. Due to rather weak recorded P-waves, only segments including S- and SS-arrivals and the initial part of surface waves are shown.

With respect to the proximity of the two hypocenters (Table 2), the propagation paths to Uppsala are practically the same. Then, the astonishing resemblence of both records, as seen in Figure 4, suggests high similarity in corresponding focal processes. The available fault-plane solutions (McKenzie, 1972; Deschamps *et al.*, 1982) both show a thrust faulting on planes striking ENE and NE for the 1954 and 1980 event, respectively.

Table 2. Source Parameters of Earthquakes Analyzed

Focal Area	Date	Epicentral Coordinates °N	°E	Focal Depth km	M
N. Algeria Orléansville	Sept 9, 1954	36.20	1.6[1]	n[2]	6.5[2]
N. Algeria El Asnam	Oct 10, 1980	36.16	1.4[3]	4[3]	7.2[3]

[1] ISS
[2] Rothé
[3] ISC

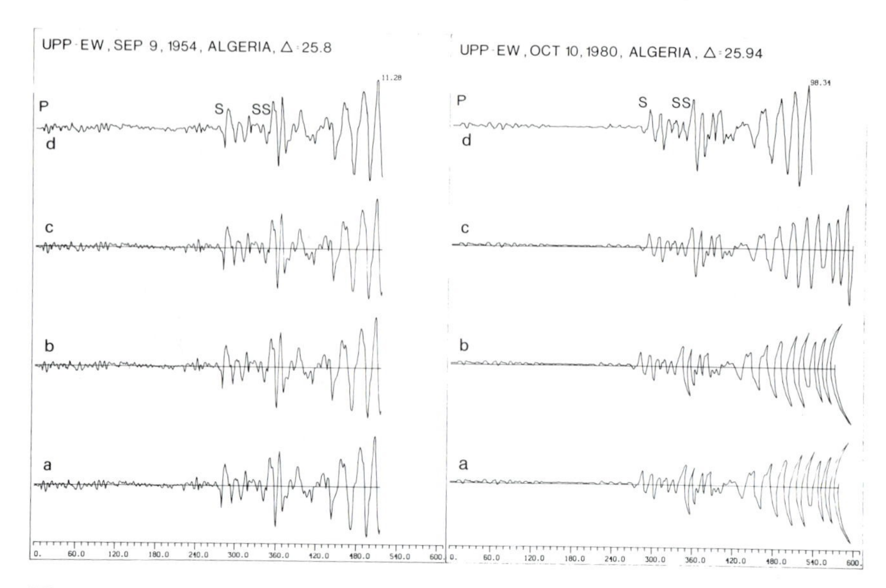

Figure 3. Uppsala Wiechert seismograms (east-west components) from two major earthquakes in northern Algeria in 1954 (left) and 1980 (right). The processing steps are as follows (from bottom to top): (a) original record digitized at non-equal time intervals; (b) correction for the arm-pivot offset introduced; (c) correction for the finite arm length introduced; (d) data interpolated and smoothed. Time in seconds is given at the bottom of the figure. For the sake of comparison, amplitudes are adjusted to the same level. Relative amplitudes are given at right edges of both top lines.

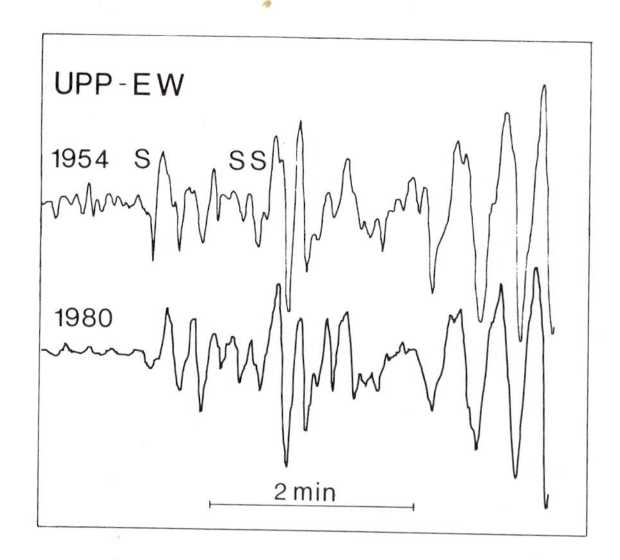

Figure 4. Comparison of corrected seismograms from the 1954 and 1980 earthquakes in northern Algeria. Maximum amplitudes are adjusted to the same level.

Obviously, the method cannot reveal all fine details of a rather complex rupture process. On the other hand, as illustrated by the Parkfield and northern Algeria examples, basic information on faulting styles may be revealed by the comparative analysis described above. It seems that this is one of a few currently available methods for studying focal processes of large earthquakes, in more active areas, from the beginning of this century when the geographical coverage by seismograph stations was rather poor and seismograms were scarce.

ACKNOWLEDGEMENTS

I wish to thank my colleague W. Y. Kim for discussing the subject and for producing the computer plots.

REFERENCES

Alsan, E., L. Tezucan, and M. Båth (1975). An earthquake catalogue for Turkey for the interval 1913-1970, Kandilli Observatory, Istanbul, and Seism. Inst., Uppsala, Report 7-75, 166 pp.

Clark, R. D. (1984). Earthquake prediction, *Geophysics: The Leading Edge of exploration*,**3**(8), 24-28.

Deschamps, A., Y. Gaudemer, and A. Cisternas (1982). The El Asnam, Algeria, earthquake of 10 October 1980: multiple-source mechanism determined from long-period records, *Bull. Seismol. Soc. Am.*, **72**, 1111-1128.

McKenzie, D. P. (1972). Active tectonics of the Mediterranean region, *Geophys. J. R. Astr. Soc.*, **30**, 109-185.

Rothé, I. P. (1969). *The Seismicity of the Earth 1953-1965*, UNESCO, 336 pp.

Singh, S. K., M. Rodriguez, and I. M. Espindola (1984). A catalogue of shallow earthquakes of Mexico from 1900 to 1981, *Bull. Seism. Soc. Am.*, **74**, 267-279.

Analysis of Historical Seismograms of Large Mexican Earthquakes ($M_S \geq 7.0$): Summary of Important Results

S. K. Singh
Instituto de Geofisica, U.N.A.M.
C.U., Mexico 04510, D.F., Mexico

ABSTRACT

Analysis of historical seismograms recorded at teleseismic distances as well as at stations of the Mexican Seismological Service has greatly improved our understanding of the tectonics and the seismic risk along the Mexican subduction zone. Important results include (1) homogeneous earthquake catalog for $M_S \geq 6.5$ since 1906, (2) strong evidence of bulge in log $N - M_S$ plot suggesting characteristic earthquakes with $M_S \simeq 7.7$, (3) complexity/simplicity of P-waveforms of large earthquakes along the Mexican subduction zone, (4) discovery that 15 January 1931 Oaxaca earthquake ($M_S = 8.0$) was an inland normal fault which probably broke the entire subducted Cocos plate, (5) seismic moments of large earthquakes ($M_S \geq 7.0$) since 1906, (6) evidence that the great Jalisco earthquake of 3 June 1932 ($M_S = 8.2$) broke the Rivera-North American plate interface showing that small, young plates subducting with small relative plate velocity are capable of generating great earthquakes, and (7) reexamination of seismic gaps and recurrence periods of large earthquakes along the Mexican subduction zone. We summarize these results which show the enormous value of the historical seismograms. Even though the quality and the quantity of historical seismograms as compared to the modern ones, is limited, there is a wealth of extremely useful information which can be extracted from them. Of great value are the seismograms from stations which have kept the old seismographs in continuous operation.

1. Introduction

The purpose of this paper is to summarize some of the important recent results related to tectonics and seismic risk in Mexico obtained from the analysis of historical seismograms. By historical seismograms we shall mean seismograms recorded prior to the installation of World-Wide Standard Seismograph Stations (WWSSN) in 1962. In much of our analysis we have used seismograms from the stations of the Mexican Seismological Network and some key stations from Europe (Uppsala, Göttingen, DeBilt, Stuttgart, and Copenhagen, among others).

2. A Catalog of Shallow Mexican Earthquakes

A homogeneous catalog is essential in the statistical analysis of earthquakes. Although a catalog of Mexican earthquakes was prepared by Figueroa (1970), the magnitudes in this catalog are not reliable because of (a) inconsistent use of an uncalibrated local and regional magnitude scale, (b) clipping of seismograms for

Historical Seismograms
and Earthquakes of the World

larger earthquakes and, (c) lack of proper calibration of seismographs. In order to prepare a homogeneous catalog we consulted Wiechert seismograms and station bulletins from Uppsala. The known magnitudes of Mexican earthquakes (M_S from Abe, 1981 for large earthquakes; M for other earthquakes listed in Gutenberg and Richter, 1954 and Rothé, 1969) were plotted as a function of zero-to-peak ground motion around 20 sec at Uppsala. Figure 1 shows the data for Uppsala. M_S is related to the ground motion by

$$M_S = \log A + 5.28, \tag{1}$$

where $A = (A_E^2 + A_N^2)^{\frac{1}{2}}$, A_E and A_N are maximum amplitude of ground motion in microns during surface-wave with period between 17 to 23 seconds. Note that no distance term is included in equation (1). This is because Δ for most events in Figure 1 was about 90°. Figure 1 also suggests that M in Gutenberg and Richter (1954) and Rothé (1969) for smaller events ($M < 7$) is significantly different from M_S. Equation (1) was used to estimate magnitudes of those events which were not listed in Abe (1981) (either because of incompleteness of Abe's catalog or because the events were smaller than $M_S = 7.0$) but for which amplitude data were available

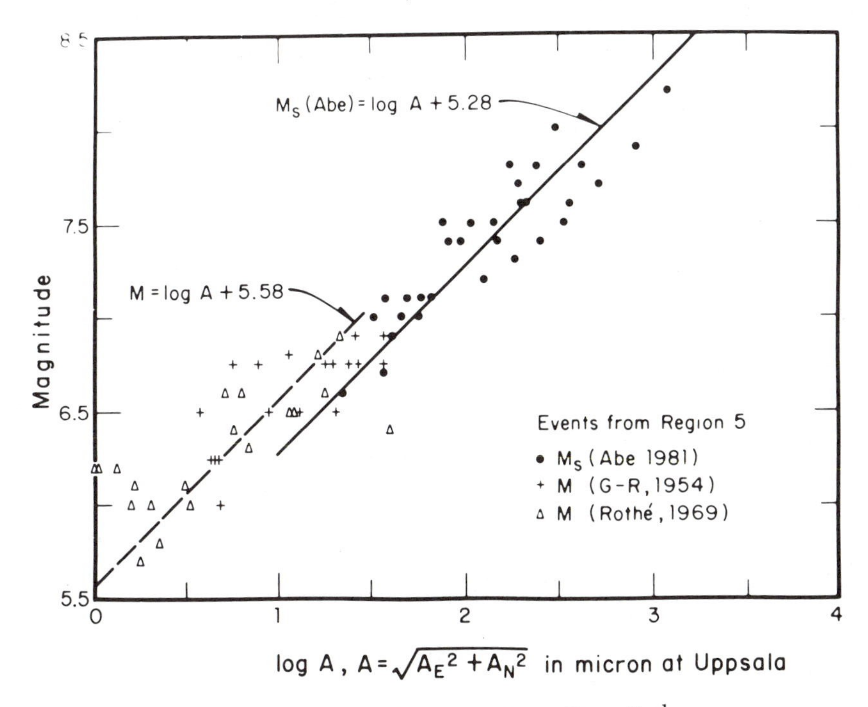

Figure 1. Magnitude vs. log A at Uppsala. $A = (A_E^2 + A_N^2)^{\frac{1}{2}}$, where A_E and A_N are ground motion during 20-sec surface waves on east-west and north-south component, respectively. Events ($H \leq 65$ km), which are from region 5 (10° - 20° N, 90.1° - 120° W), are about 90° from the station. Magnitudes are M_S from Abe (1981) (solid dots), M from Gutenberg and Richter (1954) (crosses) and Rothé (1969) (open triangles). Note that extrapolation of $M_S - \log A$ relation suggests that $M \neq M_S$ for $M < 7.0$ (From Singh *et al.*, 1984a).

from Uppsala. We also used Göttingen seismograms/bulletins to estimate magnitudes. Recall that M_S computed from (1) is 0.18 unit less that the M_S reported in Earthquake Data Report (EDR) of U.S. Geological Survey (Abe, 1981). Our catalog of shallow Mexican earthquakes (Singh *et al.*, 1984a) lists M_S so that it conforms to EDR. Tests show that it is complete for $M_S \geq 6.5$ from July 1906 to December 1981 (Singh *et al.*, 1984a).

Abe and Noguchi (1983) have further revised M_S values of large earthquakes in the period 1897-1912. This revision does not change equation (1), hence our catalog, significantly since only 3 events shown in Figure 1 have been assigned different M_S (on an average lower by 0.2 unit).

3. Statistics of Small Earthquakes and Frequency of Occurrence of Large Earthquakes: Evidence of Characteristic Earthquakes in Mexico.

Singh *et al.* (1981) found that Gutenberg-Richter (GR) relation, $\log N = a - bM_S$, was not valid for Mexico for $M_S \geq 7.0$; there was a deficiency of events for $7.0 \leq M_S \leq 7.4$ with respect to what one would expect from the GR relation. This result had great implications for seismic risk studies. It also implied that areas ruptured preferentially with certain maximum magnitude. With the completion of the new catalog the earthquake statistics were reexamined. Figure 2 shows 4 regions in which the subduction zone of Mexico was divided (Singh *et al.*, 1983). The plot of $\log N - M_S$ is shown in Figure 3 for the combined regions of Oaxaca, Guerrero, and Michoacán (Jalisco region was excluded because the seismicity in this region is related to the subduction of the Rivera plate and is expected to differ from the other 3 regions). In Figure 3, 75.5 yr data, taken from the catalog, as well as 18.5 yr data (since the installation of WWSSN) normalized to 75.5 yr, are plotted. For this

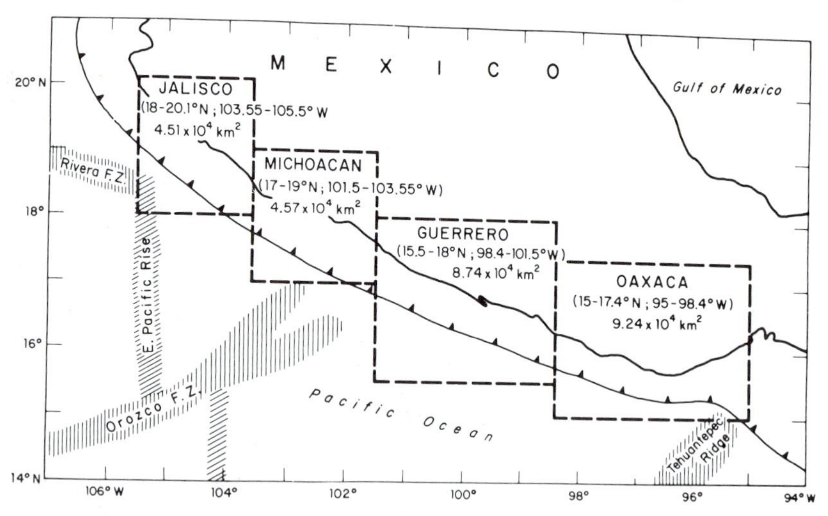

Figure 2. Regions along the Mexican subduction zone studied for the statistics of earthquakes. Seismicity of Oaxaca, Guerrero, and Michoacán regions is related to the subduction of the Cocos plate whereas the seismicity of Jalisco is mostly related to the subduction of the Rivera plate below Mexico. (From Singh *et al.*, 1983).

18.5 yr data m_b has been converted to M_S using 2 relations (shown by cross and solid dots in Figure 3):

$$M_S = 1.8\, m_b - 4.3 \quad \text{(Wyss and Habermann, 1982)},$$

$$M_S = 1.142\, m_b - 0.821 \quad \text{(Mexico; Singh } et\ al.\text{, 1983)}.$$

The data shows that the 18.5 yr data is representative of 75.5 yr data and that there is a "bulge" in log $N - M_S$ plot implying a deficiency of events in the range $6.5 \lesssim M_S \lesssim 7.4$. If the Oaxaca region is considered separately (Figure 4) the deviation of log $N - M_S$ plot from GR relation appears dramatic. In terms of seismic risk studies this observation is of paramount importance since fitting a straight line, log $N = a - b\ M_S$, to the data is grossly inadequate. The data suggests a deficiency of earthquakes in the range $6.4 \lesssim M_S \lesssim 7.4$ which is seen more clearly in Figure 5. The earthquakes in Oaxaca appear to have a characteristic magnitude of $M_S \simeq 7.7$;

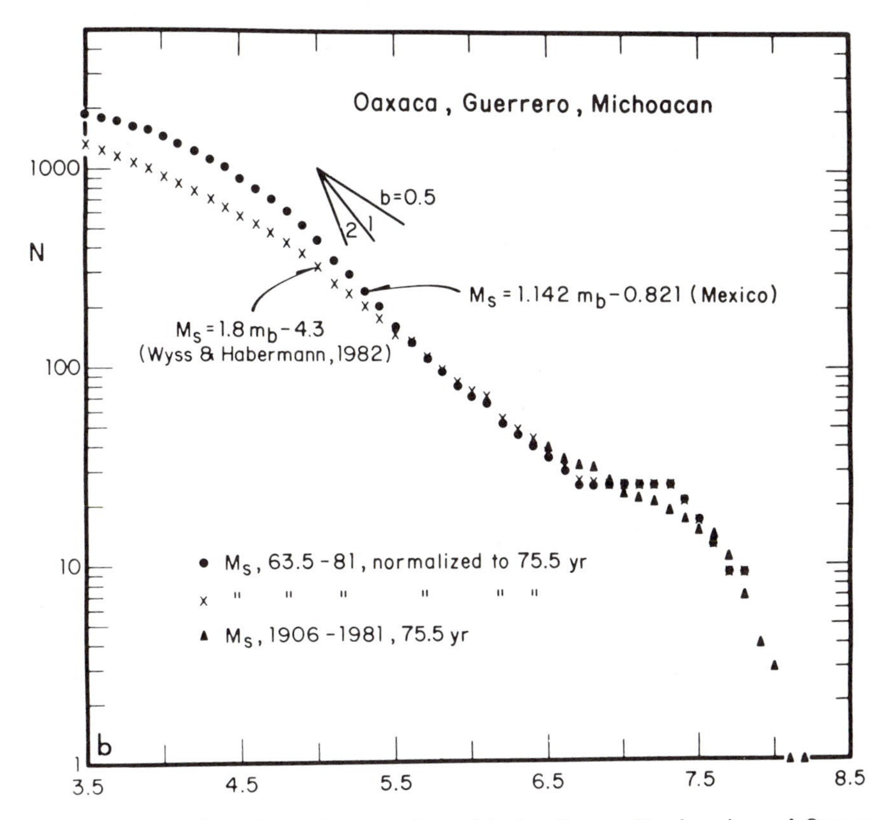

Figure 3. N (cumulative frequency) vs. M_S for the combined regions of Oaxaca, Guerrero, and Michoacán. Data from July 1963 to December 1981 (18.5 yr) have been normalized to 75.5 yr (solid circles). Observed $N(M_S)$ data ($M_S > 6.4$) from July 1906 to December 1981 (75.5 yr) are shown by solid triangles. Note the bulge in log $N - M_S$ plot. (From Singh *et al.*, 1983).

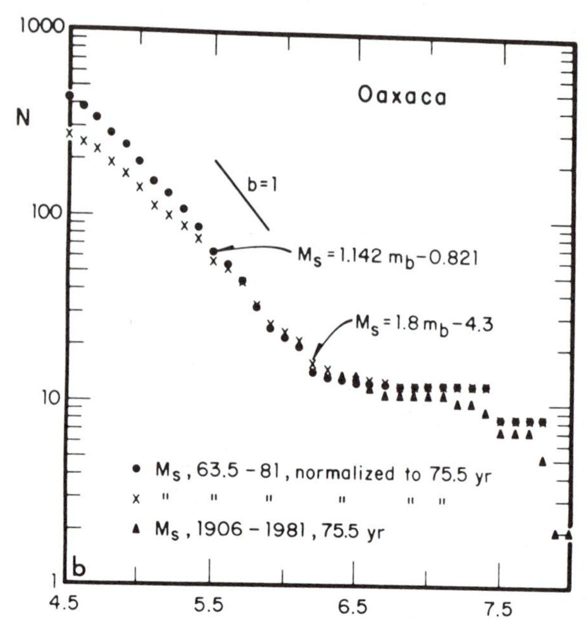

Figure 4. N vs. M_S for Oaxaca from the normalized as well as the 75.5 yr data. Symbols are the same as in Figure 3. (From Singh *et al.*, 1983).

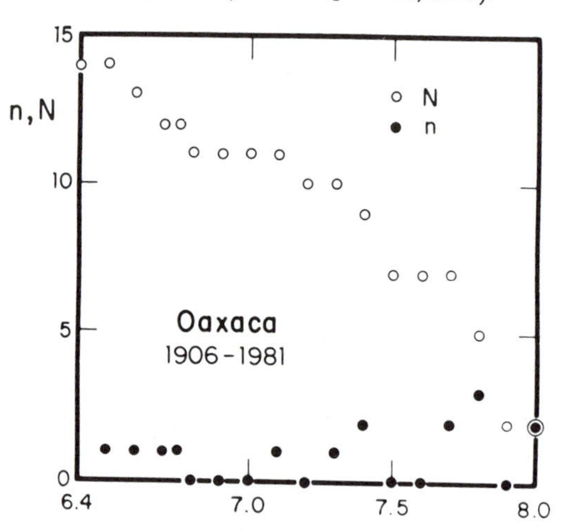

Figure 5. n and N vs. M_S for Oaxaca from the 75.5 yr data shown in greater detailed. n and N are plotted on linear scale. Peaks are seen in n at $M_S = 7.8$ and 7.4. Below $M_S = 7.4$, n is small (From Singh *et al.*, 1983).

the region ruptures with this magnitude repeatedly without giving rise to smaller magnitude earthquakes in the proportion required by the GR relation. Probably Oaxaca region provides the most clear seismological evidence in support of the idea of characteristic earthquakes (Singh *et al.*, 1983).

4. Complexity/Simplicity of P-Waveforms of Large Earthquake along the Mexican Subduction Zone

It was pointed out by Chael and Stewart (1982) and Astiz and Kanamori (1984) that P-waveforms of large subduction zone earthquakes of Mexico are very simple at teleseismic long period (LP) WWSSN stations with only few exceptions. Because of its implications to the tectonics and rupture mechanics, we decided to study the complexity/simplicity of P-waveforms of older earthquakes using historical seismograms. This presented difficulties because the response characteristics of the older seismographs (mostly Wiechert and Galitzins) are very different from LP WWSSN. A comparison of LP WWSSN seismograms at ESK of large Mexican earthquakes with the corresponding Wiechert and Galitzin seismograms at some European stations suggested that, in spite of different responses of the seismographs, qualitative and quantitative inferences regarding complexity/simplicity can be drawn from Wiecherts and Galitzins, respectively (Figure 6). Figure 7 shows Wiechert and Galitzin seismograms of large Oaxaca earthquakes recorded at European stations. With the exception of 1931 and 1950 earthquakes all earthquakes appear simple. We have now checked Galitzin seismograms from DeBilt and our previous conclusion remains valid. Note that the polarity of 1931 earthquake is inverted with respect to other earthquakes. The polarity at European stations for large thrust earthquakes in Mexico are always compressional. A dilatational first motion, then, suggests an anomalous earthquake. This led to the suggestion that the 1931 earthquake was a great (M_S = 8.0) normal or strike-slip faulting event (Singh *et al.*, 1984b).

Study of historic seismograms showed that large earthquakes in Oaxaca are mostly simple, in Guerrero and Michoacán simple or complex, whereas in Jalisco they are complex (Singh *et al.*, 1984b). [The recent great earthquake of Michoacán (19 September 1985; M_S = 8.1) consisted of 2 subevents as seen on LP WWSSN]. Synthetic and observed P-waves of some earthquakes are shown in Figure 8. Note that we have assumed a typical thrust focal mechanism based on recent studies for the synthetic modeling. Because of the location of European stations on the focal sphere, small changes in focal parameters do not affect the synthetics. Source time functions and depths estimated from just one European stations are considered reliable (Singh *et al.*, 1984a).

5. The Great Oaxaca, Mexico Earthquake of 15 January 1931: Lithospheric Normal Faulting in the Subducted Cocos Plate

As mentioned earlier the first motion at European stations for the 1931 earthquake suggested that the event was not a thrust fault (Figure 7). Location of the 1931 earthquake is shown in Figure 9. The focal mechanism constructed from many historical seismograms is shown in Figure 10 which suggests a normal faulting on a E-W plane dipping 56° to the north. Figure 11 shows synthetic and observed P-waves for Galitzin seismograms at DeBilt and LaPaz. The estimated depth of the source is about 25 to 40 km. A section through AA' in Figure 10 is shown in Figure 12. It appears that the 1931 earthquake broke the entire subducted Cocos plate. Such great normal faults generally occur near the trench (Sanriku type). The earthquake of 1931 occurred inland and destroyed the city of Oaxaca. Great normal faulting earthquake may occur elsewhere in Mexico and add another element in seismic risk estimation. Further details are given in Singh *et al.* (1985a).

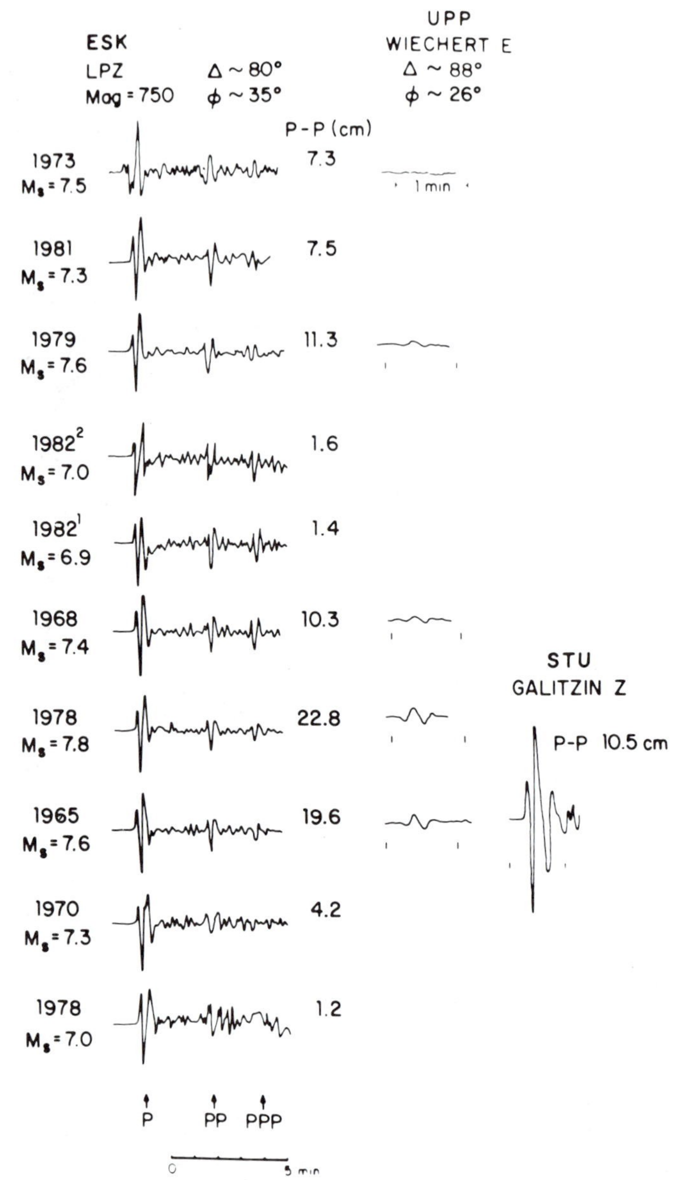

Figure 6. LPZ WWSSN seismograms from ESK for large earthquakes along the middle American trench. Note that except for the 1973 and the 1970 earthquakes, all P-waves are simple. E–W Wiechert seismograms from UPP are also shown for some of these earthquakes. The 1973 earthquake which is complex at ESK also appears complex at UPP (low-amplitude high-frequency P-waves). Four other earthquakes at UPP appear simple just as they do at ESK. Like ESK and UPP seismograms the Galitzin-Wilip seismogram for the 1965 earthquake is simple. We conclude that the simplicity or complexity of the P-waves can be studied in a qualitative and a quantitive manner from the Wiechert and the Galitzin seismograms, respectively. (From Singh *et al.*, 1984b).

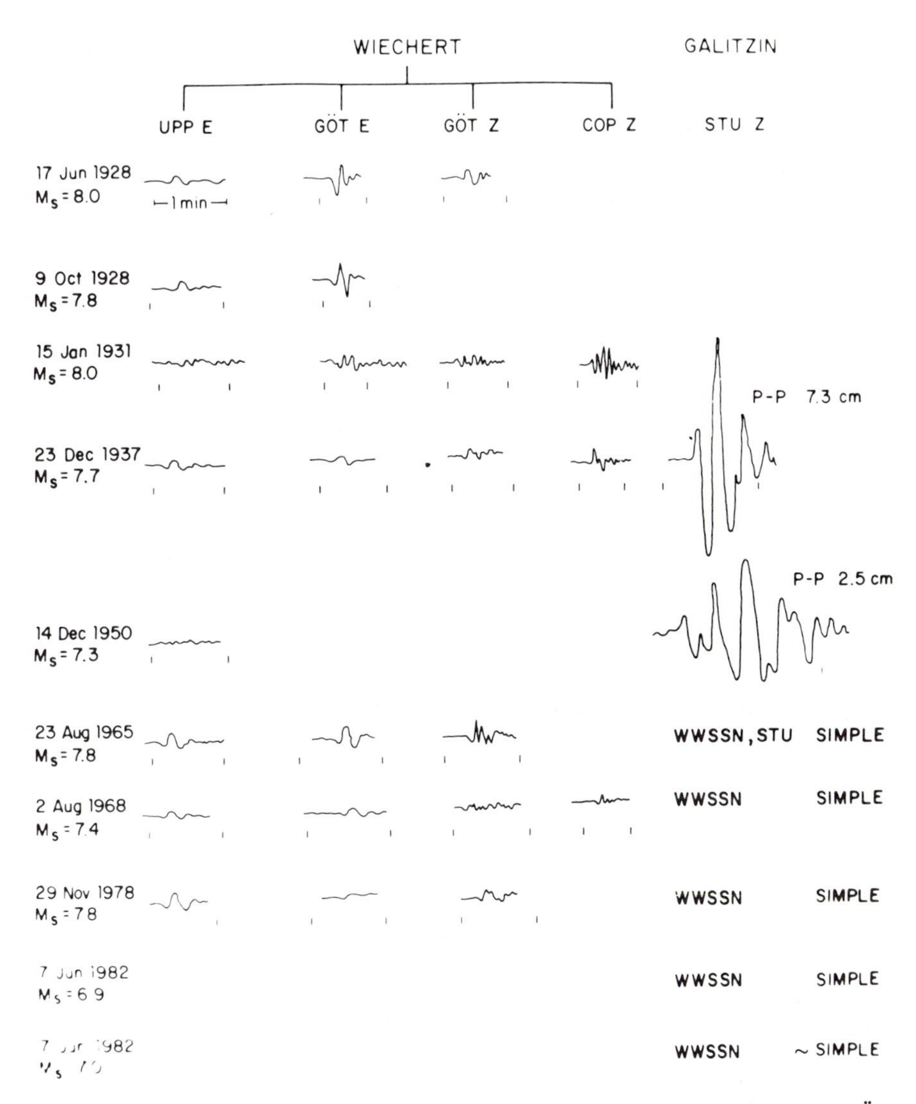

Figure 7. P-waves from Oaxaca earthquakes recorded on Wiecherts (at UPP, GÖT and COP) and Galitzin-Wilip (at STU). Note that all P-waves, except for the 1931 and the 1950 earthquakes, are simple. The earthquake of 1931 is complex with first motion at GÖT Z and COP Z being up. Compare it with first motion of other earthquakes which are down. (Polarity of these stations is reversed). Thus, the 1931 event has a drastically different focal mechanism. The 1950 earthquake is complex at UPP as well as STU. The earthquakes of 1982 are not big enough to generate visible P-waves on the Wiecherts. (From Singh *et al.*, 1984b).

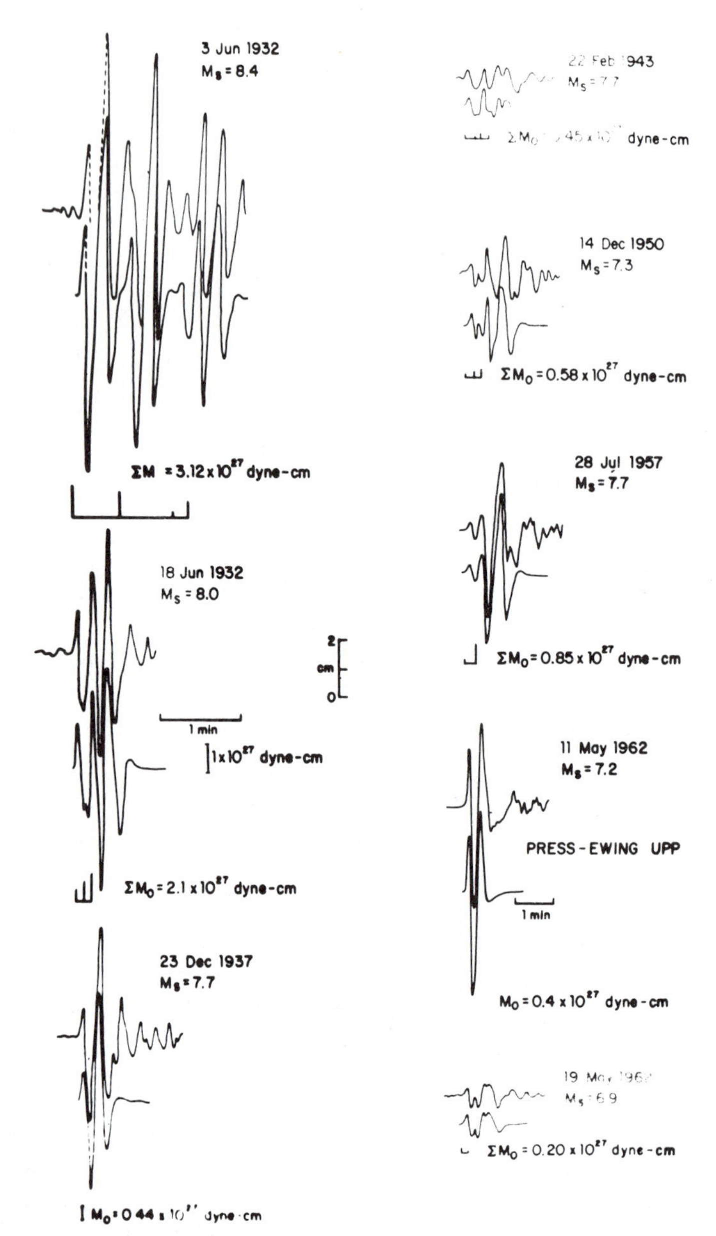

Figure 8. Observed (above) and synthetic (below) P-wave pairs of some large Mexican earthquakes. The number of subevents, their seismic moments, the time at which each acts, and the total seismic moment are indicated below each pair. The source parameters are given in detail in Singh *et al.* (1984b). Only the 1937 and the 1962 earthquakes require one source; other earthquakes are complex and require multiple sources. That part of observed trace which is not clear in the seismogram is shown by broken line. (From Singh *et al.*, 1984b).

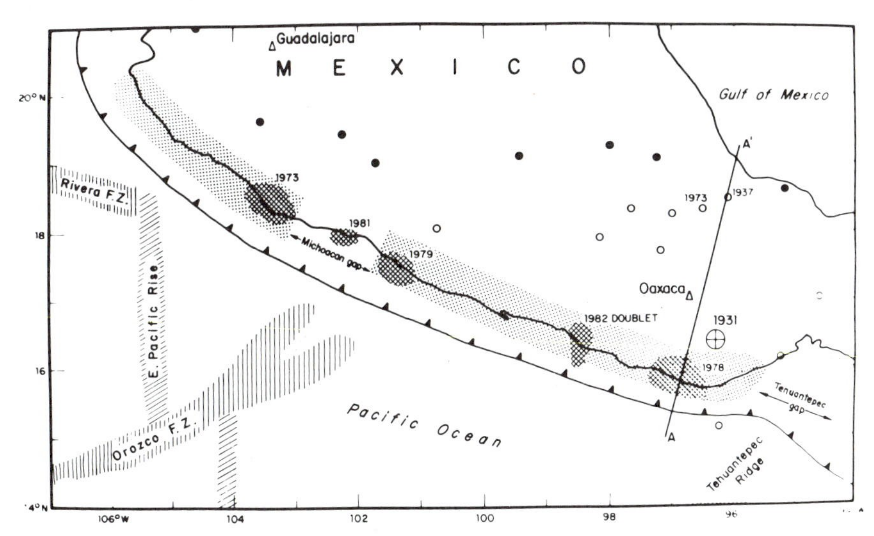

Figure 9. Location Map showing the epicenter of 1931 earthquake (circle with cross). Aftershock areas of recent large, shallow thrust earthquakes are shown cross-hatched. The dotted band (about 80 km wide) is the maximum width of strong coupling between the subducting Cocos plate and the overriding continent; this is also confirmed by aftershocks of the 19 September 1985 ($M = 8.1$) which broke the Michoacán gap (UNAM Seismology Group, 1986). Solid circles are recent volcanoes and open circles are epicenters of other normal fault earthquakes with body-wave magnitude greater or equal to 6.0. A section along A-A' is shown in Figure 12. (From Singh *et al.*, 1985a).

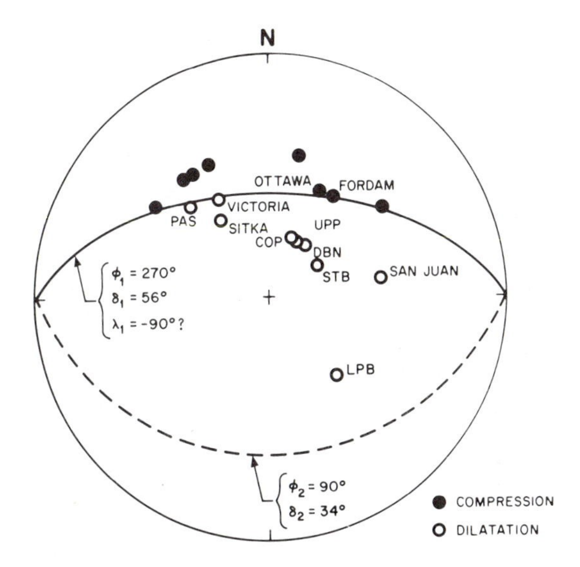

Figure 10. First motion data of 1931 earthquake on an equal-area lower-hemisphere projection. Open and closed circles represent dilatational and compressional first motions respectively. Unlabeled stations are from the Mexican Seismic Network. (From Singh *et al.*, 1985a).

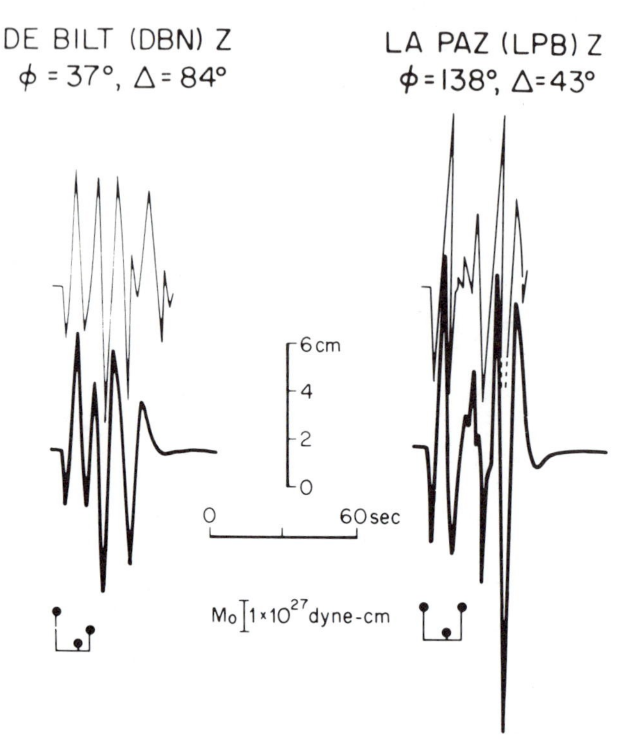

Figure 11. Observed (upper trace) and synthetic (lower trace) seismograms at DeBilt (DBN) and La Paz (LPB). Both DBN and LPB are Galitzin-Wilip seismographs (natural period of seismometer and galvanometer $\simeq$ 12 s). The largest down swing on LPB seismogram is clipped. Synthetics were generated with the focal mechanism shown in Figure 10 and required three point sources (depths 40, 25, and 25 km). Trapezoidal source time functions for each source were taken as 2, 5, and 7 s. The seismic moments (M_o) and the time delays of the sources are indicated in the figure. (From Singh *et al.*, 1985a).

6. Seismic Moments of Large Mexican Earthquakes

Historical seismograms (mostly Wiecherts from Uppsala) have been used to estimate seismic moments of old large earthquakes at a period of about 40 to 60 sec (Espíndola *et al.*, 1981; Singh *et al.*, 1982; Singh *et al.*, 1984b). Since the focal mechanism of large earthquakes and the paths to Uppsala are essentially constant, the spectral amplitudes of surface waves at 40 to 60 sec, after calibration with respect to recent earthquakes with known M_o, can be used to estimate M_o of older earthquakes. The basic assumption is that the 40 to 60 sec amplitude is on the flat part of the source spectra. This appears justified since the rupture lengths of large Mexican earthquakes are generally small ($\lesssim$ 100 km). The M_o estimates have been used in many studies. In particular, seismic potential in Acapulco-San Marcos region was quantified (Singh *et al.*, 1982). The historic seismicity and this century's earthquakes suggested the validity time-predictable model (Shimazaki and Nakata, 1980). There appears a high probability of a large earthquake in the region between now and 1993 (Singh *et al.*, 1982).

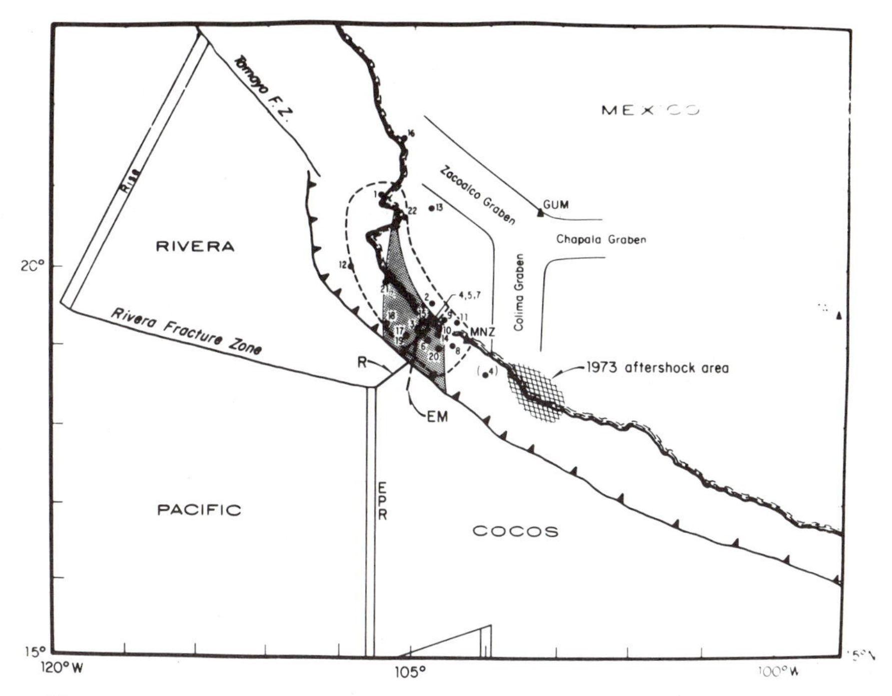

Figure 13. Tectonic setting of the Jalisco region. The boundary between Rivera and Cocos plate is uncertain; the solid line (marked R) and dashed line (marked EM) are the boundaries inferred by Reid (1976) and Eissler and McNally (1984), respectively. Stations used in the analysis are shown by solid triangles. The numbers associated with the locations of the aftershocks are in chronological order. The dashed contour is the total estimated aftershock area of the 3 June and 18 June, 1932 earthquakes. A dashed line divides this area; the NW and the SE parts roughly outline 3 June and 18 June aftershock areas, respectively. Shaded region is the early aftershock area (3 to 8 June) estimated from GUM and TAC. Aftershock area of 1973 Colima earthquake is taken from Reyes *et al.* (1979).

had only concentrated on the analysis of modern seismograms. While all historical seismograms are useful, the observatories which have been in continuous operation and have kept operating the old seismographs, in spite of many inconveniences, provide the most valuable data. Old and new seismograms from such observatories can then be directly compared and meaningful and reliable conclusions can be reached regarding older earthquakes.

ACKNOWLEDGEMENTS

The review is biased; only the results of the analysis in which the author participated are emphasized. The author thanks his many colleagues. Acknowledgement is made to the seismic observatories of Uppsala, DeBilt, Göttingen, Stuttgart, Strasbourg, Copenhagen, and La Paz for making available copies of seismograms. Part of this research was funded by Consejo Nacional de Ciencias y Tecnología (CONACYT) of Mexico.

REFERENCES

Abe, K. (1981). Magnitude of large shallow earthquakes from 1904 to 1980, *Phys. Earth Planet. Interiors*, **27**, 72-92.

Abe, K. and S. Noguchi (1983). Revision of magnitudes of large shallow earthquakes, 1897-1912, *Phys. Earth Planet. Interiors*, **33**, 1-11.

Astiz, L. and H. Kanamori (1984). An earthquake doublet in Ometepec, Guerrero, Mexico, *Phys. Earth Planet. Interiors*, **34**, 24-45.

Chael, E. P. and G. S. Stewart (1982). Recent large earthquakes along the middle American Trench and their implications for the subduction process, *J. Geophys. Res.*, **87**, 329-338.

Eissler, H. K. and K. C. McNally (1984). Seismicity and tectonics of the Rivera plate and implications for the 1932 Jalisco, Mexico earthquakes, *J. Geophys. Res.*, **89**, 4520-4530.

Espíndola, J. M., S. K. Singh, Y. Yamamoto, and J. Havskov (1981). Seismic moments of large Mexican subduction earthquakes since 1907, *EOS*, **62**, 948.

Figueroa, J. (1970). Catálogo de sismos ocurridos en la República Mexicana, *Rept No. 272, Instituto de Ingeniería*, U.N.A.M., Mexico.

Gutenberg, B. and C. F. Richter (1954). *Seismicity of the Earth and Associated Phenomena*, Princeton Univ. Press, Princeton, N.J., 310 pp.

Heaton, T. H. and H. Kanamori (1984). Seismic potential associated with the subduction in the northwestern United States, *Bull. Seism. Soc. Am.*, **74**, 933-942.

Kelleher, J., L. Sykes, and J. Oliver (1973). Possible criteria for predicting earthquake locations and their application to major plate boundaries of the Pacific and the Carribbean, *J. Geophys. Res.*, **78**, 2547-2585.

Reid, I. D. (1976). The Rivera plate: A study in seismology and tectonics, *Ph.D. Thesis*, Univ. of Calif., San Diego.

Reyes, A., J. N. Brune, and C. Lomnitz (1979). Source mechanism and aftershock study of the Colima, Mexico earthquake of January 30, 1973, *Bull. Seism. Soc. Am.*, **69**, 1819-1840.

Rothé, J. P. (1969). *The Seismicity of the Earth, 1953-1965*, UNESCO, Paris, 336 pp.

Shimazaki, K. and T. Nakata (1980). Time-predictable recurrence model for large earthquakes, *Geophys. Res. Lett.*, **7**, 279-282.

Singh, S. K., L. Astiz, and J. Havskov (1981). Seismic gaps and recurrence periods of large earthquakes along the Mexican subduction zone: A reexamination, *Bull. Seism. Soc. Am.*, **71**, 827-843.

Singh, S. K., J. M. Espíndola, J. Yamamoto, and J. Havskov (1982). Seismic potential of Acapulco-San Marcos region along the Mexican subduction zone, *Geophys. Res. Lett.*, **9**, 633-636.

Singh, S. K., M. Rodríguez, and L. Esteva (1983). Statistics of small earthquakes and frequency of occurrence of large earthquakes along the Mexican subduction zone, *Bull. Seism. Soc. Am.*, **73**, 1779-1796.

Singh, S. K., M. Rodríguez, and J. M. Espíndola, (1984a). A catalog of shallow earthquakes of Mexico from 1900 to 1981, *Bull. Seism. Soc. Am.*, **74**, 267-279.

Singh, S. K., T. Domínguez, R. Castro, and M. Rodríguez (1984b). P-waveform of large, shallow earthquakes along the Mexican subduction zone, *Bull. Seism. Soc. Am.*, **74**, 2135-2156.

Singh, S. K., G. Suárez, and T. Domínguez (1985a). The great Oaxaca, Mexico earthquake of 15 Jan., 1931: Lithospheric normal faulting in the subducted Cocos plate, *Nature*, **317**, No. 6032, 56-58.

Singh, S. K., L. Ponce, and S. P. Nishenko (1985b). The great Jalisco, Mexico earthquakes of 1932: Subduction of the Rivera plate, *Bull. Seism. Soc. Am.*, **75**, 1301-1313.

UNAM Seismology Group (1986). The September 1985 Michoacan Earthquakes: Aftershock distribution and history of rupture, *Geophys. Res. Lett.*, **13**, 573-576.

Wyss, M. and R. E. Habermann (1982). Conversion of m_b to M_S for estimating the recurrence time of large earthquakes, *Bull. Seism. Soc. Am.*, **72**, 1651-1662.

Application of Modern Techniques to Analysis of Historical Earthquakes

S. Stein and E. A. Okal
Department of Geological Sciences
Northwestern University, Evanston, IL 60201, USA

D. A. Wiens
Department of Earth and Planetary Sciences
Washington University, St. Louis, MO 63130, USA

ABSTRACT

Analysis of historical earthquakes is essential for interpretations of seismicity in regions away from recognized plate boundaries, due to long earthquake recurrence periods. We have applied a number of techniques developed for WWSSN data to a variety of instrumental records dating back to the 1930's. Basic focal mechanism constraints can be obtained from P- and S-waves polarities and S-wave polarizations. Surface wave amplitudes and body wave modeling provide further mechanism constraints and allow seismic moment estimation. Body wave modeling also yields source depth and time function information. These techniques can be combined to find a solution which simultaneously fits the different data types when only a few records are available. In one approach, the moment variance reduction technique, the model space is systematically searched for the mechanism which provides the best fit to P-, SH-, Love, and Rayleigh wave amplitudes. We present three examples where significant tectonic processes were identified from analysis of historical earthquake data. Such data first demonstrated the large magnitude strike-slip "intraplate" seismicity along the "aseismic" Ninetyeast Ridge, showed the complexity and extent of the deformation in the Chagos region, and provided the key constraints for a new tectonic model in which a diffuse plate boundary runs east from the Central Indian Ridge to the Ninetyeast Ridge, and northward along the Ninetyeast Ridge to the Sumatra Trench. Similarly, historical earthquakes along the eastern Canadian passive margin demonstrate that such margins can remain tectonically active, especially in the presence of stresses due to deglaciation or rapid sediment loading. Finally, the historical seismicity of the young lithosphere in the southeast Pacific provides important information on the complex stress release mechanism in such areas.

1. Introduction

With the advent of the WWSSN in the early 1960's and of digital networks in the late 1970's, the past 25 years have seen dramatic increase in the quantity and quality of seismological data. As a result, many investigations relying on compilations of seismicity restrict themselves to the recent data, and overlook the existence and availability of useful seismograms predating 1962, despite the fact that teleseismic instrumental recording dates back to 1889. In particular, historical seismological data (defined here as pre–WWSSN) are essential for studies of earthquakes aimed at understanding the tectonics of areas with longer seismic recurrence intervals, such as intraplate regions or slow, diffuse plate boundaries. In this application, restriction of the analysis to recent earthquakes can lead to completely erroneous results.

The purpose of this paper is to demonstrate the application of modern analysis techniques to historical seismological data. Naturally, the results are more uncertain than for recent data largely for three reasons. First, the number of available records is much less, and thus the dataset from which results are drawn is much smaller. Second, instrument responses are much less well known. Finally, the low gain of many historical seismometers prevents study of many smaller events. In general, despite these difficulties, we can obtain reasonable and valuable results.

Our examples will be drawn from oceanic "intraplate" earthquakes, as they provide excellent examples of the need for, and value of, historical earthquake studies. Typically, in these situations, the recurrence interval is long enough that post–WWSSN data alone frequently give a misleading impression. Moreover, due to their oceanic locations, cultural accounts of older earthquakes are often not available. Thus the results of historical earthquake studies, despite the uncertainties inherent in such work, are essential precisely because they provide primary tectonic information. For example, the Ninetyeast Ridge in the Central Indian Ocean was considered "aseismic", and the eastern Canadian passive margin was considered tectonically inactive, until historical earthquakes were analyzed. This effect should not be surprising; based on instrumental seismicity alone (even including pre–WWSSN records), neither the southern San Andreas fault nor the New Madrid, Missouri area would be considered zones of intense seismic activity. Only cultural accounts of the 1857 and 1811-12 earthquakes provide the key evidence for their true tectonic environment.

Less than ten years ago historical earthquake analyses could be considered unorthodox and suspect. Nonetheless, as the number of such investigations started growing, their results were important enough to draw attention. The present IASPEI symposium, on "Historical Seismograms and Earthquakes", serves proof that these endeavors are now generally accepted as an important and valuable seismological tool. Rather than attempt to summarize the many such studies, our purpose in this paper is to offer a perspective by briefly reviewing some of the sources of information and methods of analysis we have used and some of our most significant results.

2. Sources

A number of sources of information are helpful in identifying historical earthquakes of significance and conducting further studies. Perhaps the single most valuable reference, and the starting point of any study, is Gutenberg and Richter's classic *Seismicity of the Earth* (1965). We will see that references to it would have avoided several misconceptions, notably that the Ninetyeast Ridge and the Canadian passive margin were aseismic. The 1965 edition, a re-issue of the 1954 edition, discusses earthquakes through 1952. The book divides earthquakes into various regions, and classifies them as shallow, intermediate, or deep. Gutenberg and Richter personally computed all magnitudes given in the book. Geller and Kanamori (1977) critically discussed their method, and concluded that these magnitudes are basically consistent with present day M_S. The data used for locations by Gutenberg and Richter can be found in the quarterly International Seismological Summary (ISS), an indispensable source which covers the years 1913-1963 and is the predecessor to the current *Bulletin* of the International Seismological Center (ISC). Gutenberg's annotated personal copy of the ISS, and his notepads which contain his original computations for relocation and magnitudes, are available on microfiche (Seismological Society of America, 1980). Rothé's *Seismicity of the Earth* (1969) is a bilingual continuation of Gutenberg and Richter's book, covering the years 1953–1965. During 1957–1963, the information contained in the ISS becomes more succinct, and smaller events are not included. Starting in 1964, the ISC's *Bulletin* and *Regional Catalogue of Earthquakes* provide the wealth of data generated by the full development of the networks.

Several worldwide listings of seismological stations were compiled starting in the 1930's. These are very useful in identifying which stations can be sources of data; additionally, they contain crucial information on the often murky issue of instrument types and responses. Most useful are the National Research Council's *List of Seismological Stations of the World* (McComb and West, 1931) the Royal Belgian Observatory's *Liste des Stations Séismologiques Mondiales* (Charlier and van Gils, 1953), and for the United States, the U.S. Geological Survey's *Historical Survey of U.S. Seismograph Stations* (Poppe, 1979).

3. Techniques

The methods used for analysis of abundant data for modern earthquakes can generally be used for historical earthquakes, with reasonable success. We shall see that first motions, waveform modeling for both P and S body waves, and surface wave amplitude and time domain synthesis can be very valuable. Thus, it is often possible to derive reasonably well constrained mechanisms, depths, and seismic moments.

Frequently, excellent first motions can be extracted from the data, allowing the polarities of P-, SH-, and SV-waves and S-wave polarizations to be used. Figure 1 shows a long period P-wave, recorded at La Paz, Bolivia for a 1947 $M_S = 6.9$ earthquake located about 300 km northwest of the Pacific-Antarctic Ridge studied by Okal (1984). The arrival is clearly compressional. Such first motion data constrain one nodal plane well (Figure 1). The second plane, though less well determined, fits a variety of constraints including the SV/SH ratio at one station and the low amplitudes of both Love and Rayleigh waves at another. The data thus require a large component of thrust faulting on a north-south plane. The result, despite this uncertainty, is tectonically significant, in that it excludes the simplest explanation, that of a mislocated transform fault event with the expected geometry, left-lateral strike-slip on a northwest trending plane; indeed, the record shown on Figure 1 proves the point by itself.

The International Seismological Summary (ISS) offers valuable P-wave first motion information for earthquakes from about the 1930's. Though much inferior to actual examination of the data, ISS first motions can be usefully employed when a set of adjacent stations report consistent polarities. Polarities are reported using the cryptic notation "a" or "k": "a" for "anaseismic" (away from the focus, or compressional; nowadays "C"), and "k" for "kataseismic" (backwards, to the focus, or dilatational; nowadays "D").

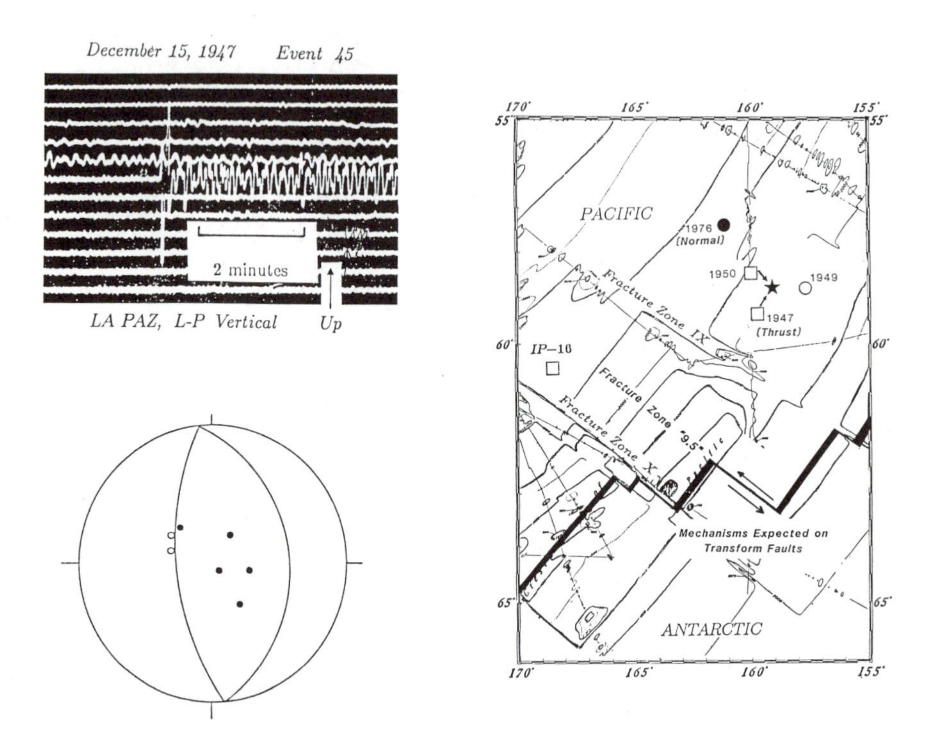

Figure 1. Sample record (top left), first motions (bottom left), and location (right) for the 1947 $M_S = 6.9$ southern Pacific earthquake [see Okal (1984) for details]. The first motion data (open circles dilatational; closed ones compressional) preclude the most likely tectonic setting, as left-lateral strike-slip would be the expected mechanism on a nearby transform fault.

Given enough data, and provided instrument responses are known, waveform modeling can be quite effective. Figure 2 shows P and SH waveforms for a 1944 M = 7.2 earthquake near the Chagos-Laccadive Ridge in the central Indian Ocean (Wiens, 1985). Synthetic body waves, generated for a thrust fault geometry derived from first motions and surface waves, match the data well and constrain the focal depth to about 12 km for P- and SH-waves at various stations. This modeling also shows the seismic moment of this event, 1.5×10^{27} dyne-cm is unusually large for an "intraplate" event.

Surface wave modeling can also be quite valuable. With abundant modern data, azimuthal coverage is generally adequate to construct an observed radiation pattern from equalized spectral amplitudes. This is hard to obtain for historical events. It is, however, often possible to employ the ratio of Love to Rayleigh waves at a small number of stations to provide a valuable constraint, especially if one nodal plane is constrained by first motions. Figure 3 shows an example for the 1939 $M_S = 7.2$ earthquake on the Ninetyeast Ridge, in the central Indian Ocean (Stein and Okal, 1978). First motions, mostly taken from the ISS, are scattered but show two clear groups which a nodal plane must bisect. Additional constraints come from surface wave ratios at two stations. Figure 3 shows a strainmeter record; after bandpass

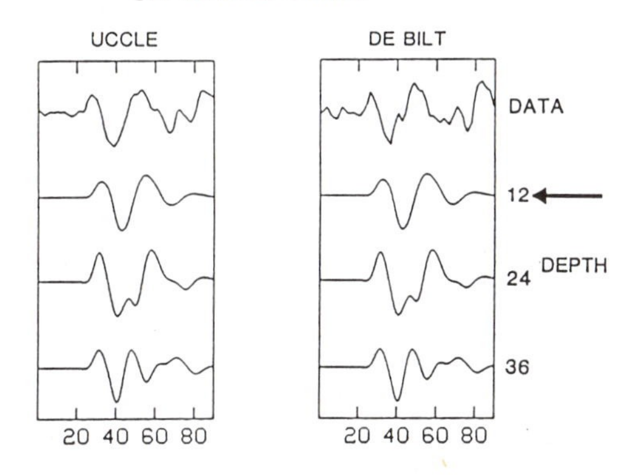

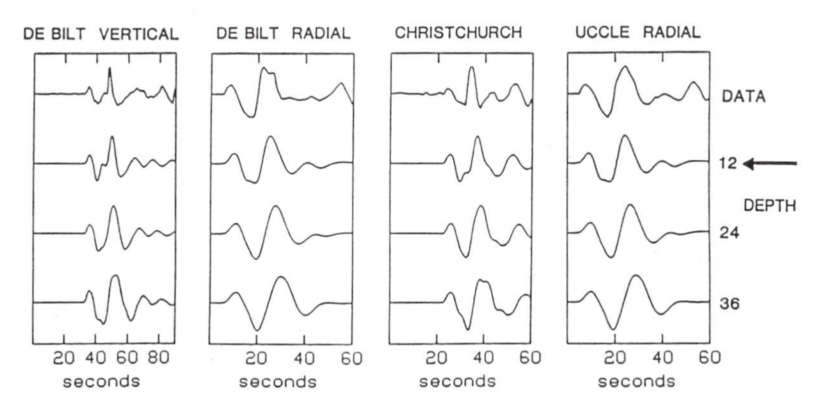

Figure 2. Synthetic body waves and data for the 1944 $M = 7.2$ Chagos-Laccadive Ridge earthquake (Wiens, 1985). The waveforms are well modeled for a focal depth of 12 km for a variety of stations and instruments.

filtering, to exclude periods below 100 s, the Love/Rayleigh ratio is about 1.5. Synthetic seismograms show that the Love/Rayleigh ratio is extremely sensitive to the slip angle on the north-south nodal plane. The best match is obtained for a slip angle of 55°, a mixture of thrust and left-lateral motion. The power of the method resides in that in many cases, the Love-to-Rayleigh ratio at a given station can span several orders of magnitude as a function of an unconstrained parameter such as slip angle, thus reducing considerably the influence of uncertainties in the gains of the various instruments used (which are inherent in the method since Love and Rayleigh waves are usually studied on different components). This is illustrated in Figure 4, for another earthquake near the Nintyeast Ridge. One fault plane is constrained by P-wave first motions, Love/Rayleigh ratios at Resolute and De Bilt uniquely determine the other nodal plane.

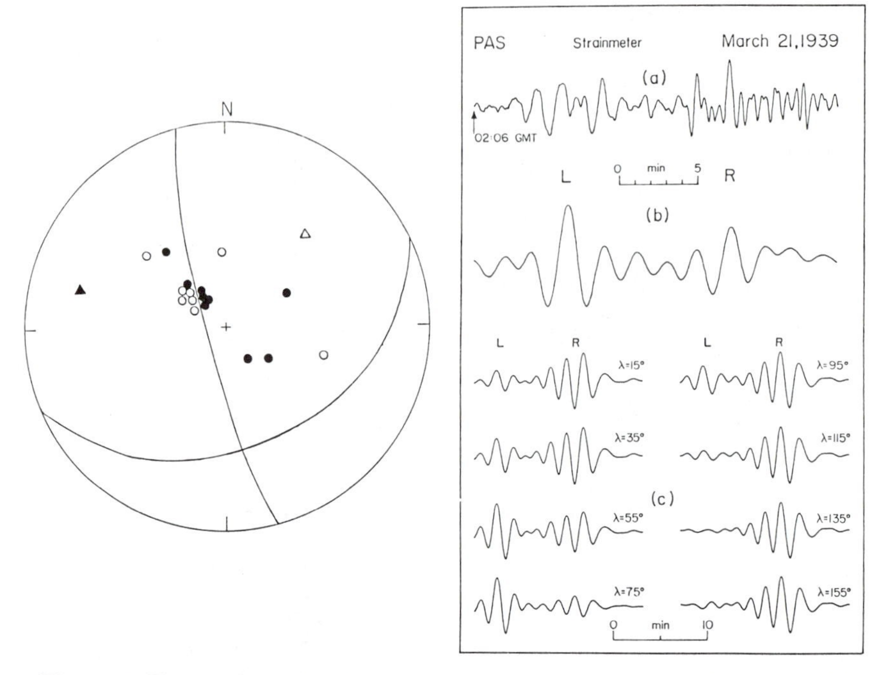

Figure 3. First motion and surface wave data for the 1939 $M_S = 7.2$ Ninetyeast Ridge earthquake (Stein and Okal, 1978). Black triangle is P axis; open triangle is T axis. Both nodal planes were obtained from surface wave data, subject to the first motion constraint (left). One surface wave datum, the Pasadena strainmeter record, is shown. The data were filtered for periods above 100 s, and modeled with various slip angles; 55 degrees yields the best fit.

Because of the limited number of records available and the often poor station distribution encountered in most studies of historical seismicity, it is desirable to analyze as many different types of data as possible in mechanism determination. Figure 5 shows an example of a particularly useful procedure, the moment variance reduction technique, applied to the 1944 Chagos earthquake mentioned previously (Wiens, 1985). The P-wave polarities are insufficient to constrain the mechanism, as all clear first motions read from actual records (large symbols) show compression. The plots on the left show the seismic moment as determined from P-, Love and Rayleigh waves at two stations of different azimuths, as a function of fault strike. The mechanism which yields the lowest variance in the seismic moments (and thus provides the best fit to the amplitudes of the various types of data), is a thrust fault with northwest striking nodal planes. (In the actual analysis, data from 4 stations were used, and the moment variance of all possible mechanisms, including those with strike-slip components, were computed. The mechanism shown produced the minimum moment variance.)

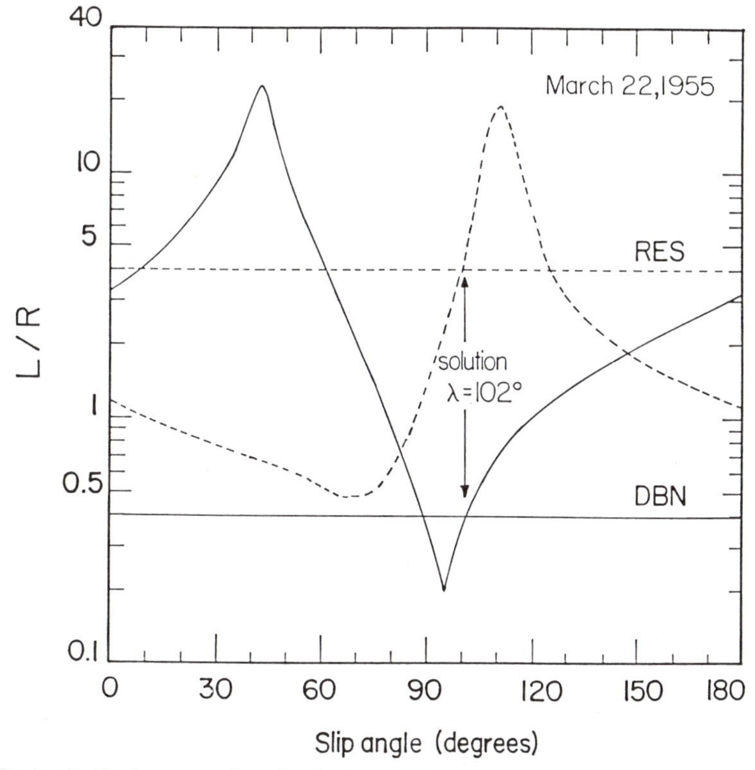

Figure 4. Surface wave data for the 1955 $M_S = 7.0$ earthquake just east of the Ninetyeast Ridge (Stein and Okal, 1978). One nodal plane is constrained by first motions. The slip angle on this plane must satisfy Love-to-Rayleigh ratios at two stations simultaneously.

4. Results

In this section, we show how seismological studies of historical earthquakes have cast decisive light on the large scale tectonic activity of three supposedly "aseismic" regions.

4.1. Indian Ocean

Tectonic models of the Indian plate are the best example of major geological results obtained from historical earthquake data. Until the late 1970's, the Indian plate was considered to be a conventional rigid plate, extending from the Central Indian Ridge to the subduction zones bordering the Pacific plate. Marine magnetic anomaly data showed that the plate was evolved from at least three previously distinct plates, now separated by linear highs known as "aseismic ridges" (Figure 6). Relative plate motion resolvable from the magnetic data along the Ninetyeast Ridge ceased 35–40 Ma ago (Sclater and Fisher, 1974; Eguchi *et al.*, 1979). Similarly, the Chagos-Laccadive Ridge was bordered by an active transform until 30 Ma ago (Fisher *et al.*, 1971).

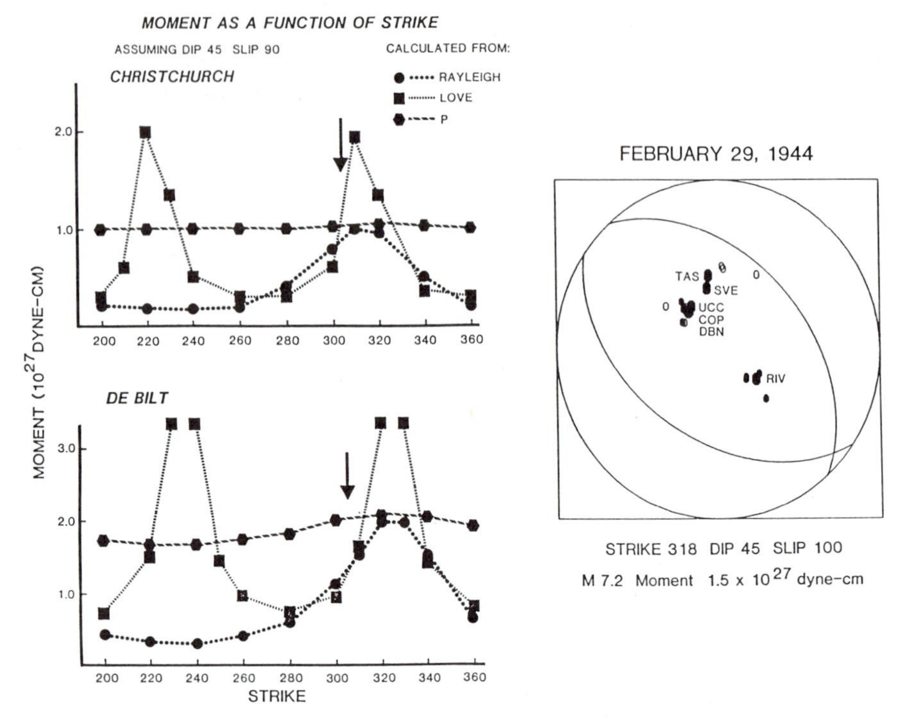

Figure 5. Mechanism determination for the 1944 $M = 7.2$ Chagos-Laccadive Ridge event (Wiens, 1985). All high quality first motions show compression; the final mechanism was constrained by minimizing the variation in seismic moments determined from body and surface waves. This process is shown for two stations.

On the other hand, until recently, present day global plate motion models (Le Pichon, 1968; Chase, 1972, 1978; Minster *et al.*, 1974) assumed a single rigid Indian plate. This assumption was reasonable given the seismicity maps in general use showing only events beginning in the 1960's (e.g., Tarr, 1974). Stein and Okal (1978) pointed out that examination of the historical seismicity yielded a very different picture. Far from being "aseismic", the Ninetyeast Ridge area is a quite active seismic zone; four magnitude seven or greater earthquakes (including one with $M_S = 7.7$) and ten magnitude six events have occurred in the general area since 1913 (Figure 6). This level of activity is much greater than along any spreading ridge and comparable to that along the southern San Andreas fault. The larger earthquakes occur along the segment of the Ninetyeast Ridge north of about 10°S. Stein and Okal (1978) pointed out that this phenomenon had, in fact, been noted by Gutenberg and Richter as early as 1941 in the first edition of *Seismicity of the Earth*:

> "A peculiarly isolated group of shocks occurs near 2°S, 89°E ... With other epicenters near 90°E north of the Equator, there is suggested a minor seismic belt following imperfectly known rises and ridges roughly north and south."

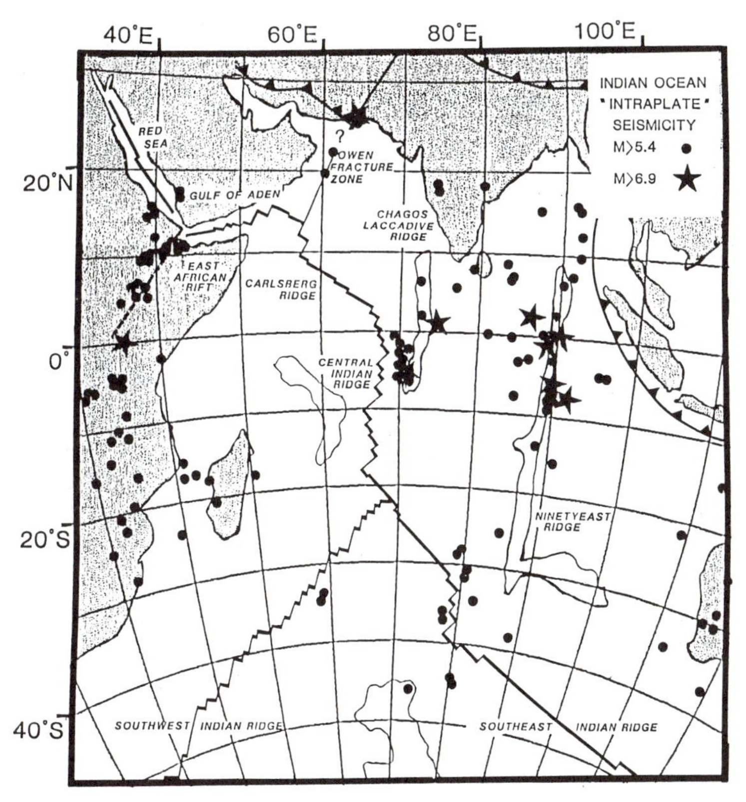

Figure 6. "Intraplate" earthquakes in the Indian Ocean region (1912-present) (Wiens *et al.*, 1985). Note the intense seismicity between the Ninetyeast and Chagos Ridges and the low level of seismicity along the Owen Fracture Zone.

Stein and Okal (1978) studied the largest historical earthquakes in the Ninetyeast Ridge area, including the 1939 $M_S = 7.2$ earthquake discussed earlier (Figure 3), and several more recent earthquakes. Based on these analyses, summarized by Figure 7, they concluded that:

> "The Ninetyeast Ridge is presently a complex zone of deformation within the Indian plate. The northern portion (3°N - 10°S) of the ridge is the active seismic zone, where both vertical and strike-slip motion occur, while further south the ridge is far less seismic ... The strike-slip motion is left lateral, which is consistent with the Indian (west) side encountering resistance due to the collision with Asia while the Australian (east) side is subducting smoothly at the Sumatra trench ... To the west the topography can be interpreted as the result of NW-SE compression which takes place largely aseismically but is observed for one large earthquake. This significant intraplate deformation may explain the difficulties that occur in attempts to close the India – Africa – Antarctica triple junction using a rigid Indian plate."

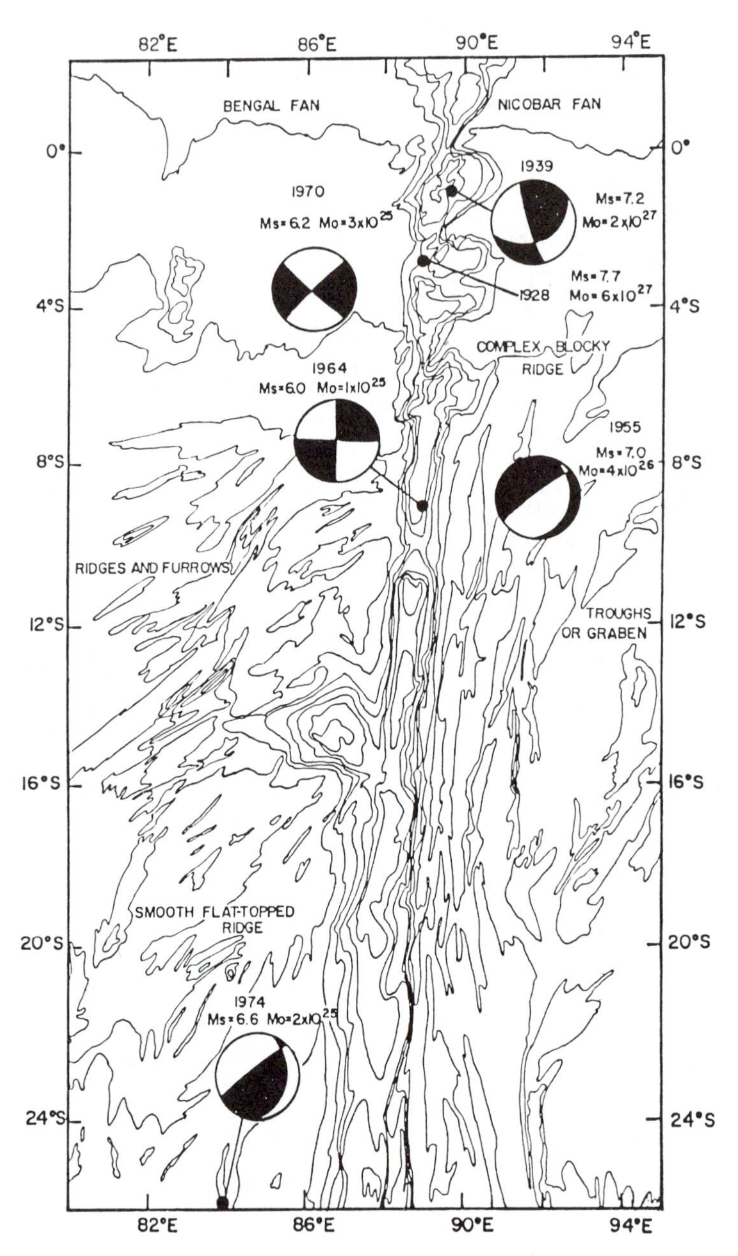

Figure 7. Results of Ninetyeast Ridge earthquake study (Stein and Okal, 1978). Left-lateral strike-slip motion occurs along the Ridge; north-south compression occurs to the west. Note that the moments of the 1928 and 1939 earthquakes are much larger than the recent events. The mechanism of the 1928 earthquake could not be reliably determined, though the data are consistent with a mechanism similar to that of the 1939 event.

They further estimated from seismic moment release that the slip rate along the Ninetyeast Ridge was about 2 cm/yr, which is greater than on some generally recognized plate boundaries. Moreover, no comparable oceanic "intraplate" seismicity occurs elsewhere. (The only other known oceanic intraplate magnitude seven earthquakes occur along passive continental margins or at sites of active volcanism like Hawaii.) The general tectonic model derived from the historical earthquakes has been confirmed by studies of recent earthquakes (Bergman and Solomon, 1985; Stein *et al.*, 1986).

An equally intriguing concentration of seismicity occurs along the Chagos-Laccadive Ridge, a parallel "aseismic" ridge to the west, where normal faulting earthquakes, including a 1965-1966 swarm (Stein, 1978) and a 1983 $M_S = 7.5$ event (Wiens and Stein, 1984), indicate N-S extension. These studies of recent seismicity seemed to indicate that the Chagos area was an isolated region of N-S tensional deformation, possibly similar to other normal faulting earthquakes found elsewhere in young oceanic lithosphere (Wiens and Stein, 1984; Bergman and Solomon, 1984). However, once again, the recent events are somewhat unrepresentative, as relocation of historical seismicity and analysis of the 1944 $M = 7.2$ thrust faulting event discussed earlier (Figures 2 and 5) show that the deformation in the Chagos area extends a considerable distance northeast of Chagos Bank and includes both thrust and normal faulting (Wiens, 1985). This finding suggests the Chagos seismicity is part of a deformation zone observed in the Central Indian Basin between the Chagos-Laccadive Ridge and the Ninetyeast Ridge, which includes unusual faulting and folding (Weissel *et al.*, 1980) and high heat flow (Geller *et al.*, 1983). Wiens (1985) proposed that the deformation observed at Chagos Bank, the Central Indian Basin, and the Ninetyeast Ridge indicates a continuous zone of deformation stretching from the Central Indian Ridge to the Sumatra Trench.

These observations provide a basis for investigating the Indian plate's internal deformation. Minster and Jordan (1978) noted that relative motion data in the Indian Ocean were poorly fit, as indicated by misclosure of the Indian Ocean triple junction, which suggests deviations from the rigid plate model used. Splitting the Indian plate along the Ninetyeast Ridge improved the fit to the data, and predicted motion in the Ninetyeast Ridge consistent with the seismological results. Stein and Gordon (1984) showed that separate Indian and Australian plates were statistically resolved, as the improved fit to plate motion data was greater than expected purely by chance, given that the new model has three additional parameters corresponding to the additional plate. Recent modeling (Cloetingh and Wortel, 1985) using the distribution of forces at the boundaries of the conventionally defined Indian plate, predicts a stress field with strong similarities to the observed seismicity, including extension in the Chagos area and large compressional stresses in the Ninetyeast area.

A recent model for Indian Ocean tectonics, which differs significantly from the conventional one, appears to explain the seismicity and plate motion data (Wiens *et al.*, 1985). The conventional model is that Australia and India are contained in a single Indian plate divided from an Arabian plate at a discrete boundary, the Owen Fracture Zone. In the new model (Figure 8), motion along the nearly aseismic Owen Fracture Zone is negligible, and Arabia and India are contained within a single Indo-Arabian plate. The Indo-Arabian plate is divided from the Australian plate by a diffuse boundary, which trends E-W from the Central Indian Ridge near

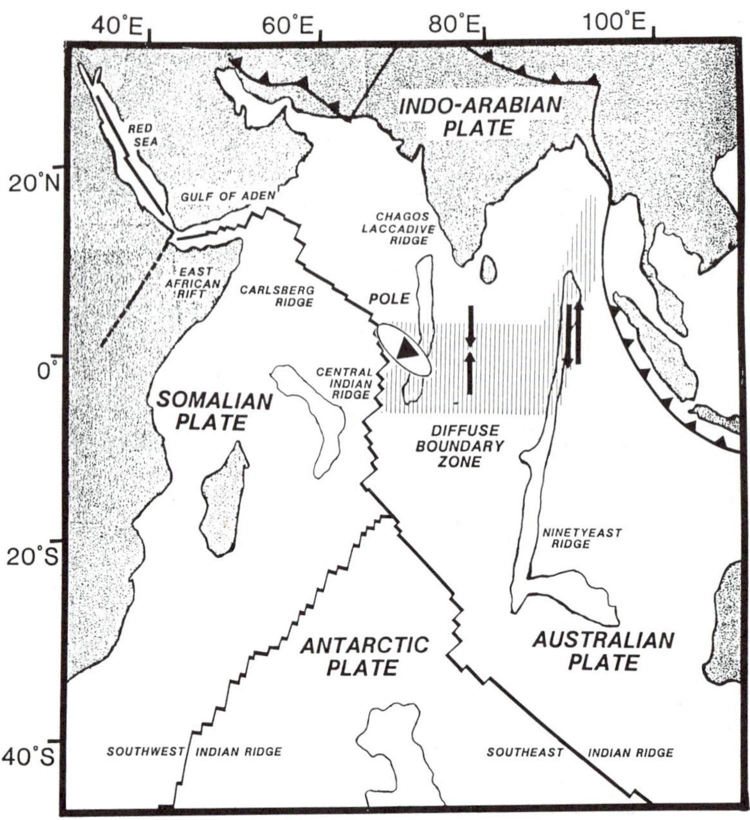

Figure 8. Schematic of the plate geometry proposed by Wiens *et al.* (1985). A diffuse boundary, with an Euler pole near the Chagos-Laccadive Ridge, separates Indo-Arabian and Australian plates. The Owen Fracture Zone is inactive. This geometry is consistent with both seismicity and plate motion data.

Chagos Bank to the Ninetyeast Ridge, and north along the Ninetyeast Ridge to the Sumatra Trench. This diffuse boundary is the zone of concentrated seismicity and deformation previously characterized as "intraplate". Relative motion data along the Carlsberg Ridge are fit significantly better by the new model than by the conventional model. The rotation vector of Australia relative to Indo-Arabia lies just east of the Central Indian Ridge, and is consistent with the about 2 cm/yr of left-lateral strike-slip observed seismologically along the Ninetyeast Ridge, with N-S compression in the Central Indian Ocean, and N-S extension near Chagos. The complexity of the deformation near Chagos is undoubtedly due to the proximity of the rotation vector. The diffuse boundary, possibly initiating in late Miocene time (Weissel *et al.*, 1980), may be related to the suturing of the Arabian plate to the Indian plate and the opening of the Gulf of Aden, or to the uplift of the Himalayas. The convergent segment of this boundary is unusual because neither a mountain range nor a subduction zone occurs along it. This segment may represent an early diffuse stage of the evolution of a convergent plate boundary; a process which may culminate with the onset of "true" subduction.

With the new model the non-closure of the Indian Ocean triple junction is reduced by 40 percent. Thus this model, though not a panacea for all problems of Indian Ocean plate kinematics, provides a simple description of motion in terms of idealized internally rigid plates, where one boundary is diffuse, not discrete. Understanding of this complex tectonic environment will improve as additional data accumulate and provide better resolution of both the diffuse boundary and motion along it. This model could not have been developed without the insight provided by historical seismicity. Consideration of only recent earthquakes underestimated the "intraplate" seismicity and resulted in plate geometry models with a single rigid Indian plate. Analysis of historical seismicity enabled the estimation of the direction and rate of motion along the Ninetyeast Ridge and established the extent and complexity of the deformation near Chagos.

4.2. Canadian Passive Margin Seismicity

A second interesting tectonic process, first observed using historical earthquake data, is intraplate seismicity at passive margins, where continental and oceanic lithosphere join. Although these areas are in general tectonically inactive, major earthquakes can occur. The type example is the active seismic zone on the eastern coast of North America (Stein *et al.*, 1979), which includes the 1929 Grand Banks (M_S = 7.2) and 1933 Baffin Bay (M_S = 7.3) earthquakes (Figure 9). Analysis of data from the 1933 earthquake, and more recent seismicity, suggested that the earthquakes were divided into thrust faulting seaward of the 1000 m isobath, and normal faulting landward. Stein *et al.* (1979) proposed that these earthquakes were associated with stresses induced by the removal of Pleistocene glacial loads extending onto the continental shelf, which reactivated the faults remaining along the continental margin from the original rifting. Simple flexure calculations predict stresses of 100-150 bars, adequate to trigger earthquakes. The only difficulty for the model, the large lithospheric thickness required by the 40 km depth estimated for the 1933 earthquake from the one available record, has been resolved by additional records which indicate a 16 km depth (Sleep *et al.*, 1986).

This general model has been accepted and extended by subsequent studies, including analysis of microseismicity in Baffin Bay (Reid and Falconer, 1982). Quinlan (1984) and Sleep *et al.* (1986) included the effect of the stress field that occurs on all passive margins due to the differing densities of continental and oceanic lithosphere (Bott and Dean, 1972; Artyushkov, 1973). This stress field is extensional in the continent and compressive seaward, in accord with the focal mechanisms. Thus, the effect of deglaciation appears to be to trigger earthquakes, as such events are observed primarily on glaciated marigins, such as Canada, Fennoscandia, and Greenland (Figure 10). This observation has obvious implications for the nature of seismic hazards along previously glaciated margins such as the northeastern United States: the largest potential for destructive earthquakes arises from the continent-ocean boundary rifting faults.

The same principal applies, to a lesser extent, on non-glaciated passive margins. Sediment loading can, in itself, generate substantial flexural stresses (Cloetingh *et al.*, 1984). Stein *et al.* (1979) suggested that, in general, these loads are much less effective at inducing earthquakes, since the sediment loads are usually in place long enough for the stress to relax. Extremely rapid sedimentation is required to induce

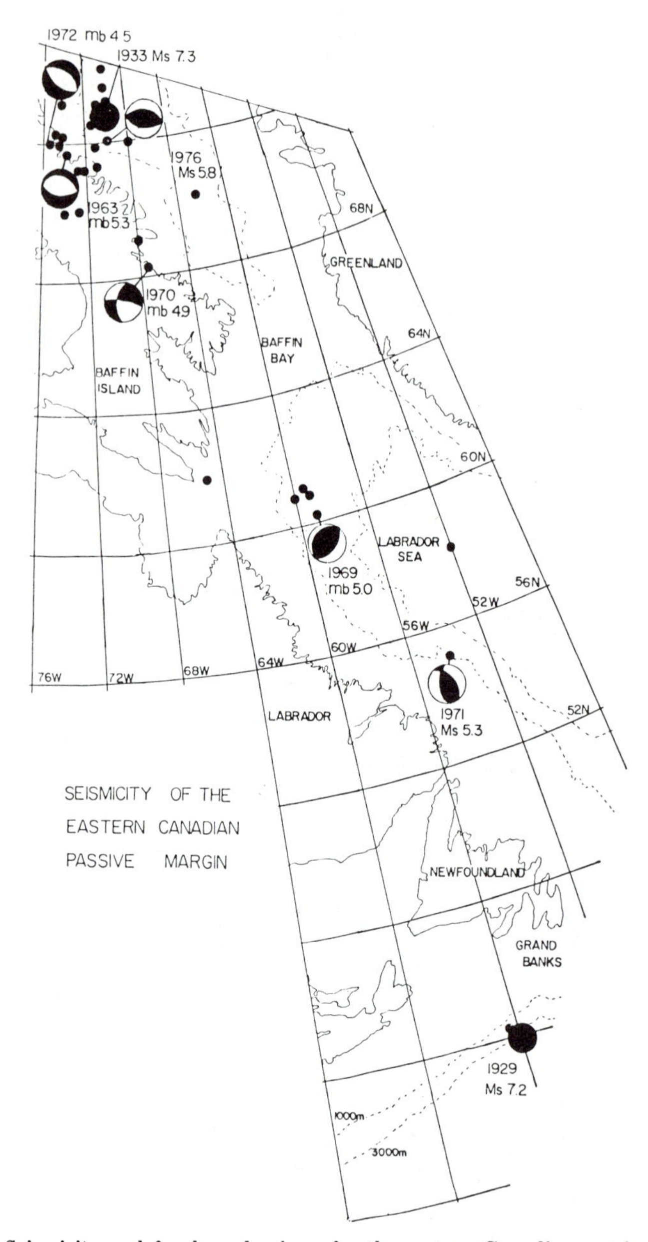

Figure 9. Seismicity and focal mechanisms for the eastern Canadian passive margin (Stein *et al.*, 1979). Earthquakes seaward of the 1000 m contour show thrust faulting; landward events show dominantly normal faulting. The mechanisms are consistent with reactivation of the continental margin rifting faults by flexure induced by deglaciation. The stress field at the margin due to differences in continent-ocean density structure also predicts the observed mechanisms.

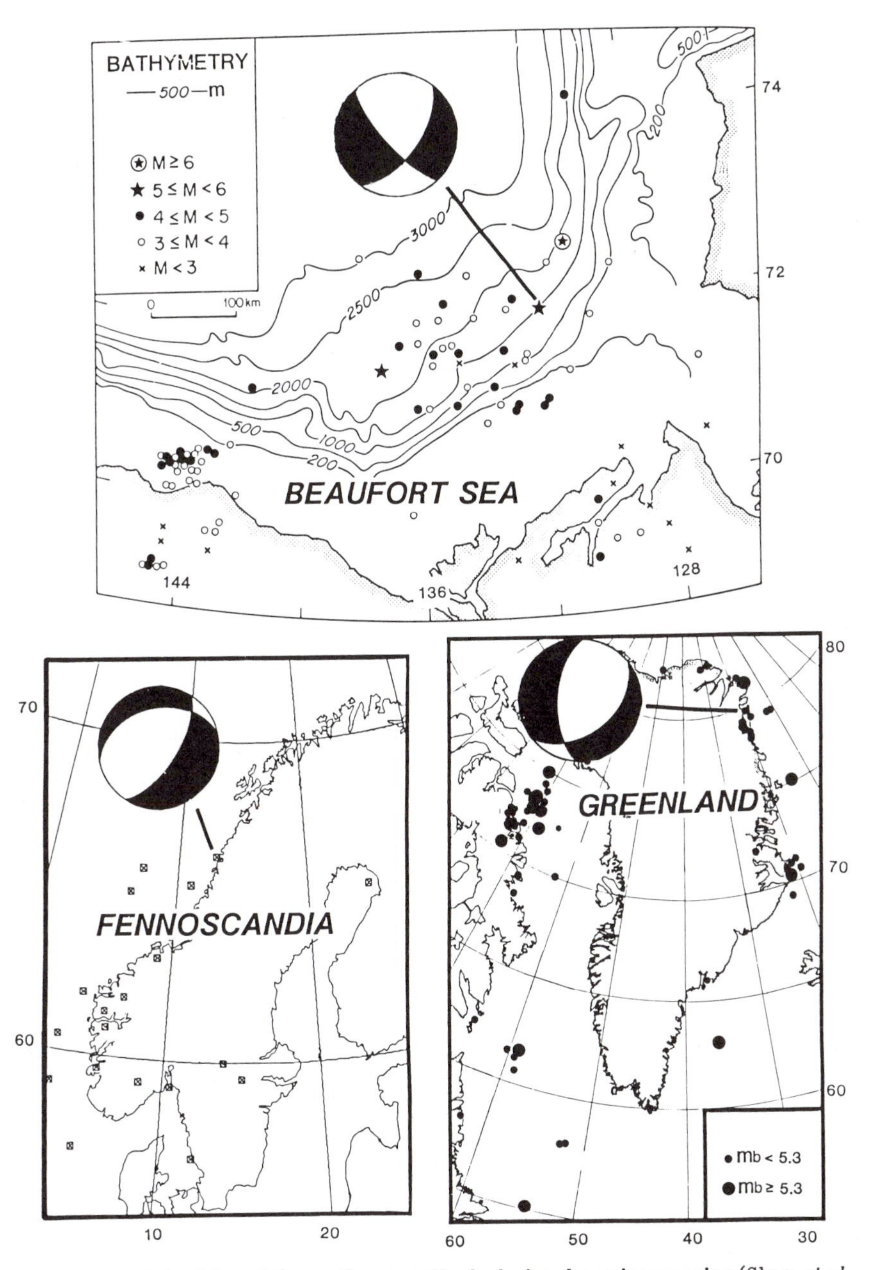

Figure 10. Seismicity of three other recently deglaciated passive margins (Sleep *et al.*, 1985); all show normal faulting inland as predicted by the eastern Canadian model. The Beaufort Sea seismicity (1920-1977) and mechanism (top) are from Hasegawa *et al.* (1979). The Fennoscandian seismicity (1497-1975) is from Husebye *et al.* (1978); the mechanism is by Bungum *et al.* (1979). The Greenland seismicity and mechanism are from Sykes (1978).

seismicity. One such case is the Gulf of Mexico; the small earthquakes in the area can be interpreted as a consequence of the rapid sediment loading history (Nunn, 1985).

As with the Indian Ocean deformation, the primary evidence for the importance of passive margin seismicity comes from the historical earthquakes. The recent seismicity is adequate to confirm its existence, but too small to convincingly demonstrate the magnitude of the effect.

4.3. Seismicity Patterns in the Eastern Part of the Pacific Plate

The easternmost part of the Pacific plate, generated at the fast-spreading East Pacific Rise in the past 20 Ma, has been the site of occasional yet intense seismic activity (see Figure 11). This seismicity is well documented by the post–WWSSN data, but only the historical earthquakes show its true dimension. As shown on Figure 11, 8 events reached or exceeded magnitude 6 between 1937 and 1962. The preferential location of these events in the young oceanic lithosphere is consistent with the observation that the rate of oceanic intraplate seismicity decreases with increasing lithospheric age (Wiens and Stein, 1983). A full study of these events is given by Okal (1984) and Okal and Wiens (1985).

At the southernmost epicenter, the 1947 earthquake (see Figure 1), the additional shock in 1950, and the probable location of the major 1949 event in its immediate vicinity, clearly reveal the area as one of preferential intraplate seismicity. The fact that the 1947 shock has a thrust faulting mechanism, as opposed to the normal faulting of a smaller event in 1976, some 200 km northwest, is a superb example of the complex regime of tectonic stress in the youngest portions of oceanic plates (Wiens and Stein, 1984; Bergman and Solomon, 1984). Once again, the historical data rules out the simplest explanation – in this case, that the lithosphere would be in a state of extension in the direction perpendicular to the ridge.

The 1955 earthquake at 25°S, 122°W, is a rare case of a large normal faulting earthquake on the flanks of a major seamount. Its similarities with shocks following documented episodes of volcanic activity suggests that the seamount may still be magmatically active.

Finally, the 1945 event further north and the recent (1984) swarm at 20°N, 116°W, further identify the fringe of the Pacific plate along the northern portion of the East Pacific Rise as a region of enhanced seismicity, particularly at the level of magnitude 6 or greater. This is in contrast to the bulk of the plate (the "F" region in Figure 11), where no magnitude 6 events are known outside of Hawaii. Okal (1984) has proposed to interpret this intriguing pattern in the framework of the reorientation which accompanied the Miocene ridge jump in the Eastern Pacific, and resulting in a geometry of particular vulnerability in relation to the intraplate tectonic stress. This picture could not have evolved without the crucial data provided by the historical earthquakes: Figure 11 shows only one event with $M_S > 6$ recorded inside the Pacific plate (Hawaii excepted) during the first 19 years of the WWSSN.

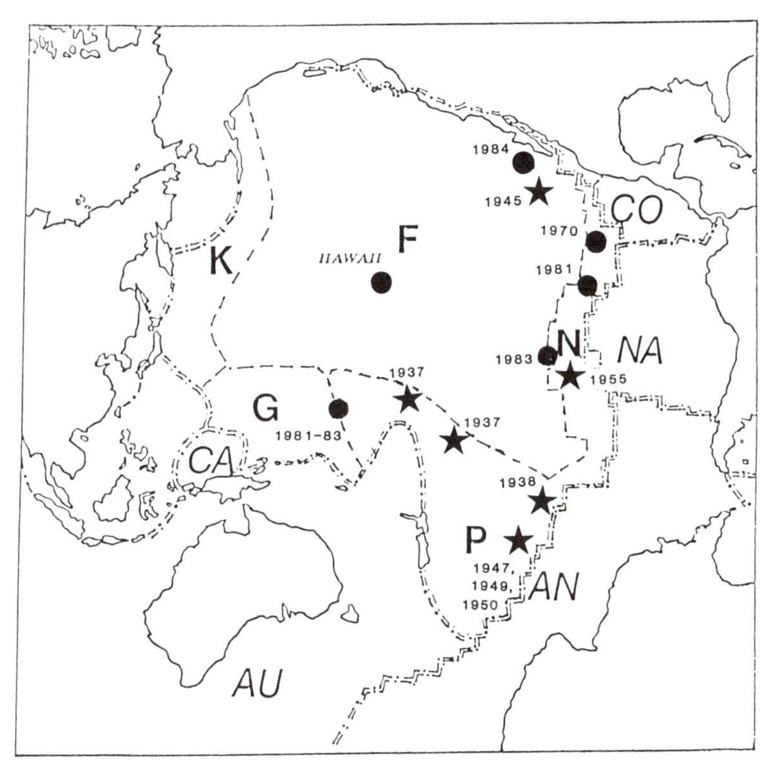

Figure 11. Map of Pacific intraplate earthquakes with at least one confirmed magnitude ≥ 6 (after Okal (1984)). Stars are historical events, dots epicenters known from post-WWSSN seismicity. This map identifies portions of the Pacific plate generated at various spreading ridges (see Okal (1984) for details). Note that magnitude 6 or greater seismicity is common outside Region "F", generated at the Farallon ridge, but that post-WWSSN seismicity gives a very incomplete picture.

ACKNOWLEDGEMENTS

This research was supported by the National Science Foundation under Grant EAR-84-07510, the Office of Naval Research under Contract N-00014-84-C-0616, and Research Initiation Funds at Washington University. We thank the reference staff of the Evanston Public Library for explaining the proper usage of "historic" and "historical".

REFERENCES

Artyushkov, E. V. (1973). Stresses in the lithosphere caused by crustal thickness inhomogeneities, *J. Geophys. Res.*, **78**, 7675-7708.

Bergman, E. A. and S. C. Solomon (1984). Source Mechanisms of Earthquakes near Mid-oceanic ridges from body waveform inversion: implications for early evolution of oceanic lithosphere, *J. Geophys. Res.*, **89**, 11415-11441.

Bergman, E. A. and S. C. Solomon (1985). Earthquake source mechanisms from body-waveform inversion and intraplate tectonics in the northern Indian Ocean, *Phys. Earth Planet. Interiors*, **40**, 1-23.

Bott, M. H. P. and D. S. Dean (1972). Stress systems at young continental margins, *Nature*, **235**, 23-25.

Bungum, H., B. K. Hokland, E. S. Husebye, and F. Ringdahl (1979). An exceptional intraplate earthquake sequence in Meloy, Northern Norway, *Nature*, **280**, 32-35.

Charlier, C. and J. M. van Gils (1953). Liste des Stations Séismologiques Mondiales, Observatoire Royal de Belgique, Uccle.

Chase, C. G. (1972). The N plate problem of plate tectonics, *Geophys. J. R. Astr. Soc.*, **29**, 117-122.

Chase, C. G. (1978). Plate kinematics: The Americas, East Africa, and the rest of the world, *Earth Planet. Sci. Lett.*, **37**, 355-368.

Cloetingh, S. and R. Wortel (1985). Regional stress field of the Indian plate, *Geophys. Res. Lett.*, **12**, 77-80.

Cloething, S. A. P. L., M. J. R. Wortel, and N. J. Vlaar (1984). Passive margin evolution, initiation of subduction and the Wislon cycle, *Tectonophysics*, **109**, 147-163.

Eguchi, T., S. Uyeda, and T. Maki (1979). Seismotectonics and tectonic history of the Andaman Sea, *Tectonophysics*, **57**, 35-51.

Fisher, R. L., J. G. Sclater, and D. P. McKenzie (1971). Evolution of the Central Indian Ridge, Western Indian Ocean, *Geol. Soc. Am. Bull.*, **82**, 553-562.

Geller, C. A., J. K. Weissel and R. N. Anderson (1983). Heat transfer and intraplate deformation in the Central Indian Ocean, *J. Geophys. Res.*, **88**, 1018-1032.

Geller, R. J. and H. Kanamori (1977). Magnitudes of great shallow earthquakes from 1904 to 1952, *Bull. Seism. Soc. Am.*, **67**, 587-598.

Gutenberg, B. and C. F. Richter (1965). *Seismicity of the Earth and Associated Phenomena*, Hafner, New York, 310 pp.

Hasegawa, H. S., C. W. Chou, and P. W. Basham (1979). Seismotectonics of the Beaufort Sea, *Can. J. Earth Sci.*, **16**, 816-830.

Husebye, E. S., H. Bungum, J. Feyn, and H. Gjøystdahl (1978). Earthquake activity in Fennoscandia between 1497 and 1975 and intraplate tectonics, *Norsk Geol. Tidsskrift*, **58**, 51-68.

Le Pichon, X. (1968). Sea-floor spreading and continental drift, *J. Geophys. Res.*, **73**, 3661-3697.

McComb, H. E. and C. J. West (1931). List of Seismological Stations of the World, *Bull. National Res. Council*, **82**, 1-119.

Minster, J.-B., T. H. Jordan, P. Molnar, and E. Haines (1974). Numerical modelling of instantaneous plate tectonics, *Geophys. J. R. Astr. Soc.*, **36**, 541-576.

Minster, J.-B. and T. H. Jordan (1978). Present day plate motions, *J. Geophys. Res.*, **83**, 5331-5354.

Nunn, J. (1985). State of stress in the northern Gulf coast, *Geology*, **13**, 429-432.

Okal, E. A. (1984). Intraplate Seismicity of the southern part of the Pacific plate, *J. Geophys. Res.*, **89**, 10053-10071.

Okal, E. A. and D. A. Wiens (1985). Major normal faulting intraplate earthquakes south of Baja California, Eos, *Trans. Am. Geophys. Union*, **66**, 307 (abstract).

Poppe, B. B. (1979). Historical Survey of U.S. Seismograph Stations, *U. S. Geol. Surv. Prof. Paper 1096*, Washington, D.C., 389 pp.

Quinlan, G. (1984). Postglacial rebound and the focal mechanisms of eastern Canadian earthquakes, *Can. J. Earth Sci.*, **21**, 1018-1023.

Reid, I. and R. K. H. Falconer (1982). A seismicity survey in northern Baffin Bay, *Can. J. Earth Sci.*, **19**, 1518-1531.

Rothé, J.-P. (1969). Seismicity of the Earth / La Sismicité du Globe, 1953-1965, *UNESCO Earth Sciences Ser.*, **1**, Paris, 336 pp.

Seismological Society of America (1980). *Seismology Microfiche Publications from the Caltech Archives, Ser. II and III*, edited by J.R. Goodstein, H. Kanamori and W. H. K. Lee, Berkeley, Calif.

Sclater, J. G. and R. L. Fisher (1974). The evolution of the eastcentral Indian Ocean, with emphasis on the tectonic setting of the Ninetyeast Ridge, *Geol. Soc. Am. Bull.*, **85**, 683-702.

Sleep, N. H., S. Stein, and G. C. Kroeger (1986). Canadian margin stress field inferred from seismicity, *J. Geophys. Res.*, submitted.

Stein, C. A., Wiens, D. A., and J. K. Weissel (1986). Seismicity and thermal constraints on the deformation of the Indo-Australian plate, *Earth Planet. Sci. Lett.*, to be submitted.

Stein, S. (1985). An earthquake swarm on the Chagos-Laccadive Ridge and its tectonic implications, *Geophys. J. R. Astr. Soc.*, **55**, 577-588.

Stein, S. and R. G. Gordon (1984). Statistical tests of additional plate boundaries from plate motion inversions, *Earth Planet. Sci. Lett.*, **69**, 401-412.

Stein, S. and E. A. Okal (1978). Seismicity and tectonics of the Ninetyeast Ridge area, evidence for internal deformation of the Indian plate, *J. Geophys. Res.*, **83**, 2233-2246.

Stein, S., N. H. Sleep, R. Geller, S.-C. Wang, and G. Kroeger (1979). Earthquakes along the passive margin of eastern Canada, *Geophys. Res. Lett.*, **6**, 537-540.

Sykes, L. R. (1978). Intraplate seismicity, reactivation of pre-existing zones of weakness, alkaline magmatism, and other tectonism postdating continental fragmentation, *Rev. Geophys. Space Phys.*, **16**, 621-688.

Tarr, A. (Ed.) (1974). World Seismicity Map, U.S. Geological Survey, Washington, D.C.

Weissel, J. K., R. N. Anderson, and C. A. Geller (1980). Deformation of the Indo-Australian plate, *Nature*, **287**, 284-291.

Wiens, D. A. (1985). Historical seismicity near Chagos: Deformation of the equatorial region of the Indo-Australian plate, *Earth Planet. Sci. Lett.*, **76**, 350-360.

Wiens, D. A., D. C. DeMets, R. G. Gordon, S. Stein, D. F. Argus, J. Engeln, P. Lundgren, D. G. Quible, C. Stein, S. A. Weinstein, and D. F. Woods (1985). A diffuse plate boundary model for central Indian Ocean tectonics, *Geophys. Res. Lett.*, **12**, 429-432.

Wiens, D. A. and S. Stein (1983). Age dependence of oceanic intraplate seismicity and implications for lithospheric evolution, *J. Geophys. Res.*, **88**, 6455-6468.

Wiens, D. A. and S. Stein (1984). Intraplate seismicity and stresses in young oceanic lithosphere, *J. Geophys. Res.*, **89**, 11442-11464.

APPENDIX

The types of studies discussed here would not have been possible without the many seismological stations whose personnel were extremely helpful in providing countless amounts of crucially important data. Such assistance in accommodating often inconvenient requests is extraordinarily valuable to the seismological community. We restrict the following list to those organizations whose data were used in the case examples described in the present paper:

Seismic Data Library, U.S. Geological Survey, Menlo Park, CA;
Seismological Laboratory, California Institute of Technology, Pasadena;
Seismographic Station, University of California, Berkeley;
Seismographic Station, St. Louis University, St. Louis, Missouri;
Lamont-Doherty Geological Observatory, Palisades, New York;
Hawaii Volcano Observatory;
Earth Physics Branch, Government of Canada, Ottawa;
Institute of Geophysics, Mexican National Autonomous University, Mexico City;
Kew Observatory, Richmond, England;
Institute of Experimental Geophysics, Trieste, Italy;
Royal Belgian Observatory, Uccle;
Royal Netherlands Meteorological Institute, De Bilt;
Royal Danish Geodetic Institute, Charlottenlund;

Geophysical Observatory, Louis-Maximilian University, Fürstenfeldbruck, Germany;

Institute of Physics of the Earth, Louis Pasteur University, Strasbourg, France;

Institute of Physics of the Earth, Academy of Sciences of the USSR, Moscow;

Kodaikanal Observatory, Kodaikanal, India;

Department of Seismology, Lwiro, Zaire;

San Calixto Observatory, La Paz, Bolivia;

Geophysics Division, Dept. of Scientific and Industrial Research, Wellington, New Zealand;

Riverview College Observatory, Riverview, Australia;

Manila Observatory, Manila, Philippines.

III. Earthquake Catalogs and Databases

Recent Activities at the International Seismological Centre

*D. M. McGregor, A. A. Hughes, and R. D. Adams**
International Seismological Centre
Newbury, RG13 4NS, Berkshire, UK
** Also at Department of Geology, University of Reading, UK*

1. Introduction

The International Seismological Centre (ISC), at Newbury, England, is a non-governmental international body, charged with the final collection, analysis and publication of standard earthquake information from all over the world. It is funded by 47 interested agencies in 44 countries, with additional support from some 20 commercial organizations representing the engineering, insurance, exploration, power generation and instrument-manufacturing professions. These Members and Associate Members are listed inside the front cover of all the Centre's publications.

The ISC continues the work of the International Seismological Summary (ISS), which was founded by Professor John Milne after his return to Britain from Japan at the end of the last century. The Centre was formed in Edinburgh in 1964, took its present constitution in 1970, and moved to Newbury in southern England in 1975.

Work at ISC is now computer dependent to a very large extent. Among the reasons for the move to Newbury was the proximity to the Rutherford and Appleton Laboratory's Computer Centre at Chilton, about 25 km away. The two IBM 360/195 computers there undertook almost all of the ISC's computing work until 1982, when the Centre purchased its own RM80/TS11 based VAX 11/750 computer system, which now has disc storage of more than 500 Mbytes. The extensive use of computers has enabled the Centre's activities to be carried out with an extremely small staff. In mid-1985 the staff totals only eight. Besides the Director, the only professional staff are two seismologists, one computer scientist and a bibliographer. The owning of its own computer has recently enabled ISC to greatly improve its services in the storage and retrieval of information on past earthquakes, as will be discussed below.

2. Seismological Analysis

The procedures used to analyse seismological data at ISC have been described in detail by Adams, Hughes and McGregor (1982). The analysis is undertaken in monthly batches and despite many calls for more speedy service, the start of analysis is delayed for 22 months to allow the information used to be as complete as possible. This delay helps ensure the receipt of data from networks in remote areas in less developed parts of the world, where there are not only difficulties of logistics in collecting data from extensive networks, but in addition the high level of activity creates a heavy burden of analysis. These difficulties are not always appreciated by seismologists working in large well-equipped institutions in developed countries.

In mid-1985 data from the latter half of 1983 are being analysed. During this period, about 1,300 individual stations submit readings to ISC, and origin estimates are received from about 35 agencies.

During analysis, the computer program first groups together origin estimates of the same earthquake from different agencies, and then associates the individual station readings with the most likely event. In a typical month up to 100,000 station readings are analysed, and initially between 1,600 and 2,000 events are identified. After mis-associations have been remedied and other discrepancies rectified, the remaining unassociated readings are searched for new events, and usually about 250 previously unlocated earthquakes are added to the file. Thus, the number of events listed each month is usually between 1,800 and 2,300. The delay of two years in the publication of the final bulletin is justified by the fact that this number of events is several times greater than that obtained by the Preliminary Determination of Epicenters Service of the U. S. Geological Survey, whose aim is to produce a rapid service for well-reported larger events, compared with ISC's wish for completeness, at the expense of speed.

3. Publications

3.1. Bulletin of the International Seismological Centre

The Bulletin is the Centre's main publication, containing all relevant information for a given month's earthquakes. For each event, the origin estimates given by other agencies are listed, followed by the refined estimate calculated by the Centre. This gives origin time, latitude, longitude and depth with computer determined standard errors. The earthquake's felt effects and any unusual scientific features are also described. Magnitudes m_b and M_S are calculated when individual stations report readings of amplitude and period, or their ratio. Since the data-year 1978 an effort has been made to obtain more long-period readings, and consequently more determinations of M_S.

For some small earthquakes located by national agencies the origin co-ordinates only can be given, but for better observed earthquakes observations from individual seismograph stations are listed with their distance, azimuth and time residual. For the largest earthquakes the number of reporting stations can reach more than 600.

The Bulletin provides seismologists with the fullest available information on all aspects of observational seismology necessary for the study of earthquake occurrence and earth structure. All issues of the Bulletin since 1964 are in fixed field format available on magnetic tape.

3.2. Regional Catalogue of Earthquakes

This is a six-monthly compilation of earthquake origin parameters and felt information, without the detailed station observations of the Bulletin. It contains lists of "major earthquakes" above certain magnitudes, reports of "felt and damaging earthquakes", "probable explosions", and the main regional catalogue in which all events in given geographic regions are listed together. This publication is that most relevant to detailed studies of seismicity, and the one most sought by institutions such as consulting engineering firms and insurance companies. It is published shortly after the issues of the Bulletin for June and December of each year. Like the Bulletin recent issues of the Catalogue are available in fixed field format on magnetic tape. It also gives world-wide lists of seismograph stations and their coordinates.

3.3. Felt and Damaging Earthquakes

This publication extracts from the Catalogue those earthquakes which have comments indicating that they were felt or caused damage. It has been published annually since the data year 1976, and continues the series "Annual Summary of Information on Natural Disasters" previously published by Unesco.

3.4. Bibliography of Seismology

The Bibliography appears twice a year, giving full references to seismological publications. These references are listed by subject and author, and citations to individual earthquakes are also given. The use of KWOC (keyword-out-of-context) indexing assures easy finding of references. The Bibliography is in demand, not only for libraries, but also for research institutions and consulting firms which wish to find references to publications relating to particular fields of research or to specific geographic areas. About 1,000 references are included in each issue. In mid-1985 the most recent issue is for items published in the first half of 1984.

4. Historical Hypocentre File

The Centre maintains a continually growing file of past earthquakes, which is available for research studies and to help with the estimation of seismic hazard. The file contains origin estimates of earthquakes dating from 1904, but naturally only major earthquakes are present for early years. The file becomes more complete with time and now contains more than 300,000 events world wide. About 20,000 new events are added each year.

4.1. Sources of Data

The main sources of data for the ISC Historical File are:

Gutenberg and Richter, *Seismicity of the Earth*	1904-1952
International Seismological Summary	1913-1963
Bureau Central International de Séismologie	1935, 1950-1963
L. R. Sykes, Catalogue of oceanic ridge events	1950-1963
U. S. Coast and Geodetic Survey	1928-1960
International Seismological Centre	1964-1982
U. S. National Earthquake Information Service, Preliminary Determination of Epicenters	1983
Telex reports from NEIS and other sources	1983 onwards.

Up to the end of 1963 there may be more than one estimate for an event, but from 1964 to 1981 only the prime estimate from the ISC Bulletin is included.

4.2. Information Stored

The file is now stored on magnetic disc on the ISC's computer. The information in a single epicentre "record" contains a set number essential fields, which may be of variable length, but must be filled, even if with a null value. The information in these fields includes such details as: agency code, source of information, date, origin Time, latitude, longitude, and depth. There is also provision for optional fields for additional information such as: associated phenomena (e.g. tsunami, volcanism); flags indicating that the event was felt, damaging etc.; intensity data; magnitude data; number of recording stations; type of event, (e.g. explosion, rock-burst etc.); and comments – any length of text may be added. This format is designed to hold the maximum possible information, without taking up storage space unnecessarily with null information.

4.3. Retrieval of Information

ISC provides a consulting service for those with scientific or commercial need for listings from this file. Events can be selected from the file by any combination or criteria relating to time, location, depth, magnitude, intensity, or locating agency. Composite estimates can also be provided, which combine origin parameters from prime estimates with magnitude determinations from other agencies.

In addition to giving a listing of events, a line printer plot can also be generated, showing the positions of events superimposed on a representation of coastlines. The earthquakes are represented by a figure which gives the magnitude of the largest event reported at each position. Where no magnitude is assigned, events are shown by an asterisk. Underlining of a symbol indicates multiple events at the same position. Another option lists estimates of expected ground acceleration and felt intensity at a specified location for each event for which a magnitude is given.

The information can also be reproduced on magnetic tape, in two formats, one of which contains all the information on the file for the chosen events, but needs some simple programming to interpret, and another which contains only the basic information in fixed fields in 80-column card images.

5. Station Bulletins

Most data arrive at ISC in machine-readable form; on magnetic tape, on punched cards, or by telex. Some readings come entered on special coding sheets for keying at the Centre, but in addition many agencies and stations send bulletins, which are stored at ISC after their data have been extracted and used. Many of the most valuable contributions to ISC analysis come in this form, with information from stations that are either remote, or in regions of high activity, or both.

Naturally, the bulletins are more complete for the later years, but there are also some more important early collections. Among these are a set from Hungary with some readings from Budapest for 1896, and more complete readings from 1905. By 1909 this network comprised seven stations. Other early bulletins are for:

Vienna	1905-1907
Hamburg	1912, 1913
Strasbourg	1913, 1915-20, from 1934
Tokyo	1922
Soviet Union (8 stations)	1926, 1927, from 1930
Kew	1927 onwards
Riverview	1927 onwards
Copenhagen	1931 onwards
La Paz, Bolivia	1931 onwards
Nanking	from 1932
Ksara, Lebanon	from 1934

These bulletins are available for study or copying.

6. Seismograms

ISC has never acted as a collecting agency for the storage of seismograms, and there are no present plans for this.

7. Future Activities

The ISC will continue to carry out its primary function of the collection of earthquake phase readings and the relocation of earthquakes for the compilation of the definitive catalogue of world seismicity.

In the late 1970's there was much discussion, fostered by both the International Association of Seismology and Physics of the Earth's Interior (IASPEI) and Unesco, about the possibility of a Global Seismic Data Bank being attached to ISC. A detailed feasibility study for this was drawn up by Dr. V. Karnik for Unesco in 1980. This ambitious plan included the collection of catalogues, bibliographies and historical data, and the revision and upgrading of early origin estimates, including where possible the assigning of magnitudes. In this way a master file of world earthquakes would be built up, that would be as complete and accurate as possible. No agency has yet been willing to fund this project, however. This proposal did not include any provision for the storing of seismograms or wave forms in computer-readable form, as it was then considered that this need was best met by national agencies.

A further suggestion that has been proposed is for the ISC to act as a Strong Motion Information Centre and Data Bank. This has also been the subject of a feasibility study undertaken by Unesco in 1975. The more widespread use of digital recording for strong ground motion now increases the need for such a centre.

The other major international sources of seismological information are the World Data Centers established by ICSU at Boulder and Moscow. These hold information from 40 major and 55 minor catalogues, as well as collections of analogue and digital seismograms. The ISC differs from these bodies, however, in that it is primarily an analysis centre for providing refined data. Any expansion or change in its activities would naturally have to be subject to the provision of adequate funding and the approval of the Centre's Governing Council of representatives of contributing Members.

For any such projects, the ISC has one great advantage. It remains an independent seismological agency fully operated and funded by international control, and it already has close and friendly links with the seismological community world-wide. The Centre particularly values its contacts with many smaller countries, which are often in areas of high earthquake activity, and whose contributions are vital to a full study of world seismicity.

REFERENCES

Adams, R.D., A.A. Hughes, and D.M. McGregor (1982). Analysis procedures at the International Seismological Centre, *Phys. Earth. Planet. Int.*, **30**, 85-93.

Flinn, E. A. and E. R. Engdahl (1965). A proposed basis for geographic and seismic regionalization, *Rev. Geophys.*, **3**, 123.

A CATALOGUE OF CHINESE EARTHQUAKES ($M \geq 6.5$) FROM 1900 TO 1948 WITH UNIFORM MAGNITUDES

Deli Cheng
Jiangsu Provincial Seismological Bureau, China

Yushou Xie
Institute of Geophysics, State Seismological Bureau, Beijing, China

Zhifeng Ding
Jiangsu Provincial Seismological Bureau, China

ABSTRACT

As the magnitude scale, the instrument characteristics, and the measurement methods are different for different time periods at various seismological organizations, the magnitudes listed in different catalogues of Chinese earthquakes are usually inconsistent. On the basis of collection and collation of instrumental data for Chinese earthquakes during the period 1900 to 1948, the authors revised the calculated magnitudes, giving a catalogue with a uniform magnitude scale. Gutenberg and Richter's formulas of 1945 and 1956 are used for surface-wave and body-wave magnitudes, respectively. As a supplement, the 1962 Moscow-Prague formula is used to determine magnitudes from short period surface waves. Statistical results show that for the period of investigation, the difference between surface-wave magnitudes obtained using the two formulas is tolerable.

Amplitude and period data from Gutenberg and Richter's unpublished original worksheets for *Seismicity of the Earth and Associated Phenomena* and all other available sources are used. A calibration curve for the Xujiahui (Zi-ka-wei) seismic station is established. An empirical formula is found for estimating magnitude from the maximum recorded distance of P waves. Regression curves are obtained for magnitudes listed in various catalogues relative to our revised magnitudes. Using this data, a new catalogue is compiled of Chinese earthquakes from 1900 to 1948 with a uniform magnitude scale. Only earthquakes with magnitudes $M \geq 6.5$ are listed in this paper.

Statistical comparison shows that the revised magnitudes for major earthquakes agree well with those of Abe (1981). Although the mean deviation in the magnitudes listed in the 1983 edition of *Catalogue of Chinese Earthquakes* [excluding those quoted directly from Hsu Mingtung's (Xu Mingtong) catalogue] is small, the scatter is rather large.

The meaning of the magnitude given in *Seismicity of the Earth and Associated Phenomena* is investigated. Our conclusion is consistent with that of Geller and Kanamori (1977). The seismicity of China and the monitoring of Chinese earthquakes by worldwide seismic stations for the period 1900 to 1948 is discussed.

1. Introduction

Since the publication of a comprehensive catalogue of worldwide earthquakes by Gutenberg and Richter (1954), Gutenberg (1956), Richter (1958), Duda (1965), Rothé (1969), and others have made supplements and revisions. As the instrument characteristics, the magnitude scale and the measurement methods are different, there is no unified standard for the magnitudes given in various catalogues, and even for different earthquakes in the same catalogue. Difficulties were encountered in the investigation of seismicity, earthquake prediction and hazard analysis based on such data, and incorrect conclusions were sometimes obtained. For example, using the catalogues given in *Seismicity of the Earth and Associated Phenomena* (Gutenberg and Richter, 1954) and *Great Earthquakes 1896-1903* (Gutenberg, 1956), it was concluded that from 1896 to 1906 the average annual energy release from earthquakes all over the world was about three times the average value for the years 1907 to 1955. Through careful study, Kanamori and Abe (1979) pointed out that this was due to the underestimation of the effective magnification for the undamped Milne seismograph in the recording of great earthquakes. When proper corrections were made (in some cases as large as 0.6 magnitude units), the abnormally high annual energy release simply did not exist.

Since the 1970s, seismologists began work on the unification of the earthquake magnitude scale. In order to compile a catalogue of global earthquake activity with a uniform magnitude scale defined by Gutenberg and Richter (G-R scale), Geller and Kanamori (1977), and Abe and Kanamori (1979) investigated in detail the magnitude scale used in *Seismicity of the Earth and Associated Phenomena.* Abe and Kanamori (1980) also studied the magnitudes of large shallow earthquakes from 1953 to 1977. Abe (1981) recalculated the magnitudes of large shallow earthquakes from 1904 to 1980 with a uniform magnitude scale. Later, Abe and Noguchi (1983a) redetermined the magnitudes of large shallow shocks from 1898 to 1917 by correcting the magnification of undamped Milne and Omori seismographs. Thereafter, by re-evaluation of the quality of surface-wave magnitudes of shallow earthquakes before 1912, using the data given in the unpublished worksheets of Gutenberg and Richter (1954), Abe and Noguchi (1983b) further revised the average effective gain of the undamped Milne seismographs and redetermined the M_S of large shallow earthquakes in the period 1897 to 1912. Thus, a catalogue of worldwide large shallow earthquakes from 1897 to 1980 was completed with uniform magnitude scale.

In 1971, the first edition of *Catalogue of Chinese Earthquakes* was published by the Central Seismological Group. Many editions were published successively, with the catalogues of the State Seismological Bureau (1977), Lee *et al.* (1978), and Gu (1983) used most often. In principle, all these catalogues use the magnitudes given in Gutenberg and Richter's *Seismicity of the Earth and Associated Phenomena* or their equivalents as standard (Gu, 1983). Actually, the data source is very complicated, the magnitude scale used is not the same, and detailed study and proper regression analysis were not carried out during the compilation. Therefore, the uniformity of magnitude in these catalogues is not satisfactory.

In compiling the fourth volume of the *Compilation of Historical Materials of Chinese Earthquakes* (Xie *et al.*, 1985-86), we revised the parameters of all earthquakes that occurred between 1900 and 1948 to prepare a catalogue with a uniform magnitude scale. In this paper, the magnitude scale used and the method of analysis are described, and the results of the revision are discussed.

2. Magnitude Scale

The Gutenberg surface-wave magnitude formula (Gutenberg, 1945a) and Gutenberg-Richter body-wave magnitude formula (Gutenberg and Richter, 1956) were used:

$$M_S = \log A + 1.656 \log \Delta + 1.818 + S \tag{1}$$

$$m_B = \log\left(\frac{A}{T}\right) + Q(\Delta, h) + C. \tag{2}$$

In formula (1): A is the vector sum of maximum horizontal displacements in μm of surface waves with period of 20 ± 3 sec for earthquakes with epicentral distance between 15° and 130° (a factor of $\sqrt{2}$ is multiplied when only one component is available); Δ is the epicentral distance, in degrees; and the station correction (S) is taken from Gutenberg (1945a). In using formula (1), Gutenberg also considered the influence of focal depth, radiation pattern and wave path. He pointed out that formula (1) was obtained from earthquakes with focal depth (h) from 20 to 30 km; and that for $h \leq 40$ km, the correction does not exceed 0.2. Thus, in this paper, formula (1) is ordinarily used for shallow shocks with $h \leq 40$ km; and for $40 < h \leq 100$ km, a magnitude correction, ΔM_S, according to Bäth (1977) is used:

h (km)	40	50	60	70	80	90	100
ΔM_S	+0.15	+0.20	+0.30	+0.35	+0.45	+0.50	+0.55

When the average value of many stations was used, the influence of azimuth and wave path were not considered. Nevertheless, it is necessary to point out that according to Gutenberg's investigation, for seismic waves passing through the Pacific Basin and that outside of it, the difference of magnitude may attain 0.5.

For the years 1900 to 1948, no strict specification could be given for the amplitude A in formula (1). Generally, one half of the wave crest-to-trough value of the surface-wave maximum was taken, which is slightly larger than the weighted average of the three consecutive amplitudes around the maximum taken by Gutenberg (Bäth, 1981).

The body-wave magnitude formula (2) was given by Gutenberg for unified magnitude. The Q-value resulted from the modification of the shallow and deep focus body-wave magnitude scales (Gutenberg, 1945b, 1945c) by using more observational data. Thus, the m_B-value calculated by using formula (2) differs slightly from that given in Gutenberg and Richter (1954), indicated by M_B in this paper. The 1945 values of Gutenberg (1945b, 1945c) were used for the station correction C. The maximum amplitude of the body wave was taken as amplitude A, usually 10 sec after the first impulse on seismograms of broad-band intermediate-period seismographs.

For earthquakes with focal depth $h \leq 40$ km, both M_S and m_B were determined. For $40 < h \leq 100$ km, m_B was usually given and M_S with depth correction whenever possible. For medium and deep focus earthquakes, only m_B is given.

For surface waves with periods less than 17 sec, the Moscow-Prague formula (Kárník *et al.*, 1962), indicated by M_V in this paper, was used:

$$M_V = \log\left(\frac{A}{T}\right) + 1.66 \log \Delta^\circ + 3.3 + S_V \quad (2^\circ \leq \Delta \leq 160^\circ). \tag{3}$$

Here, the vector sum of amplitudes of the horizontal components of surface waves was taken for A. When the epicentral distance is less than 5°, care is taken to differentiate between surface wave and S wave.

Although the magnitudes calculated using formulas (1) and (3) were obtained from seismic waves in different frequency bands, statistical results show they are mutually consistent for earthquakes during the period 1900 to 1948, as shown in Figure 1. The average deviation of M_V from M_S is +0.02 and the standard deviation is ± 0.18. The station corrections are given in Table 1. As data are limited, the standard deviation is comparatively large. The corrections for different time periods are not always the same, and in some cases the differences are relatively large.

3. Data Analysis and Magnitude Determination

During the period 1900 to 1948, seismic stations were established, instruments were improved, and the determination of seismic parameters was steadily improved. Thus, the magnitude scales used in different earthquake catalogues differ greatly. In this paper, the Gutenberg-Richter magnitude scale (Gutenberg, 1945a, Gutenberg and Richter, 1956) is taken as the standard. Amplitude and period data given by different seismic stations were used to calculate earthquakes magnitudes. For earthquakes with no amplitude data, magnitudes given in different catalogues were extrapolated to our standard scale through careful investigation.

3.1. From the amplitude and period data of the 82 Chinese earthquakes given in Gutenberg and Richter's unpublished original worksheets for *Seismicity of the Earth and Associated Phenomena*, M_S and m_B were calculated using formulas (1) and (2), respectively. The results show that: for $h \leq 40$ km, the magnitude values in that book are calculated using formula (1); for $40 < h \leq 100$ km, mainly m_B was calculated, and in rare occasions M_S; and for $h > 100$ km, all magnitudes are m_B (as shown in Figure 2). In Figure 2, the average deviation of body-wave magnitude $\overline{\Delta M} = 0.07$. It may be due to the difference between m_B and M_B. This conclusion agrees with that of Abe (1981). For class "d" earthquakes in Gutenberg's book, no reliable magnitudes were assigned. Results of our calculation show that their mean value is 5.6 ± 0.22.

3.2. In addition to Gutenberg's worksheets, amplitude and period data from other seismic stations were used. For earthquakes before 1935, the most important data used are from the stations listed in Table 1. As shown in Table 1, there was a large change in the Osaka station correction around 1920, and in the Strasbourg station correction around 1930. Undamped Omori seismographs were used at Osaka in its earlier stage. According to Abe (1979), the station correction is −0.1, which agrees with the −0.06 value for the years 1902 to 1920, as shown in Table 1. After 1934, data from Tashkent, Vladivostok, Moscow, Pulkovo, Sverdlovsk, Irkutsk, Baku, and other U.S.S.R. stations were used frequently. Their station corrections are shown

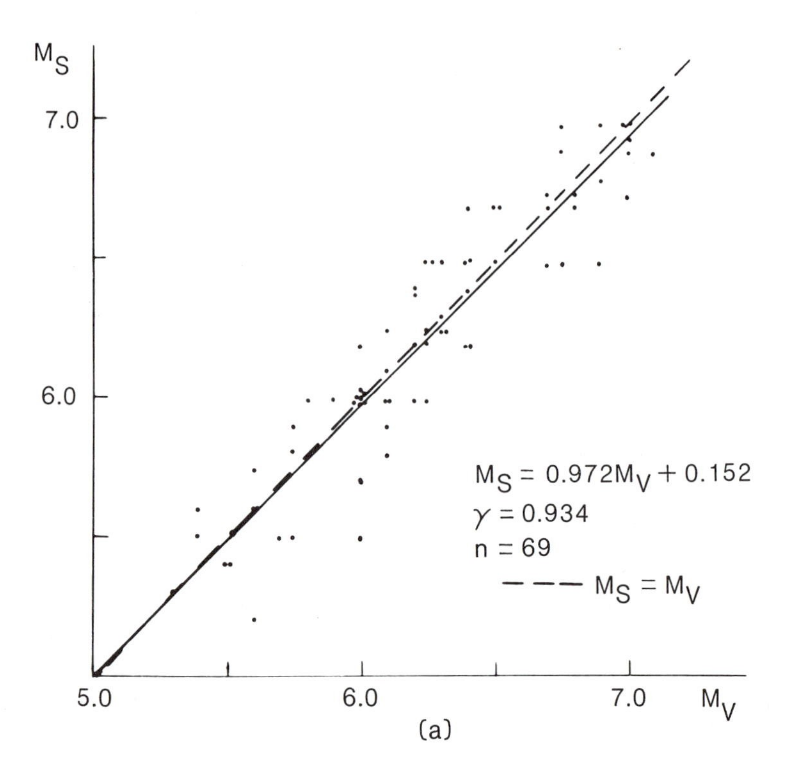

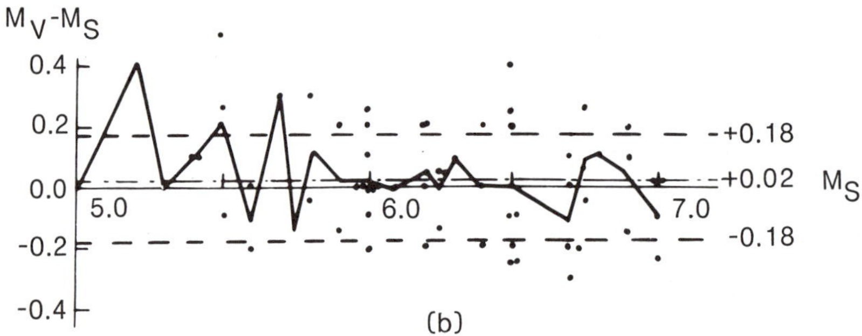

Figure 1. (a) M_S vs. M_V (Moscow-Prague) curve; (b) Deviation of M_V from M_S.

in Table 1. After 1940, maximum amplitudes without corresponding periods were listed in the "Bulletin of Seismic Stations of U.S.S.R." For $\Delta \geq 15°$, magnitudes were calculated using formula (1) with the amplitude multiplied by $\sqrt{2}$. When $\Delta < 15°$, the following formula (Kondorskaya and Shebalin, 1977) supplemented formula (3):

$$M_S = \log A + 1.32 \log \Delta, \tag{4}$$

where Δ is in km.

Table 1. Station Corrections, S_V

Station	Correction S_V	Standard deviation, σ	Number of observations	Period (year.month)
Osaka	−0.057	0.25	37	1902 - 1920
	−0.23	0.27	21	1921 - 1930
Kobe	+0.20	0.38	14	1914 - 1927
Strasbourg	+0.00	0.15	12	1921 - 1930
	−0.24	0.22	18	1931 - 1940
	−0.46	0.26	10	1941 - 1948
Tashkent	−0.12	0.24	85	1934 - 1940
Vladivostok	+0.16	0.30	69	1934 - 1940
Moscow	+0.004	0.17	69	1935.04 - 1940
Pulkovo	−0.11	0.21	75	1934 - 1940
Sverdlovsk	+0.13	0.20	51	1934 - 1940
Irkutsk	−0.014	0.19	28	1934.06 - 1940
Baku	−0.06	0.17	24	1934 - 1940
Hongkong	+0.054	0.37	35	1937.08 - 1940.02
Jiufeng (Chiufeng)	−0.025	0.28	8	1931 - 1935
Uppsala	−0.23	0.33	10	1941 - 1948
Prague	−0.24	0.22	24	1941 - 1948

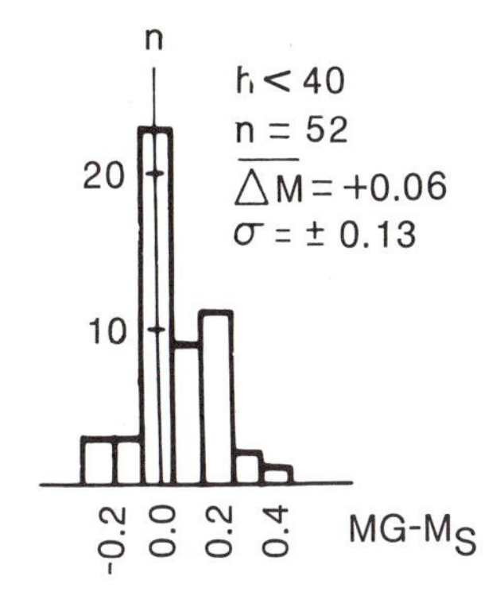

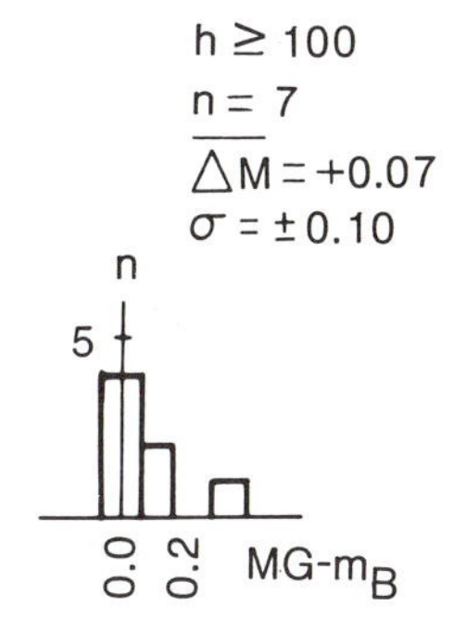

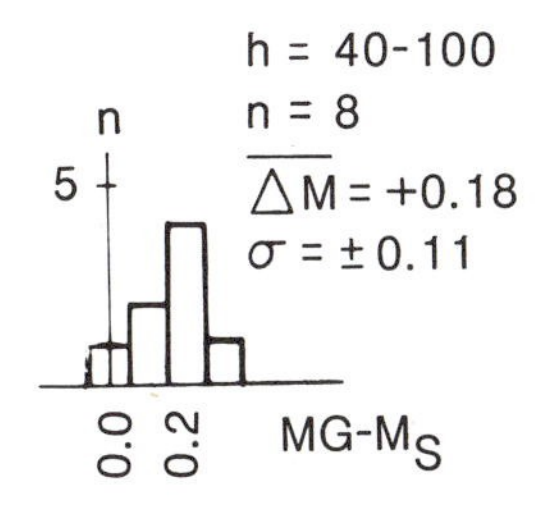

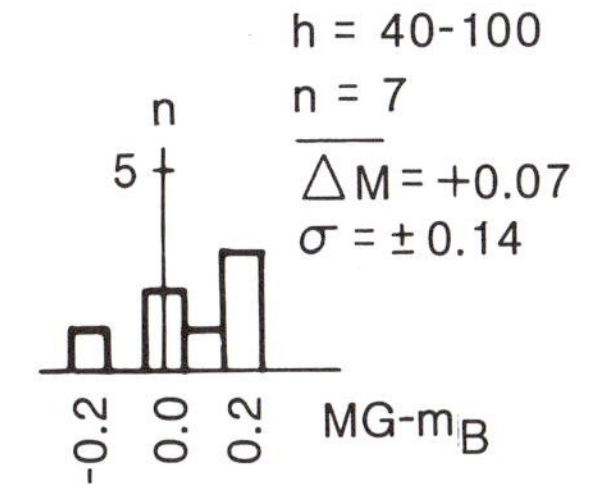

Figure 2. Statistical comparison of MG (Gutenberg, 1954) with M_S and m_B.

3.3. The Xujiahui (Zi-ka-wei) seismic station has published bulletins since 1909. Mr. Sun Quinxuan revised the bulletin by reanalyzing twice as many original seismograms in 1979 and 1982. The calibration curve is shown in Figure 3 and the formulas for magnitude determination are:

$$M_S = \log\left(\frac{A}{T}\right) + 3.83 \log \Delta^\circ + 1.80 \quad (5^\circ \leq \Delta \leq 13^\circ) \tag{5a}$$

$$M_S = \log\left(\frac{A}{T}\right) + 2.29 \log \Delta^\circ + 2.53 \quad (13^\circ < \Delta \leq 40^\circ), \tag{5b}$$

where A is the maximum displacement (in μm) of a single horizontal component of the ground motion. The calibration curve is composed of two segments with obviously different slopes, and may be caused by the fact that the maximum amplitude at different epicentral distances correspond to different types of seismic waves. The scatter in the data points gives a r.m.s. error of $\pm$ 0.3.

3.4. For earthquakes without amplitude data, magnitudes listed in related catalogues were used and their values were extrapolated from the G-R magnitude scale.

3.4.1. For earthquakes in the western part of China, mainly U.S.S.R. catalogues were used. For earthquakes before 1962, magnitudes were revised by using formula (3) in *New Catalogue of Strong Earthquakes in USSR* (Kondorskaya and Shebalin, 1977). Statistical results show that for Chinese earthquakes the magnitudes given are near the G-R magnitudes, with a correction of +0.08.

3.4.2. In Richter's (1958) catalogue, unified magnitude based upon surface waves, i.e. modified magnitude scale, is used. Statistical results show that the magnitude value is 0.22 too high, and thus can not be used.

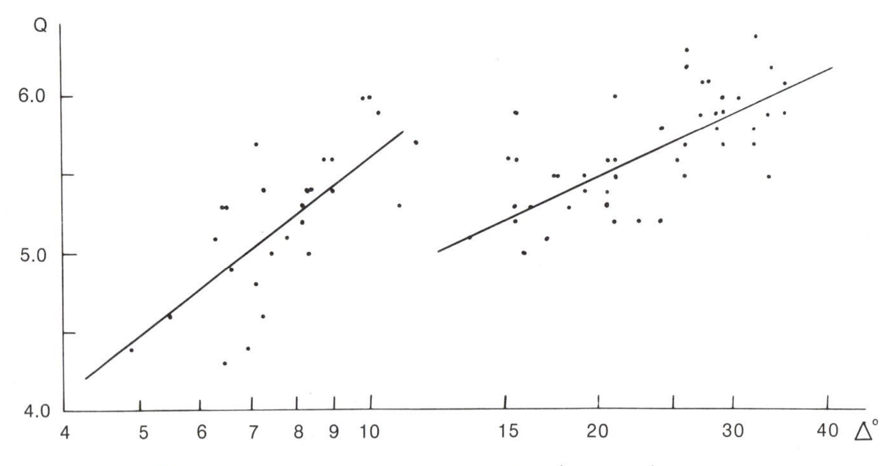

Figure 3. Calibration curve of Xujiahui (Zi-ka-wei) station.

3.4.3. The magnitudes given in Duda's (1965) catalogue are quite complicated. Most of the values were adopted from Gutenberg (1954, 1956) and Richter (1958). For Chinese earthquakes, excluding those adopted directly from Gutenberg (1954), statistical results show that the magnitude value is 0.25 too high on the average ($n = 30,\ \sigma = \pm 0.25$).

3.4.4. For the Taiwan region, the earthquake catalogues used are: *History of Taiwan earthquakes* (Taipai Observatory, 1936), *Catalogue of major earthquakes which occurred in the vicinity of Japan (1885-1950)* (Central Meteorological Observatory of Japan, 1952), and *Seismicity of Taiwan and some related problems* (Hsu, 1971). In *History of Taiwan earthquakes*, earthquakes are divided into "obvious," "moderate," "small," and "local" according to their strength. Statistical study shows that they correspond to 7, $6\frac{1}{4}$, $5\frac{1}{2}$, and $4\frac{3}{4}$ of G-R magnitude scale respectively. In *Catalogue of major earthquakes which occurred in the vicinity of Japan (1885-1950)*, magnitude values for earthquakes before 1930 were converted from Kawasumi's magnitude, M_K. They differ greatly from the G-R scale values and were not adopted. For earthquakes after 1931, average magnitudes obtained from data of six Japanese stations were added and can be used for reference.

Hsu's (1971) catalogue is composed of two parts. In Appendix II, the magnitudes for earthquakes before 1935 are mostly Kawasumi's magnitude or Hsu's estimation. Statistics show that for the years before 1925, the magnitudes given differ greatly from the G-R scale with a correction of +0.43 ($n = 12,\ \sigma = \pm 0.34$); for 1925 to 1935, the deviation is smaller, with a correction of +0.1 ($n = 9,\ \sigma = \pm 0.17$). In Appendix I, the magnitudes of earthquakes after 1936 were determined from observational data by Hsu using complicated magnitude scales. For earthquakes between 1936 and 1948, the result of least square regression is

$$M_S = 1.49\, M_H - 2.97 \qquad (n = 63, \gamma = 0.80, \sigma = \pm 0.29), \tag{6}$$

where M_H is Hsu's estimation of magnitude. It can be seen from Figure 4 that $M_H = 5.2$, $M_S = 4\frac{3}{4}$; and when $M_H < 6$, the correction is negative.

3.4.5. In the International Seismological Summary, magnitude, amplitude and period data are not given. For these earthquakes, we estimate their magnitudes from the maximum distance of P-wave observation, Δ_P. Considering the probable difference in P-wave attenuation for earthquakes occurring in mainland China and in Taiwan Province and its neighbouring seas, they were treated separately (Figure 5). For $\Delta_P < 105°$, the following emperical formulas were obtained:

$$M_S = 0.0119\, \Delta_P{}^{\circ} + 4.89 \tag{7}$$

for mainland China, and

$$M_S = 0.0179\, \Delta_P{}^{\circ} + 4.23, \tag{8}$$

for Taiwan Province and its neighboring seas ($\sim 5 < M_S < \ \sim 6$).

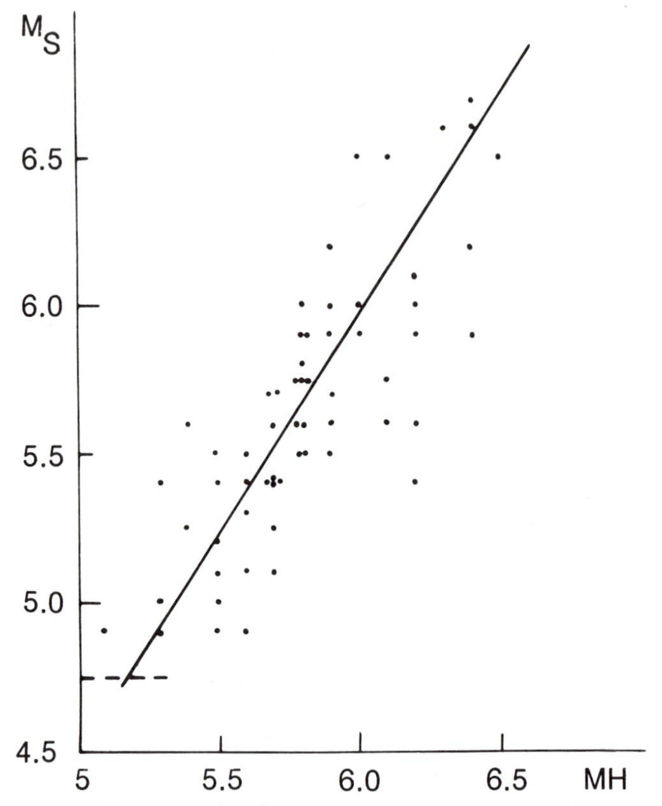

Figure 4. M_S vs. M_H (Hsu) curve of Taiwan earthquakes (1936-1948).

4. Results and Discussion

The magnitudes of the 1,006 Chinese earthquakes that occurred between 1900 and 1948 were calculated, checked, or extrapolated, according to the given G-R magnitude scale, the data analysis method and magnitude determination. The revised magnitudes were listed in the fourth volume of *Compilation of Historical Materials of Chinese Earthquakes* (Xie and Cai, 1985-86). Table 2 lists the parameters for earthquakes with M_S (or m_B) ≥ 6.5, together with their magnitudes from several popular catalogues for comparison.

4.1. Figure 6 shows the statistical comparison of our unified magnitudes with those given in several other catalogues. The magnitude values listed in this paper agree well with that in Gutenberg's table (1954), with a mean deviation $\overline{\Delta M} = +0.01$ and $\sigma = \pm 0.11$. This deviation seems to be related to the fact that supplementary data were added and formula (3) was used in the treatment of data from near stations. In comparison with Abe's results (1981, 1983a), $\overline{\Delta M} = +0.02$, with $\sigma = \pm 0.06$. Our magnitude is a little higher than that given in *New Catalogue of Strong Earthquakes in the USSR* (Kondorskaya and Shebalin, 1977), with $\overline{\Delta M} = -0.08$ and $\sigma = \pm 0.15$; and differs most from that of *Atlas of Earthquakes in USSR* (Savarenskii *et al.*, 1962), with $\overline{\Delta M} = -0.27$ and $\sigma = \pm 0.22$.

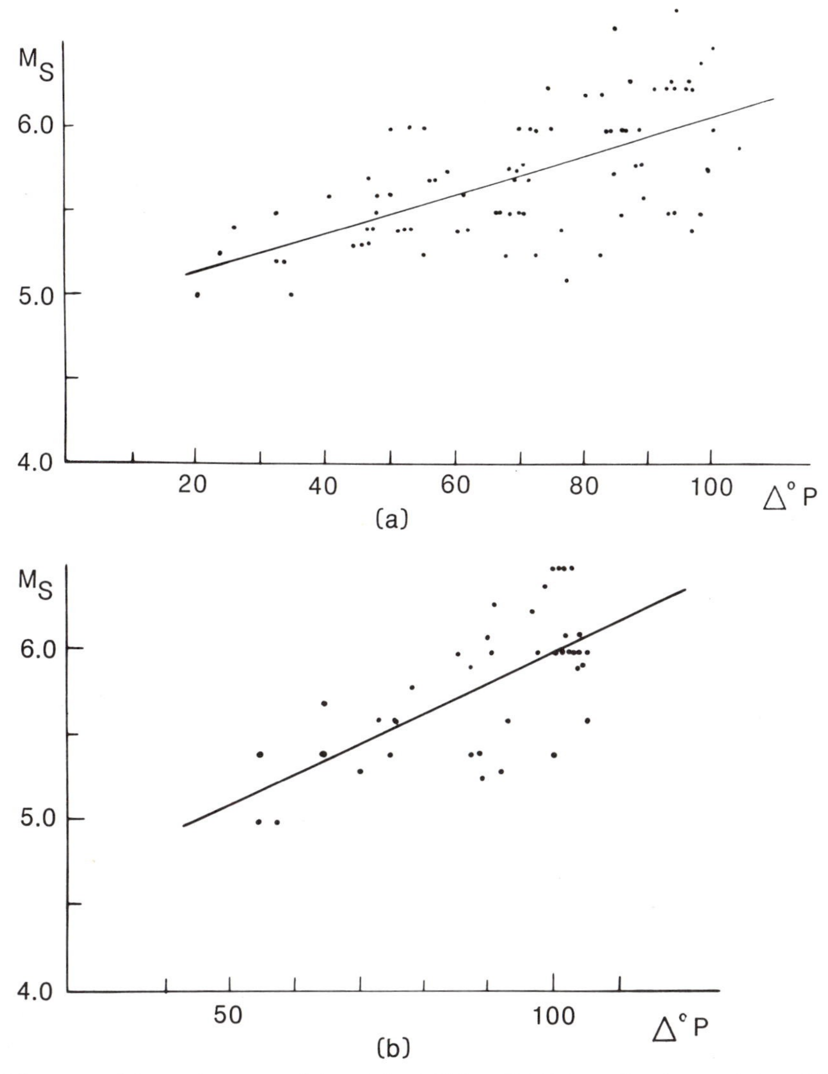

Figure 5. M_S vs. Δ_P curve: (a) mainland China; (b) Taiwan Province and its neighbouring seas.

The deviations between our magnitudes and those of the 1971 and 1983 editions of *Catalogue of Chinese Earthquakes* are: $\overline{\Delta M} = -0.04$ with $\sigma = \pm 0.25$ and $\overline{\Delta M} = -0.03$ and $\sigma = \pm 0.23$ respectively. Although the mean deviations are small, the σ values are large, reflecting the large scatter. It should be pointed out that the magnitude values directly quoted from Hsu (1971) in the 1983 catalogue were not considered in the statistics. Otherwise, the deviation would be larger.

Table 2. Catalogue of Chinese Earthquakes with $M \geq 6.5$ (1900-1948)

Date yr/mo/da	Time h:m:s	Epicenter °N	°E	h km	M_S	m_B	L	Magnitude M C1	C2	C3	S2	G	R	D	H	A1	A_B	A2
1902/07/03	23:36:45	43.2	129.6	(20)	6.7		Jl				6.6							
1902/08/22	11:01	39.8	76.2	40	7.7		Xi	$8\frac{1}{4}$	$8\frac{1}{4}$	$8\frac{1}{4}$	8.1		8.6	8.6		8.2		7.7
1902/08/31	(05:48)	37	78		6.8		Xi				6.9					7.0		6.8
1902/11/04	19:33:30	36	96		6.9		Qi									6.9		
1902/11/21	15:03	23	121		6.8		Ta									7.0		6.8
1904/04/24	14:39	23.5	120.5		(6.5)		Ta	$5\frac{1}{2}$	$5\frac{1}{2}$	6					6.1			
1904/08/30	19:42	30	101		6.8		Si	6*	6*	6*						7.5	7.6	6.8
		31.2*	100.9*															
1905/08/25	17:46:45	43	129	470		6.6	Jl	$6\frac{3}{4}$	$6\frac{3}{4}$	$6\frac{3}{4}$		$6\frac{3}{4}$						
1906/03/02	14:15:15	43.0	80.0		(7)		Xi							7.3				
1906/03/17	06:42	23.6	120.5		6.8		Ta	$6\frac{3}{4}$	$6\frac{3}{4}$	$6\frac{3}{4}$				7.1	7.1	7.1		6.8
1906/04/14	03:18	23.4	120.4		$(6\frac{3}{4})$		Ta	6	6	$6\frac{1}{2}$					6.6			
1906/04/14	07:52	23.4	120.4		$(6\frac{1}{2})$		Ta		5.7	$6\frac{3}{4}$					5.8			
1906/12/23	02:21:00	43.5	85.0		7.2		Xi	8	8	8	7.5	7.9	8.3	8.3		7.9	7.5	7.2
		43.9*	85.6*															
1908/01/11	11:35	23.7	121.4		6.8		Ta		7.3	$6\frac{1}{2}$				7.3	7.3	7.2		6.8
1908/02/10	02:13	26.0	100.0		(7.3)		Yu							7.3				
1908/08/20	17:53	32.0	89.0		7.0		Xz							7.0		7.0		
1909/04/15	03:54	25.0	121.5	80	(6.9)	7.1	Ta	$7\frac{1}{4}$	$7\frac{1}{4}$	7.3		7.3		7.3	7.3		7.1	
1909/11/21	15:36	24.4	121.8		7.0		Ta		7.3	7.3				7.3	7.3			
1910/01/08	22:49:30	35	122		6.7		Ji	$6\frac{3}{4}$	$6\frac{3}{4}$	$6\frac{3}{4}$		$6\frac{3}{4}$						
1910/04/12	08:22:13	25.5	122.5	200		7.6	Ta	$7\frac{3}{4}$	$7\frac{3}{4}$	$7\frac{3}{4}$		$7\frac{3}{4}$	8.3	8.3	8.3		7.6	
1910/06/17	13:28	21.0	121.0		6.8		Ta		7.0	$6\frac{1}{2}$					7.0			
1910/07/12	15:36:12	37	76	120		6.7	Xi	$6\frac{3}{4}$	$6\frac{3}{4}$	$6\frac{3}{4}$		$6\frac{3}{4}$						
1910/09/01	08:45	22.7	121.7		$(6\frac{3}{4})$		Ta			$6\frac{1}{2}$				7.1		7.0		7.0
1910/09/01	22:21	24.1	122.4		$(6\frac{3}{4})$		Ta			$6\frac{1}{2}$				7.3				6.8
1910/11/14	15:34:31	24.5	122.0		$(6\frac{1}{2})$		Ta			$6\frac{1}{4}$				7.0				
1911/10/15	07:24	31.0	80.5		6.9		Xz	$6\frac{3}{4}$	$6\frac{3}{4}$	$6\frac{3}{4}$		$6\frac{3}{4}$						
1912/12/25	02:07	24.0	121.6		6.5		Ta			$6\frac{1}{4}$								
1913/01/08	06:50	24.0	121.6		$6\frac{3}{4}$		Ta			$6\frac{1}{2}$								
1913/12/21	23:37:48	24.5	102.0		$7\frac{1}{4}$		Yu	$6\frac{1}{2}$	$6\frac{1}{2}$	7				7.2		7.2		
		24.2*	102.5*															
1914/03/28	18:44:48	25	99	100	$(6\frac{3}{4})$	6.7	Yu					6.9						
1914/07/06	14:37:30	24	122	60		$6\frac{3}{4}$	Ta	$6\frac{3}{4}$	$6\frac{3}{4}$	$6\frac{3}{4}$		$6\frac{3}{4}$						
1914/08/05	06:41:36	43.5	91.5		7.3	7.0	Xi	$7\frac{1}{2}$	$7\frac{1}{2}$	$7\frac{1}{2}$		$7\frac{1}{2}$		7.3		7.3	7.1	
1914/10/09	10:39:10	35	78		6.6		Xi	$6\frac{1}{2}$		$6\frac{1}{2}$								
1915/01/06	07:26:42	25	123	160		7.3	Ta		$7\frac{1}{4}$	$7\frac{1}{4}$		$7\frac{1}{4}$		$7\frac{1}{4}$	7.3		7.3	
1915/12/03	10:39:19	29.5	91.5		7.0		Xz	7	7	7				7.1		7.0		
1915/12/17	15:04:47	42.0	79.5		$(6\frac{3}{4})$		Xi			$6\frac{1}{2}$	6.7							
1916/08/28	14:39:42	30	81		7.3	7.5	Xz			7.5		7.5		7.7		7.3	7.5	
1916/08/28	15:27	23.7	120.9		$(6\frac{3}{4})$		Ta	6	6	$6\frac{1}{2}$					6.4			
1917/01/05	00:55	23.9	120.9		6.5		Ta	$6\frac{1}{4}$	$6\frac{1}{4}$	$6\frac{1}{4}$					5.8			
1917/01/07	02:08	23.9	120.9		$(6\frac{1}{2})$		Ta	$5\frac{3}{4}$	$5\frac{3}{4}$	$5\frac{3}{4}$					5.6			
1917/01/24	08:48:26	31.4	116.5		$(6\frac{1}{2})$		An	$6\frac{1}{4}$	$6\frac{1}{4}$	$6\frac{1}{4}$								
		31.3*	116.2*															

continued

Table 2. *Continued*

Date yr/mo/da	Time h:m:s	Epicenter °N	°E	h km	M_S	m_B	L	Magnitude M C1	C2	C3	S2	G	R	D	H	A1	A_B	A2
1917/07/04	08:38:20	25	123		7.3		Ta		7	7				7.7		7.3		
1917/07/04	13:36:30	25	123		($6\frac{3}{4}$)		Ta		$6\frac{1}{2}$	$6\frac{1}{2}$				7.2				
1917/07/31	07:54:05	29	104		$7\frac{1}{4}$		Si	$6\frac{1}{2}$	$6\frac{1}{2}$	$6\frac{3}{4}$				7.5		7.5		
		28.0*	104.0*				Yu											
1917/07/31	11:23:10	42.5	131.0	460		7.4	Jl	$7\frac{1}{2}$		7.5		7.5		7.5			7.4	
1918/02/10	04:46:26	43	130	450		$6\frac{1}{2}$	Jl	$6\frac{1}{2}$	$6\frac{1}{2}$	$6\frac{1}{2}$		$6\frac{1}{2}$						
1918/02/13	14:07:13	24	117		7.4	7.2	Fu	$7\frac{1}{4}$	$7\frac{1}{4}$	7.3		7.3		7.3		7.4	7.2	
		23.5*	117.2*				Gu											
1918/04/10	10:03:54	43.5	130.5	570		7.0	Jl	$7\frac{1}{4}$	$7\frac{1}{4}$	7.2		7.2		7.2			7.0	
1918/12/01	10:35:04	39.0	73.0		6.6		Xi			$6\frac{1}{2}$								
1919/07/24	10:03:20	40	76		6.6		Xi	$6\frac{1}{2}$	$6\frac{1}{2}$	$6\frac{1}{2}$	6.7							
1919/08/26	03:55:15	32	100		6.5		Si	$6\frac{1}{4}$	$6\frac{1}{4}$	$6\frac{1}{4}$								
1919/12/21	04:38	22.8	121.7		7.0		Ta	7.0	7.0	7		7		7.0		7.0		
1919/12/21	05:38:55	23.0	121.7		$6\frac{3}{4}$		Ta											
1920/06/05	12:21:28	23.5	122.7		8.0	7.7	Ta	8.0	8.0	8		8.0	8.3	8.3	8.3	8.0	7.8	
1920/10/20	18:02:27	25.1	120.8		$6\frac{3}{4}$		Ta	$6\frac{1}{2}$	$6\frac{1}{2}$	$6\frac{1}{2}$								
1920/10/21	03:15:51	23.7	120.1		$6\frac{1}{2}$		Ta	$6\frac{1}{4}$	$6\frac{1}{4}$	$6\frac{1}{4}$								
1920/12/16	20:05:53	36.8	104.9	20	8.6	7.9	Ga	$8\frac{1}{2}$	$8\frac{1}{2}$	$8\frac{1}{2}$		$8\frac{1}{2}$	8.6	8.6		8.6	7.9	
		36.5*	105.7*				Ni											
1920/12/25	19:33:13	36.6	105.2		7.0		Ni	7	7	7								
1921/03/19	16:19:45	24.0	116.5		6.5		Gu	$6\frac{1}{4}$	$6\frac{1}{4}$	$6\frac{1}{4}$								
1921/04/02	17:36:45	23.3	122.0		6.7		Ta	$6\frac{1}{2}$	$6\frac{1}{2}$	$6\frac{1}{2}$								
1921/12/01	18:49:40	33.7	122.0		6.5		Ji		$6\frac{1}{4}$	$6\frac{1}{2}$								
1922/09/02	03:16:06	24.5	122.0		7.6	7.5	Ta	$7\frac{1}{2}$	$7\frac{1}{2}$	7.6		7.6		7.6	7.6	7.6	7.5	
1922/09/05	01:53:35	24.0	120.0		($6\frac{3}{4}$)		Ta	$6\frac{1}{2}$	$6\frac{1}{2}$	$6\frac{1}{2}$								
1922/09/15	03:32	24.6	122.3		7.2		Ta	$7\frac{1}{4}$	$7\frac{1}{4}$	7.2		7.2		7.2	7.2	7.2		
1922/09/17	17:58	24.2	122.2		6.9		Ta	$6\frac{3}{4}$	$6\frac{3}{4}$	$6\frac{3}{4}$								
1922/10/15	07:47	24.6	122.3		6.7		Ta	$6\frac{3}{4}$	$6\frac{3}{4}$	$6\frac{3}{4}$					5.9			
1922/10/17	00:01:32	39.5	91.0		$6\frac{1}{2}$		Xi	$6\frac{1}{2}$	$6\frac{1}{2}$	$6\frac{1}{2}$		$6\frac{1}{2}$						
1922/10/27	22:23	24.5	122.2		$6\frac{1}{2}$		Ta	$6\frac{1}{2}$	$6\frac{1}{2}$	$6\frac{1}{2}$								
1923/03/24	20:40:06	31.5	101.0		7.3		Si	$7\frac{1}{4}$	$7\frac{1}{4}$	7.3		7.3		7.3		7.3		
		31.3*	100.8*				Si											
1923/07/01	15:54:55	22.0	100.5		($6\frac{1}{2}$)		Yu	$6\frac{1}{2}$	$6\frac{1}{2}$	$6\frac{1}{2}$								
		23.0*	101.0*															
1923/07/02	10:32	23.1	122.0		6.5		Ta	$6\frac{1}{2}$	$6\frac{1}{2}$	$6\frac{1}{2}$								
1923/11/19	05:30	24.2	122.5		($6\frac{3}{4}$)		Ta			$6\frac{3}{4}$								
1923/11/22	15:21	24.2	122.5		($6\frac{1}{2}$)		Ta			6								
1924/01/27	12:22:12	20	121		($6\frac{1}{2}$)		Ta			$6\frac{1}{2}$								
1924/07/03	12:40:10	36.8	83.8	7	7.2		Xi	$7\frac{1}{4}$	$7\frac{1}{4}$	$7\frac{1}{4}$		7.2		7.2		7.2		
1924/07/12	03:44:44	37.1	83.6	20	7.3	7.3	Xi	$7\frac{1}{4}$	$7\frac{1}{4}$	7.2		7.2		7.2		7.2	7.3	
1924/10/09	04:32:57	30	90		$6\frac{3}{4}$	$6\frac{1}{2}$	Xz	$6\frac{1}{2}$	$6\frac{1}{2}$	$6\frac{1}{2}$		$6\frac{1}{2}$						
1925/03/16	22:42:17	25.7	100.4	20	7.0	7.0	Yu	7	7.0	7		7.1		7.1		7.0	7.0	
1925/04/17	03:53	20.4	120.2		7.1	7.1	Ta	7.0	7	7.1		7.1		7.1	7.1	7.1	7.1	
1925/12/22	13:05:30	21	101.5		$6\frac{3}{4}$		Yu			$6\frac{3}{4}$		$6\frac{3}{4}$						

continued

Table 2. *Continued*

Date yr/mo/da	Time h:m:s	Epicenter °N	°E	h km	M_S	m_B	L	Magnitude M C1	C2	C3	S2	G	R	D	H	A1	A_B	A2
1926/08/03	11:41:30	22.0	121.0		$6\frac{3}{4}$		Ta	$6\frac{1}{2}$	$6\frac{1}{2}$	$6\frac{1}{2}$								
1927/05/18	05:44:16	44	131	430		$6\frac{1}{2}$	He	$6\frac{1}{2}$	$6\frac{1}{2}$	$6\frac{1}{2}$		$6\frac{1}{2}$						
1927/05/23	06:32:47	37.7	102.2	20	7.9	7.9	Ga	8	8	8		8.0	8.3	8.3		7.9	7.9	
		37.6*	102.6*															
1927/08/25	02:09:00	23.0	120.5	60		$6\frac{3}{4}$	Ta	$6\frac{3}{4}$	$6\frac{3}{4}$	$6\frac{3}{4}$		$6\frac{3}{4}$			6.5			
1927/09/23	21:54:13	42.3	85.6	33	6.7		Xi	$6\frac{3}{4}$	$6\frac{3}{4}$	$6\frac{3}{4}$	6.6	$6\frac{3}{4}$						
1929/08/19	10:44	24.2	122.5		6.7		Ta	$6\frac{3}{4}$	$6\frac{3}{4}$	$6\frac{3}{4}$		$6\frac{3}{4}$						
1930/04/29	02:34:37	25.8	98.6	20	6.5		Yu	$6\frac{1}{4}$	$6\frac{1}{4}$	$6\frac{1}{4}$								
1930/07/14	03:27:21	38.1	98.2	32	6.5		Qi	$6\frac{1}{2}$	$6\frac{1}{2}$	$6\frac{1}{2}$		$6\frac{1}{2}$						
1930/08/21	04:54	24.6	122.0	40	$6\frac{3}{4}$		Ta	7	7	$6\frac{3}{4}$								
1930/09/22	07:04:14	25.8	98.4	20	6.5		Yu	$6\frac{1}{2}$	$6\frac{1}{2}$	$6\frac{1}{2}$								
1930/12/21	22:51	23.4	120.5		(6.6)		Ta											
1930/12/22	07:52	23.3	120.4		$(6\frac{1}{2})$		Ta		$5\frac{1}{2}$	$5\frac{1}{2}$					6.5			
1930/12/22	08:08	23.3	120.4		$(6\frac{1}{2})$		Ta	$5\frac{3}{4}$	$5\frac{3}{4}$	$5\frac{3}{4}$					6.5			
1931/08/11	05:18:43	47.1	89.8	20	7.9	7.6	Xi	8	8.0	8	7.8	8.0	7.9	7.9		7.9	7.6	
		46.8*	89.9*															
1931/08/18	22:21:04	47.4	90.0	20	7.3	7.1	Xi	$7\frac{1}{4}$	$7\frac{1}{4}$	$7\frac{1}{4}$	6.7	7.2		7.2		7.3	7.1	
1931/09/21	18:27:20	19.8	113.1	20	$6\frac{3}{4}$		Gu	$6\frac{3}{4}$	$6\frac{3}{4}$	$6\frac{3}{4}$		$6\frac{3}{4}$						
1932/12/25	10:04:27	39.7	96.7	8	7.7	7.3	Ga	$7\frac{1}{2}$	$7\frac{1}{2}$	7.6		7.6		7.6		7.7	7.4	
		39.7*	97.0*															
1933/02/13	10:49:15	46.3	90.5	20	$6\frac{1}{4}$		Xi	$6\frac{1}{2}$	$6\frac{1}{2}$	$6\frac{1}{2}$	6.2	$6\frac{1}{2}$						
1933/04/19	14:44:36	24.3	121.5		$6\frac{3}{4}$		Ta	$6\frac{1}{2}$	$6\frac{1}{2}$	$6\frac{1}{2}$		$6\frac{1}{2}$						
1933/08/25	15:50:30	31.9	103.4	20	7.4	7.3	Si	$7\frac{1}{2}$	$7\frac{1}{2}$	$7\frac{1}{2}$		7.4		7.4		7.5	7.3	
		32.0*	103.7*															
1933/09/26	02:51:27	38.3	86.9	58	$6\frac{3}{4}$		Xi	$6\frac{3}{4}$	$6\frac{3}{4}$	$6\frac{3}{4}$		$6\frac{3}{4}$						
1934/02/14	11:59:34	17.5	119.0		7.6	7.6	Zh			7.6		7.6	7.9	7.9		7.6	7.6	
1934/12/15	09:57:40	31.3	89.0	22	7.1	7.2	Xz	7.0	7.0	7		7.1		7.1		7.1	7.2	
1935/04/21	06:01:54	24.3	120.8		7.1	6.8	Ta	7	7	7.1		7.1		7.1	7.1	7.1	6.8	
1935/09/04	09:37:42	22.2	121.3	20	7.2	7.1	Ta	$7\frac{1}{4}$	$7\frac{1}{4}$	7.2		7.2		7.2	7.2	7.2	7.1	
1936/02/07	16:56:27	35.4	103.4	18	6.7		Ga	$6\frac{3}{4}$	$6\frac{3}{4}$	$6\frac{3}{4}$		$6\frac{3}{4}$						
		35.2*	103.4*															
1936/04/27	07:59:11	28.9	103.6	33	6.8	6.7	Si	$6\frac{3}{4}$	$6\frac{3}{4}$	$6\frac{3}{4}$		$6\frac{3}{4}$						
		28.7*	103.7*															
1936/05/16	15:05:44	28.5	103.6	20	6.9	6.9	Si	$6\frac{3}{4}$	$6\frac{3}{4}$	$6\frac{3}{4}$		$6\frac{3}{4}$						
1936/08/22	14:51:35	22.3	120.8		7.2	7.1	Ta	$7\frac{1}{4}$	$7\frac{1}{4}$	7.2		7.2		7.2	7.1	7.3	7.1	
1937/01/07	21:20:41	35.5	97.6	20	7.6	7.2	Qi	$7\frac{1}{2}$	$7\frac{1}{2}$	$7\frac{1}{2}$		7.6		7.6		7.7	7.2	
1937/08/01	04:35:48	35.4	115.1	20	6.8		Sh	7	7.0	7		6.9						
		35.2*	115.3*															
1937/08/01	18:41:05	35.3	115.2	20	6.7		Sh	$6\frac{3}{4}$	$6\frac{3}{4}$	$6\frac{3}{4}$		$6\frac{3}{4}$						
		35.3*	115.4*															
1937/11/16	05:37:34	35	78	100	$(6\frac{3}{4})$	$6\frac{1}{2}$	Xi	$6\frac{1}{2}$		$6\frac{1}{2}$		$6\frac{1}{2}$						
1937/12/08	16:32:11	22.9	121.2	22	7.0	6.9	Ta	7	7	7		7.0		7.0	7.0	7.0	7.0	
1937/12/14	02:53	23.8	121.3		6.6		Ta	$6\frac{3}{4}$	$6\frac{3}{4}$	$6\frac{3}{4}$					6.4			
1937/12/17	17:32	22.8	121.5		6.5		Ta	$6\frac{3}{4}$	$6\frac{3}{4}$	$6\frac{3}{4}$					6.5			

continued

Table 2. *Continued*

Date yr/mo/da	Time h:m:s	Epicenter °N	°E	h km	M_S	m_B	L	Magnitude M C1	C2	C3	S2	G	R	D	H	A1	A_B	A2
1938/04/01	06:31:10	20.0	120.5	60	(6.3)	$6\frac{1}{2}$	Ta			$6\frac{1}{2}$		$6\frac{1}{2}$						
1938/05/23	16:21:53	18.0	119.5	80	(7.1)	7.0	Do			7.0		7.0		7.0			7.0	
1938/09/07	12:03:18	23.8	121.5		7.0	6.6	Ta	7	7.0	7		7.0		7.0	7.0	7.0	6.6	
1938/12/07	07:00:53	22.8	120.8		7.0	7.1	Ta	7	7	7		7.0		7.0	7.0	7.0	7.1	
1940/07/10	13:49:55	44	131	580		7.3	He	$7\frac{1}{4}$	$7\frac{1}{4}$	7.3		7.3		7.3			7.3	
1941/01/21	20:42:01	27.5	91.9	180		$6\frac{3}{4}$	Xz	$6\frac{3}{4}$	$6\frac{3}{4}$	$6\frac{3}{4}$		$6\frac{3}{4}$						
1941/01/27	10:30:16	26.5	92.5	180		$6\frac{1}{2}$	Xz			$6\frac{1}{2}$		$6\frac{1}{2}$						
1941/05/16	15:14:32	23.6	99.4	20	6.9	$6\frac{3}{4}$	Yu	7	7	7		6.9						
		23.7*	99.4*															
1941/12/17	03:19:42	23.3	120.3	21	7.0	7.1	Ta	7	7	7		7.1		7.1	7.1	7.2	7.1	
1941/12/18	04:29	23.4	120.4		6.6		Ta		6.4	6.4					6.4			
1941/12/26	22:48:09	22.7	99.9	47	7.0	7.0	Yu	7	7	7		7.0		7.0		7.0	7.0	
		22.2*	100.1*															
1942/09/24	11:38:58	23.9	121.7		6.5		Ta	$6\frac{1}{2}$	$6\frac{1}{2}$	$6\frac{1}{2}$					6			
1943/04/05	09:56:01	39.3	73.3	(20)	6.5		Xi			$6\frac{1}{2}$	6.3	$6\frac{1}{2}$						
1943/11/24	21:17:13	22.5	122.0		6.8	7.1	Ta	7	7	6.9		6.9			6.3			
1943/12/02	13:09	22.5	121.5	40	6.6		Ta	$6\frac{1}{2}$	$6\frac{1}{2}$	$6\frac{1}{2}$					6.1			
1944/02/06	01:20	23.8	121.4	5	6.6		Ta		$6\frac{1}{2}$	$6\frac{1}{2}$					6.4			
1944/03/10	06:12:58	44	84		7.1	7.0	Xi	$7\frac{1}{4}$	$7\frac{1}{4}$	7.2		7.2		7.2		7.1	7.0	
1944/09/28	00:25:07	39.1	75.0	20	6.8	7.1	Xi	7	7	7	6.7	7.0		7.0		6.8	7.1	
1944/10/18	02:36:56	31.4	83.3	33	6.7		Xz	$6\frac{3}{4}$	$6\frac{3}{4}$	$6\frac{3}{4}$		$6\frac{3}{4}$						
1944/10/29	08:11:30	31.3	83.4		6.6		Xz	$6\frac{3}{4}$	$6\frac{3}{4}$	$6\frac{3}{4}$		$6\frac{3}{4}$						
1944/12/19	22:09:04	39.7	124.3	64	(6.5)		Li	$6\frac{3}{4}$	$6\frac{3}{4}$	$6\frac{3}{4}$		$6\frac{3}{4}$						
1946/01/11	09:33:29	44.0	129.5	580		6.9	He	$7\frac{1}{4}$	$7\frac{1}{4}$	7.2		7.2		7.2			6.9	
1946/12/05	06:47	23.1	120.3		6.6		Ta	$6\frac{3}{4}$	$6\frac{3}{4}$	$6\frac{3}{4}$					6.3			
1946/12/19	10:57:19	24.8	122.5	100		6.8	Ta	$6\frac{3}{4}$	$6\frac{3}{4}$	$6\frac{3}{4}$		$6\frac{3}{4}$			6.7			
1947/02/10	12:02:02	31.8	85.4	20	$6\frac{1}{2}$		Xz	$6\frac{1}{4}$	$6\frac{1}{4}$	$6\frac{1}{4}$								
1947/03/17	16:19:41	33.3	99.5	79	7.6	7.4	Qi	$7\frac{3}{4}$	$7\frac{3}{4}$	7.7		7.7		7.7		7.6	7.4	
1947/07/29	21:43:33	28.6	93.6	114	7.6	7.5	Xz	$7\frac{3}{4}$	$7\frac{3}{4}$	7.7		7.7	7.9	7.9		7.5	7.5	
1947/09/27	00:01:57	24.8	123.0	110		7.4	Ta	$7\frac{1}{2}$		7.4		7.4		7.4	7.4		7.4	
1948/03/03	17:10:02	18.8	119.0	74	7.2	7.4	Ta	$7\frac{1}{4}$	$7\frac{1}{4}$	7.2		7.2		7.2		7.2	7.3	
1948/05/23	15:11:21	29.5	100.5		7.3	7.4	Si	$7\frac{1}{4}$	$7\frac{1}{4}$	7.3		7.3		7.3		7.3	7.4	
		29.7*	100.3*															

Notes:

Time (120° E) = GMT + 8 hours; * = estimated from macroseismic observations.

Magnitude M: C1 = Catalogue of Chinese Earthquakes, 1971; C2 = Catalogue of Chinese Earthquakes, 1977; C3 = Catalogue of Chinese Earthquakes, 1983; S2 = New Catalogue of Strong Earthquakes in U.S.S.R., 1977; G = Gutenberg, 1954; R = Richter, 1958; D = Duda, 1965; H = Hsu, 1971: A1 = Abe, 1981,1983a; A_B = Abe, 1981, 1983a (body wave magnitude); A2 = Abe, 1983b.

L = Location; Location abbreviations: An = Anhui; Do = Dongshaqundao; Fu = Fujian; Ga = Gansu; Gu = Guangdong; He = Heilongjiang; Ji = Jiangsu; Jl = Jilin; Li = Liaoning. Ni = Ningxia; Qi = Qinghai; Sh = Shandong; Si = Sichuan; Ta = Taiwan; Xi = Xinjiang; Xz = Xizang (Tibet); Yu = Yunnan; Zh = Zhongshaqundao.

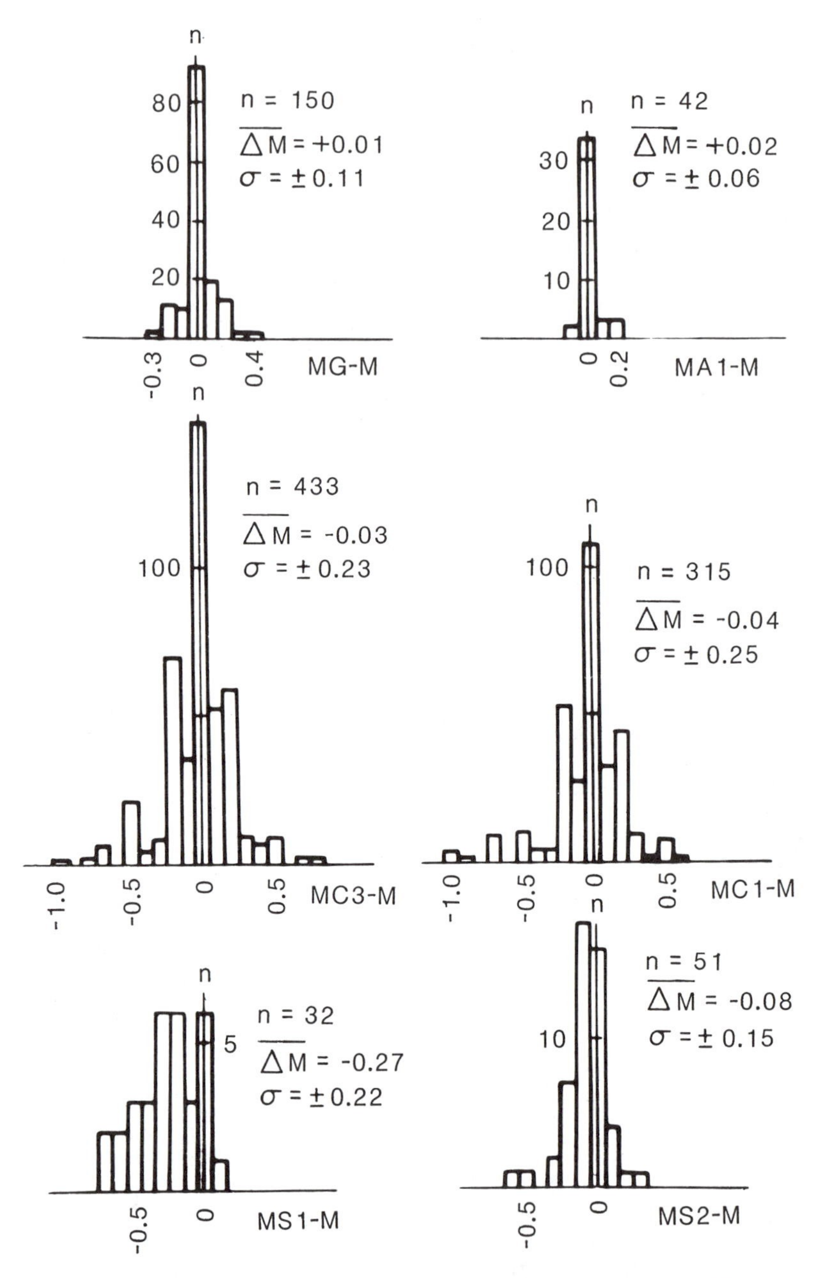

Figure 6. Statistical comparison of our catalogue with other catalogues: MG = Gutenberg, 1954; MA1 = Abe, 1981, 1983a; MS1 = Atlas of Earthquakes in USSR, 1962; MS2 = New Catalogue of strong Earthquakes in USSR, 1977; MC1 = Catalogue of Chinese Earthquakes, 1971; MC3 = Catalogue of Chinese Earthquakes, 1983.

4.2. From Table 2, it can be seen that after careful revision the magnitudes of some strong earthquakes differ greatly from their previously adopted values. For example: for the August 22, 1902 Xinjiang earthquake, the M_S calculated using data from Gutenberg's worksheet was 8.2. After calibration, it was revised to 7.7. The epicenter of the August 31, 1902 Xinjiang earthquake was redetermined. According to Gutenberg's worksheet, its M_S is 7.4. Abe (1983a) changed it to 7.0 by correcting the effective amplification of the undamped seismograph, and later he further revised it to 6.8, which agrees well with the revised magnitude of USSR. The magnitude of the August 30, 1904 Sichuan earthquake was estimated to be 6 from macroseismic observations. By using the data of Gutenberg's worksheet and 10 other stations, $M_S = 7.5$ was obtained. It was further revised to 6.8 after calibration. In addition, the epicenter determined from instrumental data is 1° away from the macroseismic location. The calculated M_S for the December 23, 1906 Xinjiang earthquake was 7.9. After calibration, it was reduced to 7.2. The magnitude of the December 21, 1913 Yunnan earthquake was estimated to be between $6\frac{1}{2}$ and 7 from macroseismic observations in 1971 and 1983 editions of the *Catalogue of Chinese Earthquakes* respectively. The result of our calculation based upon instrumental data is $M_S = 7\frac{1}{4}$, which is very close to Abe's revised magnitude of 7.2. In the two editions of the Chinese catalogues, the magnitude of the July 31, 1917 Yunnan earthquake (the instrumentally determined epicenter is at Sichuan) was given between $6\frac{1}{2}$ and $6\frac{3}{4}$ respectively. Abe corrected it to 7.5. According to the data from several seismic stations, we obtained $M_S = 7\frac{1}{4}$. Similar circumstances exist for the Taiwan earthquakes, although the correction is less. As the instrumental data increased gradually after 1918, such large corrections of magnitude became rare. However, the accuracy and reliability were improved through careful revision. For example: the magnitude of the May 23, 1927 Gansu earthquake was taken as 8.0 by Gutenberg. Among the magnitudes calculated by using formula (1) from the data of 15 stations, only one attained 8.1, with all the others lower than 8. The average value is 7.8, near Abe's value of 7.9. We revised it to 7.9. Based upon the data of 10 seismic stations, we revised the magnitudes of the two August 1, 1937 Shandong earthquakes to 6.8 and 6.7 respectively by using formula (3).

4.3. During the period from 1900 to 1948, the number of stations was insufficient, the seismogram quality was poor, the station distribution was not optimal, and the data collected were insufficient. The fact that the frequency-magnitude curve (Figure 7) bends below $M_S = 6.5$ indicates that many earthquakes with magnitudes less than 6.5 were not recorded. The b-value for the straight line portion of the curve is 0.84. It is close to the value $b = 0.90 \pm 0.02$ obtained by Gutenberg for worldwide shallow shocks, but differs greatly from the value given in *Report of Seismic Zoning of China* (State Seismological Bureau, 1982).

4.4. Figure 8 shows the variation in annual frequency, energy and cumulative strain release for earthquakes with $M \geq 6.5$. Obviously, the seismicity had a maximum in 1920, when the $M_S = 8.6$ Haiyuan earthquake and the $M_S = 8.0$ Taiwan earthquake occurred. The year 1902 was the next largest and includes the $M_S = 7.7$ Atush, Xinjiang shock. In addition, there are small jumps in the years 1906, 1927, 1931, and 1947.

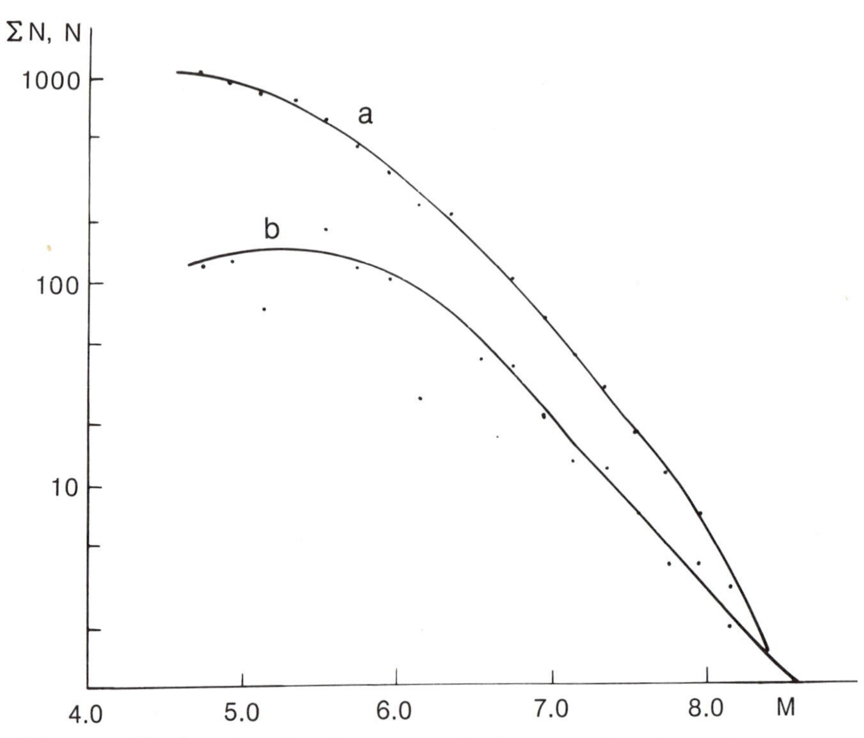

Figure 7. Number of earthquakes vs. magnitude curves of 1900-1948 Chinese earthquakes: (a) Cumulative number of earthquakes with magnitude $\geq M$; (b) Number of earthquakes with magnitude M.

REFERENCES

Abe, K. (1979). Instrumental magnitudes of Japanese earthquakes 1901-1925, *J. Seism. Soc. Japan*, Ser. II, **32**, 341-353.

Abe, K. (1981). Magnitude of large shallow earthquakes from 1904 to 1980, *Phys. Earth Planet. Interiors*, **27**, 72-92.

Abe, K. and H. Kanamori (1979). Temporal variation of the activity of intermediate and deep focus earthquakes, *J. Geophys. Res.*, **84**, 3589-3595.

Abe, K. and H. Kanamori (1980). Magnitudes of great shallow earthquakes from 1953 to 1977, *Tectonophysics*, **62**, 191-203.

Abe, K. and S. Noguchi (1983a). Determination of magnitude for large shallow earthquakes, 1898-1917, *Phys. Earth Planet. Interiors*, **32**, 45-59.

Abe, K. and S. Noguchi (1983b). Revision of magnitudes of large shallow earthquakes, 1897-1912, *Phys. Earth Planet. Interiors*, **33**, 1-11.

Bäth, M. (1977). Teleseismic magnitude relations, *Ann. Geofis.*, **30**, 299-327.

Bäth, M. (1981). Earthquake magnitude – recent research and current trends, *Earth-Science Rev.*, **17**, 315-398.

Central Meteorological Observatory Japan (1952). Catalogue of major earthquakes which occurred in the vicinity of Japan (1885-1950), Suppl. Seism. Bull. for the year 1950, Tokyo, Japan.

Central Seismological Group (1971). *Catalogue of Chinese Earthquakes*, Science Press, Beijing, (in Chinese).

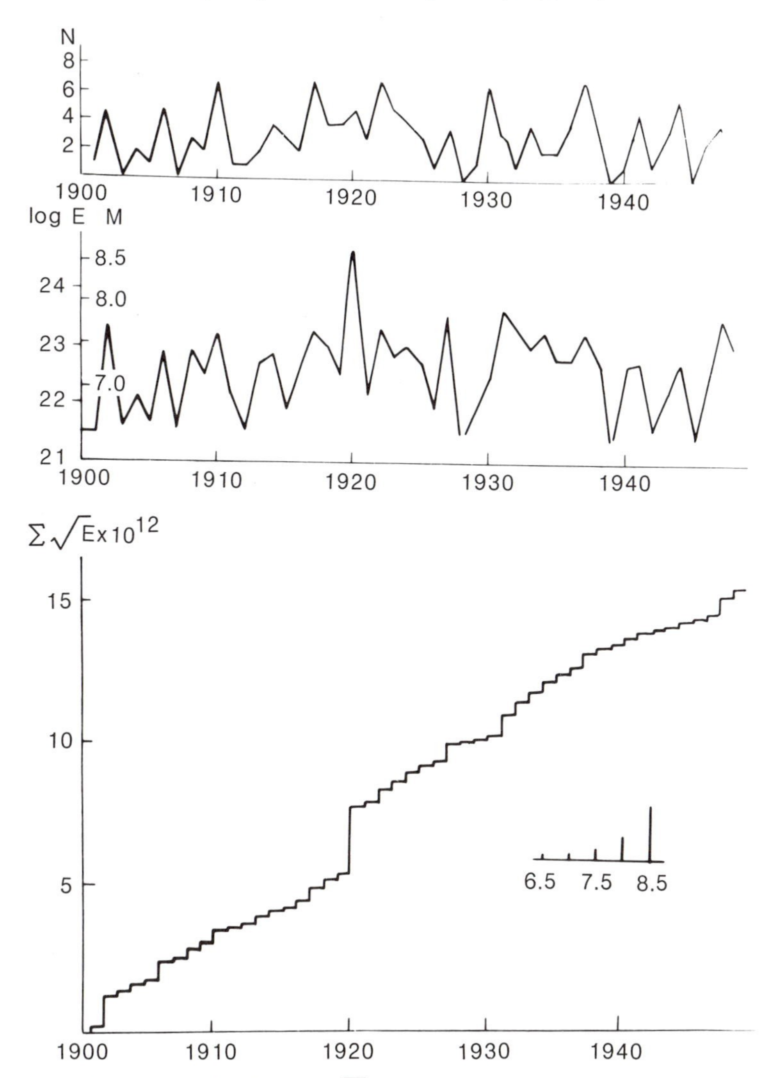

Figure 8. Variations of N, E, and $\Sigma\sqrt{E}$ with time of Chinese earthquakes (1900-1948).

Duda, S. J. (1965). Secular seismic energy release in the Circum-Pacific belt, *Tectonophysics*, **2**, 409-452.

Geller, R. J. and H. Kanamori (1977). Magnitudes of great shallow earthquakes from 1904 to 1952, *Bull. Seism. Soc. Am.*, **67**, 587-598.

Gu Gongxu, (1983). *Catalogue of Chinese Earthquakes*, Science Press, Beijing, (in Chinese).

Gutenberg, B. (1945a). Amplitudes of surface waves and magnitudes of shallow earthquakes, *Bull. Seism. Soc. Am.*, **35**, 3-12.

Gutenberg, B. (1945b). Amplitudes of P, PP and S and magnitude of shallow earthquakes, *Bull. Seism. Soc. Am.*, **35**, 57-69.

Gutenberg, B. (1945c). Magnitude determination for deep-focus earthquakes, *Bull. Seism. Soc. Am.*, **35**, 117-130.

Gutenberg, B. (1956). Great earthquakes 1896-1903, *Trans. Amer. Geophys. Union*, **37**, 608-614.

Gutenberg, B. and C. F. Richter (1954). *Seismicity of the Earth and Associated Phenomena*, Princeton University Press, New Jersey.

Gutenberg, B., and C. F. Richter (1956). Magnitude and energy of earthquakes, *Ann. Geofis.*, **9**, 1-15.

Hsu Ming-Tong (Xu Mingtong) (1971). Seismicity of Taiwan and related problems, *Bull. Inter. Inst. Seism. Earthq. Eng.*, **8**, 41-160.

Kanamori, H. and K. Abe (1979). Reevaluation of the turn-of-the-century seismicity peak, *J. Geophys. Res.*, **84**, 6131-6139.

Kárník, V. *et al.* (1962). Standardization of the earthquake magnitude scale, *Studia geoph. et geod.*, **6**, 41-48.

Kondorskaya, N. V. and N. V. Shebalin (1977). *New Catalog of Strong Earthquakes in the USSR from Old Time to 1975*, Nauka, Moscow (in Russian).

Lee, W. H. K., F. T. Wu, and S. C. Wang (1978). A catalog of instrumentally determined earthquakes in China (magnitude $\geq$ 6) compiled from various sources, *Bull. Seism. Soc. Am.*, **68**, 383-398.

Richter, C. F. (1958). *Elementary Seismology*, Freeman, San Francisco, 768 p.

Rothé, J. P. (1969). *The Seismicity of the Earth, 1953-1965*, UNESCO, Paris.

Savarenskii, E. F., S. L. Soloviev, and D. A. Kharin, editors (1962). *Atlas of Earthquakes in USSR*, Nauka, Moscow (in Russian).

State Seismological Bureau (1977). *Concise Catalogue of Chinese Earthquakes*, Seismological Press, Beijing, (in Chinese).

State Seismological Bureau (1981). Report of Seismic Zoning of China, Seism. Press, Beijing (in Chinese).

Taipei Observatory (1936). History of Taiwan earthquakes, Report of Shinchu-Taichuang Strong Earthquake, 147-160, Taipei Observatory, Taipei, (in Japanese).

Xie, Yushou and Cai Meibiao (1985-86). *Compilation of Historical Materials of Chinese Earthquakes*, Vol. 4, (Part 1, 1985; Part 2, 1986), Science Press, Beijing.

Data Bases on Historical Seismicity: Structure, Quality of Information, and Applications

A. D. Gvishiani
Institute of Physics of the Earth
Academy of Sciences, Moscow, USSR

Wilbur A. Rinehart
National Geophysical Data Center
NOAA, Boulder, CO 80303, USA

ABSTRACT

The earthquake data bases of the National Geophysical Data Center, National Oceanic and Atmospheric Administration, USA, and the Institute of Physics of the Earth, Academy of Sciences, USSR, are described. Together, these two data bases form one of the largest collections of earthquake catalogs in the world. Information on historical seismicity is a significant part of this collection.

The main files are regularly supplemented with new earthquake determinations, taking into account priorities of their quality. A significant problem is verification and correction of errors in the computer data bases. To solve this problem, we have developed special logical algorithms. Using these algorithms, errors have been found in 10% of the records in the computer-readable earthquake catalogs. Some examples of applications of historical seismic data to earthquake-prone areas are given.

1. Introduction

This paper describes two approaches to the storage, updating, and application of large collections of computer-readable earthquake catalogs. The first approach was developed at the National Geophysical Data Center (NGDC), National Oceanic and Atmospheric Administration (NOAA), USA. The second approach was developed at the Institute of the Physics of the Earth (IPE), USSR. These Centers maintain contact with each other and exchange data on a regular basis. They participate in joint research efforts in the fields of seismological data bases and optimizing the data base structure for data applications.

The seismological data base archived by NGDC in Boulder, Colorado, known as the NOAA earthquake data base, is one of the largest collections of computer-readable earthquake catalogs in the world. This data base is distributed by magnetic tape with a fixed record format of 86 characters. Events are arranged on the tape in either chronological or geographical order. It consists of the following 11 files: (1) Preliminary Determination of Epicenters; (2) worldwide historical catalogs; (3) California and W. Nevada catalogs; (4) regional catalogs of the USA; (5) regional catalogs of Alaska; (6) regional catalogs of Hawaii; (7) regional catalogs of Canada; (8) regional catalogs of Europe; (9) regional catalogs of USSR; (10) regional catalogs of Asia; and (11) unclassified, local and regional catalogs. This data base was used for selecting the seismograms to be photographed for the Historical Seismogram Filming Project, begun in 1977 under the auspices of the U.S. Geological Survey. In the first stage, Dr. Gordon Stewart of the California Institute of Technology prepared a list of significant earthquakes from the time when seismometers were first used at observatories throughout the world (about 1900). The final list of earthquakes for the Historical Seismogram Filming Project contained data on earthquakes that have a magnitude sufficient to be recorded by the relatively low-gain seismometers of the 1900 to 1960 time period. Their locations are shown on a world map in Figure 1. Seismograms of the events included in this list were microfilmed, forming the base of the NGDC archive of historical seismograms.

In the seismological data base maintained by NGDC, a unified format is used. The major data fields are: (1) source code; (2) year-month-day; (3) hour-minute-second; (4) latitude; (5) longitude; (6) depth; (7) m_b magnitude; (8) intensity I_o; (9) M_S magnitude; (10) geological effects codes; (11) cultural effects codes; (12) Flinn-Engdahl codes; (13) magnitude reported by other sources; (14) International Declared Event (IDE); (15) number of stations and quality indicator; (16) source authority for time and coordinates; and (17) local magnitude – type and source code. This format makes the earthquake data available using FORTRAN 77 retrieval programs.

Another large earthquake computer-readable data base was independently developed beginning in 1975 by the Institute of the Physics of the Earth (IPE), Academy of Sciences, USSR. A significant part of this data base is information on historical seismicity. Data exchange between IPE and NGDC gave an opportunity to fill in the gaps in both data bases and to have sufficient and complete information about earthquakes in all regions of the world. The NOAA data base was supplemented by the earthquake hypocenter data file of the USSR. In return, the worldwide catalog of the NOAA data base is regularly provided by NGDC to update the IPE data base.

As a result of the exchange of data, NGDC prepared a map (Figure 2) of the strong earthquakes in the territory of the USSR from ancient times to 1977.

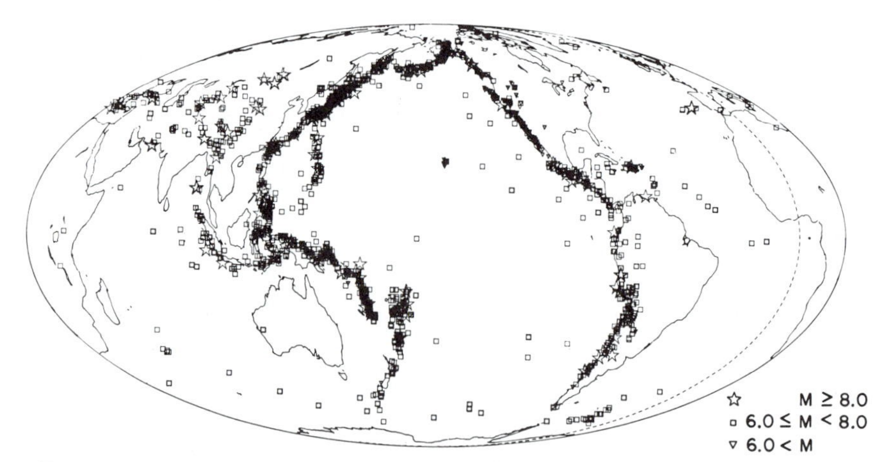

Figure 1. 1900-1960 earthquake locations proposed for the historical microfilming project.

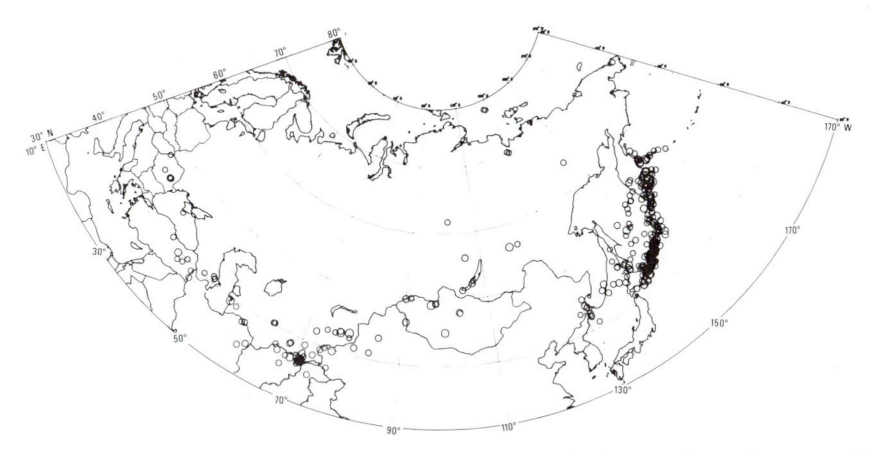

Figure 2. Catalog of strong earthquakes in USSR from ancient times to 1977 ($M > 6.75$).

2. Quality Control

The usage of seismological data for prediction and other research requires continuous updating and verification. NGDC (Rinehart, *et al.*, in press) and IPE (Gvishiani and Zheligovsky, in press) have developed independent procedures for this activity. The logical scheme for verifying and updating the NOAA data base is shown in Figure 3.

At IPE, the research applications begin with seismicity files that are compiled from data of different sources (Shebalin, in press). The data base contains more than 70 catalogs, including the historical seismicity stored in sequential files, including: (1) worldwide catalogs – NOAA, ISS-ISC, Abe, Gutenberg and Richter, etc.; (2) New Catalog of Strong Earthquakes in the Territory of the USSR from Ancient Times to 1984; (3) republic and regional catalogs of the USSR; and (4) regional catalogs – USA, Japan, Italy, France, Spain, Canada, Turkey, New Zealand and others. The reliability of the earthquake information increases with the time delay between occurrence of the event and publication of its parameters. (Immediate earthquake reports and bulletins are not as reliable as later catalogs.) Thus, it is necessary to create from the existing sources a file that contains for every time period the most reliable earthquake data.

Figure 4 shows how this problem is solved at IPE. Priorities of reliability are assigned to different sources according to the amount of primary information and the time period between occurrence and publication. The lower the priority number, the greater the reliability of source. This system of updating assures that IPE's data base includes, for every time period, only the earthquake parameters from the source of highest priority. In this file, the delay of the data entry for certain regional data catalogs is no longer than ten days after receipt of published listings.

Thus, the data in the current IPE catalog have been verified, as illustrated in Figure 5, using an analytical approach to the geophysical data. Applying this approach to most of the IPE computer catalogs has led to the discovery of errors in up to 10% of the records in some data files.

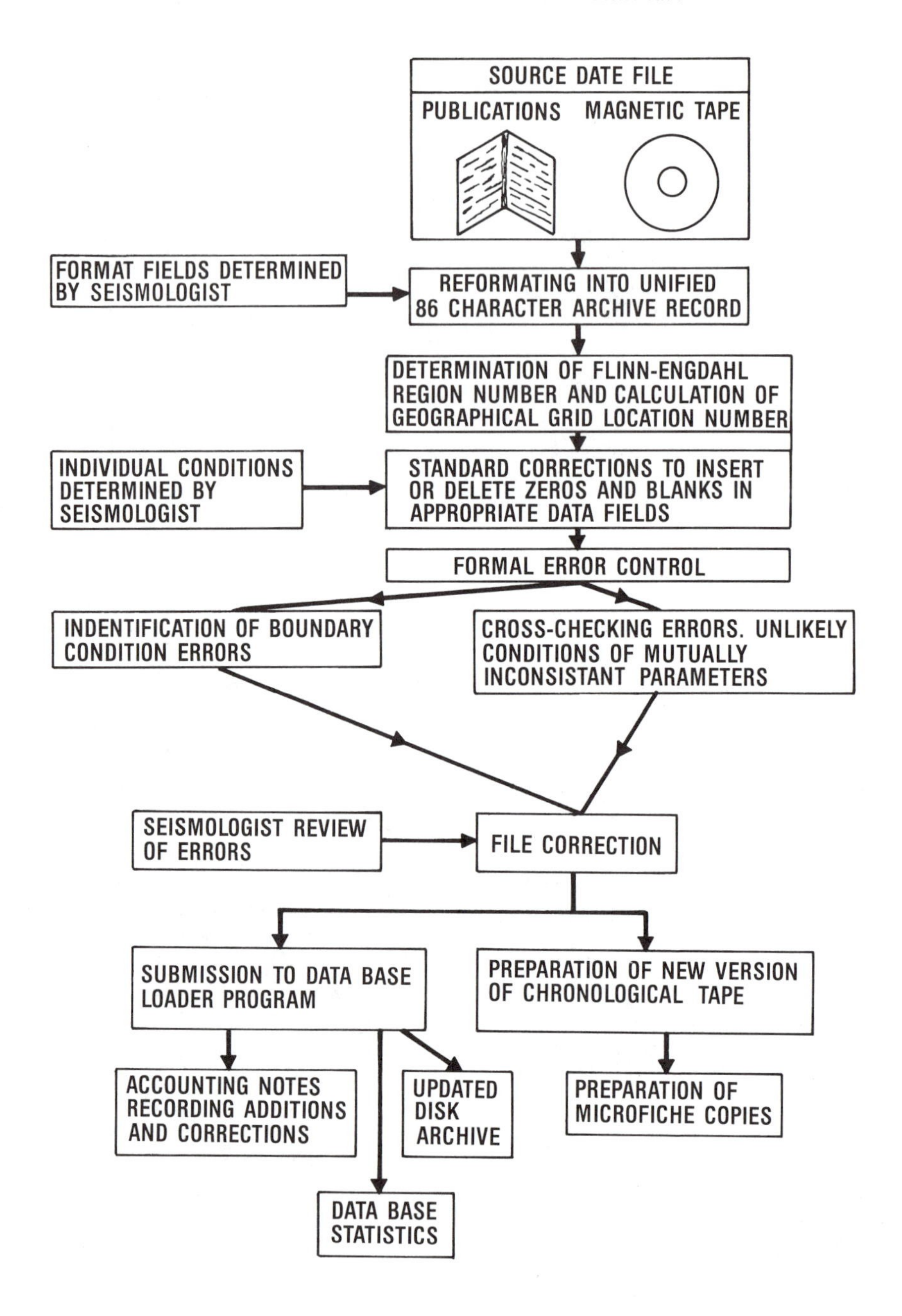

Figure 3. Quality control, updating and maintenance of the NOAA earthquake data base.

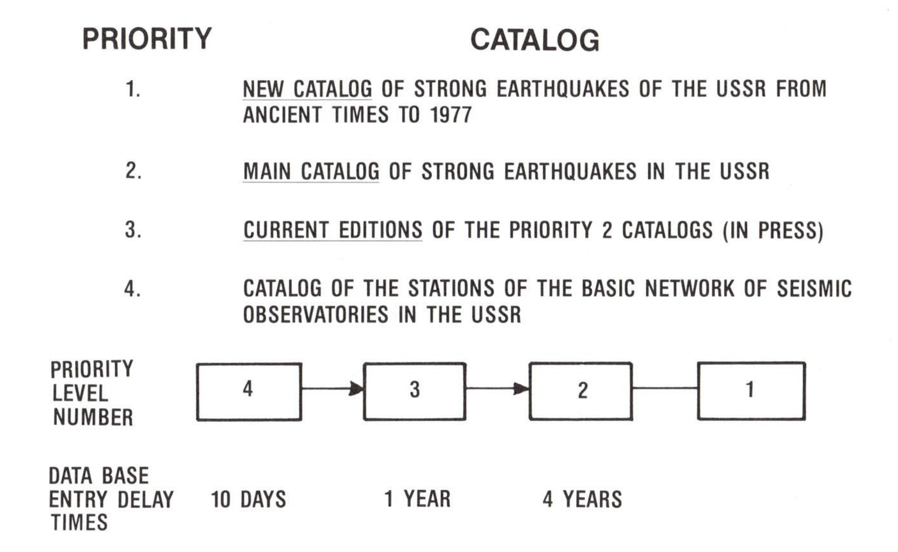

Figure 4. Current data file of strong earthquakes in the USSR.

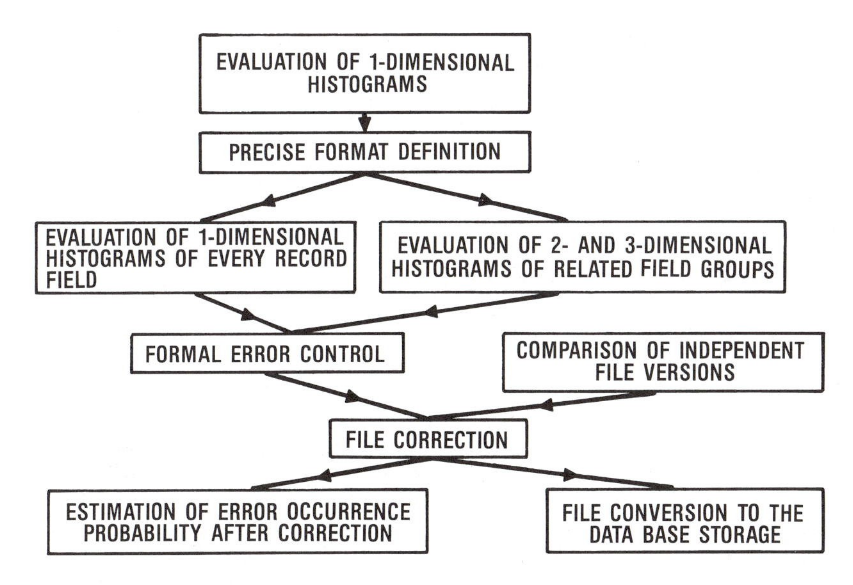

Figure 5. Logical scheme of analytical approach to verification of geophysical data (V. Zheligovsky).

k —NUMBER OF ERROR RECORDS BEFORE CORRECTION PROCESS
P_i —PROBABILITY TO FIND AN ERROR IN THE ith VERIFICATION, $i=1, 2$
n_i —NUMBER OF DISCOVERED ERRORS IN THE ith VERIFICATION, $i=1, 2$
n —NUMBER OF DISCOVERED ERRORS IN BOTH VERIFICATIONS

$$K \approx \frac{n_1 n_2}{n} \; ; P_1 \approx \frac{n_1}{k} \; P_2 \approx \frac{n_2}{k}$$

PROBABILITY OF THE EXISTENCE OF AN ERROR IN THE VALUE OF THE PARAMETER AFTER THE VERIFICATION PROCESS

$$q = 1 - P_1 - P_2 + P_1 P_2$$

WITH THE PROBABILITY

$$P(\ell) = C_k^{\ell} \, q^{\ell} \, (1-q)^{k-\ell}$$

WHERE

$$C_k^{\ell} = \frac{k(k-1) \ldots\ldots\ldots\ldots (k-\ell+1)}{\ell!}$$

FOR ℓ ERRORS IN THE FILE

Figure 6. Estimation of the probability of an error after data base correction.

According to the method represented in Figure 6, Gvishiani and Zheligovsky (in press) obtained a 90% probability that the number of erroneous records in the file (which currently contains more than 7,000 strong earthquakes in the territory of the USSR) is not higher than 10. This number of errors, which is small compared to the size of the file, is unusually small for computer-readable catalogs.

3. Applications

The main goal of creating the seismological data bases in NGDC and IPE is to provide data for scientific and engineering research and for development of worldwide and regional epicenter maps.

The number of users of NOAA and IPE data bases increases each year. The data from computer-readable earthquake catalogs are widely used in scientific and engineering research. An important example of such scientific application of the NOAA and IPE data bases is evidenced by the recognition of areas prone to strong earthquakes. In this research, as objects of recognition, we recognize the morphostructures near which strong earthquake epicenters are a priori clustered. Using the descriptions of these objects as geological and geophysical parameters, pattern recognition algorithms make an attempt to answer the question: "Is a strong earthquake possible in the vicinity of the given object?" Strong earthquakes are those of magnitude larger than a chosen threshold $M \geq M_o$, where M_o depends on the territory under consideration.

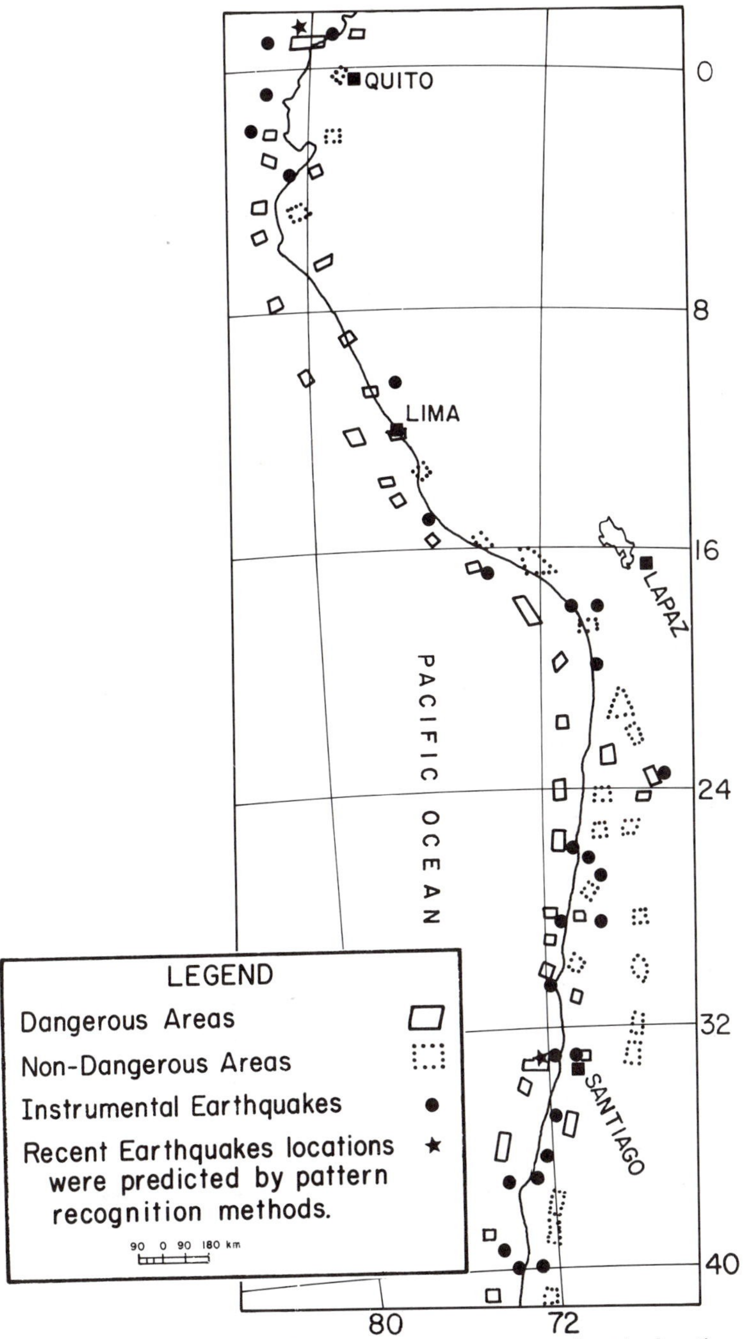

Figure 7. Pacific coast of South America where recent earthquake locations were predicted by pattern recognition methods.

The pattern recognition methods used for this problem and the results of its applications were described in many papers (see, for example, Gelfand, *et al.*, 1976). The regions of strong earthquakes which were investigated by these methods are: Pacific Belt ($M \geq 8.0$); Andes ($M \geq 7\frac{3}{4}$); Kamchatka ($M \geq 7\frac{3}{4}$); Pamir-Tienshen ($M \geq 6.5$, $M \geq 7.0$); Anatolia and Balkans ($M \geq 6.5$, $M \geq 7.0$); California ($M \geq 6.5$); Apennines ($M \geq 6.0$); Alps ($M \geq 5.0$); and Pyrenees ($M \geq 5.0$).

Noninstrumental historical epicenters are not used in this process of pattern recognition. Thus, the catalogs of historical seismicity data represent independent information, which gives an opportunity to evaluate the reliability of the pattern recognition results. It is clear that if an object is recognized as a high seismic danger, then the existence of an epicenter of a strong historical earthquake supports the reliability of the pattern recognition results. For example, on the Pacific coast of South America, as shown in Figure 7, the locations of the historical epicenters agree well with the result of the pattern recognition procedures. After the paper with these results was published (Gabrielov, *et al.*, 1982), two strong earthquakes occurred in the Andes region. The first one, on December 12, 1979, on the border of Ecuador and Colombia, had a magnitude of 7.9 and a depth of 62 km. The second one, the recent earthquake of March 4, 1985, which killed more than 100 people and caused extensive damage, had its epicenter near Santiago, Chile, and a magnitude of 7.75. In Figure 7, the locations of both earthquakes predicted by the pattern recognition results are shown.

4. Conclusion

The recognition of zones of high seismicity is one of the many applications of the earthquake catalog data bases. For these applications, the advantages of comprehensive collections of seismicity data in two major world centers is obvious. The creation of such collections is the goal of the joint NGDC-IPE project. We appeal to the participants of this symposium to support this effort.

REFERENCES

Gabrielov, A. M., A. D. Gvishiani, and M. P. Zhidkov (1982). Normalized morphostructural zoning of the mountainous zones of the Andes, mathematical model of the earth's structure, and forecasting of earthquakes (in Russian), *Comput. Seism.* (Moscow), **14**, 38-56.

Gelfand, I. M., Sh. I. Guberman, V. I. Keilis-Borok, L. Knopoff, F. Press, E. Ya. Ranzman, I. M. Rotwain, and A. M. Sadovsky (1976). Pattern recognition applied to earthquake epicenters in California. *J. Phys. Earth Planet. Interiors*, **11**, 227-283.

Gvishiani, A. D., and V. A. Zheligovsky (in press). Logical scheme of the analytical approach to error control of geophysical data files, Solid Earth Series Rept., World Data Center A (Boulder, CO).

Rinehart, W. A., K. D. Woerner, and S. Godeaux (in press). Quality control in the NOAA database, Solid Earth Series Rept., World Data Center A (Boulder, CO).

Shebalin, P. (in press). Methods of computer integration of different earthquake catalogs, Solid Earth Series Rept., World Data Center A (Boulder, CO).

Compilation and Processing of Historical Data: Summary Remarks

by V. Kárník
Geophysical Institute, Czechoslovak Academy of Sciences
Prague, Czechoslovakia

Every study on earthquake hazard and seismic risk started with the inventory of information available on past seismic events and on their geological and geophysical setting. As a matter of fact, the results of any study depend on the quality and completeness of input information.

Earthquake generating processes are rather slow; and for their understanding a data sample covering at least several centuries is required, if we have to answer the basic questions about the location of potential seismogenic zones and about the frequency of occurrence of events at different magnitude levels.

Statistical models employed in earthquake hazard and seismic risk analyses must introduce various simplifying assumptions, for instance, a constant rate in earthquake activity, independency of events, etc. Observations, however, show that the activity varies with time at different "periods", that the centres of activity migrate within an active belt, that events cannot be independent, etc. Consequently, the time window of a data sample must be as long as possible to allow for detection of possible time variations. Thus, there is a need for searching in historical records and even in geological deposits for any trace of past earthquake activity.

Another requirement is for as many earthquake parameters as possible defining the earthquake generating process. In this respect we depend on the capabilities of the observation techniques as well as on the development of theoretical and laboratory investigations.

Basic parameters defining a seismic event are: date, origin time, coordinates of the focus (the point from which the initial P waves propagate; usually specified by longitude, latitude and focal depth), and size of the event in terms of seismic energy ($\log E$), Magnitude (M) or macroseismic intensity in the epicenter (I_o). This is the minimum information; naturally, it is desirable to determine other quantities describing the event, such as the stress drop ($\Delta\sigma$), or seismic moment (M_o), dimensions and orientation of the focus (fault plane), direction and size of the movement along the fault, shaken area, damage assessment, accompanying geological and geophysical phenomena, etc. Additional parameters are, however, available only for a rather limited number of events in a particular area and cannot be as yet processed statistically.

For seismic events prior to 1900, even the determination of basic parameters is a problem because of the decreasing completeness and accuracy of information if we go back in history. Earthquake information is dispersed in chronicles, in various treaties on natural disasters, in administrative records, in old newspapers, etc. The basic problem is the reliability or accuracy of the sources, and every investigator must check it by consulting historians and other experts. The tendency of some chroniclers to attract attention by exaggerating or manufacturing information is known. The ideal situation is when we can confront several sources of information, which is the case only for some large events. Often seismologists have to consider brief sentences like "the town X was destroyed by an earthquake on ... ". It depends now on the density of population, on the period of peace or war, on the source and other aspects whether the information can be classified as reliable or doubtful.

In some cases storm effects, landslides or subsidences are reported as seismic phenomena. Another source of error is the wrong transcription of names of localities, or the case of some localities having identical or very similar names; as a consequence, earthquake epicentres have been moved to wrong places. "New" earthquakes can originate simply by listing twice an event reported with the date given in different sources according to the Julian or the Gregorian calendars. Another source of similar manufacturing of earthquakes are errors in transcribing the dates, e.g. Jan. – June, VI – XI, etc. An opposite phenomenon may occur because of a long period of war, foregin occupation of a country, plague, or other reason for which the records either were not made or were destroyed, which results in an artificial interval of quiescence. It is imperative to work with the original reports as much as possible, but this is not easy because some old sources are not accessible to an investigator or have been destroyed or require special knowledge of language.

Sources giving only the evaluation of basic parameters by the author are of less value than those containing also the names of all affected localities and the description of seismic effects. Although macroseismic scales have been introduced to convert systematically the description of earthquake effects into grades, there have been always discrepancies between the assessments made by different investigators for the same event, even if identical macroseismic scales were used. Examples can be quoted from several European catalogues.

The important role of historical data is most evident in regions of medium or low seismicity, where a sample covering less than 100 years can rarely reveal the main features of earthquake activity. The published epicentre maps corresponding to different periods of time provide the simplest overview. For instance, in Central Europe, including the Alps and the Carpathians, the twentieth century sample would not help to discover such zones of extreme seismic energy release such as the epicentres near Nice, Basel, Villach, Komárno, Eger, and others. The epicentre map of the areas for the twentieth century shows only a few large events ($M = 5\text{-}6$), with the exception of the well defined East-Alpine belt. If historical events are added, other belts are emerging and seismo-tectonic relations are more evident.

The Western Carpathians were relatively quiet in the twentieth century, except the event of I_o = VIII in the Little Carpathians in 1906; in the central part of the Carpathians earthquakes did not exceed I_o = VI. However, historical sources (chronicles) mention a destructive earthquake in Central Slovakia – in 1443; stone buildings such as churches and castles partly collapsed, apparently killing several tens of people. The exact location of the epicentre is unknown; however, the scanty historical information proves that in the central part of the Western Carpathians, medium size earthquakes can occur ($M = 6$, I_o = VIII-IX). Other examples are: the region of Komarno on the Danube, where one active period lasting 80 years was observed (1763-1843); the destructive earthquakes in the middle ages in the Carnic Alps, etc. In most regions of Central Europe we know only about one release of a large amount of seismic energy during the historical period. Such a situation introduces the serious problem of how to estimate an average recurrence rate of extreme events for the Central European earthquake zones.

Historical information is normally not suitable for statistical processing because its homogeneity can be rarely guaranteed. However, its main importance is in evidencing the long term variation of earthquake activity and the existence of foci which become active after a time span of several hundred years or more, sometimes in regions which would not be considered as potentially active using other evidence. The occurrence of large, damaging events of very low probability of occurrence is the phenomenon which must be taken into account, particularly when considering earthquake hazard for important structures, such as nuclear power plants. In this respect, every effort must be made for improving historical information.

THE U.S. GEOLOGICAL SURVEY'S DATABASES OF SEISMIC EVENTS

J. N. Taggart
U. S. Geological Survey
Box 25046, Mail Stop 967
Denver Federal Center, Denver, CO 80225, USA

W. H. K. Lee and K. L. Meagher
MS-977, U. S. Geological Survey
Menlo Park, CA 94025, USA

ABSTRACT

Lists of seismic events are now on databases at the U.S. Geological Survey's National Earthquake Information Center in Golden, Colorado, and in Menlo Park, California. These controlled-access computer files describe the location, size, focal characteristics, and effects of earthquakes and large explosions. Some also refer to sources of information, point to associated observations, estimate parametric uncertainties and indicate the precision of the data. The databases thus make it easy to access large quantities of data for many purposes. For example, various database parameters are used to plot seismicity maps, delineate active faults and aftershock zones, evaluate the probabilities of earthquake recurrence, evaluate the rate at which regional strain energy is released, compare the published hypocenters of different agencies, compare estimates of magnitudes or seismic moments, plot focal mechanisms, and evaluate regional stress patterns.

The databases are organized into lists of global, national, regional and local earthquake events. The global, regional, and local event lists are archived with associated observational data for future research applications. Each database comes with interactive computer codes that prompt the user to select the type, range, and output format of the data. Other interactive codes are used by database administrators to add, modify, or delete individual event listings, and to sort, merge, correct, or tag duplicate event listings among the entire contents of the database. Thus, the databases are not static, but are continually modified when new data become available.

1. Introduction

During the past century, scientists and lay observers have recorded nearly 1 million earthquakes throughout the world. The quantity of data on seismic events is staggering, even discounting the several hundred thousand microearthquakes recorded locally or regionally that are included in this total. Seismologists of the U.S. Geological Survey (USGS) have developed computer databases that list most of these seismic events so that they can be studied efficiently – a necessity for further research. The databases are on computers at the National Earthquake Information center (NEIC) in Golden, Colorado, and in Menlo Park, California.

Most of the earthquake databases are on-line computer files and associated software that now or can soon be accessed by all users of USGS computers. The NEIC database of global earthquakes, for example, is routinely accessed on data telephones by USGS seismologists in Menlo Park and in Reston, Virginia. It may also be possible in the future to make these databases available to non-USGS users, if problems of funding, computer availability and computer security can be resolved.

In October 1979, the authors began a "National Earthquake Catalog Project". The goal of this project is to consolidate the various earthquake lists and catalogs into a uniform U.S. national earthquake catalog with references and estimates of parameter uncertainties. Work in Menlo Park concentrates on California earthquakes and earthquakes in the rest of the United States are handled in Golden. The present paper briefly describes (1) databases that are developed for this project, and (2) databases that we utilize but are developed by other projects in the U.S. Geological Survey and other institutions.

2. Database Content

The USGS currently has five principal databases. Three of the principal databases contain computed or assigned parameters from three categories of earthquakes: (1) the local-regional database mostly contains small events recorded by USGS and cooperating seismograph networks; (2) the national database contains all published hypocenters of earthquakes recorded in or near the U.S.; and (3) the global database contains published hypocenters of earthquakes larger than about m_b 3.0 recorded throughout the world.

The above three databases contain the origin time, location, depth, and source reference for the included events. Magnitudes, if determined, are also listed. Most of the earthquakes have been located with arrival-time data from seismograph stations, but a few thousand events, including nearly all historically important earthquakes before 1895, have been located only from published descriptions of damages and intensities. In all, intensity or damage information is available for perhaps 40,000 earthquakes from 1177 B.C. to the present.

The fourth principal database, and by far the largest, is the archival files. These files comprise magnetic tapes of digital waveform data recorded by the Global Digital Seismograph Network (GDSN), and the USGS California Seismic Networks, several magnetic tapes of phase data edited from the International Seismological Centre Bulletins for 1964-present, results of special studies by USGS seismologists, and global lists of fault plane and moment tensor solutions for many earthquakes larger than m_b 5.3 that have occurred since 1977. The waveform data (1980-1984) from the Global Digital Seismograph Network have been edited and reduced from about 1825 network-day tapes to 92 event tapes with earthquakes $\geq m_b$ 5.5. Although magnetic tapes of digital waveform data from recent earthquakes have been drastically edited, it is not now feasible to keep these data on line at USGS. The edited phase data, (which currently totals 200 megabytes and is increasing by about 1 megabyte per month), may be placed on-line at the NEIC computer as soon as the necessary disk storage becomes available. Digital waveform data recorded by the USGS California Seismic Networks are systematically collected since 1984 at a rate of about 100 megabytes of data per day. These waveform data are kept off-line on

6250 BPI magnetic tapes, but can be loaded back on the computer. The associated phase data are collected at a rate of about 0.5 megabyte of data per day, and are kept on-line in the USGS Menlo Park computer.

The fifth principal database contains non-instrumental descriptions of California earthquakes which are compiled from existing literature. Information are extracted and arranged chronologically by earthquake dates. This effort is very similar to that described by Usami (1985) and Xie (1985), except that in our case, we put the data (mostly text descriptions) on computer for easier access, cross checking, and re-arrangement.

3. Database Format

The systematic collection of seismic events for the databases began with older event lists (pre-1964), which were verified against published values and corrected where necessary. Events after 1963 did not have to be verified or corrected because these data were calculated on digital computers and stored on magnetic tape essentially free of transcription errors. The event lists were then reformatted to that of one or more database files, and, finally, these files were merged into operational databases.

Individual records in all of the Geological Survey's database files consist of ASCII-coded character strings, which require very little decoding time on the computers. Although records could have been stored as strings of binary constants, which would require less than half the space of the character strings, they would have required much more decoding time.

Figure 1 shows records of the same earthquake in three of the primary database formats. The local-regional databases are on-line files with fixed record lengths of 80 characters (figure 1a) that accommodate output from the HYPO71, HYPOELLIPSE, HYPOINVERSE and related computational hypocenter programs. About half of the data in these databases comprise confidence ellipsoid parameters and semi-quantitative estimates of the quality of the hypocenters.

The global database (figure 1b) is on-line indexed files with fixed record lengths of 100 characters. The national database (figure 1c) contains on-line primary indexed files with fixed record lengths of 168 characters. The national database also contains secondary files with intensity (36 characters), additional magnitude (84 characters), and hypocenter statistics (80 characters) data for some events, which are indexed as in the primary file. The date-origin-time indices in the global and national databases help the user to begin at any event in the files. Thereafter, the records are checked sequentially to determine whether various selection criteria have been matched.

Figure 2 shows the selection criteria for events in the global and national databases. The geographic search modes are mutually exclusive, and with two exceptions the event parameters must match all of the specified options for an event to be selected. The first exception to this requirement is that events without recorded depths will be included if a range of shallow depths is selected. The second exception is that events without magnitudes that occurred before 1963 will be included when a range of magnitudes less than M 6.0 is selected.

```
(a)  LOCAL - REGIONAL DATABASES

 DATE   TIME     LAT     LONG     DEPTH MAG  N GAP DMIN   RMS   TR PL ERH1 TR PL ERH2 ML MD    ERZ Q PM NS
790228 2127072 60N385 141W359    127   71  20  66  60     42    0  0  13  90  0  13            25  B  S  0

(b)  GLOBAL DATABASE

AGENCY  DATE      TIME  R    LAT      LONG    DEPTH NP   RMS    mb   Ms ZH     MAG          MAG      INT  EFFECTS IDE F-E  N  C P
GS    49790228 21270720G  60.642 -141.598  12.7              6.4  7.1Z  6.90MLPMR   7.52MWGS    7 D...TS.... X   19  20

(c)  NATIONAL DATABASE

       (PRIMARY FILE)
  DATE       TIME        UNC S  REF PR ID L F-E GRID     LAT       LONG    C    N    UNC     REF  DEPTH  R    UNC    NP    REF
919790228 2127072        2.0 s   GS  1 AK    19        60.6417 -141.5983     20  0.05     GS  12.7        10.0    0     GS

TYPE MAG  UNC  RMS  N  REF  TYPE MAG  UNC  RMS  N  REF ZH NM PM TD TE HP MO    BINARY FLAGS     BINARY FLAGS
mb   6.2  0.3      99  ISC  Ms   7.1  0.3              ISC  Z  4  2  3  1 21 23  100000001100100 000000000000011

        (SECONDARY FILE: HYPOCENTER STATISTICS)
  DATE      TIME       ERH1  AZ1 P1  ERH2  AZ2 P2   ERZ  AZ3 P3 RMS  GAP DMIN NF CO UFAC QQQ NS
919790228 2127072       1.3    0  0   1.3   90  0   2.5    0 90 0.42  66   60    68       B   0

        (SECONDARY FILE: ADDITIONAL MAGNITUDES)
  DATE      TIME      TYPE MAG  UNC  RMS  N  REF  TYPE MAG  UNC  RMS  N  REF  TYPE MAG  UNC  RMS  N  REF
919790228 2127072      ML   6.9  0.5            PMR  Mw   7.52 0.1            627

        (SECONDARY FILE: INTENSITY DATA)
  DATE      TIME     INT  E UNC C  REF    AREA
919790228 2127072     7.  D 1.0    622  5000(00)
```

Figure 1. Internal formats of three U.S. Geological Survey seismic event databases for the St. Elias, Alaska, earthquake on 28 February 1979 at 21:27::07.2 UTC. For clarity, the formats have been expanded by adding one or two blank spaces between the parameters. The national database contains all of the parameters in the local-regional and global databases, plus additional counters, control characters, uncertainty estimates, references, and secondary files when needed. The symbols are abbreviated parameter labels, which are explained in the user manuals for each database.

4. Database Development

The USGS database files go through as many as five developmental stages in which the data are collected, verified, corrected, merged, and prioritized. The principal sources from which the data were collected are shown in Figure 3. The sources include numerous regional catalogs and even a few global catalogs of historical earthquakes that had been published before instrumental seismology was well established (for example, see Milne, 1911). Other sources include monthly weather reports (1874-1924) of the U.S. Army Signal Corps, reports of explorers and mapping expeditions, private journals, and early newspaper articles. Data from these sources for the more significant felt or damaging earthquakes in the United States and its territories were assembled and published in *Earthquake History of the United States* (latest revision by Coffman, von Hake, and Stover, 1982). *Abstracts of Earthquake Reports for the Pacific Coast and the Western Mountain Region* (1932-1973) and *United States Earthquakes* (1928-current), respectively issued quarterly and annually by the U.S. Coast and Geodetic Survey and successor organizations, are important sources of information on felt or damaging earthquakes in the United States.

The USGS national and global databases contain calculated and computed hypocenter parameters from several published sources. The monthly bulletin (1913-1917) of the Seismological Committee, British Association for the Advancement of Science, is the earliest source of the calculated hypocenter parameters in the databases. These data continued to be published (covering 1918-1963) in the *International Seismological Summary* (ISS), enhanced by the development of improved travel times (Jeffreys and Bullen, 1940) that are still used throughout the world. The monthly

GEOGRAPHIC SEARCH MODES

GLOBAL (all), or
RANGE OF COORDINATES, or
FLINN-ENGDAHL REGIONS, or
DISTANCE FROM A POINT.

SEARCH OPTIONS

FILE SELECTION
- Choice of agencies
- All agencies

RANGE OF DATE - ORIGIN TIME
- Beginning time = index key

RANGE OF DEPTHS
RANGE OF INTENSITIES
RANGE OF MAGNITUDES
- All magnitudes
- Selected magnitude

FLAGGED EVENTS
- Events with moment tensor solutions
- Events with fault plane solutions
- *Felt, damaging, or disastrous events
- *Events with isoseismal maps
- *Events with associated diastrophism
- *Tsumagenic events
- *Events with associated ground effects
- *International Data Exchange events

IRREGULAR SHAPED GRID ZONE
DUPLICATE ENTRY ELIMINATOR
- Order of agency preference
- Origin time difference
- Distance difference

OUTPUT FORMAT OPTIONS

GLOBAL DATABASE OUTPUT
@GLOBAL DATABASE OUTPUT WITH PAGE HEADERS
*LOCAL - REGIONAL DATABASE OUTPUT
*ONE LINE OUTPUT
*ONE LINE OUTPUT WITH PRECISIONS
*ONE LINE OUTPUT WITH PRECISIONS, UNCERTAINTIES
*DISPLAY COMPLETE DATABASE RECORDS
*DUPLICATE DATABASE

@ Global database only
* National database only

Figure 2. Selection criteria for events in the national and global databases. For example, a user might enter the national database and request a one line per event listing of earthquakes in Flinn-Engdahl region 7 (Andreanof Islands, Alaska), which: 1) were published by the ISC, 2) occurred between 01 January 1957 and 31 December 1979, 3) had depths less than 50 km, 4) had M_S magnitude > 6.4, and 5) were felt or caused damage.

MAGNETIC TAPE FILES

International Seismological Centre – Historical Hypocenter File
- International Seismological Centre, Bulletin, 1964 – current
- International Seismological Summary, 1918 – 1963
- Bur. Central International de Seismologie, Month. Bull., 1950 – 1963
- Gutenberg and Richter: Seismicity of the Earth, 1904 – 1952
- Japan Meteorological Agency, 1921 – current

National Oceanic and Atmospheric Administration – Hypocenter Data Files
- China Earthquakes, 1177 BC – 1976
- Earthquake History of the United States, 1534 – 1980
- United States Earthquakes, 1928 – current
- Seismological Bulletin, 1933 – 1950 and 1954 – 1966
- PDE Monthly Listing, 1963 – current

Regional Files
- USGS State Seismicity Files, earliest – current
- Alaska (USGS), 1786 – current
- Northern California (U. Cal. Berkeley), 1910 – 1974
- Central and Northern California (USGS), 1966 – current
- Southern California (Caltech, USGS), 1932 – 1981
- California (Calif. Dept. of Mining and Geology), 1900 – 1974
- Western United States (University of Utah), 1900 – 1981
- Central Mississippi Valley Earthquake Bulletin, 1975 – current
- Northeastern U.S. Seismic Network, 1975 – current
- Southheastern U.S. Seismic Network, 1977 – current

SOURCES OF ADDITIONAL DATA

Abstracts of Earthquake Reports for the Pacific Coast and Western Mountain Region, 1932 – 1973
Archives of Religious Orders
Newspapers
Personal Journals and Diaries
Published Histories of Regions
Published Research Results
Published National Catalogs
Published State Catalogs
U.S. Army Signal Corps Monthly Weather Reports, 1874 – 1924
Unpublished Files and Catalogs (USGS)

Figure 3. Principal sources of earthquake listings in the U.S. Geological Survey databases. The dates represent the range in time of earthquake occurrences listed in the source.

bulletins of the Bureau Central International de Seismologie also are an important source of hypocenters and arrival times, especially for the 1950's when they contain more hypocenters than any other publication. The International Seismological Centre (ISC), successor to the ISS, has the internationally-supported responsibility for the final publication (1964-current) of standard earthquake information from all over the world. The ISC's historical hypocenter file contains data on magnetic tape from the above sources, and data contributed by agencies and universities throughout the world.

Some of the database files were created by merging lists of events from two or more sources. As a result, events commonly are multiple-listed, perhaps with slight differences in parameters, or even with exact duplications of data. The exact duplications are easily identified by simple comparisons and the redundant listings can be deleted, but the listings that differ slightly must either be winnowed or prioritized by a more complicated logic.

The preferred reference is usually, but not always, the one that is the most complete. The ISC Catalogue is the most complete reference for global earthquake listings for 1964 to the present, although there are many regions where the uncertainties in the ISC data are quite large. The listings of the International Seismological Summary (ISS) may be the most complete global reference for 1914-1953. The monthly bulletins of the Bureau Central International de Seismologie (BCIS) certainly are the most complete global reference for 1954-1963 when many other agencies were still having difficulty processing the large volumes of data. Finally, the catalog of Gutenberg and Richter (1954) is a favored reference for data on large earthquakes that occurred during the years 1904-1952.

Special research investigations of many earthquakes have yielded hypocenters and magnitudes that have much smaller uncertainties than those of the originally published values. For example, important reevaluations of M_S magnitudes have been published that allow better comparisons of large earthquakes throughout this century (Abe, 1981; Abe and Noguchi, 1983; Lienkaemper, 1984). Also, Dewey and Gordon (1983) and Gordon (1985) use calibration events, relative locations, and carefully identified phases to locate earthquake hypocenters more accurately in the United States. The data from these and similar investigations are included as preferred listings in the USGS global and national databases.

In many regions the listings of national agencies are more accurate than the listings of agencies that specialize in global data. The Japan Meteorological Agency, for example, locates shallow earthquakes in and near Japan more accurately than either the ISC or the USGS. In general, shallow hypocenters estimated with local network data are much more accurate than hypocenters estimated with data from more distant stations. The listings of national agencies and local networks are added to the USGS databases as soon as they are made available on magnetic tape.

Although it is unlikely that lists of computer-generated hypocenters contain enough errors to justify checking them against published values, the verification of pre-1962 listings is a necessary part of the development of the USGS database files. The most common error in the early data arises where the origin times are converted from local to Universal Coordinated Time. After files from different sources are merged, it is important to identify the duplications that result from the time conversion errors, so that a preferred listing can be selected. Typographic and transpositional errors can affect any parameter, but are somewhat less common than the time conversion errors.

Special interactive programs have been written to correct errors in several of the USGS databases. These interactive programs permit any designated parameter to be replaced by another value. At the end of an error-correcting session, the "before" and "after" versions of modified listings are printed to confirm the corrections. In order to protect the integrity of the databases, however, the editing programs are available only to the database administrators or their assistants.

The computers used for the databases in the U.S. Geological Survey in Golden and in Menlo Park are primarily VAX-type minicomputers which have a limited capacity for large scale data retrieval and processing. Consequently, we also developed a separate earthquake data archiving and retrieval system utilizing the large scale computer facility at the Stanford Linear Accelerator Center (SLAC). The design of this system is described in Lee *et al.* (1983), its implementation is presented in Crane *et al.* (1984), and a guide to its use is given in Crane *et al.* (1985). The SLAC computer and the VAX-type computers are accessible via high-speed commercial data transfer networks, which are linked to computers at major universities in the U.S. and abroad.

5. Conclusion

In recent years many new parameters have been added to those already used to describe earthquakes. Among the new parameters are moment tensor components, seismic moment, stress drop, average displacement, rupture length (and width), rupture area, fault plane and stress axes descriptors, M_w magnitude, M_{Lg} magnitude, M_d magnitude, magnitude statistics, and ground motion descriptors. The global database format contains alphanumeric flags, which, when set, indicate the existence of some of the above data. The primary format of the national database accommodates many of these parameters, and binary flags point to secondary files in the database that contain much of the remaining new data. Furthermore, the national database can be expanded as necessary to acccommodate new event parameters that may be routinely determined in the future.

REFERENCES

Abe, K. (1981). Magnitudes of large shallow earthquakes from 1904 to 1980, *Phys. Earth Planet. Interiors*, **27**, 72-92.

Abe, K. and S. Noguchi (1983). Revision of magnitudes of large shallow earthquakes, 1897-1912, *Phys. Earth Planet. Interiors*, **33**, 1-11.

Coffman, J. L., C. A. von Hake, and C. W. Stover (1982). *Earthquake history of the United States*, revised edition with supplement (through 1980): Washington, D.C, U.S. Government Printing Office, 208+50a p.

Crane, G. R., W. H. K. Lee, and J. T. Newberry (1984). Earthquake data archiving and retrieval system: reference manual, *U.S. Geol. Surv. Open-File Rept. 84-840*, 159 p.

Crane, G. R., W. H. K. Lee, and M. E. O'Neill (1985). Earthquake data archiving and retrieval system: user's manual, *U. S. Geol. Surv. Open-File Rept. 85-368*, 24 p.

Dewey, J. W. and D. W. Gordon (1983). Seismicity of the eastern United States and adjacent Canada, 1925-1976, U.S. Geological Survey, written commun., 1985.

Gordon, D. W. (1985). Revised instrumental hypocenters and correlation of earthquake locations and tectonics in the Central United States, U.S. Geological Survey, written commun., 1985.

Gutenberg, B. and C. F. Richter (1954). *Seismicity of the Earth*, New York, Hafner Publishing Company, 310 p.

Jeffreys, H. and K. E. Bullen (1940), *Seismological Tables*, London, British Association for the Advancement of Science, Gray Milne Trust, 50 p.

Lee, W. H. K., D. L. Scharre, and G. R. Crane (1983). A computer-based system for organizing earthquake-related data, *U. S. Geol. Surv. Open-File Rept. 83-518*, 28 p.

Lienkaemper, J. J. (1984). Comparison of two surface-wave magnitude scales: M of Gutenberg and Richter (1954) and M_S of "Preliminary Determination of Epicenters", *Bull. Seism. Soc. Am.*, **74**, 2357-2378.

Milne, J. (1911). Catalogue of destructive earthquakes, British Association for the Advancement of Science, Report of 81st Annual Meeting, Appendix No. 1, 649-740.

Usami, Tatsuo (1987). Study of historical earthquakes in Japan, *this volume*, p. 276-288.

Xie, Yushou, (1987) Historical materials on Chinese earthquakes and its seismological analysis, *this volume*, p. 162-170.

A Catalog of Large Earthquakes ($M \geq 6$) and Damaging Earthquakes in Japan for the Years 1885–1925

Tokuji Utsu
Earthquake Research Institute
University of Tokyo, Tokyo 113, Japan

1. The Old Catalog

In 1885, the Central Meteorological Observatory (CMO) in Tokyo began to collect macroseismic and instrumental data on earthquakes throughout Japan. Since then, the CMO successively published the bulletins containing the occurrence times, locations of epicenters, and radii of shaken areas of moderate to large earthquakes in Japan. The catalog of Japanese earthquakes compiled by the CMO (1952) lists the occurrence times, epicenters, radii of felt areas of earthquakes between 1885 and 1950 taken from the previous bulletins of CMO, together with the magnitudes given by Kawasumi in 1952. The epicenters, focal depths, and magnitudes of earthquakes which occurred after 1926 were redetermined by the Japan Meteorological Agency (JMA – the successor of CMO) in 1958 and 1982 (JMA, 1958, 1982), but no revision was made for earthquakes before 1925.

The old CMO catalog is inaccurate in several respects. No consideration was given to the depth of focus. Most deep earthquakes which occurred before the discovery of the deep earthquake were mislocated by up to 1000 km. Magnitudes were considerably overestimated for many earthquakes especially those which occurred before 1913. Some fairly large earthquakes are missing in the list. In view of these situations, I undertook a compilation of a new catalog of Japanese earthquakes for the years 1885-1925.

2. Compilation of a New Catalog

Most of the data collected by the CMO before September 1, 1923 were lost in the fire caused by the great Kanto earthquake on that date. To prepare a new catalog, we had to re-collect the old data. Data on some large earthquakes are found in the CMO bulletins and some seismological papers. Most of the written data-files stored in the JMA stations were microfilmed about 20 years ago. The Geophysical Institute of the University of Tokyo stores voluminous written data-sheets collected from many CMO stations before 1925, some of which are unavailable at the reporting stations and the main office of JMA now. Some data can be obtained only at individual JMA stations.

During this work, some old seismograms were examined by myself. Many seismograms were missing or stored in bad condition. It was often difficult to access required seismograms quickly, even if they were preserved in the storeroom of a station.

Time-keeping in seismic observation during the period 1885-1925 was not sufficiently accurate for the determination of hypocenters from arrival times of P-waves. S–P times could be used in the later stage of the period. In the earlier stage, few S–P times were available, and the epicenters were estimated by the use of the distribution of seismic intensities. I constructed isoseismal maps of about 2000 earthquakes from 1885 to the present. By comparing the patterns of isoseismals of earthquakes of known and unknown hypocenters, we can estimate the location of earlier earthquakes. But this method sometimes fails for deep earthquakes, which exhibit anomalous intensity distributions due to the effect of the descending Pacific plate or Philippine Sea plate beneath the Japanese Islands. Actually, few deep earthquakes have been located for the period before 1900.

The magnitudes were calculated by the same method as presently adopted by the JMA. The maximum amplitudes measured by old-fashioned seismographs, many of which have no damping devices, may be systematically different from those measured by modern instruments. However, no appreciable systematic difference is evident between my determinations and the magnitudes given by Gutenberg and Richter (1954) or those by Abe (1979). For earthquakes before about 1900 for which few data on the maximum amplitudes were available, the magnitude were estimated by the comparison of the distributions of seismic intensities. Consequently, the magnitudes of so-called low-frequency earthquakes occurring before about 1900 may be considerably smaller. The low-frequency earthquake has a relatively small felt area for its instrumental magnitude determined from the amplitudes of medium or long-period seismic waves (Utsu, 1980).

Figure 1 shows an example of the hypocenter and magnitude determination for the new catalog. The old CMO catalog indicates the epicenter located off the Pacific coast of eastern Honshu, but the S—P times at several stations clearly indicate that this is a deep earthquake beneath the Sea of Japan. This earthquake (08:06, Jan. 21, 1916) is listed neither in Gutenberg and Richter (1954) nor in the Bulletin of the British Association.

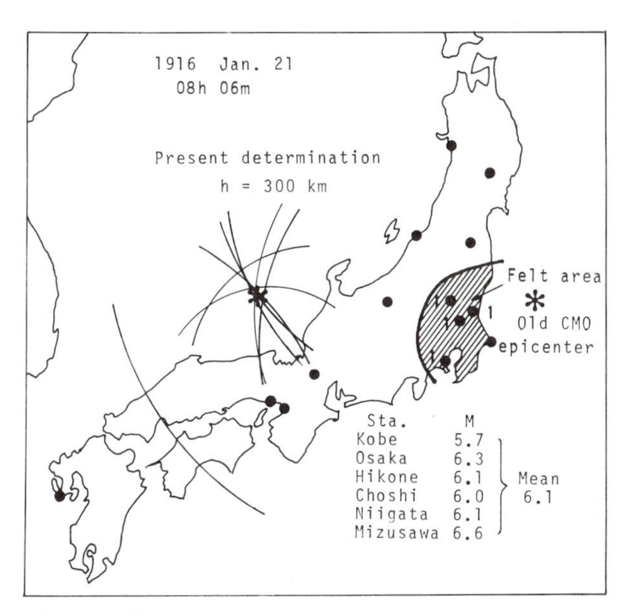

Figure 1. An example of the redetermination of hypocenters and magnitudes for the new catalog. The radii of arcs correspond to the S—P times for h = 300 km from the standard travel-time tables presently used in JMA.

The first version of the new catalog was published in 1979 (Utsu, 1979), to which some corrections and additions were made in 1982 (Utsu, 1982a). The revised catalog is reproduced in Table 1 in a simplified form. In the original form, the accuracy of epicenter and magnitude determination is indicated by using various notations, but these are neglected in Table 1 to save space. The last column of Table 1 (headed D) gives the degree of damage as judged by the following criteria:

0 – No damage.
1 – Slight damage (cracks on walls and ground).
2 – Light damage (broken houses, road, etc.).
3 – 2-19 persons killed or 2-999 houses totally destroyed.
4 – 20-199 persons killed.
5 – 200-1999 persons killed.
6 – 2000-19,999 persons killed.
7 – 20,000 or more persons killed.
x – Damage inseparable from that caused by the previous earthquake.

In compiling the new catalog, I intended to include all earthquakes of $M \geq 6$ which were felt in Japan and all earthquakes which caused damage in Japan. "Japan" means the present territory of Japan. Table 1 contains 534 felt earthquakes of $M \geq 6$ and 178 damaging earthquakes of which 63 have magnitudes below 6.0. In addition, if unfelt earthquakes are found in the area enclosed by the solid lines in Figure 2 (hereafter called region A), they are included in the catalog. Only two earthquakes are in this category. In total, Table 1 lists 599 earthquakes.

3. Seismicity of Japan for the Past 100 Years

It is easy to compile a similar catalog for the years from 1926 to the present, since the hypocenters and magnitudes of earthquakes which occurred from 1926 through 1960 were redetermined by the JMA (JMA, 1982). The focal parameters of earthquakes in and after 1961 have been published in monthly bulletins of JMA. By using the catalogs of JMA, Gutenberg and Richter (1954), ISS, ISC, etc., I have extended the present catalog to the end of 1984. Thus the extended catalog covers just 100 years from 1885 through 1984. A total of 1812 earthquakes are listed in this catalog. Table 2 lists the number of earthquakes classified according to the damage rating.

Figures 2 and 3 show the distribution of seismicity for the past 100 years in space and time, respectively. Data on deep earthquakes ($h \geq 100$ km) before 1900 are not plotted because of insufficiency of the data.

Figure 4 shows the magnitude-frequency relation for shallow earthquakes. Four formulas representing the relation have been fitted to the data using the maximum likelihood method (Utsu, 1984). The straight line G indicates the well-known Gutenberg-Richter formula ($b = 0.94$). The three curves U, L, and M indicate the following formulas:

$$\log n = a - bM + \log (c - M) \qquad \text{Utsu (1971)}$$

$$\ln N = A - B \exp (\alpha M) \qquad \text{Lomnitz-Adler and Lomnitz (1979)}$$

$$\log N = A + \tfrac{1}{k} \log (c - M) \qquad \text{equivalent to Makjanić (1980)}$$

Table 1. List of Earthquakes of $M \geq 6$ and Damaging Earthquakes in the Region of Japan from 1885 through 1925

Y	M	D	h	m	Lat °N	Long °E	Dep km	*M*	D
1885	2	8	17	0	37.5	142.0	s	6.0	0
1885	3	20	4	0	35.5	139.0	s	6.0	0
1885	6	11	0	20	40.5	143.0	s	6.9	0
1885	6	14	16	40	35.0	141.0	s	6.5	0
1885	7	28	20	30	42.0	142.0	s	6.0	0
1885	9	5	10	50	43.0	146.5	s	6.5	0
1885	9	8	14	20	42.5	145.0	s	6.0	0
1885	9	26	3	0	34.0	139.0	s	6.8	0
1885	9	27	20	30	34.0	139.0	s	6.7	0
1885	10	30	11	30	39.5	143.0	s	6.2	0
1885	12	7	4	2	36.5	141.0	s	6.3	0
1885	12	19	9	26	36.5	141.0	s	6.0	0
1886	4	12	20	44	38.5	142.0	s	6.3	0
1886	7	2	3	33	38.5	142.0	s	6.3	0
1886	7	22	15	57	37.1	138.5	s	5.3	2
1887	1	15	9	51	35.5	139.3	s	6.2	2
1887	4	29	1	55	32.0	132.0	s	7.1	0
1887	5	28	15	50	37.5	142.0	s	6.4	0
1887	5	28	16	10	37.5	142.0	s	6.2	0
1887	7	22	11	27	37.5	138.9	s	5.7	2
1887	9	5	6	23	35.8	140.5	s	6.3	1
1888	2	4	15	50	41.5	142.5	s	7.1	0
1888	4	29	1	0	36.6	140.0	s	6.0	1
1888	11	23	17	3	39.0	143.0	s	6.5	0
1889	1	15	8	31	41.0	144.0	s	6.0	0
1889	2	17	21	9	35.5	139.7	s	6.0	1
1889	3	30	21	42	37.0	141.5	s	6.6	0
1889	5	12	1	42	35.4	136.8	s	5.9	1
1889	7	28	14	45	32.8	130.7	s	6.3	4
1889	9	30	16	50	28.0	130.0	s	6.0	1
1889	12	31	4	5	40.0	130.0	d	7.8	0
1890	1	7	6	43	36.5	138.0	s	6.2	2
1890	4	16	12	34	34.2	139.3	s	6.8	1
1890	4	16	19	56	34.0	139.0	s	6.3	0
1890	4	16	21	42	34.0	139.0	s	6.0	0
1890	11	17	0	31	41.5	142.5	s	6.3	0
1891	4	7	0	49	39.0	143.0	s	6.7	0
1891	4	21	1	49	37.0	142.0	s	6.5	0
1891	5	4	23	16	39.0	142.0	s	6.2	0
1891	7	21	11	19	37.0	141.5	s	7.0	0
1891	10	15	22	6	33.2	131.8	s	6.3	2
1891	10	27	21	38	35.6	136.6	s	8.0	6
1891	10	28	1	38	35.5	136.5	s	6.0	0
1891	10	29	15	15	35.5	136.5	s	6.0	0
1891	12	23	20	35	35.4	138.9	s	6.5	2
1892	1	3	7	21	35.3	137.1	s	5.5	2
1892	6	2	22	10	35.7	139.9	s	6.2	2
1892	7	3	10	10	43.0	146.0	s	6.0	0
1892	9	6	20	42	35.7	137.0	s	6.1	1
1892	10	22	10	9	42.0	142.5	s	6.0	0
1892	12	9	1	42	37.1	136.7	s	6.4	2
1892	12	10	16	34	37.0	136.7	s	6.3	3
1893	6	3	17	27	43.5	148.0	s	7.7*	0
1893	6	13	10	42	42.5	145.5	s	6.9	1
1893	9	6	17	25	31.4	130.5	s	5.3	2
1894	1	4	13	9	31.4	130.5	s	6.3	2
1894	1	10	9	45	35.4	136.7	s	6.3	2
1894	2	24	19	18	41.5	142.0	s	6.8	0
1894	3	14	9	15	41.5	142.5	s	6.0	0
1894	3	22	10	23	42.5	146.0	s	7.9	3
1894	3	29	8	53	42.5	146.0	s	6.0	0
1894	4	25	0	15	42.5	146.0	s	6.8	0
1894	6	20	5	4	35.7	139.8	s	7.0	4
1894	8	8	14	19	32.8	131.0	s	6.3	2
1894	8	29	10	55	38.0	142.0	s	6.6	0
1894	10	7	11	30	35.6	139.8	s	6.7	1
1894	10	22	8	35	38.9	139.9	s	7.0	5
1894	11	27	16	5	41.5	142.5	s	7.1	0
1894	12	1	9	37	39.0	142.0	s	6.3	0
1895	1	18	13	48	36.1	140.4	s	7.2	3
1895	8	27	13	42	32.8	131.0	s	6.3	2
1896	1	9	13	17	36.5	141.0	s	7.3*	1
1896	1	9	20	52	36.5	141.0	s	6.0	0
1896	1	10	2	25	36.5	141.5	s	6.3	0
1896	2	23	10	42	36.5	141.0	s	6.1	0
1896	3	6	14	52	36.5	141.0	s	6.0	0
1896	4	1	16	41	37.5	137.3	s	5.7	2
1896	4	11	14	0	36.5	141.0	s	6.0	0
1896	4	19	10	56	40.0	140.0	d	6.5	0
1896	6	15	10	32	39.5	144.0	s	8.5*	7
1896	6	15	19	16	39.5	144.0	s	7.5*	0
1896	6	15	23	1	39.5	144.0	s	7.5*	0
1896	6	17	3	43	39.5	144.0	s	7.5*	0
1896	7	29	8	44	36.5	141.0	s	6.1	0
1896	8	1	2	49	37.5	141.5	s	6.5	1
1896	8	23	6	56	39.7	140.8	s	5.5	1
1896	8	31	7	37	39.6	140.7	s	6.4	0
1896	8	31	8	6	39.5	140.7	s	7.2	5
1896	9	5	14	7	39.5	144.0	s	6.5	0
1896	11	18	2	6	43.0	146.5	s	6.7	0
1897	1	16	15	49	36.2	139.9	s	5.6	1
1897	1	16	20	36	36.7	138.3	s	5.2	2
1897	2	7	7	35	40.0	139.0	s	7.5*	0
1897	2	19	20	50	38.1	141.9	s	7.4	3
1897	2	19	23	47	38.0	142.0	s	7.0*	0
1897	3	27	10	49	39.5	143.5	s	6.3	0
1897	4	30	7	2	36.7	138.3	s	5.4	2
1897	5	23	12	22	39.0	142.8	s	6.9	1
1897	7	22	9	31	37.0	141.8	s	6.8	0
1897	7	29	13	45	37.0	142.0	s	6.0	0
1897	8	5	0	10	38.3	143.3	s	7.7	2
1897	8	5	23	48	38.0	143.0	s	6.3	0

(*continued*)

Table 1. (*Continued*)

Y	M	D	h	m	Lat °N	Long °E	Dep km	*M*	D
1897	8	12	1	50	37.5	142.0	s	6.1	0
1897	8	16	7	50	39.6	143.6	s	7.2	0
1897	10	2	12	45	38.0	141.7	s	6.6	1
1897	12	4	0	18	40.0	143.5	s	6.2	0
1897	12	26	7	41	37.0	141.5	s	6.2	0
1898	2	13	14	58	36.2	139.8	s	5.6	1
1898	4	2	21	9	35.4	138.4	s	5.9	2
1898	4	3	6	48	34.6	131.2	s	6.2	2
1898	4	22	23	37	38.6	142.0	s	7.2	2
1898	5	25	18	0	37.0	138.9	s	6.1	2
1898	8	10	12	57	33.6	130.2	s	6.0	2
1898	8	11	23	36	33.6	130.2	s	5.8	1
1898	9	1	9	0	24.5	124.7	s	7.0	2
1898	10	7	2	0	41.5	142.5	s	6.0	0
1898	11	13	2	33	35.3	136.7	s	5.7	1
1898	11	14	7	5	42.5	146.0	s	6.5	0
1898	12	3	16	45	32.7	131.1	150	6.7	1
1898	12	15	16	47	41.5	142.5	s	6.0	0
1899	3	7	0	55	34.1	136.1	s	7.0	3
1899	3	22	10	23	39.0	142.5	s	6.5	0
1899	3	24	4	5	31.8	131.1	100	6.4	1
1899	3	31	14	1	35.6	136.6	s	5.5	1
1899	4	15	10	25	36.3	141.0	s	5.8	1
1899	5	8	3	29	42.8	146.2	s	6.9	2
1899	7	10	22	12	43.0	146.0	s	6.5	0
1899	7	11	7	39	48.0	146.0	d	7.5	0
1899	8	3	9	52	38.0	142.0	s	6.0	0
1899	11	10	11	59	43.0	146.0	s	6.5	0
1899	11	10	17	41	40.0	143.0	s	6.5	0
1899	11	24	18	43	31.9	132.0	s	7.1	2
1899	11	24	18	55	32.7	132.3	s	6.9	2
1900	1	18	7	46	44.5	148.5	s	6.7	0
1900	1	31	19	22	48.0	146.0	d	7.5	0
1900	2	13	4	28	40.0	143.0	s	6.0	0
1900	3	12	1	34	38.3	141.9	s	6.4	1
1900	3	21	15	55	35.8	136.2	s	5.8	3
1900	4	24	23	16	27.0	126.5	s	7.0	0
1900	5	11	17	23	38.7	141.1	s	7.0	3
1900	5	31	8	43	35.7	136.6	s	5.3	1
1900	8	5	4	21	37.3	141.7	s	6.6	0
1900	8	29	2	32	41.2	142.8	s	6.8	0
1900	9	24	3	32	39.0	143.0	s	6.0	0
1900	11	5	7	42	33.9	139.4	s	6.6	2
1900	11	9	17	55	33.8	138.8	s	6.4	0
1900	11	24	7	57	43.5	148.0	s	7.0	0
1900	12	25	5	9	43.0	146.0	s	7.1	1
1901	1	13	22	41	42.3	143.8	s	6.8	1
1901	4	5	23	30	44.5	149.0	s	7.3	0
1901	4	22	18	10	35.5	139.5	120	6.0	0
1901	5	13	20	11	36.0	141.0	s	6.0	0
1901	5	14	6	51	44.5	149.0	s	6.6	0
1901	6	15	9	34	39.0	143.0	s	7.0	1
1901	6	24	7	2	28.0	130.0	s	7.5*	1
1901	6	24	13	40	28.0	130.0	s	6.5	0
1901	8	9	9	23	40.5	142.5	s	7.2	3
1901	8	9	18	33	40.6	142.3	s	7.4	x
1901	8	9	20	0	40.5	142.5	s	6.3	0
1901	8	11	11	31	40.5	142.5	s	6.0	0
1901	8	29	12	16	40.5	142.5	s	6.3	0
1901	9	30	10	19	40.2	141.9	s	6.9	1
1901	9	30	10	44	40.0	142.0	s	6.2	0
1901	12	31	15	20	41.0	142.0	s	6.1	0
1902	1	17	19	38	41.6	141.9	s	6.5	0
1902	1	30	14	1	40.5	141.3	s	7.0	3
1902	1	31	1	42	41.6	142.2	s	6.6	0
1902	2	20	15	38	41.4	141.8	s	6.7	0
1902	3	25	5	35	35.9	140.5	s	5.6	1
1902	5	2	11	31	39.0	144.0	s	7.0	0
1902	5	8	2	19	30.5	131.5	s	6.6	0
1902	5	25	11	29	35.6	139.0	s	5.4	1
1902	5	28	9	1	42.8	144.8	s	6.5	1
1902	6	13	0	22	42.5	144.0	s	6.3	0
1902	7	1	8	19	40.0	143.5	s	6.3	0
1902	7	8	14	5	41.0	143.0	s	6.2	0
1902	8	7	9	22	40.5	143.5	s	6.1	0
1902	11	21	7	3	22.5	121.5	s	7.0	0
1902	12	10	20	6	31.0	130.0	s	5.3	1
1902	2	3	12	14	34.0	137.0	d	6.5	0
1903	3	21	10	36	33.8	132.2	s	6.2	1
1903	6	7	9	5	25.0	122.0	s	6.0	0
1903	7	6	4	55	35.0	136.5	s	5.7	1
1903	8	10	4	40	36.2	137.6	s	5.5	2
1903	10	10	16	41	31.8	132.0	s	6.2	1
1904	3	18	13	42	42.7	146.1	s	6.8	1
1904	5	7	19	23	37.1	138.9	s	6.1	2
1904	6	5	18	40	35.3	133.2	s	5.4	1
1904	6	6	2	51	35.3	133.2	s	5.8	1
1904	6	7	8	17	39.0	135.0	350	7.2	0
1904	7	1	13	27	42.8	146.4	s	6.4	1
1904	8	22	13	0	42.1	145.5	s	6.3	0
1904	8	24	20	59	30.0	131.0	s	7.4	0
1904	12	17	7	2	41.2	142.7	s	6.3	0
1904	12	24	2	46	38.8	142.5	s	6.0	0
1904	12	27	22	47	31.0	138.0	d	6.7	0
1905	6	2	5	39	34.1	132.5	s	7.2	3
1905	6	2	10	55	34.0	132.5	s	6.0	1
1905	6	7	5	39	34.8	139.3	s	5.8	2
1905	6	26	16	10	40.0	143.3	s	6.2	0
1905	7	6	16	21	37.4	141.8	s	7.1	0
1905	7	6	22	17	42.0	132.0	d	6.5	0
1905	7	11	15	37	22.0	143.0	450	7.3	0
1905	7	23	8	26	37.1	138.4	s	5.2	1
1905	8	25	9	46	43.0	131.0	500	6.8	0
1905	9	1	2	45	45.0	143.0	250	7.0	0

(*continued*)

Table 1. (*Continued*)

Y	M	D	h	m	Lat °N	Long °E	Dep km	*M*	D
1905	10	3	23	14	41.2	141.8	s	6.1	0
1905	10	24	3	46	34.0	138.0	250	6.1	0
1905	12	8	3	8	34.1	132.6	s	6.1	0
1905	12	8	4	25	34.0	132.4	s	6.2	0
1905	12	23	2	37	38.5	141.8	s	5.9	1
1905	12	26	3	11	36.5	141.2	s	6.0	0
1906	1	21	13	49	34.0	137.0	350	7.6	1
1906	2	4	6	24	37.8	142.0	s	6.0	0
1906	2	23	9	49	34.8	139.8	s	6.3	1
1906	2	24	0	14	35.5	139.8	s	6.4	1
1906	3	13	13	27	32.5	132.2	s	6.4	1
1906	3	16	22	42	23.6	120.4	s	6.9	0
1906	4	5	2	50	36.9	141.6	s	6.0	0
1906	4	8	17	37	36.3	142.0	s	6.1	0
1906	4	13	19	17	23.6	120.4	s	7.1	0
1906	4	13	23	51	23.6	120.4	s	6.9	0
1906	4	20	12	48	35.9	137.2	s	4.9	1
1906	4	20	19	38	35.9	137.2	s	5.9	1
1906	5	4	23	9	33.9	135.3	s	6.2	1
1906	9	7	18	52	34.0	141.0	s	7.0*	0
1906	10	12	0	56	40.0	140.5	s	5.6	1
1906	10	12	1	4	40.0	140.5	s	5.4	1
1907	1	4	16	46	40.4	143.6	s	6.3	0
1907	2	6	8	37	29.0	139.0	400	6.9	0
1907	3	1	6	16	42.2	143.6	s	6.1	0
1907	3	10	13	3	32.9	130.7	s	5.4	1
1907	3	26	11	21	38.0	135.0	350	6.7	0
1907	4	23	0	57	36.8	141.6	s	6.0	0
1907	5	4	8	36	28.0	141.0	200	6.9	0
1907	5	22	22	54	38.4	143.0	s	6.3	0
1907	5	25	14	2	50.5	148.0	600	7.3	0
1907	7	5	15	46	43.7	145.5	100	6.7	1
1907	8	13	19	50	40.0	139.0	s	6.1	0
1907	11	21	17	17	35.8	139.2	s	6.0	0
1907	12	2	13	53	40.1	142.3	s	6.7	1
1907	12	23	1	13	43.8	145.0	150	6.9	1
1908	1	11	3	34	23.0	121.1	s	6.7	0
1908	1	15	12	56	37.3	141.8	s	6.9	0
1908	1	17	16	5	36.2	141.1	s	6.0	0
1908	2	5	12	7	38.3	142.5	s	6.0	0
1908	3	2	15	36	43.5	147.0	s	6.3	0
1908	4	16	3	27	31.7	130.6	s	4.0	1
1908	4	19	7	58	42.0	134.0	450	7.3	0
1908	5	3	0	51	43.0	146.5	s	6.5	0
1908	5	12	20	22	33.9	138.9	s	6.0	0
1908	6	7	20	49	28.0	140.0	400	6.0	0
1908	6	17	1	28	33.8	139.7	120	6.3	0
1908	6	27	14	21	36.0	142.5	s	6.1	0
1908	7	1	7	28	24.0	122.0	s	6.1	0
1908	9	25	13	31	42.3	141.5	100	6.1	0
1908	11	4	20	55	27.0	141.0	d	6.1	0
1908	11	22	7	15	41.7	142.2	s	6.4	0
1908	12	28	8	8	35.6	138.7	s	5.8	1
1909	3	10	23	55	30.5	130.5	100	6.5	0
1909	3	12	23	14	34.7	141.7	s	6.2	0
1909	3	12	23	19	34.5	141.5	s	6.7	1
1909	3	13	2	36	35.0	142.0	s	6.0	0
1909	3	13	14	29	34.5	141.5	s	7.5	2
1909	3	17	22	26	30.5	137.5	450	6.7	0
1909	3	22	20	4	35.0	141.5	s	6.6	0
1909	4	14	19	53	25.0	122.5	s	7.2	0
1909	7	2	20	54	35.6	139.8	s	6.1	1
1909	8	14	6	31	35.4	136.3	s	6.8	4
1909	8	29	10	27	26.0	128.0	s	6.2	2
1909	9	10	18	8	28.0	129.0	100	6.6	0
1909	9	16	19	39	42.0	142.0	s	6.8	1
1909	10	3	14	1	32.0	138.0	300	6.3	0
1909	11	10	6	13	32.3	131.1	150	7.6	3
1909	11	21	7	36	25.5	122.0	s	7.0	0
1910	1	6	19	55	24.0	123.0	s	6.2	0
1910	1	21	23	25	42.0	145.0	s	6.0	0
1910	2	12	18	9	33.0	138.0	350	7.3	0
1910	4	12	0	22	25.0	123.0	200	7.6	1
1910	5	4	15	19	28.5	139.5	400	6.3	0
1910	5	9	9	53	36.5	142.0	s	6.0	0
1910	5	10	13	56	36.5	142.0	s	6.1	0
1910	5	12	3	22	36.4	142.1	s	6.0	0
1910	5	22	6	25	44.0	148.0	s	7.1	0
1910	6	9	11	48	30.0	139.0	400	6.7	0
1910	6	26	15	59	32.0	138.0	300	6.0	0
1910	7	5	18	33	27.0	129.0	s	6.0	0
1910	7	24	6	49	42.5	140.9	s	5.1	2
1910	9	8	2	50	44.2	141.6	s	5.3	1
1910	9	26	10	26	36.8	141.5	s	5.9	1
1910	10	13	14	56	36.2	141.0	s	6.3	0
1911	2	17	20	14	31.9	131.5	s	5.6	1
1911	2	18	14	45	35.4	136.3	s	5.5	1
1911	2	23	11	14	27.0	128.0	s	7.0*	0
1911	3	24	3	18	24.0	123.0	s	6.8	0
1911	6	9	22	20	34.0	135.0	s	6.0	0
1911	6	15	14	26	28.0	130.0	100	8.0	3
1911	8	8	14	25	28.0	130.0	s	6.2	0
1911	8	21	22	48	32.9	131.0	s	5.7	1
1911	9	6	0	54	46.0	143.0	350	7.1	1
1911	11	8	14	12	34.5	140.5	s	6.5	0
1911	11	15	13	46	36.0	137.5	280	6.1	0
1911	11	21	7	36	39.0	136.0	300	6.6	0
1911	12	6	8	31	36.1	139.7	s	6.0	0
1912	1	3	19	4	42.0	144.0	s	6.1	0
1912	1	8	21	21	37.2	141.2	s	6.1	0
1912	3	10	11	10	46.0	145.0	300	6.2	0
1912	4	18	7	37	38.6	142.0	s	5.8	1
1912	5	20	7	54	27.5	141.0	400	6.4	0
1912	5	30	15	30	35.4	139.9	s	6.0	0

(*continued*)

Table 1. (*Continued*)

Y	M	D	h	m	Lat °N	Long °E	Dep km	*M*	D	Y	M	D	h	m	Lat °N	Long °E	Dep km	*M*	D
1912	6	8	4	41	40.5	142.0	s	6.6	1	1915	10	13	19	43	39.0	144.0	s	6.2	0
1912	7	13	14	31	35.5	140.5	s	6.2	0	1915	10	14	16	28	39.0	144.0	s	6.1	0
1912	7	15	22	46	36.4	138.5	s	5.7	1	1915	10	14	18	40	39.0	144.0	s	6.3	0
1912	7	24	23	24	31.0	137.5	450	6.6	0	1915	10	15	16	55	39.0	144.0	s	6.0	0
1912	8	17	14	22	36.4	138.3	s	5.1	2	1915	10	16	15	21	39.0	144.0	s	6.1	0
1912	12	8	23	50	39.0	143.5	s	6.6	0	1915	11	1	7	24	38.3	142.9	s	7.5	0
1912	12	31	14	30	35.0	138.0	200	6.3	0	1915	11	1	7	50	38.0	143.0	s	6.7	0
1913	1	19	23	47	46.0	151.0	150	7.2	0	1915	11	1	9	1	38.0	143.0	s	7.0	0
1913	2	20	8	58	41.8	142.3	s	6.9	1	1915	11	1	15	43	38.0	143.0	s	6.2	0
1913	3	3	20	2	28.0	129.0	150	6.6	0	1915	11	4	3	13	38.0	143.0	s	6.4	0
1913	4	2	23	54	32.0	132.0	s	6.7	0	1915	11	16	1	38	35.4	140.3	s	6.0	2
1913	4	13	6	40	32.0	132.0	s	6.8	1	1915	11	18	4	4	37.7	143.1	s	7.0	0
1913	5	21	20	36	36.0	141.1	s	6.1	0	1915	12	6	20	58	41.0	145.0	s	6.5	0
1913	5	29	10	14	36.1	141.0	s	6.4	0	1916	1	21	8	6	37.0	135.0	300	6.1	0
1913	6	6	2	41	34.0	142.0	s	6.1	0	1916	1	25	11	38	45.0	144.0	250	6.9	0
1913	6	29	8	23	31.6	130.3	s	5.7	1	1916	2	1	7	37	29.5	131.0	s	7.4	0
1913	6	30	7	45	31.6	130.3	s	5.9	2	1916	2	6	10	56	45.0	151.0	s	6.7	0
1913	7	31	22	6	41.8	142.5	s	5.7	1	1916	2	22	9	12	36.5	138.5	s	6.2	3
1913	10	3	0	17	40.0	144.0	s	6.1	0	1916	3	6	9	12	33.5	131.6	s	6.1	1
1913	10	11	9	10	40.0	144.0	s	6.9	0	1916	3	18	0	58	41.5	144.5	s	6.6	0
1913	10	12	17	5	40.0	144.0	s	6.6	0	1916	3	25	23	52	24.0	124.0	s	6.5	0
1913	12	15	2	2	35.5	140.0	s	6.0	1	1916	4	14	2	11	32.5	142.0	s	6.6	0
1914	1	8	10	54	34.5	141.0	s	6.1	0	1916	4	21	11	31	32.5	141.8	s	7.1	0
1914	1	12	9	28	31.6	130.6	s	7.1	4	1916	5	14	23	56	38.3	142.1	s	6.0	0
1914	2	7	6	50	41.0	143.0	s	6.8	0	1916	7	16	18	16	39.5	144.0	s	6.8	0
1914	3	14	19	59	39.5	140.4	s	7.1	4	1916	8	5	22	52	34.0	133.4	s	5.7	1
1914	3	27	17	50	39.2	140.4	s	6.1	3	1916	8	8	4	25	36.4	141.2	s	6.3	0
1914	5	23	3	38	35.3	133.2	s	5.8	1	1916	8	21	14	33	36.4	141.2	s	6.2	0
1914	7	4	17	48	29.0	128.0	200	7.0	0	1916	8	27	22	43	37.2	141.1	s	6.8	0
1914	9	11	16	53	44.0	148.0	s	6.5	0	1916	8	28	7	28	23.9	120.5	s	7.2	0
1914	10	16	22	9	44.0	148.0	s	6.0	0	1916	9	15	7	1	34.4	141.2	s	7.0	1
1914	11	15	13	29	37.1	138.1	s	5.7	1	1916	10	28	3	20	46.0	145.0	d	6.5	0
1914	11	28	10	45	29.0	131.0	s	6.9	0	1916	11	22	14	4	40.5	135.0	300	6.1	0
1914	12	22	8	57	42.3	140.0	180	6.4	0	1916	11	24	4	4	39.2	142.7	s	6.6	0
1914	12	25	18	18	39.0	143.0	s	6.1	0	1916	11	26	5	8	34.6	135.0	s	6.1	3
1915	1	5	23	26	25.1	123.3	150	7.4	1	1916	12	28	21	41	32.3	130.5	s	6.1	1
1915	2	28	18	59	23.0	124.0	s	7.4	0	1917	1	30	15	40	35.2	139.0	s	4.5	1
1915	3	8	15	29	38.2	141.9	s	6.8	0	1917	2	21	15	47	35.7	135.8	350	6.2	0
1915	3	17	18	45	42.1	143.6	s	7.0	3	1917	3	15	0	14	39.5	144.5	s	6.9	0
1915	4	5	20	25	39.5	143.0	s	6.0	0	1917	4	21	3	53	39.6	144.8	s	6.3	0
1915	4	6	5	32	36.5	142.0	s	6.2	0	1917	5	17	19	7	35.0	138.1	s	6.3	3
1915	4	24	17	9	36.2	141.2	s	6.4	0	1917	5	31	6	6	35.0	141.5	s	6.1	0
1915	5	1	5	0	47.5	154.5	s	8.0	0	1917	6	14	13	22	37.5	141.7	s	6.1	0
1915	5	27	17	26	36.1	142.0	s	6.0	0	1917	7	4	0	38	25.0	128.0	s	7.4	0
1915	6	4	21	59	40.0	143.5	s	6.7	0	1917	7	4	5	36	25.0	128.0	s	6.9	0
1915	6	19	16	1	35.5	139.0	s	5.9	1	1917	7	29	14	32	41.0	144.0	s	7.3	0
1915	6	27	15	29	44.0	148.0	s	6.5	0	1917	7	31	3	23	42.0	131.0	500	7.6	0
1915	7	8	22	21	37.1	142.2	s	6.4	0	1917	8	23	5	14	42.6	146.0	s	6.0	0
1915	7	14	12	13	31.9	130.8	s	5.0	1	1917	11	15	15	2	40.2	143.7	s	6.0	0
1915	8	6	13	13	44.0	150.0	s	7.1	0	1917	12	6	11	39	40.0	143.0	s	6.3	0
1915	10	8	15	36	32.8	139.1	200	6.9	0	1917	12	17	22	18	33.3	136.8	s	6.0	0
1915	10	12	21	30	39.0	144.0	s	6.8	0	1918	1	30	21	18	45.0	136.0	350	7.8	0

(*continued*)

Table 1. (*Continued*)

Y	M	D	h	m	Lat °N	Long °E	Dep km	*M*	D
1918	2	9	20	46	42.0	131.5	500	6.7	0
1918	2	13	6	7	24.0	117.0	s	7.5	0
1918	4	2	3	33	32.0	132.4	s	6.3	0
1918	4	10	2	3	43.5	130.5	600	7.5	0
1918	5	25	22	30	44.2	141.6	s	5.8	1
1918	5	31	8	47	43.2	147.8	s	6.3	0
1918	6	26	13	46	35.4	139.1	s	6.3	1
1918	7	25	20	50	36.2	142.3	s	6.7	0
1918	9	7	17	16	45.5	152.0	s	8.0	2
1918	9	13	9	8	37.5	141.4	s	6.1	0
1918	9	22	13	50	45.0	150.0	s	6.7	0
1918	11	8	4	38	44.5	150.5	s	7.7	0
1918	11	10	17	59	36.5	137.9	s	6.1	3
1918	11	11	7	4	36.5	137.9	s	6.5	x
1918	12	13	21	33	40.5	144.5	s	6.2	0
1918	12	31	5	31	34.5	137.5	250	6.0	0
1919	3	10	21	21	26.0	128.0	s	6.2	0
1919	3	26	13	37	30.5	139.0	350	6.2	0
1919	3	28	22	40	36.9	138.4	s	5.4	1
1919	5	3	0	52	40.9	145.0	s	7.4	0
1919	6	1	6	51	26.0	125.0	200	6.8	0
1919	6	6	13	24	33.5	134.7	s	6.0	0
1919	7	16	4	10	43.0	147.0	s	6.3	0
1919	7	21	23	51	41.5	142.5	s	6.1	0
1919	8	3	18	8	36.0	142.5	s	6.7	0
1919	8	7	16	32	40.0	144.0	s	6.2	0
1919	9	12	14	54	39.7	144.7	s	6.1	0
1919	10	11	13	17	41.1	142.9	s	6.3	0
1919	10	31	23	34	34.8	132.9	s	5.8	1
1919	12	20	0	28	37.9	142.3	s	6.3	0
1919	12	20	19	33	22.5	122.5	s	7.1	0
1919	12	20	20	37	22.5	122.5	s	7.3	0
1920	1	13	18	30	40.1	139.0	s	6.0	0
1920	1	17	18	42	40.1	139.0	s	6.2	0
1920	2	7	15	6	40.6	143.0	s	6.7	0
1920	2	22	17	35	47.5	146.0	350	7.5	0
1920	4	11	23	5	47.5	153.0	100	6.9	0
1920	4	15	12	13	33.3	138.2	300	6.4	0
1920	5	6	9	40	43.0	131.5	500	6.5	0
1920	5	12	21	53	34.9	139.3	100	6.4	0
1920	6	5	4	21	24.0	121.7	s	7.8	0
1920	7	3	14	19	36.2	139.2	100	6.0	0
1920	7	20	12	18	34.0	141.0	s	6.2	0
1920	9	16	15	8	41.6	142.1	s	6.5	0
1920	9	20	20	27	41.0	143.0	s	6.1	0
1920	10	18	8	11	44.5	148.5	s	7.1	0
1920	11	8	17	38	36.1	142.0	s	6.3	0
1920	12	2	23	39	36.4	141.8	s	6.0	0
1920	12	19	20	11	37.3	141.0	s	6.8	0
1920	12	27	9	21	35.2	139.0	s	5.7	1
1921	1	2	7	6	44.5	149.5	s	6.8	0
1921	1	9	18	56	36.1	142.0	s	6.1	0
1921	3	3	3	2	37.7	141.5	s	6.9	0
1921	3	4	12	51	29.5	139.0	450	6.4	0
1921	3	15	4	32	32.5	139.0	250	6.0	0
1921	4	2	9	36	23.0	123.0	s	7.2	0
1921	4	18	17	58	32.6	132.1	s	5.5	1
1921	6	22	11	23	42.3	143.8	s	6.3	0
1921	7	4	14	18	26.0	141.5	200	7.0	0
1921	8	9	10	38	42.2	145.1	s	6.1	0
1921	8	22	4	5	35.7	142.5	s	6.3	0
1921	9	26	21	16	30.5	140.0	200	6.3	0
1921	9	27	16	20	40.5	139.0	s	6.5	0
1921	10	12	7	52	45.3	148.6	100	6.6	0
1921	12	8	12	31	36.0	140.2	s	7.0	1
1922	1	22	22	5	37.5	141.5	s	6.5	1
1922	2	26	8	56	43.0	147.0	s	6.0	0
1922	3	16	18	31	36.4	141.7	s	6.2	0
1922	4	26	1	11	35.2	139.8	s	6.8	3
1922	4	27	9	15	36.5	142.0	s	6.0	0
1922	5	9	3	28	36.0	140.0	s	6.1	1
1922	5	15	20	21	40.3	144.2	s	6.5	0
1922	6	3	4	56	36.5	141.4	s	6.1	0
1922	6	18	12	17	34.0	137.3	s	6.1	0
1922	7	5	20	20	38.3	142.0	s	6.5	0
1922	7	11	14	13	22.0	143.0	s	6.6	0
1922	9	1	19	16	24.5	122.2	s	7.6	0
1922	9	14	19	31	24.5	122.2	s	7.3	0
1922	10	14	23	45	24.5	122.0	s	6.9	0
1922	10	24	21	21	46.5	152.0	100	7.7	0
1922	12	7	16	50	32.7	130.1	s	6.9	4
1922	12	8	2	2	32.7	130.1	s	6.5	3
1922	12	8	22	33	40.2	143.2	s	6.8	0
1922	12	31	7	19	45.0	151.0	s	7.2	0
1923	1	14	5	51	36.1	139.9	s	6.1	2
1923	3	21	8	28	46.0	151.0	s	6.0	0
1923	4	23	3	16	27.0	126.0	s	7.2	0
1923	5	26	3	12	36.0	141.8	s	6.4	0
1923	5	31	5	55	36.5	142.0	s	6.2	0
1923	6	1	17	24	35.9	142.0	s	7.3	0
1923	6	1	20	14	36.0	142.0	s	7.1	0
1923	6	6	17	36	36.5	142.5	s	6.2	0
1923	6	29	10	47	28.0	141.0	400	6.4	0
1923	7	2	2	31	23.5	122.5	s	6.7	0
1923	7	13	11	13	30.6	131.2	s	7.1	2
1923	7	13	23	55	30.6	131.2	s	6.6	0
1923	7	20	16	50	32.5	138.0	300	6.2	0
1923	8	12	10	6	26.0	128.0	s	6.8	0
1923	9	1	2	58	35.1	139.5	s	7.9	7
1923	9	1	3	1	35.0	139.5	s	6.5	0
1923	9	1	3	7	35.0	139.5	s	6.0	0
1923	9	1	3	14	35.0	139.5	s	6.0	0
1923	9	1	3	17	35.0	139.5	s	6.4	x
1923	9	1	3	23	35.4	139.4	s	6.6	x

(*continued*)

Table 1. (*Continued*)

Y	M	D	h	m	Lat °N	Long °E	Dep km	*M*	D	Y	M	D	h	m	Lat °N	Long °E	Dep km	*M*	D
1923	9	1	3	34	35.0	139.5	s	6.0	0	1924	5	31	12	4	36.4	141.6	s	6.4	0
1923	9	1	3	36	36.0	140.0	s	6.0	0	1924	6	30	15	44	45.0	147.5	150	7.6	0
1923	9	1	3	39	35.2	139.7	s	6.6	x	1924	7	22	14	23	23.7	122.0	s	6.8	0
1923	9	1	3	48	35.4	139.8	s	7.0	x	1924	8	6	14	22	35.5	140.6	s	6.3	0
1923	9	1	4	13	35.0	140.0	s	6.4	0	1924	8	12	18	18	33.9	135.2	s	5.9	1
1923	9	1	4	20	35.1	139.8	s	6.2	0	1924	8	14	18	2	36.2	141.6	s	7.1	0
1923	9	1	4	30	35.2	139.0	s	6.3	x	1924	8	14	23	27	36.3	142.2	s	6.7	0
1923	9	1	5	23	35.4	139.0	s	6.7	x	1924	8	17	1	45	35.6	142.1	s	6.3	0
1923	9	1	6	19	35.5	140.5	s	6.5	0	1924	8	17	2	10	36.0	142.5	s	6.6	0
1923	9	1	7	38	35.5	138.9	s	6.8	1	1924	8	25	14	31	36.2	142.0	s	6.7	0
1923	9	1	13	52	35.0	140.0	s	6.0	0	1924	8	28	23	50	32.4	132.6	s	6.2	0
1923	9	2	0	59	35.2	139.5	s	6.0	0	1924	9	18	1	8	36.3	140.2	s	6.6	0
1923	9	2	2	46	34.9	140.2	s	7.3	1	1924	11	25	17	26	45.5	142.0	300	6.7	0
1923	9	2	9	27	34.9	140.5	s	7.1	0	1924	12	27	11	22	44.0	146.0	150	7.4	0
1923	9	2	9	49	35.3	140.6	s	6.3	0	1924	12	28	22	54	43.0	147.0	s	7.0	0
1923	9	2	13	9	35.3	139.1	s	6.5	0	1925	1	18	12	5	47.0	154.0	s	7.4	0
1923	9	2	14	16	35.3	139.1	s	6.2	0	1925	1	28	4	5	43.5	147.5	s	6.9	0
1923	9	4	22	23	34.8	140.3	s	6.1	0	1925	2	1	5	24	43.5	148.0	s	6.4	0
1923	9	7	17	32	34.4	139.8	s	6.2	0	1925	2	2	13	29	42.5	147.5	s	6.2	0
1923	9	9	17	11	34.8	139.2	s	5.9	1	1925	2	2	19	25	43.0	148.0	s	6.2	0
1923	9	17	3	39	31.0	140.0	150	6.4	0	1925	2	2	19	48	43.0	148.0	s	6.8	0
1923	9	26	8	24	34.8	139.4	s	6.7	1	1925	2	2	22	12	43.0	147.5	s	6.3	0
1923	10	3	15	54	35.4	139.1	s	6.4	0	1925	2	6	17	11	35.6	141.4	s	6.0	0
1923	10	5	13	5	35.4	139.1	s	6.0	0	1925	2	20	1	2	45.5	150.5	s	7.0	0
1923	10	9	11	22	39.5	141.0	120	6.2	0	1925	3	27	4	16	30.0	139.0	500	6.3	0
1923	11	3	16	19	29.4	130.8	s	6.8	0	1925	4	19	15	46	33.0	138.0	350	6.8	0
1923	11	4	20	45	35.7	139.2	s	6.3	0	1925	4	19	20	41	38.6	142.2	s	6.3	0
1923	11	5	21	27	29.1	130.9	s	7.1	0	1925	5	15	18	25	30.5	138.5	400	6.0	0
1923	11	6	19	18	29.2	130.4	s	6.5	0	1925	5	20	11	4	30.5	142.5	s	6.0	0
1923	11	17	20	40	36.2	141.0	s	6.3	0	1925	5	22	9	40	30.5	142.0	s	6.3	0
1923	11	23	2	32	35.4	139.5	s	6.2	0	1925	5	23	2	9	35.6	134.8	s	6.8	5
1923	11	25	17	3	23.5	123.2	s	6.4	0	1925	5	25	16	22	35.6	134.8	s	6.3	0
1923	12	4	23	40	33.3	134.0	s	6.5	0	1925	5	27	2	29	37.5	134.0	400	6.9	0
1923	12	27	14	39	36.3	141.1	s	6.4	0	1925	6	2	5	18	41.3	142.0	s	6.4	0
1924	1	14	20	50	35.5	139.2	s	7.3	3	1925	6	23	4	43	42.5	142.3	100	6.2	0
1924	1	14	21	5	36.2	140.5	s	6.0	0	1925	7	3	19	20	35.5	133.3	s	5.8	1
1924	2	2	22	25	35.4	141.2	s	6.3	0	1925	7	6	16	46	35.4	136.5	s	5.8	1
1924	4	3	2	30	32.0	138.5	350	6.0	0	1925	8	10	0	37	33.3	130.9	s	4.4	1
1924	5	22	18	8	42.7	142.3	100	6.3	0	1925	10	20	9	41	28.0	140.0	400	6.4	0
1924	5	28	9	52	48.0	147.0	450	7.4	0	1925	11	10	14	44	39.0	143.5	s	6.0	0
1924	5	31	12	2	36.4	141.6	s	6.3	0										

Data and time are in GMT. Letters "s" and "d" mean shallow ($h < 100$ km) and deep ($h \geq 100$ km), respectively. Asterisks (*) indicate magnitudes estimated mainly from tsunami data and/or amplitudes of long-period seismic waves at distant stations. The last column headed "D" gives the damage rating (see text).

where n is the frequency of earthquakes of magnitude M, and N is the cumulative frequency ($n = -dN/dM$). The logarithm of likelihood function is -247.04 for U, -247.19 for L, and -246.76 for M. Therefore the Makjanić formula is the best in this particular case, but the difference among the three 3-parameter formulas is very small.

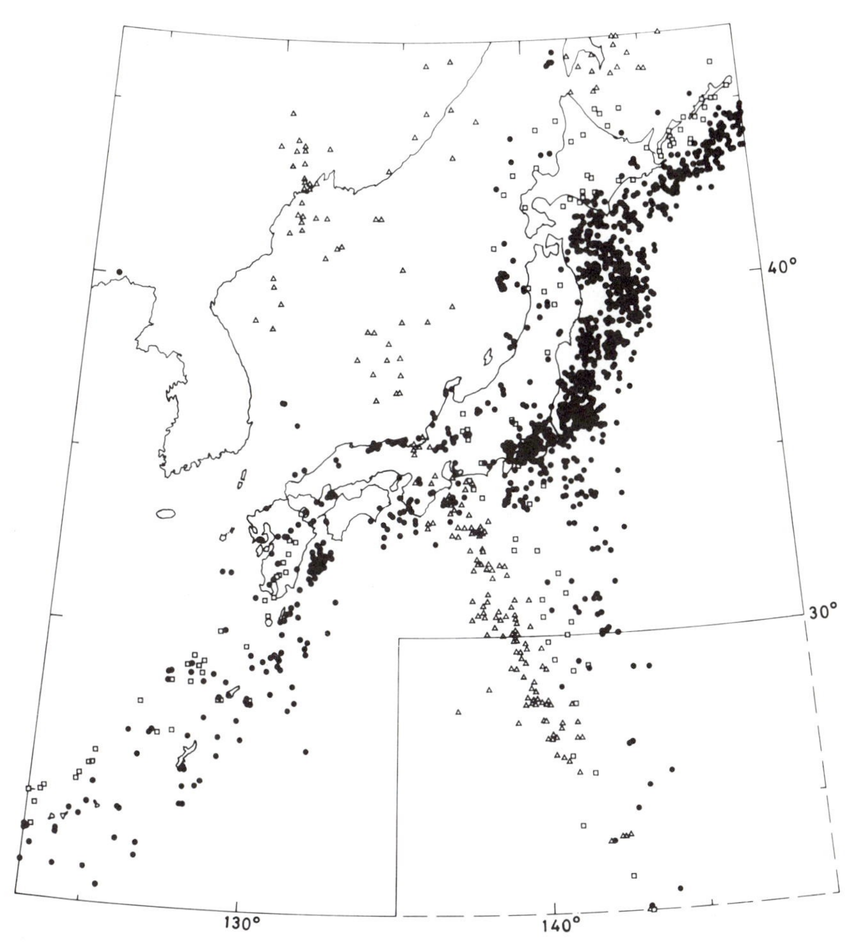

Figure 2. Distribution of epicenters of earthquakes of $M \geq 6$. Solid circles represent shallow earthquakes ($h < 100$ km) for the years 1885-1984. Squares and triangles represent deep earthquakes (100 km km $\leq h < 300$ km and $h \geq 300$ km, respectively) for the years 1901 - 1984.

Table 2. Number of Earthquakes Accompanied by Damage (cf. p. 152) in Japan, 1885-1984

Damage	1	2	3	4	5	6	7	x	Total
Number	259	104	60	16	8	4	2	20	473

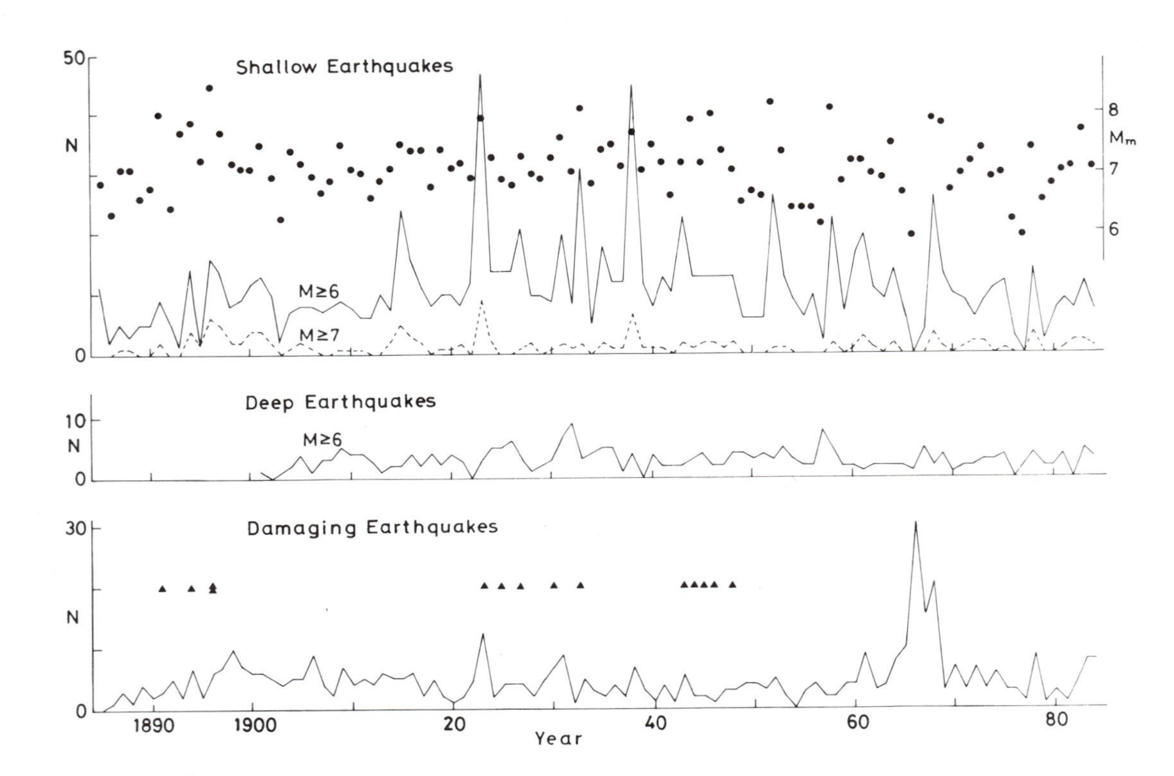

Figure 3. Variation of the number of earthquakes per year, 1885-1984. *Top*: Shallow earthquakes of $M \geq 6$ and $M \geq 7$ ($h < 100$ km) in region **A** (region enclosed by solid lines in Figure 2). Circles represent the magnitude of the largest earthquake in each year. *Center*: Deep earthquakes of $M \geq 6$ ($h \geq 100$ km) in region A. Data before 1900 are not shown. *Bottom*: Earthquakes which caused damage in Japan. Triangles indicate the time of earthquakes which killed more than 200 persons. High rates around 1966 are mostly due to the Matsushiro swarm.

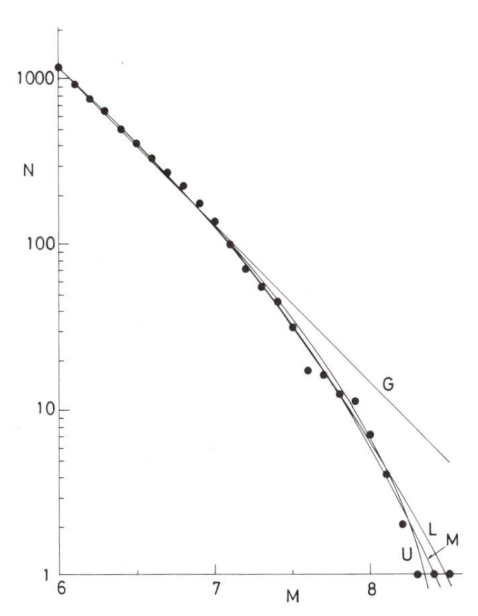

Figure 4. Magnitude vs. cumulative frequency relationship for shallow earthquakes in region A in the years 1885-1984. Curves G, U, M, and L represent Gutenberg and Richter, Utsu, Makjanić, and Lomnitz-Adler and Lomnitz formulas fitted to the data using the maximum likelihood method, respectively.

REFERENCES

Abe, K. (1979). Instrumental magnitudes of Japanese earthquakes, 1901-1925, *Zisin* (J. Seism. Soc. Japan), Ser. 2, **32**, 341-354 (in Japanese).

Central Meteorological Observatory (1952). The magnitude catalogue of major earthquakes which occurred in the vicinity of Japan (1885-1950), *Seism. Bull. CMO*, for 1950, 99-183.

Gutenberg, B. and C. F. Richter (1954). *Seismicity of the Earth and Associated Phenomena*, 2nd Ed., 310 pp., Princeton University Press.

Japan Meteorological Agency (1958). Catalogue of major earthquakes which occurred in and near Japan (1926-1956), *Seism. Bull. JMA*, Suppl. Vol. 1, 91 pp.

Japan Meteorological Agency (1982). Catalog of relocated major earthquakes in and near Japan (1926-1960), *Seism. Bull. JMA*, Suppl. Vol. 6, 109 pp.

Lomnitz-Adler, J. and C. Lomnitz (1979). A modified form of the Gutenberg-Richter magnitude-frequency relation, *Bull. Seism. Soc. Am.*, **69**, 1209-1214.

Makjanić, B. (1980). On the frequency distribution of earthquake magnitude and intensity, *Bull. Seism. Soc. Am.*, **70**, 2253-2260.

Utsu, T. (1971). Aftershocks and earthquake statistics (III), *J. Fac. Sci. Hokkaido Univ.*, Ser. VII, **3**, 379-441.

Utsu, T. (1979). Seismicity of Japan from 1885 through 1925, *Bull. Earthq. Res. Inst. Univ. Tokyo*, **54**, 253-308 (in Japanese).

Utsu, T. (1980). Spatial and temporal distribution of low-frequency earthquakes in Japan, *J. Phys. Earth*, **28**, 361-384.

Utsu, T. (1982). Seismicity of Japan from 1885 through 1925 (correction and supplement), *Bull. Earthq. Res. Inst. Univ. Tokyo*, **57**, 111-117 (in Japanese).

Utsu, T. (1984). Comparison of four formulas for the distribution of recurrence times of earthquakes and comparison of five formulas for the distribution of earthquake magnitudes, Programme and Abstracts, *Seism. Soc. Japan*, 1984 No. 2, 189 (in Japanese).

Historical Materials of Chinese Earthquakes and Their Seismological Analysis

Yushou Xie
Institute of Geophysics, State Seismological Bureau
China

ABSTRACT

Science and culture developed early in China. Its seismicity is high and historical material is abundant. The earliest record goes back to the 23rd century B.C. The Seismological Committee of Academia Sinica compiled the *Chronological Table of Chineses Earthquakes* in 1956. In 1978, we organized hundreds of seismologists, historians, archaeologists, archivists and librarians from various institutions and localities to revise it. More than 20,000 varieties of literatures, documents and materials including archives written in ancient Tibetan and Manchurian, archives of the Republic of China, the Customs archives, rare books and local annals, peculiar literatures and stone inscriptions, abundant materials of macroseismic investigation, documents of neighbouring countries, bulletins of global and regional networks of seismic stations and catalogues of earthquakes were looked up. Through analysis and textural research, the false were eliminated and the true were retained. As a result, a series of five volumes of *Compilation of Historical Materials of Chinese Earthquakes* arranged chronologically from the 23rd century B.C. to 1980 A.D. totalling about 5,000,000 Chinese characters was compiled. Two volumes were published and the others will be published before the end of 1986.

A catalogue of earthquakes from 1900 to 1980 with $M_S \geq 4.7$ was added. The origin time and the location of focus were adjusted with computer whenever possible. In order to unify the magnitude scale, Gutenberg and Richter formulas of 1945 and 1956 were used as standards for the calculation of surface wave and body wave magnitudes respectively. Correlations of magnitudes listed in various catalogues relative to our revised magnitudes were obtained for regression purposes. For earthquakes of 1949-1980, the unification of magnitude scale was not well done, and we are now revising them.

The analysis of macroseismic materials of historical earthquakes is more complicated. The conversion of calendars and the identification of localities have been done by paleoastronomers and historical geographers respectively. For the estimation of magnitudes, regions with different rates of intensity attenuation were delineated. We are now planning to get different sets of emperical relations between instrumentally determined magnitude and macroseismic parameters, such as maximum intensity, isoseismal pattern, and extent of isoseismals with various intensities, for individual regions from recent earthquakes with both instrumental and macroseismic information. On this basis, we believe, the magnitude of historical earthquakes can be better estimated.

1. Introduction

Historical materials are indispensable for the prolongation of time span in the study of seismicity. In this paper, instrumental data, as well as historical records, are included. The historical documents in China are investigated. The compilation of historical materials is described. The use of them in the late decades is briefly reviewed. Finally, the unification of magnitude scale for instrumentally determined earthquakes of the 20th century and the method to be used for the seismological analysis of macroseismic records of historical earthquakes are given.

2. Historical Documents in China

Science and culture developed early in China, and we have long recorded history. It was in the 2nd century B.C. that Chang Heng invented the first seismoscope of the world. Since the appointment of an official historian in the Ying dynasty (13-11th century B.C.), not only political affairs, but also such natural phenomena as astronomical, meteorological and seismological events were recorded. The invention of printing around the 10th century greatly facilitated the preservation of historical materials. After the Yuan dynasty, i.e., since the 14th century, local records or annals of different provinces, districts and counties became popular. Thus, the seismological materials increased greatly both in number and in contents.

Ancient Chinese usually correlated natural phenomena with social and political affairs. The occurrence of rare natural phenomena, such as earthquakes, comets, etc., and the appearance of rare animals and plants such as phoenix, glossy garnoderma, etc. were considered to be auspicious sign or ill omen. Under the control of such thought of "God's will", the emperor required the officials of the central government, as well as the local authorities to report such things, officials and people paid attention to and recorded them in different ways and by different means. As a result, some of them were handed down till now.

Which item of our recorded historical earthquakes is the first reliable one is a debatable problem. "Taiping Yü Lan" quoted an item from the classic "Mo-tze": "While the San-Miao was going to be destroyed, the Earth quaked, fountains sprang". This is a tale of about 2221 B.C., before the invention of Chinese characters in the Shang Dynasty (16-11th century B.C.). The classic "Mo-tze" was written in the 4th century B.C., it only recorded legends popular at that time. Therefore, it is considered to be not very reliable. Nevertheless it seems to be the earliest written description of an earthquake in the world. Another classic "Chronicle on Bamboo" inscribed on bamboo at the time of the Warring States (5th–3rd century B.C.) recorded: "In the 35th year of Emperor Shun (about 1831 B.C.) Mount Taishan quaked"; "In the 10th year of Emperor Jie of Xia Dynasty (about 1767 B.C.) five stars ran out of orbits, stars fall like rain, the Earth quaked, Rivers Yihe and Luohe exhausted"; and "In June, the third year of Emperor Yi of Shang Dynasty (about 1189 B.C.), Zhou (northeast of Qishan in Shaanxi Province) quaked". As the history before Shang Dynasty is not well known, these items are also ancient legends.

The "Lu's Chronicle" written in 239 B.C. recorded: "In June, the 8th year of Emperor Wen-wang of Zhou Dynasty (about 1177 B.C.), the Emperor sickened for 5 days and the Earth quaked in the whole country". Some historians claim that it was written less than a thousand years after the event, at a time before the burning of still earlier books by the first Emperor of Qin Dynasty in 212 B.C., and is therefore more reliable. Another item in the same book recorded: "In the second year of Emperor You (about 780 B.C.), all the three rivers (Jinhe, Weihe and Luohe) quaked.... In this year, the three rivers exhausted, Mount Qishan slided." This item is more concrete in its contents and is generally considered to be reliable.

Besides, we have collections of seismic materials since early days, for example: a chapter in "Taiping Yü Lan", written by Li Fang in 977 A.D., collected 45 items of earthquakes between the 11th century B.C. and 618 A.D.; the "Chapter on Strange Things" in "General Reference to Literatures and Documents", selected by Ma Duanlin in the 13th century, collected 268 items of earthquakes between the 11th century B.C. and the 13th century; the "Chapter on Telluric Abnormal Phenomena" in "Collection of Ancient and Modern Books", written by Jiang Jianjiun and published in 1725, recorded 654 items of earthquakes, landslides and ground fissures from the 11th century B.C. to 1722 A.D.

Our historical materials contain not only earthquake phenomena, but also accompanying phenomena before, during and after their occurrences, the influence of local conditions on the destructiveness of earthquakes, anti-seismic measures of buildings, methods to safeguard lives, and philosophy of earthquake generation, e.g. in an imperial edict of Emperor Kangxi in the Qing Dynasty written in 1720, it was said that earthquake is caused by the bursting of gas buried underground, thus seismicity is high in North China where the overburden is thick. Of course, they can not all be expected to be true. Especially the explanations are not always scientific. But, through systematic and objective analysis, they can inspire our train of thoughts on earthquake prediction and mitigation of seismic hazards.

3. Compilation of Historical Materials

Since the founding of our People's Republic, the assessment of seismic risk of construction sites is of urgent need. As we had only a few seismic stations before 1949, instrumental data are far from sufficient to solve such a problem. Under the unified leadership of the Central Government, specialists from related organizations and scientific institutes organized expeditions and collected large amount of seismic materials from field work, as well as literatures and historical documents.

On this basis, the Seismological Committee of Academia Sinica organized historians and seismologists to further search for new materials. Within two years, they read more than 8,000 various sorts of literatures and collected more than 10,000 items of historical materials concerning 880 destructive earthquakes. In 1956, the *Chronological Table of Chinese Earthquakes* (Academia Sinica, Seismological Committee, 1956) was published. This book of nearly two million words was highly valued by seismologists. They deemed that it has profound significance in the study of seismicity and mitigation of seismic hazards. On the other hand, people also discovered that there are imperfections in the materials collected and the editorial work.

Since the publication of the *Chronological Table*, seismological study has made great development in our country. Seismologists, historians, archeologists, archivists, engineers and librarians did a lot of collection, sorting and research work on historical earthquakes, and published various collections of historical records of earthquakes. Seismological organizations have been formed in all provinces, metropolis and autonomous regions of our country. Many of them have collected historical materials of earthquakes in their own regions.

The great Tangshan Earthquake of 1976 attracted general attention to earthquakes. The Academy of Sciences, the Academy of Social Sciences and the State Seismological Bureau of China deemed it necessary and also possible to revise the *Chronological Table of Chinese Earthquakes.* An editorial board was established in 1978. All provinces, metropolis and autonomous regions also organized their respective editorial groups. Hundreds of seismologists, historians, archaeologists, archivists, and librarians from various institutes and localities took part in this work.

In the *Chronological Table*, more than 1,600 counties among the 2,000 and more counties in our country have materials of historical earthquakes. Most of them are distributed in the eastern part of our country. For the bordering regions, historical records are rare. For instance, in Xizang (Tibet) Autonomous Region (Figure 1), only 30 earthquakes between 1893 and 1954 were recorded, of which 13 were destructive. Recently, the Seismic Office and the Archives of Xizang Autonomous Region looked up different sorts of historical documents which were never used before; such as, the archives of local authorities and monasteries written in ancient Tibetan, the Buddhistic classics, histories, biographies, the lunar calendar used by the Zang (Tibetan) nationality, gazettes, instrumental data and reports of expeditions, etc. Through careful selection, translation and verification, more than 800 items, totalling about 600,000 Chinese characters, representing 437 earthquakes between 641 and 1980 were collected. For Taiwan Province, only 84 earthquakes between 1661 and 1951 were collected in the *Chronological Table.* In the new compilation, the number of earthquakes collected increased more than ten times. Even for the densely populated eastern China, where culture and economy were well developed, we added many materials in the new edition. For example, nine destructive earthquakes were supplemented for Jiangsu Province in the Lower Yangtze Valley. Several of them occurred in the vicinity of Liyang county (Figure 1), where two destructive earthquakes with magnitudes $5\frac{1}{2}$ to 6 occurred in 1974 and 1979, successively. In an investigation of the 1303 and the 1695 earthquakes, both occurred in southern Shanxi Province, out of 2,386 stone tablets, 47 were inscribed with records of earthquakes. Besides, in northern Hainan Island, remains of villages and graveyards can be seen at ebb tide, testifying the sinking of lowland near the coast during the 1605 earthquake as recorded in historical documents.

To check the materials concerning historical earthquakes is a difficult and painstaking work. For example: the explosion of a powder magazine in Beijing in 1626 was formerly misunderstood as a local destructive earthquake from historical records. An item from local record was quoted as: "In Tang Dynasty (870 A.D.) Sichuan was shaken strongly." It was thus explained as "A big earthquake occurred in Sichuan". But, tracing the original document, it wrote: "The city Chengdu in Sichuan was encircled by the local minority nationality troops, arrows fall like rain, Sichuan was strongly shaken". Really, it is the morale of the residents that was shaken by the fear of the war and not the ground shaken by an earthquake. In the *Chronological Table of Chinese Earthquakes*, there is an item saying: "In Xichang County earth shaked slightly". The whole story as recorded in "History of Ming Dynasty" reads: "In Xichang County, a fire ball like a wheel with a tail flew from the north and disappeared in the south, with a noise like thunder, the earth shaked slightly." It is clear that the shaking was caused by a comet.

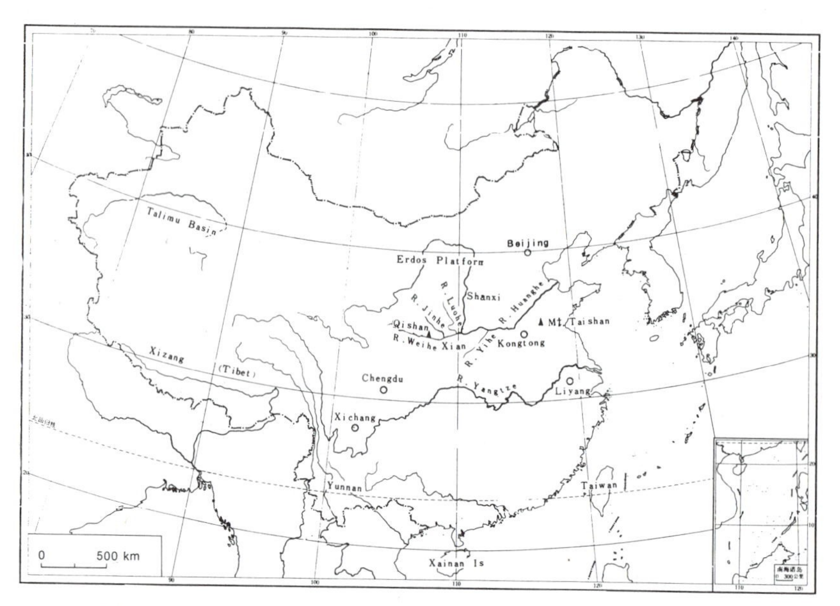

Figure 1. Map Showing Location of Places Mentioned in the Text.

Locations are often difficult to pin down because of obscure description or changed place names. The cooperation of historical geographers helped us a great deal on this matter. For example, as all the three provinces – Shanxi, Shaanxi and Henan – have places named Kuntung, where is the real epicenter of the Kuntung earthquake of 466 B.C. is a long disputed problem. After thorough study, it is finally put at Henan Province.

As we use a lunar calendar and the calendar system in various dynasties changed frequently, accurate dates are sometimes difficult to establish. With the help of paleoastronomers, we have already corrected more than 200 errors and mistakes on the data of historical earthquakes in the former catalogues. Both lunar calendar and Gregorian calendar are listed. For dates before 1582, the Julian calendar is added for reference.

More than 20,000 kinds of literatures, documents and materials including archives written in ancient Tibetan, archives written in Manchurian, archives of the Republic of China, the Customs archives, rare books and local annals, peculiar literatures and stone inscriptions, abundant materials of macroseismic investigation, documents of neighbouring countries, bulletins of global and regional networks of seismic stations and catalogues of earthquakes were looked up. Through analysis and textural research, the false were eliminated and the true were retained. Errors and mistakes in the *Chronological Table of Chinese Earthquakes* published in 1956 were corrected and its content is greatly enriched. About 50,000 items of historical materials and more than 20,000 sets of instrumental earthquake parameters were collected. As a result, a series of five volumes of *Compilation of Historical Materials of Chinese Earthquakes* (Xie *et al.*, 1983-1987) arranged chronologically from the 23rd century B.C. to 1980 A.D. totalling around 6,500,000 Chinese characters was compiled.

The first and the fifth volumes were published in 1983. The second and the fourth volumes were published in 1985 and 1986. The manuscript of the third volume has been sent to the publishing house recently.

At the same time, the provinces, metropolis and autonomous regions with high seismic activity compiled materials of historical earthquakes of their respective regions in more detail. Besides, monographs of individual great earthquakes were published. The sum total much exceeds ten million words.

4. Use of Historical Materials in China

In order to use the abundant material of historical earthquakes in modern seismological study, the macroseismic data should be converted into seismic parameters.

Chiefly on the basis of field observations and historical materials, with reference to instrumental data, foreign seismic reports and other scales of seismic intensity, a new seismic intensity scale with characteristic Chinese criterions was compiled (Hsieh, 1957) for use in intensity assessment. The scale consists of 12 degrees, each corresponds approximately to the same degree in other 12-degree scales, such as the MM or the MSK scale. The destruction of buildings and structures is taken to be the main criteria of intensity evaluation. We tried our best to select typical Chinese buildings and structures as criteria, and took that type of masonry building, which is familiar in European countries and is built in large amount in new China as a bridge linking Chinese and western buildings.

Our colleagues published a *Catalogue of Chinese Earthquakes* in 1960 (Lee *et al.*, 1960). It contains about 10,000 earthquakes occurred between 1189 B.C. and 1955 A.D. Among them, 1,180 have magnitudes greater or equal to 4.7. As the data are not complete, these earthquakes are not evenly distributed in either time or space. There are only 131 earthquakes between 1189 B.C. and 1500 A.D., mostly distributed along the Yellow River Valley and in the south-western part of our country; 405 events between 1501 and 1900 A.D., generally east of 100° E meridian; and 618 quakes from 1901 to 1955, mostly west of 100° E meridian.

For earthquakes with instrumentally determined parameters, macroseismic data were used simply for reference. When there were no instrumental data, they tried to draw isoseismals and took the center of the meizoseismic region as the epicenter. For earthquakes with insufficient data, the site of the county which suffered most was taken as the location of the epicenter.

The magnitudes of historical earthquakes were deduced from their epicentral intensities. From the 35 earthquakes distributed all over the country with both instrumentally determined magnitudes and macroseismic materials, an empirical equation between the magnitude, M, and the "epicentral intensity, I_0" was obtained ($M = 0.58\,I_0 + 1.5$), in which focal depth was not considered.

In 1971, a revised and enlarged edition of the *Catalogue of Chinese Earthquakes* (Central Seismological Group, 1971) was published. It contains 2,257 earthquakes with magnitudes greater or equal to 4.7 occurred between 1177 B.C. and 1969 A.D. Besides its prolongation in time span, supplementation of materials and correction of errors and mistakes, the instrumentally determined earthquake foci were redetermined with electronic computer whenever possible.

Several concise catalogues with earthquake parameters only (see e.g. State Seismological Bureau, 1977, and Lee *et al.*, 1978) and a *Collection of Isoseismals of Chinese Earthquakes* (State Seismological Bureau, 1979) were published later. The latest edition of *Catalogue of Chinese Earthquakes* (Gu, 1983) was published at the end of 1983. The method of determination of magnitude remains the same.

From materials of historical earthquakes, we investigated the division of seismic active zones and their temporal variations of seismic activity (see e.g. Shih *et al.*, 1974). Generally speaking, the cycle of seismic activity in the eastern part of China mainland is comparatively long. It is of the order of hundreds of years, or even more than 1,000 years. For example, in the Valley of the River Weihe, where the most disastrous historical earthquake with a death toll of more than 830,000 people occurred in 1556, tens of felt to destructive earthquakes occurred in the eastern part of the graben during the few decades before and after this earthquake, but it has remained quiet since 1568. As this region was an important political, cultural and economic center and has long recorded history, we believe that no destructive earthquake could be overlooked during this period. Since the establishment of a regional network of seismic stations in the late 1970's, seismicity in the whole graben has been very low. Only several small earthquake swarms were recorded in the central part of the graben, where strong earthquakes rarely occurred in history. Earthquakes with magnitude around 3 occurred once in several years. On the other hand, the cycle of activity in Yunnan and Xinjiang is only several tens of years.

That seismic activity is closely related to geological conditions was noticed long ago. Strong earthquakes concentrate in regions with active Neotectonic movements and almost no destructive earthquakes occur in stable platforms, such as Talimu Basin and Erdos Platform. For example, the central eastern part of Yunnan Province is one of the most active seismic regions in China. Geologically, it is characterized by two sets of surface faults with obvious Neotectonic movements: one striking nearly meridionally, the other trending N 60° W. Almost all epicenters of the 46 medium to strong earthquakes occurred before 1940 are located on known fault zones. Their isoseismals are usually elongated in form with their major axes parallel to the local surficial faults (see Hsieh, 1958). Thereafter, several destructive earthquakes occurred in this region. Generally speaking, they agreed well with the above-mentioned correlation between seismic activity and geological conditions.

As a reference for the assessment of seismic risk of construction sites, "The Map of Seismicity of China" (Lee, 1957) was compiled in the mid-1950's. The following two principles were applied: in a given region, earthquakes of the same intensity as those which occurred previously may recur; and regions with similar geological conditions can also be regarded as of similar seismicity. Such a map simply reflects the maximum possible intensity of different regions, without any indication of time. In the 1970's, our colleagues compiled a new map of seismic zoning valid for the coming 100 years.

5. Seismological Analysis of Historical Materials

In the *Compilation of Historical Materials of Chinese Earthquakes*, all historical materials were quoted directly from their original records. Thus, seismological analysis is indispensable for it to be used conveniently by seismologists. The conversion of calendars and the identification of localities have been done by paleoastronomers and historical geographers, respectively. The estimation of magnitudes of historical earthquakes with macroseismic data only is much more complicated.

A catalogue of instrumentally determined parameters of earthquakes of the 20th century is included in the *Compilation.* Considering the detecting capability of the networks of seismic stations and the precision of magnitude determination, 4.7 is taken as the lower limit of magnitude. The origin time and the location of earthquake foci were adjusted with computer whenever possible. For the revision of magnitudes, all available amplitude and period data were collected and collated. In order to unify the magnitude scale, Gutenberg and Richter's formulas of 1945 (Gutenberg, 1945) and 1956 (Gutenberg and Richter, 1956) were used as standards for the calculation of surface wave and body wave magnitudes respectively. As a supplement, the 1962 Moscow-Prague formula (Kárník *et al.*, 1962) was used in determining magnitudes from surface waves with shorter periods. Calibration curves of Xujiahui (Zi-ka-wei) seismic station was established. An emperical formula for the estimation of magnitude from maximum distance of registration of P-waves was found. Regression curves of magnitudes listed in various catalogues relative to our revised magnitudes were obtained. On these basis, a new catalogue of Chinese earthquakes from 1900 to 1948 with uniform magnitude scale was compiled. It will be published in volume IV, part 2 of the *Compilation.* In the 1949-1980 catalogue published in volume V of the *Compilation*, the unification of the magnitude scale was not well done, and we are now revising it.

Although affected by local conditions, macroseismic effects of earthquakes basically reflect the intensity of ground shaking. The medium through which a seismic wave propagates influences its dynamic and kinematic properties as well as the attenuation of seismic intensity. To inquire into the regional differences and to give a quantitative description of the attenuation of intensity with distance, we used the isoseismals given in volume V of the *Compilation* to find the intensity attenuation factor "k" of different earthquakes. The regionalization map delineated with "k" as the criterion agrees well with those of intensity attenuation, Q-value, tectonic and crustal structure zonations given by others.

The assessment of seismic intensities is sometimes not so simple as it appears to be. This is especially true for very high intensities. Besides the usual insufficiency of macroseismic phenomena observed in meizoseismic regions, the main difficulty is the indefiniteness of the criteria at very high intensities. At times, the assessment of intensity in the meizoseismic region is not convincing.

Historical materials were written by scholars with various styles of writing and great differences in detail. In most cases, only scattered descriptions of casualties and damages in individual living centers were given. As a matter of fact, vast amount of ancient historical materials simply recorded: "the Earth quaked" in a certain region. Thus, only in rare occasions can isoseismals be drawn.

In recent years, strong earthquakes occurred successively in most of our major seismic regions. Both macroseismic materials and instrumentally determined parameters of these earthquakes were obtained. We are now investigating the emperical relations between magnitude and pattern of isoseists, maximum intensity, felt range, rate of attenuation of intensity, length of surficial faulting, etc. of regions with different intensity attenuation factors, to be used in the estimation of magnitudes of historical earthquakes occurred in the corresponding regions. On this basis, we believe, the magnitude of historical earthquakes can be determined more reliably.

REFERENCES

Academia Sinica, Seismological Committee (1956). *Chronological Table of Chinese Earthquakes*, Science Press, Beijing, 2 volumes (in Chinese).

Central Seismological Group (1971). *Catalogue of Chinese Earthquakes*, Science Press, Beijing, 2 volumes (in Chinese).

Gu, Gongxu (1983). *Catalogue of Chinese Earthquakes*, Science Press, Beijing (in Chinese).

Gutenberg, B. (1945). Amplitudes of surface waves and magnitudes of shallow earthquakes, *Bull. Seism. Soc. Am.*, **35**, 3-12.

Gutenberg, B. and C. F. Richter (1956). Magnitude and energy of earthquakes, *Ann. Geofis.*, **9**, 1-15.

Hsieh (Xie), Y. S. (1957). A new scale of seismic intensity adapted to the conditions in Chinese territories, *Acta Geophysica Sinica*, **6**, 35-47 (in Chinese with English abstract).

Hsieh (Xie), Y. S. (1958). The seismicity and surface faulting of central eastern Yunnan, *Acta Geophysica Sinica*, **7**, 31-40.

Kárník, V., *et al.* (1962). Standardization of the earthquake magnitude scale, *Studia geoph. et geod.*, **6**, 41-48.

Lee, S. P., *et al.* (1957). The map of seismicity of China, *Acta Geophysica Sinica*, **6**, 127-158 (in Chinese with English abstract).

Lee, S. P., *et al.* (1960). *Catalogue of Chinese Earthquakes*, Science Press, Beijing, 2 volumes (in Chinese with English abstract).

Lee, W. H. K., F. T. Wu, and S. C. Wang (1978). A Catalogue of instrumentally determined earthquakes in China (magnitude ≥ 6) compiled from various sources, *Bull. Seism. Soc. Am.*, **68**, 383-398.

Shih, C. L., W. L. Huan, H. L. Tsao, H. Y. Huan, Y. P. Liu, and W. K. Huang (1974). Some characteristics of seismic activity in China, *Acta Geophysica Sinica*, **17**, 1-13 (in Chinese with English abstract).

State Seismological Bureau (1977). *Concise Catalogue of Chinese Earthquakes*, Seismological Press, Beijing.

State Seismological Bureau, Seismic Intensity Regionalization Group (1979). *Collection of Isoseismals of Chinese Earthquakes*, Seismological Press, Beijing.

Xie, Yushou and Meibiao Cai (1983-1987). *Compilation of Historical Materials of Chinese Earthquakes*, Science Press, Beijing, 5 volumes (in Chinese).

IV. Individual Historical Earthquakes

The Anatolian Earthquake of 17 August 1668

N. N. Ambraseys and C. F. Finkel
Dept. of Civil Engineering, Imperial College of Science and Technology
London SW7 1BU, England, UK

1. Introduction

One of the largest earthquakes associated with the North Anatolian fault zone occurred on 17 August 1668, causing heavy damage within a narrow band about 100 km wide and 600 km long, running along the fault zone from Bolu in the west to near Erzincan in the east (see Figure 1). Virtually one-third of the entire known length of the individual fault zone broke during this one shock, an event somewhat analagous to the 1939 Erzincan earthquake of magnitude 7.7 associated with a 350 km fault rupture in the same zone between Amasya and Erzincan. The great length of the epicentral region and the associated fault-breaks provide an illustration of the very large magnitude of the earthquake of 1668, which was preceded by widely-felt damaging foreshocks at its western end, and was followed by aftershocks that continued for six months. This earthquake was the largest last event of a series which affected Anatolia for almost six centuries, and was most probably responsible for the relative quiescence of the fault zone that has ensued for almost two and a half centuries.

Existing earthquake catalogues for Anatolia are very incomplete for the period before about 1930, and of little value for the study of the long-term seismicity of the region. The indiscriminate use of such catalogues often leads to the false impression that prior to the sequence of large earthquakes that began in 1938, the North Anatolian fault zone was, for all practical purposes, quiescent (see, for instance, Toksöz *et al.*, 1979, Figure 2). The same applies to Central Anatolia, to the Border Zone (East Anatolian fault zone) and to its extension into the Dead Sea fault where, although recent seismicity has been very low, large historical earthquakes are known to have occurred with disastrous consequences (Ambraseys and Melville, 1986).

2. Sources of Information

Earthquake catalogues either ignore the Anatolian earthquake of 17 August 1668, or refer vaguely to a series of earthquakes that occurred in different parts of Asia Minor during the period 3 July to 13 September 1668 (Hoff, 1840; Perrey, 1850; Mallet, 1850). Their source of information, the Dressdnische Gelehrte Anzeigen (1756), is an extract from the contemporary Theatrum Europaeum (1668), which appears to have had access to reliable local information, most probably deriving from correspondence with European merchants living in different parts of western Anatolia. Such information is remarkable for its lack of exaggeration. However, these early catalogues fail to identify not only the position of most of the places mentioned in their source as having been affected by the earthquake, but also the exact date of the main event.

With the exception of Stepanian (1942), authors since Mallet add nothing but confusion, which distorts the effects of this earthquake. Thus Sieberg (1932), without quoting his source of information and without any good justification, splits the

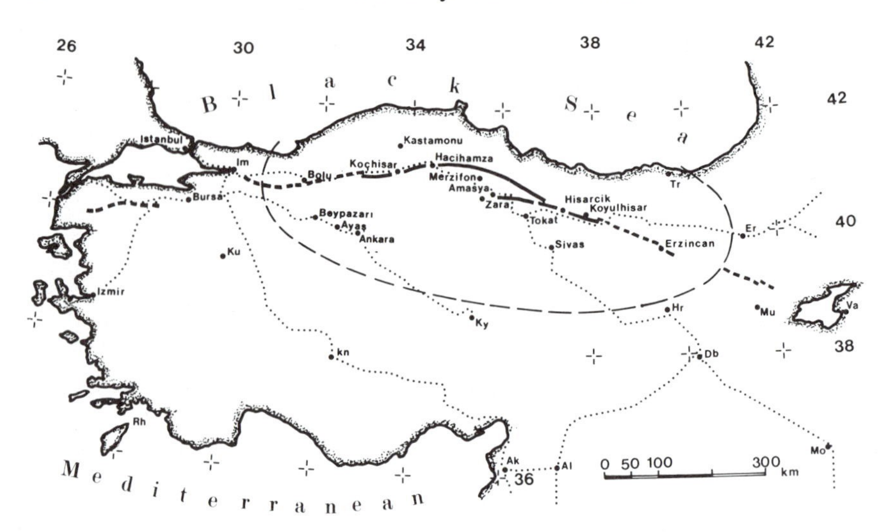

Figure 1: Location map of the earthquake of 17 August 1668. Place names given in full are known or assumed sites affected by the earthquake (see Appendix 2). Heavy broken lines show the location and extent of fault breaks associated with 20th century earthquakes in the North Anatolian fault zone; solid lines show the location and probable extent of faulting associated with the earthquake of 1668. Dotted lines show main caravan routes in the seventeenth and eighteenth centuries (Taeschner, 1926; Taeschner, 1960); main staging posts are shown by two letters:

Tr = Trabzon	Mu = Mus	kn = Konya	Ky = Kayseri
Er = Erzurum	Al = Aleppo	Ku = Kutahya	Va = Van
Hr = Harput	Mo = Mosul	Im = Izmit	Rh = Rhodes
Db = Diyarbakir	Ak = Antakya		

Note the bias toward information from towns of the main caravan routes. Dashed contour shows approximate limits of epicentral area with I = VI (MSK) with r_{vi} = 270 km.

sequence of events into three distinctly different major earthquakes: one in Kastamonu and Bolu on 18 August, another in Konya on 13 September, and he adds a third catastrophic earthquake in Izmir on 10 July 1668, which is, in fact, the earthquake of 1688, one of the many erroneous entries in this catalogue. Later catalogues follow Sieberg, although different works by others are quoted, and the details are not always consistent. Pinar and Lahn (1952) copy Sieberg and Calvi (1941), a secondary source, producing four large earthquakes with epicenters located in different parts of Turkey. Ergin *et al.* (1967) merely copies Pinar and Lahn, adding nothing new, and Soysal *et al.* (1981), more recently, copies all the previous secondary sources including Ambraseys (1975), to produce seven distinct earthquakes with epicenters widely dispersed in Anatolia.

The primary (that is, contemporary or near-contemporary) sources at our disposal, show that there was only one major event during the period 3 July to 13 September 1668, preceded and followed by damaging shocks, the main shock occurring on 17 August, and affecting the region between Bolu and Erzincan. For the limits of the epicentral region to the west, we have letters written by merchants in late August and early September 1668 from Beypazari and Izmir respectively, confirming that the shock felt in these places and in Istanbul occurred during the same period as that damaging Ankara, and overwhelming Tokat (State Papers; see Appendix 1). We have the account of the English cleric, Thomas Smith (1684), who visited the western part of the region affected by the earthquake about a year after the event, and heard details of the shock affecting Ankara and Beypazari from a Scottish merchant living in the latter town. Smith's informant was possibly one of the correspondants of Theatrum Europaeum, and perhaps, also, the author of the letter from Beypazari in August 1668: if so, this would deprive us of an independent source in Smith.

The Armenian minor chroniclers in Hakobyan (1951) are also contemporary sources of information from Tokat and Erzincan, and probably from the area of Sivas, that greatly enhance the picture of the extent of the earthquake, pinpointing the limits of the epicentral region to the east. They distinguish quite clearly between the effects of the August earthquake in Anatolia, and those of the earthquake of 14 January 1668 in the Caucasus (Nikonov, 1982a, b). These sources are supplemented by the near-contemporary Katip Celebi (1734) [although this work is always cited under the name of Katip Celebi, the section which includes the year 1668 was written after his death by Mehmet Seyhi Efendi], and by information given in an early twentieth century history of Amasya (Hüsameddin Hüseyn 1910) [the author tells us (vol. 1, p. 6) that his history is based in part on local chronicles, court records, and other contemporary sources].

3. The Anatolian Earthquake of 17 August 1668

Early in July 1668, shocks began to be felt in Ankara that continued intermittently until 20 July, causing considerable concern among the inhabitants of the area. After that date their intensity began to increase, when on 12 August, between 3 and 4 p.m. a violent shock, that was felt as far as Istanbul and Izmir, caused the collapse of several houses and some hundreds of chimneys in Beypazari, killing seven people. The principal minaret of the town also fell. There is no evidence that other regions of Anatolia were affected by this shock, which, however, was followed by intermittent tremors, some of which were felt in both Ankara and Beypazari, and forced the people to camp in open spaces in the towns or in the countryside (Theatrum, 1668; Dressdnische, 1756; State Papers; Smith, 1684).

These tremors recurred with increasing intensity until 15 August, when at 3 p.m. a strong earthquake in Ankara not only destroyed masonry walls, houses, stables, and part of the town wall, but also shattered the castle on the hill above the town. Only two people were caught in the town and buried, all those who had escaped to the open being saved (Theatrum, 1668; Dressdnische, 1756). At Beypazari this shock added to the damage already caused by the previous earthquakes, and elsewhere it caused great concern.

On 17 August, there followed another violent earthquake, stronger than any that had gone before, which affected a very large part of Anatolia. In Ankara the earthquake consisted of a series of powerful jolts occurring at intervals of 3 to 4 minutes, as a result of which the cliffs of the hill above the town began to break up, hurling down huge boulders weighing up to a tonne. In the town, the ground opened up in places and many houses, already damaged by the foreshocks, were ruined. Those few people still in Ankara fled, leaving behind only the garrison at the fort (Theatrum, 1668; Dressdnische, 1756).

However, with the exception of a village between Beypazari and Ayas which was all but wiped out, damage was not so great here or in Bursa as in their adjacent districts to the north, where destruction was almost complete. The town of Bolu was almost totally destroyed, with 1,800 people losing their lives, among this number being about 60 christian merchants. The effect of the earthquake was very severe to the east in Hacihamza, most of which was razed to the ground, with only a small fort remaining standing. In Amasya and its neighboring villages the earthquake caused great destruction: the ground was "cleaved", an allusion to faulting. The walls of the castle, stone dwellings, and religious and commercial buildings, including many parts of the *bedesten* (central secure part of bazaar), were demolished with much loss of life. The mosque and *medrese* (theological school) in the Kuba district of the town were ruined. The Sultan Bayazid mosque, completed in 1486, sustained severe damage, with an arch and the domes collapsing. One of its minarets was sheared off and rotated, while part of the other fell. The mosque and *mescid* (small mosque) of Emir 'Imadüddin Sulu Bey were totally destroyed, as well as the gate of the Square, and the domes of many other mosques in the town. Continuing shocks kept the survivors camping in the open for many days. In the region of Amasya, at Zara, and at Kochisar, much of the destruction was brought about by large scale ground deformations, the description of which implies extensive faulting (Theatrum, 1668; Dressdnische, 1756; Stepanian, 1942; Katip Celebi, 1734; Hüsameddin, 1910).

The town of Tokat was badly damaged, the town walls and the castle being demolished. Many mosques in the town, including those of Bahadir and Avlik(?), and many other public buildings, with the exception of the Armenian church, collapsed. Houses were destroyed in the town and in 30 settlements in the Tokat district, but casualties were not excessive here. Tremors recurred here for many weeks, causing the people to abandon their homes for open country. Merzifon, Hisarcik (mod. Asarcik) and Koyulhisar were also ruined, and in the latter places caravanserais collapsed. There were reports of 6,000 deaths, among which were some 50 Persian and Anatolian merchants (State Papers; Hakobyan, 1951; Stepanian, 1942; Theatrum, 1668; Dressdnische, 1756).

Damage extended to Erzincan, where a number of houses collapsed, and to Kastamonu where many buildings were destroyed causing many casualties, although the majority of these were cases of heavy injury. Other towns which we could not identify (see Appendix 2) were laid waste, such as Coujam, Listrien, Derben, and sites along the Black Sea coast. Between Stammas and Marannoy a stable of horses and mules collapsed. To the south, most probably near the Armenian monastery of Nishan, in the region of Sivas, a town and several villages were engulfed (Hakobyan, 1951; Theatrum, 1668; Dressdnische, 1756).

The earthquake was strong enough to be felt in Izmir, Istanbul, and possibly in Alexandria, but it is not mentioned in the accounts of Europeans who were in Thrace, Larissa and Jerusalem at the time. By late August there had been more than 200 shocks felt in Ankara in the preceding 37 days, while foreshocks and aftershocks lasted for a total of 47 days in that town, and for 15 days at Beypazari, until they ceased being reported after 13 September 1668. In Tokat, aftershocks continued for six months altogether. After the earthquake it took a year to rebuild some of the public buildings, such as those in Amasya (al-'Umari, 1796; Colier, 1671; Simopoulos, 1976; Troilo, 1676; State Papers; Smith, 1684; Theatrum, 1668; Dressdnische, 1756; Hakobyan, 1951; Hüsameddin, 1910).

4. Discussion

Although it is impossible at this stage to assess the effects of this earthquake on the towns of Anatolia in greater detail, it is quite clear that the event may have been comparable to, if not larger than, the 1939 earthquake in the extent of the area affected and the number and duration of foreshocks and aftershocks experienced.

From Figure 1, we notice that part of the area affected was on the main caravan route through northern Anatolia (Taeschner, 1926; Taeschner, 1960), which for much of its length ran along the North Anatolian fault zone, extending from Bolu in the west to past Koyulhisar, to Erzincan in the east, a distance of 680 km, as is suggested not only by the considerable damage to Bolu in the west and to a lesser extent to Erzincan in the east, but also by the ground deformation reported from the regions of Kochisar, Zara and Amasya that coincide with more recent fault breaks (Ketin, 1969). It would appear that the area affected extended into the mountainous rural area of Sivas and further to the east toward Erzincan, as well as to the north along the Black Sea coast.

The surface-wave magnitude of this earthquake may be assessed from the magnitude fault length relationships for earthquakes in the Middle East (Ambraseys, 1986). Assuming a 400 km long rupture and average dislocation of the order 10^{-5} of the rupture length, we find that $M_S = 8.0$. Alternatively we may use the average radius of the isoseismal VI (MSK) shown in Figure 1, i.e. $r_{vi} = 270$ km, and Equation (1) in Ambraseys and Melville (1986). This would give $M_S = 7.8$. Thus, as a first approximation, the magnitude of the 1668 North Anatolian earthquake should have been close to 8.0. The casualties, in excess of 8,000, would have been much greater had the people not been warned by the protracted foreshock period. It is rather strange that this apparently major event seems not to be mentioned by many contemporary Ottoman sources for this period, which is otherwise relatively well-documented. The search for additional details of this event goes on.

ACKNOWLEDGEMENTS

This research project is supported by a Natural Environment Research Council grant for the study of the historical seismicity of the eastern Mediterranean region.

REFERENCES

Ambraseys, N. N. (1975). Studies in historical seismicity and tectonics, Geodynamics Today, Royal Soc. Publ., London.

Ambraseys, N. N. (1986). Magnitude-fault length relationships for earthquakes in the Middle East, this volume.

Ambraseys, N. N. and C. P. Melville (1986). An analysis of the eastern Mediterranean earthquake of 20 May 1202, this volume.

d'Aramon, see Chesneau below.

Burnaby, F. (1877). On horseback through Asia Minor, v. 2, 307 pp., London.

Calvi, V. S. (1941). Erdbebenkatalog der Türkei and einiger Benachbarter Gebiete, Report n.276 of M.T.A. Enstitüsü (unpub.), Ankara. (not seen)

Chesneau, Jean (1887). Le voyage de Monsieur d'Aramon ambassadeur pour le roy en Levant, Recueil de Voyages et de Documents, Histoire de la Geographie, v. 8, pp. 64, 65, 71, Annot. Ch. Schefer. Paris.

Colier, Justin (1671). Journal de Monsieur Colier, Resident à la Porte ... etc., Translated from Flemish by V. Minutoli, Geneva.

Curzon, G. (1966). Persia and the Persian question, v. 1, p. 246, London.

Dressdnische Gelehrte Anzeigen (1756). Nachrichten von Erdbeben, 189-191, Dresden.

Ergin, K., U. Güclü, and Z. Uz (1967). Türkiye ve Civarinin Deprem Kataloğu (Milattan sonra 11 yilindan 1964 sonuna kadar), Publication no. 24 of I.T.U. Maden Fakültesi Arz Fiziği Enstitüsü, Istanbul.

Hakobyan, V. (1951-1956). Manr Zhamanakagrut'yunner XIII-XVIII DD (Minor Chronicles), v. 1, p. 157, 162; v. 2, p. 519, Erivan.

Hoff, K. von (1840). Chronik der Erdbeben und Vulkan-Ausbrüche, ... etc., Gesch. Ueberlieferung nachgew. natürl. Veränder. Erdoberfläche, w. 2, 3, 4, Gotha.

Howel, Th. (1791). A journal of a passage from India, p. 115, London.

Hüsameddin Hüseyn (1910-1928). Amasya Tarihi, v. 1, 79, 130-31, 156, 158; v. 4, p. 152, Istanbul.

Katip Celebi (1734). Takvim üt-Tevarih, Istanbul.

Ketin, I. (1969). Ueber die nordanatolische Horizontal-verschiebung, *Bull. Miner. Res. Explor. Inst. Turkey*, **72**, 1-28, Ankara.

Mallet, R. (1850-1858). Report on the facts of earthquake phaenomena, Rep. British Assoc. Adv. Sci., London.

Newberie, John (1625-1626). Two Voyages of Master John Newberie ... etc., Hakluytus Posthumus or Purchas his Pilgrimes ... etc., v. 2, p. 1418, London.

Nikonov, A. (1982a). Sil'neishee zemletriasenie bal'shogo Kavkaza 14 ianvar 1668, *Fizika Zemli*, **9**, 90-106.

Nikonov, A. (1982b). O zemletriasenii 1668 g na vostochnom Kavkaze, *Fizika Zemli*, **9**, 123-127.

Otter, Jean (1747). Voyage en Turquie et en Perse; ... etc., v. 2, p. 348, Paris.

Perrey, A. (1850). Sur les tremblements de terre ressentis dans la Peninsular Turco-Hellenique et en Syrie, *Mém. Acad. Roy. Sci. Belgique*, **23**, pt. 1, 1-75, Bruxelles.

Pinar, N., Lahn, E. (1952). Türkiye Depremleri izahli Katalogu, Publication series 6, n. 36 of Bayindirlik Bakanliği Yapi ve Imar Isleri.

Poullet, Sieur du (1668). Nouvelles Relations du Levant ... deuxième partie, ... etc., v. 2, p. 33, Paris.

Schefer, C. (1887). p. 65, note 2. See Chesneau, Jean above.

Sieberg, A. (1932). Erdbebengeographie, in B. Gutenberg's *Handbuch der Geophysik*, v. 4, Berlin. (sub-annum.)

Simopoulos, K. (1976). Xenoi taxidiotes stin Hellada 333 mCh - 1700, v. 1, 580-634, Athens.

Smith, Thomas (1684). An Account of the city of Prusa in Bithynia, ... etc., *Philosophical Transactions*, 14, n. 155, 442-443, London.

Soysal, H., S. Sipahioğlu, D. Kolcak, and Y. Altinok (1981). Türkiye ve cevresinin tarihsel deprem kataloğu, Project n.TBAG 341 of TÜBITAK, Istanbul.

State Papers, SP 97/19 fol. 34r, Public Record Office, London.

Stepanian, V. (1942). Istoricheskii obzor zemletriaseniiakh v Armenii i v prilegaiushchikh raionakh. Zakavkazskaia Konfer. Antiseism. Stroyte. Izdatel. Arm. F.A.N., Biuro Antiseismicheskogo Stroitelstva, v. 1, p. 66, Erivan.

Taeschner, F. (1926). Die Verkehrslage und das Wegenetz Anatoliens im Wandel der Zeiten, in Dr. A. Petermanns Mitteilungen, v. 72, pp. 202-206, Gotha.

Taeschner, F. (1924-1926). Das anatolische Wegenetz nach osmanischen Quellen, 2 volumes, Leipzig.

Taeschner, F. (1960). article "Anadolu", *Encyclopaedia of Islam* (second edition), v. 1, pp. 474-79 plus map, Leiden and London.

Tavernier, Jean Baptiste (1681). Les six voyages de Jean Baptiste Tavernier en Turquie, en Perse et aux Indes, ... etc., v. 1, pp. 7, 71, Paris.

Theatrum Europaeum (1617-1721). v. 10/2 pp. 973-974, Ed. J. B. Abelin, Frankfurt.

Toksöz, N., A. Shakal, and A. Michael (1979). Space-time migration of earthquakes along the North Anatolian Fault Zone and seismic gaps, *Pageoph.*, **117**, 1258-1270.

Tournefort, Joseph Pitton de (1717). Relation d'un Voyage du Levant fait par ordre du Roy, ... etc., v. 2, p. 434, Paris.

Troilo, Franz Ferdinand von (1676).... orientalische Reise-Beschreibung ... etc., pp. 577-578, Dresden.

al-'Umari, Yasin (1796). Al-athar al-jaliya fi 'l-hawadith al-ardiya, fol. 216 v., Manuscript, Iraq Academy, Baghdad.

APPENDIX 1

A caravan cannot be trusted to do more than about 40 km a day on average. However, post riding by day alone and resting at night can accomplish 190 km between dawn and leaving the saddle. Chapar-riding in Persia and eastern Turkey does 100 km a day quite comfortably (see, for instance, Curzon, 1966). Thus, for the news of the destruction of Tokat to reach Izmir, a distance of 1050 km on the caravan route via Bolu and Bursa, it should have taken an average of 26 days of caravan travel depending on the season of the year. Poullet covered this distance in 21 days in 1658, and Tavernier gives 35 days in 1632. Tournefort noted in 1701, that a camel caravan would take 40 days to cover the distance from Tokat to Izmir by the direct route without passing through Ankara or Bursa (which gives a slow daily average of under 30 km per day), but that a mule caravan would take only 27 days. A post courier would have taken only a week to 10 days to cover the same distance with ease. News of the event could thus have reached Izmir well within the eighteen day period between the earthquake of 17 August and 4 September, the date on which news of the event was sent to Istanbul (Poullet, 1668; Tavernier, 1681; Tournefort, 1717).

APPENDIX 2

The names of some of the towns and villages mentioned in Theatrum Europaeum and Dressdnische Gelehrte Anzeigen are so corrupted in their translation into German or Armenian transcription that their position is very difficult to determine at the present day. According to the order in which they are mentioned in the reports in these two sources, we have arrived at the following locations, aided by Taeschner's concordance of Anatolian place-names (Taeschner, 1924).

ANGORI is clearly Ankara, as BOLLE is Bolu, CASTOMME is Kastamonu, BEYZAZAR is Beypazari, and AMIAS is Ayas. STAMMAS suggests Amasya, derived probably from the Greek Istamasia. The final syllable of MARANNOY appears to be a corruption of the Turkish -köy for village, but this place remains unidentified. The English traveller Newberie (Newberie, 1625) passed through SARDAEL in 1582, transcribing it as Searradella, which is the modern Zara, and should not be confused with a larger place of the same name to the northeast of Sivas. CESARIA/COSARIA is modern Ilgaz, known in earlier times as Kochisar. It has been variously transcribed in European travellers' accounts as Cosizar (Tavernier, 1681), Cagiassar (d'Aramon), Cojaste (Howel, 1791) and Kodjehisar (Otter, 1747). The locations of COUJAM and LISTRIEN remain in doubt, as does that of DERBEN, a common place name in its Turkish form of Derbend, indicating a pass or defile. ISARNO may be Asarcik, formerly Hisarcik, which lies to the northwest of Resadiye, a rather tenuous identification: d'Aramon knew this place as Assarguict. CACHETTE, may be tentatively equated with Koyulhisar, d'Aramon's Coyouassar: the caravanserai destroyed here may well have been the Haci Murad Han, lying a little distance from the town itself. NABUZZIO, described as lying in a valley between two mountains, may be identified as Haci Hamza on the Kizilirmak River. d'Aramon referred to this town as Cabouziac, and Schefer notes that it took its name from Haci Hamza (Schefer, 1887).

The rather imperfect sense of geography at this time, witnessed, for example, in the statement that Nabuzzio and Cachette were a mere day's journey apart, is further shown in the belief that the final place mentioned as having suffered destruction, the residence of an Armenian patriarch, lay in the neighborhood of the Taurus mountains. This place, whose name is not given, may, in reality, be identifed as near the 500-year old Monastery of the Cross, lying 2 miles for Sivas, the home of the Armenian Bishop (Burnaby, 1877).

An Analysis of the Eastern Mediterranean Earthquake of 20 May 1202

N. N. Ambraseys
Dept. of Civil Engineering, Imperial College of Science and Technology
London SW7 2BU, England, UK

C. P. Melville
Faculty of Oriental Studies, University of Cambridge
Cambridge CB3 9DA, England, UK

1. Introduction

A large earthquake was widely felt in the Middle East around daybreak on the morning of 20 May 1202. Contemporary sources indicate that the shock was felt from Lesser Armenia, parts of Anatolia and northwest Iran down to Qus in upper Egypt, and from Sicily in the west to Iraq and Mesopotamia in the east, i.e. within an area of average radius about 1,200 km (Figure 1). Along with this very large felt area, an ensuing seismic sea-wave and aftershock sequence, the earthquake was associated with extensive and serious damage in Syria and to a lesser extent in Cyprus, with great loss of life. Comparison with 20th century earthquakes in the eastern Mediterranean suggests that this was a shallow, large magnitude ($M_S = 7.6$) event.

We have chosen to study this event not only to try to resolve the problems connected with it, and for its large size, but also because it occurred in what appears today to be a seismically quiescent region in a densely populous part of the Middle East. Reflecting some uncertainties contained in contemporary accounts of the earthquake, particularly regarding its date, this event is found in different guises in various catalogues, often appearing more than once under different entries. A thorough review of primary sources is necessary to clarify the effects and scale of the shock. More recent earthquakes in the area also need to be re-examined before they can be used to calibrate this and other early events in the Middle East.

The analysis that follows provides a detailed account of our investigations, rather than a mere summary of the findings. It is important that all the direct and circumstantial evidence that combine to build a picture of an early earthquake should be presented and discussed.

2. Analysis of Historical Earthquake Data

Our procedure for the study of the historical seismicity of Iran, the Middle East and certain regions in Europe and Africa, has been to collect macroseismic data from original sources and review them gradually as they accumulate. Likely sources of information include local histories and documentary material, general chronicles, diaries, private correspondence and travel narratives. A considerable amount of such material is available in libraries throughout Europe, including official archives of diplomatic and political correspondence. European and local newspapers and special studies of past earthquakes also provide useful macroseismic data (see for instance Ambraseys and Melville, 1982, for Iran and 1983, for the Yemen; Melville, 1983, for the UK and 1984, for the Northern Red Sea; Ambraseys, 1985a, for Scandinavia).

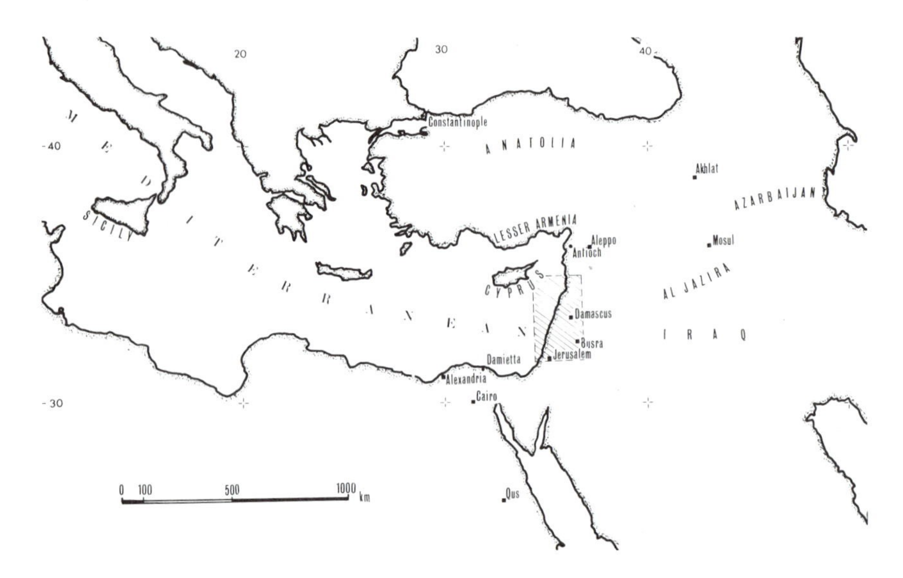

Figure 1. Map showing the area over which the earthquake of 20 May 1202 was felt. Ceuta is situated on the Moroccan coast opposite Gibraltar, 3,800 km from the epicentral region. The shaded inset shows the location of Figure 2.

The next stage is the analysis of all the data retrieved, including both the macroseismic and instrumental data recovered for 20th century events. Following the procedures put forward by Ambraseys, *et al.* (1983), this requires the unification of instrumental data in terms of location and magnitude, which will provide a consistent quantitative yardstick for the assessment of long-term seismicity. Examples of this kind of reduction of macroseismic data into a uniform body of information are found in Melville (1985) and Ambraseys and Jackson (1985) for the UK, Ambraseys (1985b) for Northwest Europe, and Ambraseys and Melville (1982) for Iran, where extensive field work was necessary to resolve ambiguities in the epicentral areas of important historical earthquakes and to delineate local tectonics. Similar procedures have been followed by Muir Wood *et al.* (1987), Vogt (1984a,b) and in particular in Italy (ENEA-ENEL, 1981; Postpichl, 1985). The work of these groups is distinguished by the fact that, regardless of the details of the evaluation procedures adopted, the basic data have been established from primary sources of information. Some general points may be made arising from previous experience of this work.

2.1. Location

In addition to early 20th century events for which only felt information exists, there are a considerable number of early teleseismic locations of offshore events and earthquakes with epicenters in sparsely populated parts of the Middle East, Europe and Africa, which can be examined. Up to 1915, these are recorded mainly on Milne seismographs, with readings reported in the Shide Circulars (no. 1-27) or early issues of the International Seismological Summary (ISS). These instruments do not record body waves well and the best results are obtained by using the time of the reported maximum phase, presumed to be a surface wave travelling with a velocity of about 2.9 to 3.1 km/sec. Simple graphical location methods have been used. For some later events, readings of P- and S-phases from other instruments are also used to support those of phase maxima. Such relocations are not accurate by modern standards, but an earthquake's epicenter may usually be estimated within a few degrees and felt information may be found to help resolve ambiguities. When the existence of such an event is established in a certain region and teleseismic amplitudes and periods are available, a magnitude can be estimated even if the position is unreliable to within several degrees, since calibration functions do not change rapidly with distance in the teleseismic range (Ambraseys and Adams, 1986).

2.2. Magnitudes

For earthquakes of the period 1898 to 1914, surface-wave magnitudes may be estimated using trace maxima recorded by undamped Milne seismograms. This method has been applied to a number of geographically restricted regions in the Middle East and North Africa with excellent results, along much the same lines as discussed by Abe (1987). For the majority of early earthquakes in Europe and the Middle East, data from the more advanced instruments are available from 1903, so that the need to resort to the undamped Milne trace amplitudes for magnitude determination is minimal.

Magnitudes are often given without indications as to whether they are local (M_L), surface (M_S) or body wave (m) magnitudes and the indiscriminate amalgamation of such values from existing catalogues leads to a body of non-homogeneous data. Similarly, the use of empirical correlations to convert m or M_L into M_S and vice versa is quite unacceptable in areas such as the Middle East and Southeast Europe, where the seismotectonic characteristics of the region vary so rapidly and where the values of m, M_L and M_S reported by local stations are based on procedures that have changed many times during the last few decades.

For events of the pre-instrumental period, surface wave magnitudes may be estimated quite consistently using a set of macroseismic observations such as the radii of the isoseismals of a given intensity grade (see for example Equation (1) below). The use of uniformly processed macroseismic data may also lead to an estimate of focal depth (Ambraseys, 1985b). In practice, however, there are often considerable problems in assessing intensities and felt areas from historical information, particularly in areas of low seismicity or low population density and also the further back in time one goes, since the available documentary sources become both fewer and more laconic.

2.3. Intensities

Whatever the limitations imposed on the type of data available by the particular cultural characteristics of a given age or area, intensities should be assessed uniformly by one or more observers, either on a scale such as the MSK scale or using a simpler one designed for early events in a specific region. Conversion from one intensity scale to another, using empirical formulae, has been found to lead to unacceptably large errors and should be avoided.

Many of these issues are relevant to the identification and analysis of a large earthquake that occurred in the Eastern Mediterranean on 20 May 1202.

3. Previous Catalogues

Early catalogues either ignore the 1202 earthquake (Coronelli, 1693; Seyfart, 1756; Dressdnischer, 1756; Berryat, 1761; Huot, 1837) or mention an earthquake on 30 May 1202 (perhaps converting to the New Style calendar), which caused great damage to the Christian-held territory of Tyre, Tripoli, 'Arqa and Acre (Manetti, 1457; Batman, 1581; Beuther, 1601). None of these authors quote their source of information.

Among later authors, Hoff (1840) quotes early editions of Muslim writers (Abu 'l-Faraj, Abu 'l-Fida and Hajji Khalifa), all secondary sources, but separates the events into two distinct earthquakes in 1201 and 1204. Hoff is followed by Mallet (1852, p. 30), who notes however that non-Arab authors have 13th, 20th or 30th May 1202, probably referring to the sources cited by Perrey (1850, p. 18). Perrey quotes Baronius's *Annalium Ecclesiasticorum*, where an account similar to that of Robert of Auxerre's, but dated 30 May 1202, is found (ed. O. Raynaldo, Rome 1646, xiii, 89); also Ralph of Coggeshall and Robert of Auxerre's accounts (ed. Dom Bouquet, *Rec. des Hist. de Gaule et de la France*, xviii, 97, 265-6), cf. below. The chronicle published in L. Muratori, *Rerum Italicarum Scriptores*, XXVI, 85 is probably Sozomenus's 15th century universal history. The chronicle published in Luc d'Achery, *Spicilegium*, xi, 478 has not yet been examined. One or other of these latter two must give the date 13 May 1202. This results in three earthquakes, occurring in 1201 or 1202, 20 May 1202 (Old Style) and 1204, which affected parts of Syria, Mesopotamia and Egypt, though Perrey notes that the first two are probably the same.

Authors since Mallet add nothing but further chronological confusion, which in turn obscures the size and effects of the earthquake. Willis (1928), on the authority of secondary works (Tholozan, 1879; Arvanitakis, 1903; Blankenhorn, 1905; Vigouroux, 1912) increases the number of earthquakes in the period to four, i.e. in 1201, 1202, 1203, and 1204, and adds another two shocks in 597 and 600 A.D., which are in fact the Muslim years covering the period 1201-1204 that he misinterprets from Sprenger (1843), cf. Ambraseys (1962). Ambraseys' own early work (1961), which is an uncritical translation of one of the many manuscripts of al-Suyuti's treatise, passes on the errors in the text. Ambraseys' studies on the seismicity of Iran and adjacent regions (1968, 1974) are very incomplete and occasionally misleading, but likewise continue to be used uncritically. Willis is followed by later authors, such as Sieberg (1932a,b), Kallner-Amiran (1951), Plassard and Kagoj (1968) and others. Some produce two major earthquakes out of the 20 May 1202 event (Ben-Menahem, 1979), others five (Alsinawi and Ghalib, 1975), with an average of three destructive shocks given by Poirier and Taher (1980), the only 20th century authors who have the credit of using an extensive number of Arabic sources.

4. Sources of Information.

Owing to the Crusader presence in the Levant, information on the effects of the earthquake is available from both Christian and Muslim authors. Both sets of data naturally refer most particularly to the territory belonging to the respective sides, but to a large degree they complement each other. It is clear that most of the chronological confusion surrounding the event is caused by the uncritical use of Muslim chronicles. It is also remarkable that almost no use has been made of western sources, which are far more accessible to most European authors and unambiguously resolve the dating of the earthquake. These works, though largely ignored by earthquake cataloguers are of course well known to the historians of the Crusades (e.g. Röhricht, 1898).

The political context of the earthquake is briefly outlined in Mayer (1972, 1984) and more fully in Cahen (1940), Runciman (1971) and Setton (1969), where detailed reference is made to the narrative sources available. The Crusader states had been greatly reduced by Saladin's campaign of 1187 and only partially reconstituted by the Third Crusade. Regarding the non-Muslim accounts, it is unfortunate that the main political and military developments at this time were not taking place in the Levant at all, but lay in the preparations for the ill-fated Fourth Crusade. The focus is not therefore so clearly on events in the east, where the Crusader states were on the defensive and greatly reduced in their sphere of operations. Most of the relatively few places retained by the Christians are mentioned in European accounts, all in the truncated kingdom of Jerusalem and the county of Tripoli, on or near the coastal strip. No details are given in Christian sources of wider effects in the Syrian hinterland. Similarly, no details are given of the shock further north, in the principality of Antioch, beyond the indications that it was not so severe there.

The two letters from the Hospitaller and Templar Grand Masters published in Mayer (1972) contain the fullest occidental accounts and refer particularly to the possessions of their respective Orders. Very few additional details are found in other sources (among them the references to Jubail in various texts of the *Annales de Terre Sainte*). As demonstrated by Mayer, the near contemporary account of Robert of Auxerre (d. 1212) has many points of similarity with Philip du Plessis' description. Variations of date occur in the Christian sources, but not concerning the year: William of Nangis (d.ca 1300) has 30 May, three days before Ascension (which was in fact on 23 May in 1202); Felix Fabri (fl. 1480) has 30 March. The Barletta manuscript (Kohler, 1901, p. 401) appears to read 3 March. Most of these sources are telegraphic, containing only general summary information.

Arabic sources from the Muslim areas surrounding the Christian states naturally present a wider view and provide the most information. Just as both the contemporary European letters date the earthquake Monday, 20 May 1202, so too do a comparable pair of Arabic letters from Hamah and Damascus. These were received by 'Abd al-Latif b. al-Labbad al-Baghdadi, who was in Cairo at the time of the earthquake and wrote his account in Ramadan 600/May 1204, two years after the event. Both he and the letters he transcribes give the date as early on the morning of Monday 26 Sha'ban 598 hijri [Muslim calendar] (= 21 May 1202, which was a Tuesday) or 25 Pashon [Coptic calendar] (= Monday 20 May). A discrepancy of one day is common in converting the Muslim calendar. As noted above, the latter date is confirmed by the contemporary European accounts. Abu Shama, quoting the testimony of al-'Izz Muhammad b. Taj al-umana' (d. 643/1245), also has Monday, 26 Sha'ban 598 or 20 Ab [Syriac calendar] (= August (sic.) 1202).

There can thus be no doubt that the correct Muslim year is 598 H., which runs from 1 October 1201 to 19 September 1202. Unfortunately other later Arabic texts contain variations in the date of the earthquake and in cases split its effects into accounts of separate events in different years. The most influential of these alternative texts is that of Ibn al-Athir of Mosul (d. 1233), who has a general account of the earthquake felt throughout Mesopotamia and in Egypt, Syria and elsewhere, dated Sha'ban 597 H., which is a year early. He is clearly referring to the same event. His account is followed almost verbatim in the Syriac Chronicle of Bar Hebraeus [Abu 'l-Faraj] (d. 1286), and in greatly abbreviated form by Abu 'l-Fida (d. 1331), under 597 H. Another early source, Abu 'l-Fada'il of Hamah (ca. 1233) has a brief notice of the shock under 597 H. It is of interest that he does not refer to the shock in Hamah, but mentions that it destroyed most of the towns belonging to the "Franks". Reconciling these accounts is no problem, simply an error of one year has occurred.

A greater problem is introduced when Ibn al-Athir has another, shorter but similar, account of the (same) earthquake under the year 600 H. (10 September 1203 - 28 August 1204), without specifying the month. He says the shock destroyed the walls of Tyre and also affected Sicily and Cyprus. This "second" earthquake is once more reported by Bar Hebraeus and Abu 'l-Fida. A similar account, but adding new information that the shock was felt in Sabta (Ceuta), is given by Ibn Wasil (d. 1298). Since Ibn Wasil was a native of Hamah, it is surprising that he does not have independent local information, also that he does not have any reference to the shock under 597 or 598 H.

It is not clear why Ibn al-Athir should duplicate his account under the dates 597 and 600 H., but it is perhaps sufficient to note that this sort of duplication is not uncommon in both European and Islamic medieval chronicles. Within this repetition, there may be some echo of a strong aftershock or a prolonged period of seismic activity. News from Sicily and Cyprus clearly took longer to arrive than information from Syria.

Two separate notices are also found in the chronicle of Sibt b. al-Jauzi (d. 1256), this time under 597 and 598 H. The first account, under Sha'ban 597 H., echoes that of 'Abd al-Latif, while mentioning a few additional places. The date, however, is the one given by Ibn al-Athir. Sibt b. al-Jauzi supports this date by saying (p. 480) that <u>after</u> these earthquakes in 597/1201, died both 'Imad al-Din [the historian whose work he had earlier quoted for an account of the famine in Egypt that year] and the author's own grandfather [the historian Ibn al-Jauzi]. It is generally accepted that both men did indeed die in 597/1201 and thus <u>before</u> the earthquake. This is awkward to explain, but the author is probably trying to rationalize two conflicting pieces of chronological data. He is not so much dating the deaths by reference to the earthquake, as accommodating the false date that he has accepted for the earthquake within the sequence of other events that year. Under the correct year, 598 H., he has a much briefer account, describing damage to the castles at Hims and Hisn al-akrad. He says the shock extended to Cyprus and destroyed what was left of Nablus (i.e. after the first earthquake). This implies two shocks. On the other hand, Sibt b. al-Jauzi's second account is not unlike Ibn al-Athir's second account (under 600 H.), and may again simply be an attempt to accommodate the conflicting dates. It is significant that Sibt b. al-Jauzi has no report of an earthquake under 600 H.

Abu Shama, who quotes Sibt b. al-Jauzi's accounts under 597 and 598 H. in turn (pp. 20, 29), in both cases cites the additional testimony of al-'Izz b. Taj al-umana', a descendant of Ibn 'Asakir and continuator of the latter's Biographical history of Damascus (Cahen, 1940). It is clear that the first part of Sibt b. al-Jauzi's 597 H. account also follows al-'Izz. Under 598 H., al-'Izz records the effect of the shock in north Syria and in Damascus, with some minor details additional to those provided by 'Abd al-Latif.

Al-Suyuti summarizes the dating confusion found in his sources, by entering the earthquake under 597 H. (quoting al-Dhahabi, *'Ibar* and Sibt b. al-Jauzi); 598 H. (quoting Sibt b. al-Jauzi) and 600 H. (citing Ibn al-Athir). Later sources add no details. It is worth noting that the Aleppo author, Ibn al-'Adim (d. 1262), makes no reference to the earthquake under any of the years found elsewhere.

Despite the conspicuous duality of accounts in almost all Muslim sources, probably reflecting protracted aftershock activity, there remains no evidence of more than one principal earthquake. Apart from the silence of contemporary occidental and oriental authors, 'Abd al-Latif was in a position to record separate earthquakes in both 597 and 600 H. had they occurred. The amalgamation of these several accounts therefore removes much of the mystery surrounding the 598/1202 earthquake, and allows a coherent identification of its effects and felt area.

5. The Earthquake of 20 May 1202

Many sources speak of strong effects and significant damage along the Mediterranean littoral of Syria, affecting both the "Franks" and "Saracens" (Abu 'l-Fada'il, fol. 113a-b; Hethum Gor'igos, p. 480, Ibn al-Furat, p. 240). Specifically, both Acre and Tyre, the two main Christian centres, were severely damaged, with heavy loss of life (Figure 2). Contemporary letters (Mayer, 1972) speak of damage to walls and towers in both cities, including the palace at Acre. The house of the Templars in Acre (in the Southwest of the city, see Enlart, 1928, p. 23) was however fortunately spared. All but three towers and some outlying fortifications were destroyed in Tyre, along with churches and many houses. The English chronicler, Ralph of Coggeshall (d. 1228) says most of Tyre and one third of Acre were overthrown (p. 141-2). Muslim sources largely confirm this, 'Abd al-Latif stating that the greatest part of Acre and one third of Tyre were destroyed. Intensities in Tyre may be assessed higher than those in Acre, respectively around IX and VIII (see Table 1). Funds were made available for both cities to be reconstructed (*L'Estoire d'Eracle*, p. 245; Sanuto, p. 203), though no specific indication is available of the extent of these repairs (Enlart, p. 4, Deschamps, 1939, p. 137).

Inland from the Christian territories, in Shamrin (Samaria) and Hauran, damage was equally severe. It was reported that Safad was partially destroyed, with the loss of all but the son of the garrison commander; also Hunin (Chastel Neuf), Baniyas (Paneas) and Tibnin (Toron). At Bait Jann (Bedegene), not even the foundations of walls remained standing, everything having been "swallowed up". Two possibilities present themselves for the identification of Bait Jann out of the three noted by de Sacy in 'Abd al-Latif, p. 446, both being known to the Crusaders (see Dussaud, 1927, pp. 7, 391). The first is 10 km west of Safad and the second on the road between Damascus and Baniyas, see Ibn Jubair, p. 300, who described it as situated in between the mountains. The context in which Bait Jann is mentioned by 'Abd al-Latif allows either alternative to be acceptable, but the second is preferred here

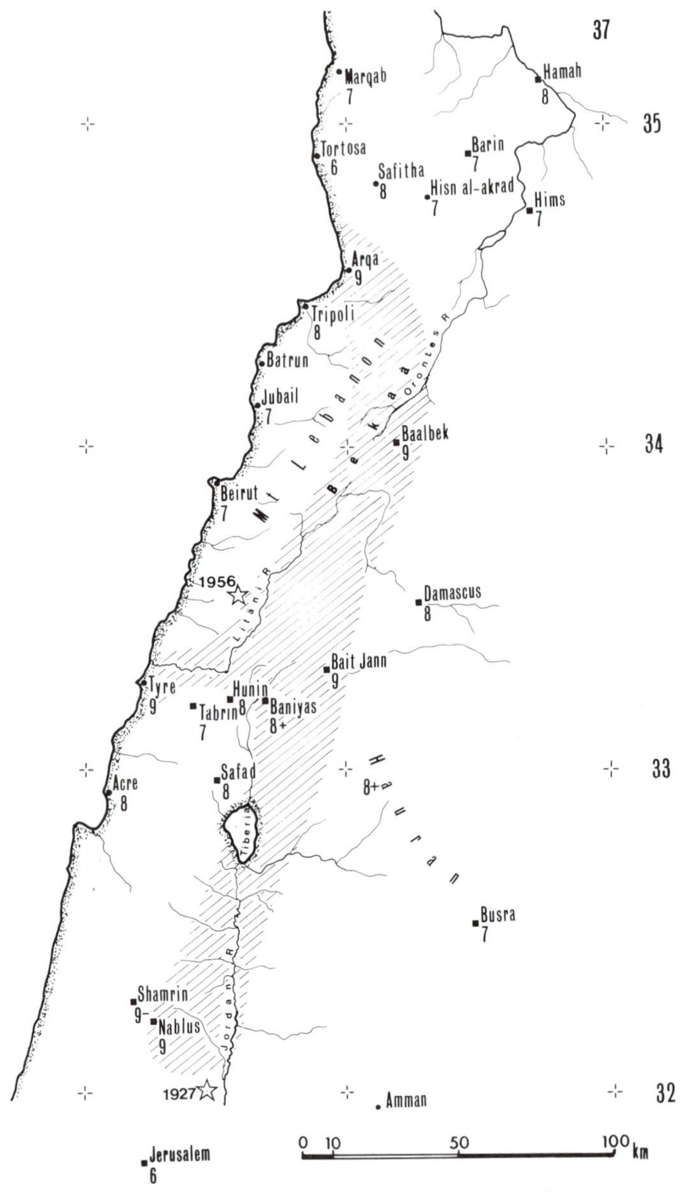

Figure 2. The epicentral region of the earthquake of 20 May 1202. Sites shown are localities affected by the earthquake and numbers indicate the intensity experienced on the MSK scale, see Table 1. Dots mark Christian-held localities, squares indicate Muslim possessions. The shading shows the approximate extent of the meizoseismal region, associated with the main shock and its aftershocks. Open stars show epicenters of the more important 20th century events discussed in the text. The location of sites is based mainly on Dussaud (1927).

Table 1. Places Where the Earthquake of 20 May 1202 Was Reported

Places mentioned	Intensity (MSK)	Territory
Acre ('Akka)	VIII *	Christian
Akhlat (Khilat)	(IV)	Muslim
Aleppo (Halab)	(V)	M
Alexandria (al-Iskandariyya)	(V)	M
Anatolia (Bilad al-rum)	(IV)	M
Antioch (Antakiya)	(V)	C
'Arqa ('Irqa, Irqata)	IX	C
Arsum ('Arima ?)	IX	C (Temp.)
Azarbaijan	(IV)	M
Baalbek (Ba'labakk)	IX	M
Bait Jann (Betegen)	IX	M
Baniyas (Paneas)	VIII+	M
Barin = Ba'rin (Montferrand)	VII	M
Batrun (Botron)	VII ?	C
Beirut (Bairut)	VII *	C
Busra (Bosra)	VII-VIII	M
Cairo (al-Qahira)	V	M
Ceuta (Sabta)	(III) ?	M
Constantinople	IV	C
Cyprus (Qubrus)	VII ?	C
Damascus (Dimashq)	VIII *	M
Damietta (Dumyat)	(V)	M
Hamah (Epiphania, Hamath)	VIII	M
Hauran	VIII	M
Hims (Homs, Emessa)	VII	M
Hisn al-akrad (Krak des Chevaliers)	VII *	C (Hosp.)
Hunin (Chastel Neuf)	VIII	M
Iraq	(IV)	M
Jerusalem (al-Quds)	VI	M
Jubail (Gibelet, Byblos)	VII	C
Lesser Armenia	IV	M
Marqab (Margat)	VII	C (Hosp.)
Mesopotamia (al-Jazira)	(IV)	M
Mosul (al-Mausil)	IV-V	M
Mount Lebanon (Jabal Lubnan)		
Nablus (Nabulus)	IX	M
Qus	(IV)	M
Safad (Saphet)	VIII	M
Safitha (Chastel Blanc)	VIII *	C (Temp.)
Shamrin (Samaria)	VIII+	M
Sicily (Siqilliya)	(IV)	C
Tabrin = Tibnin (Toron)	VII	M
Tortosa (Tartus)	VI	C (Temp.)
Tyre (Sur)	IX *	C
Tripoli (Tarablus)	VIII	C

Alternative European and Arabic spellings are given. Locations in the epicentral region are shown in Figure 2. Intensity estimates in brackets are based on felt reports only. An asterisk (*) indicates that subsequent reconstruction or repairs are mentioned (see text).

because the location was better known as marking the boundary between Muslims and Franks before the conquests of Saladin (cf. Deschamps, 1939, p. 146). In Nablus there was total destruction except for some walls in the "Street of the Samaritans", while in Hauran province most of the towns were so badly damaged that they could not be readily identified ('Abd al-Latif, p. 417, Sibt b. al-Jauzi, p. 478). One of the villages around Busra is said to have been completely destroyed, perhaps by landslides (Ibn al-Athir, xii, 112). This region must be near the epicentral area of the earthquake, and there is some hint in the account of Sibt b. al-Jauzi (p. 510) of additional damage in an aftershock at Nablus (see above), where intensity IX (MSK) may be assigned.

To the south of this area, Jerusalem suffered relatively lightly, according to the information available to 'Abd al-Latif (pp. 415, 417), at intensities not exceeding VI. His account indicates that further north, however, Damascus was strongly shaken. A large number of houses are reported to have collapsed and beside the destruction in town, major buildings near the citadel were damaged. The Umayyad mosque lost its eastern minaret and 16 ornamental battlements along its north wall. One man was killed in the collapse of the Jirun (eastern) gate of the mosque. The lead dome of the mosque was split in two and one other minaret fissured (cf. Le Strange, 1890, p. 241). The adjacent Kallasa mosque was ruined, killing a North African and a Mamluk slave (Abu Shama, p. 29, quoting al-'Izz). This building had been founded in 1160 by Nur al-Din and restored by Saladin in 1189 after its destruction by fire (Elisséeff, 1967, p. 294). West of the mosque, Nur al-Din's hospital was completely destroyed. People fled for the open spaces. The shock in Damascus was of long duration and old men could not recall such a severe one having occurred before ('Abd al-Latif, p. 416-417). Previous destructive earthquakes had occurred in 1157 and 1170. Another slight shock was felt early the following morning (Abu Shama, p. 29), and aftershocks continued for at least four days ('Abd al-Latif, p. 417).

Further north, houses are said to have collapsed at Jubail (Gibelet), recently recovered by the German Crusade (1197) which restored the landlink between the Kingdom of Acre and the County of Tripoli (*Annales de Terre Sainte*, p. 435, *Chronique de Terre Sainte*, p. 16). The walls of Beirut, also regained in 1197, are said to have been repaired around this time following earthquake damage (variant readings in *L'Estoire d'Eracle* (p. 244-5) incorrectly under A.D. 1200; likewise Ernoul, p. 338). The fact that the Prince of Batrun, a Pisan, granted his compatriots remission of taxation early in 1202 indicates that this town too suffered damage (Muralt, 1871, p. 264). The extent of the destruction is not easy to assess in these places. The walls of Jubail were dismantled by Saladin in 1190 and were probably not rebuilt after the Christian takeover. Wilbrand of Oldenborg who visited Jubail in 1211 found only a strong citadel, and a similar situation in Beirut and Batrun (p. 166-7; cf. Rey, 1871, p. 121). There is therefore the danger that damage from the extensive military operations in the period before and during the Third Crusade is misreported as earthquake damage, and even if not, some of these castles may have been rendered more vulnerable by acts of warfare. Inland, however, rockfalls in Mount Lebanon overwhelmed about 200 people from Baalbek who were gathering rhubarb; Baalbek itself was destroyed despite its strength and solidity ('Abd al-Latif, p. 416).

In the County of Tripoli, the Christian sources disagree slightly on the degree of damage to Tripoli itself, though both main accounts refer to heavy loss of life (Mayer, 1972). Ibn al-Athir (xii, 112) also refers to the heavy damage there, suggesting intensities not less than VIII. Other strongholds were severely shaken: the castle of 'Arqa (Arches) was completely ruined and deserted villages in the area were taken to indicate heavy loss of life (Philip du Plessis: but perhaps simply the flight of the inhabitants, since famine and sickness were also rife). It may be noted that Rey (1871, p. 92) cites 'Abd al-Latif and Robert of Auxerre concerning an earthquake in Sha'ban 597 (sic.)/20 May 1202 which destroyed Jebel 'Akkar and Chastel Blanc, falsely equating "Archas" with 'Akkar, which the occidentals called Gibelcar. The destruction of 'Arqa is also mentioned by Arab writers ('Abd al-Latif, p. 417, Abu Shama, p. 29). Philip du Plessis records the complete destruction of the castle at "Arsum", which is not satisfactorily identified but perhaps refers to 'Arima. Mayer (1972, p. 304) is reluctant to identify Arsum but points to the possibility of Arsuf, near Caesarea. Support for this is found in the account of the pilgrimage of Wilbrand of Oldenborg, who in 1212 found the small ruined town of "Arsim" (Arsuf) on his way to Ramla (p. 184). As Mayer mentions, however, the letter seems to refer rather to a place in Tripoli, and 'Arima is suggested on the grounds: 1) that it probably belonged to the Templars; and 2) it was one of the few strongholds retained by the Christians in the truce that ended the Third Crusade (Setton, 1969, i, 664). It is situated a few miles SSW of Chastel Blanc. Philip further reported that the greater part of the walls of the Templar stronghold Chastel Blanc (Safitha) had fallen and the keep weakened to such an extent that it would have been better had it collapsed completely. 'Abd al-Latif (p. 417) also mentions the destruction of the castle. The castle keep was probably rebuilt using existing materials (Deschamps, 1977, pp. 257-258). Tortosa (Tartus) however and the Templar citadel there seem largely to have been spared, notably the Cathedral of Notre Dame (Berchem and Fatio, 1914, p. 323, Enlart, 1928, p. 397).

The Grand Master of the Hospitallers (Geoffrey of Donjon) wrote that their strongholds at Margat (Marqab) and Krak were badly damaged but could probably still hold their own in the event of attack. Damage to Krak (Hisn al-akrad) is also mentioned in the account of Sibt b. al-Jauzi (p. 510). In the same vicinity, but in Muslim hands, the castle of Barin (Montferrand), despite its compactness and fineness, was also damaged ('Abd al-Latif, p. 416).

There is little additional evidence to help assess the intensities indicated by these reports. Studies of military architecture (e.g. Rey, 1871, Deschamps, 1934, 1977) on the whole use documentary evidence of earthquakes to support the chronology and identification of building phases at the castles, rather than documentary or archaeological evidence of rebuilding to indicate the extent of earthquake damage. Indeed, it is interesting that Deschamps, unaware of the reports of earthquake damage at Marqab in 1202, makes no reference to this specific period as being one of substantial building at the castle (Deschamps, 1977, p. 282-284), whereas in the case of Krak damage done by the earthquake is thought to have been responsible for some of the reconstruction work analysed (Deschamps, 1934, p. 281). Even so, the fact that the knights of Krak were frequently on the offensive in the next few years after 1203, and were joined by the knights from Marqab, is thought to indicate that both castles were "already in a perfect state of defense". These raids may rather suggest that attack was the best form of defense. Nevertheless, the circumstantial testimony by Geoffrey can be taken at face value and is supported by the fact that Marqab successfully resisted a counter-attack by al-Malik al-Zahir, amir of Aleppo, in 601/1204-1205 (Ibn Wasil, iii, 165). Both Marqab and Krak were visited in 1211 by Wilbrand of Oldenborg and seemed to his probably unprofessional gaze to be very strong, the latter housing 2000 defendants (p. 169-170). Few details are available about Barin, which was finally dismantled in 1238-1239 (Deschamps, 1977, p. 322). It seems unlikely that intensities exceeding VII were experienced at any of these strongholds.

In neighboring Muslim territory, the shock was experienced at similar intensities in Hims (Homs, Emessa), where a watchtower of the castle was thrown down (Sibt b. al-Jauzi, p. 510) and Hamah, where the earthquake was experienced as two shocks, the first lasting "an hour" and the second shorter but stronger. Despite its strength, the castle was destroyed, along with many houses and other buildings. Two further shocks were felt the following afternoon ('Abd al-Latif, p. 416). Considerable damage to houses in both towns is implied by Ibn al-Athir (xii, 112).

Further north, the earthquake is said to have been felt in Aleppo and other regional capitals (Sibt b. al-Jauzi, p. 478), and also in Antioch, though less strongly (Geoffrey of Donjon). It was also reported in Mosul and throughout the districts of Mesopotamia, as far as Iraq, though without destruction of houses. Azarbaijan, Armenia, parts of Anatolia and the town of Akhlat are said to have experienced the earthquake (Ibn al-Athir and Sibt b. al-Jauzi, *loc. cit.*).

In the south, the shock was felt throughout Egypt from Qus to Alexandria. Sibt b. al-Jauzi (p. 478, probably quoting al-'Izz) says that the shock came from al-Sa'id and extended into Syria; al-Sa'id being the region south of Fustat (Old Cairo) down to Aswan (Yaqut, iii, 392). In Cairo, the shock was of long duration and aroused sleepers, who jumped from their beds in fear. Three violent shocks were reported, shaking buildings, doors and roofs. Only tall or vulnerable buildings were particularly affected, and those on high ground, which threatened collapse ('Abd al-Latif, p. 414-5). Such a strong shock was considered unusual for Egypt and must have been at least intensity V. The details provided indicate that Egypt experienced long-period shaking at a large epicentral distance. A lesser shock was felt at about midday the same morning, probably the one reported from Hamah at midday on Tuesday 27 Sha'ban (21 May).

In Cyprus, under Frankish rule since 1191, the earthquake damaged churches and other buildings and was strongly felt (*Annales* 5689, fol. 108b; 'Abd al-Latif, p. 415; Ibn al-Athir, xii, 130). Damage to buildings is not however very well attested and it is noteworthy that most of the "Cypriot Chronicles" refer only to damage on the mainland. In the words of the Arabic authors, the sea between Cyprus and the coast parted and mountainous waves were piled up, throwing ships up onto the land. Eastern parts of the island were flooded and numbers of fish were left stranded ('Abd al-Latif, p. 415; Ibn Mankali in Taher, 1979). The significance of this seismic sea-wave is discussed below.

The earthquake is said to have been felt as far as Sicily (Ibn al-Athir, xii, 130) and Ceuta (Ibn Wasil, iii, 161), but this still lacks confirmation in the annals of the Muslim west, dominated at this period by the Almohads. No details have been recovered of the shock in the western Mediterranean area. It is very likely that the shaking reported on or after 1 March 1202 felt in and around Constantinople was from the earthquake of 20 May (Nicetas, p. 701).

The loss of life caused by this earthquake and its aftershocks is difficult to estimate. A figure frequently quoted in Arab sources is 1,100,000 dead (e.g., al-Dhahabi, iv, 296; al-Suyuti, p. 47) for the year 597-598 H. (A.D. 1201-1202). This specifically includes those dying of famine and the epidemic consequent on the failure of the

Nile floods, graphically described by 'Abd al-Latif, who notes 111,000 [sic.] deaths in Cairo alone between 596 and 598 H. (p. 412). More realistically, the figure of 30,000 casualties is given, primarily, it would seem, in the Nablus area (Sibt b. al-Jauzi, p. 478). No reliance can be placed on such figures, but the fact that the main shock occurred at dawn, when most people were in bed, without noticeable foreshocks, probably contributed to a high death toll.

Aftershocks were reported from Hamah, Damascus, and Cairo, for at least four days ('Abd al-Latif, p. 417, Abu Shama, p. 29), one of which, apparently felt in Cairo and Hamah, must have been a large event. There remains the possibility that the aftershock sequence was terminated with a destructive shock that totally destroyed what was left of Nablus, but it seems preferable to consider both reports by Sibt b. al-Jauzi as referring to the same one shock. Whatever the exact sequence of events, the cumulative effects of the earthquake were clearly catastrophic. Most of the sites affected in the epicentral region (see Figure 2) must have needed total reconstruction or major repairs (cf. Table 1), although in most cases the evidence is circumstantial, not specific.

6. Discussion

From the foregoing it appears that the 1202 earthquake was a shallow, large magnitude multiple event. This is attested by: (1) the large area over which the shock was felt; (2) the long-period effects observed at large epicentral distances; (3) the fact that the main shock was followed by aftershocks at least one of which was very widely felt; (4) by a seismic sea-wave generated between Cyprus and the Syrian coast; and finally (5) by the observation that in the epicentral region the earthquake was experienced as more than one shock.

From Figure 1 we notice that the radius of the felt area, which includes Sicily but not Ceuta, was about 1,200 km. From Figure 2 we see that the epicentral region, within which intensities exceeded VIII (MSK), forms a narrow inland strip about 250 km long and 40 km wide that extends from Nablus in the south to 'Arqa in the north. The number of sites at which intensities can be assessed (shown in Figure 2) is obviously insufficient to allow the construction of a proper isoseismal map (but cf. Sieberg, 1932b). However, it would appear that the maximum effects of the earthquake were experienced away from the coast, in the upper Jordan and Litani valleys as well as the upper reaches of the Orontes river, in the vicinity of Baalbek. Several thousand people perhaps perished in this area. Without further details, it is difficult to indicate more precisely the exact location and extent of the epicentral region. The vague details of severe damage in the Hauran district may suggest that the rupture zone was wider than shown on Figure 2. Since most of the aftershocks were reported from the north (Hamah), we may conjecture that the event nucleated in the south, near Nablus, and that it was completed by a second rupture that originated in the Tyre-Baalbek segment of the meizoseismal area. Apart from the statement that large-scale landslides occurred in Mount Lebanon, there is no indication that this event was associated with faulting.

The 1202 earthquake may however be compared with the earthquake sequence between June 1759 and January 1760, which affected almost exactly the same epicentral region. Preceded by strong foreshocks on 10 June, the main shock on 30 October 1759 completely destroyed the region of the Litani and upper Orontes valleys. A violent aftershock on 25 November extended the damage to Safad in the south, with a cumulative epicentral region somewhat smaller than that of the 1202 event. Aftershocks continued well into January 1760 and damage in Tyre, Tripoli and Damascus was as serious as in 1202. One important aspect of the 1759 earthquake, which is much better documented, is that we know it was associated with a 95 km long fault-break in the Bekaa, on the west side of the valley, in places many metres wide (Archives Nationales, 1759). It is not possible to assess the tectonic effects of the 1202 earthquake, which was much larger than the multiple shock of 1759, but it may have been comparable in location and in its extent of faulting.

7. Calibration with 20th Century Earthquakes

No earthquake of comparable felt area has occurred in the Middle East during the present century that can be used to calibrate the magnitude of the 1202 event. The nearest is the eastern Mediterranean earthquake of 26 June 1926, which had an offshore epicenter near Rhodes and a radius of perceptibility of about 900 km as shown in Figure 3 (Shebalin and Kárník, 1974). This earthquake was originally located by the International Seismological Summary (ISS) at 36.0° N, 28.0° E, with a shallow depth, but because of its relatively large felt area it was relocated at intermediate depth between 100 and 150 km and assigned a body wave magnitude between 7.7 and 8.0 (Gutenberg and Richter, 1965; Shebalin and Kárník, 1974; Alsan, *et al.* 1975; Makropoulos, 1978). It has also been suggested that its radius of perceptibility was as large as 1,600 km (references in Kárník, 1968), a value inferred from the fact that Gassmann (1926) reports the event felt in Switzerland. However, Galanopoulos (1953) and Wyss and Baer (1981) show that these relocations cannot be trusted and that the 1926 earthquake was a relatively shallow event. The felt foreshocks of 1 and 27 April, 26 and 27 June and 5 July, of magnitude $M_S \geq 5$, all located as shallow events, imply a shallow source. This is supported by a revised radius of perceptibility of the main shock of just under 900 km (Figure 3), which is based on a re-examination of Sieberg's data (1932a,b), with the use of additional information (Mihailovič, 1928; Critikos, 1928; Press Reports). The recent evaluation of Italian data (Margottini, 1982) in particular shows that the shock reported in Switzerland should be attributed to a local earthquake. Moreover, using the method put forward by Ambraseys (1985b) and the seven isoseismal radii listed by Shebalin and Kárník (1974), our estimate of the macroseismic focal depth of the 1926 earthquake is only 35 km with a corresponding absorption coefficient of $3 \times 10^{-3}\,\mathrm{km}^{-1}$ and an n-value close to the Airy phase, i.e. $n = 1.3$. For this focal depth the recalculated surface wave magnitude from 25 station estimates was found to be 7.0 ($\pm$0.2), while the body wave magnitude from 7 station estimates was 7.5 ($\pm$ 0.3).

With an epicentral area entirely offshore, the effects of the 1926 earthquake were obviously not representative of its magnitude. The shock killed 13 people and destroyed or damaged beyond repair about 4,000 houses, mainly in Rhodes, the nearby coast of Turkey and in Crete. A few old and delapidated houses in Cairo, Alexandria and the Jordan Valley suffered damage; minarets and belfries in Turkey and Greece swayed dangerously. The 1926 earthquake thus had a radius of perceptibility at $I = II^{+}$ (MSK) of about 900 km, a shallow depth and a surface wave magnitude of 7.0.

The largest 20th century earthquake in the Jordan Valley, on 11 July 1927, was again of shallow depth but a much smaller event, with a surface wave magnitude

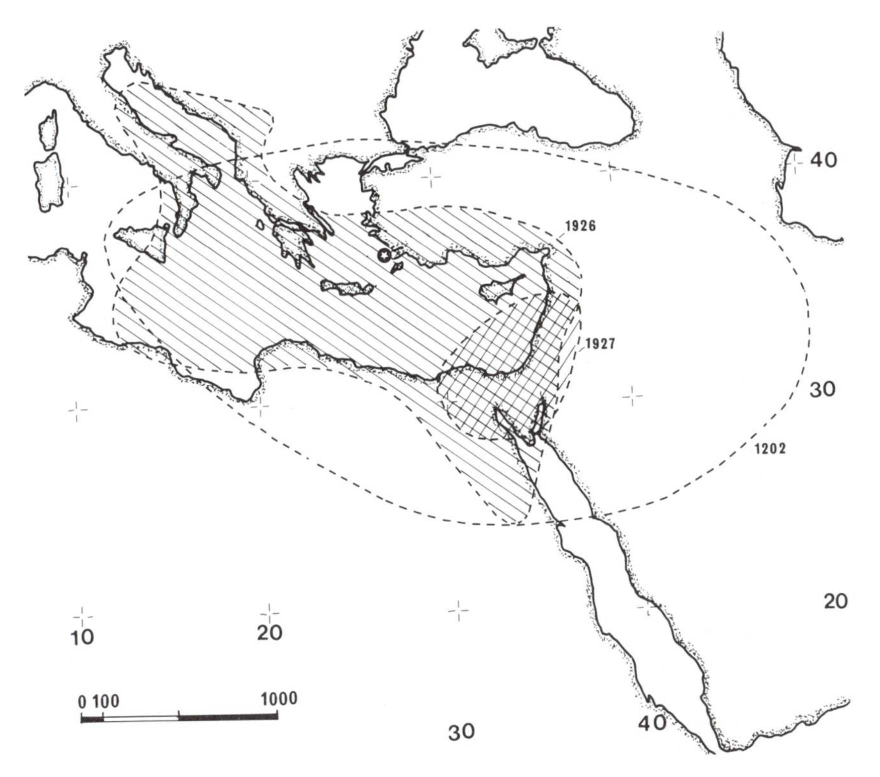

Figure 3. Comparison of the felt areas of the earthquakes of 20 May 1202, 26 June 1926 ($M_S = 7.0$, $m = 7.5$) and 11 July 1927 ($M_S = 6.0$, $m = 6.4$).

of only 6.0 ($\pm$ 0.1) from 20 station estimates ($m = 6.4$). The shock was centered at 32.0° N and 35.4° E. It destroyed hundreds of houses on both sides of the Jordan River, killing 361 people and injuring about 1,000. Detailed studies of this event have been published by Sieberg (1932a, b), Berloty (1927) and more recently by Vered and Striem (1976). This earthquake must be described as one which excited widespread interest and sympathy rather on account of where it happened than because of its special violence, and its effects have been overestimated. Our re-evaluation of its effects, based on official reports compiled by the British and French authorities, shows that Sieberg's intensity assessments are greatly exaggerated and that the radius of perceptibility he deduces, of 600 km, contains large areas from which negative reports predominate (see Berloty, 1927, p. 80-81). On Figure 3 we show an area of radius only 300 km where the shock was felt at intensity III (MSK), and on Figure 2 the epicenter of this event is shown with respect to the epicentral region of the 1202 earthquake. The 1927 earthquake was followed by a long series of aftershocks which terminated on 22 February 1928 with a damaging shock of $M_S = 5.2\,(\pm 0.2)$ that was centered in the Judean Hills.

The earthquake of 16 March 1956 was a smaller, double shock consisting of two events of $M_S = 5.0\,(\pm 0.1)$ and $5.1\,(\pm 0.1)$, both shocks having $m = 5.6$, occurring within ten minutes of each other. The epicentral region was situated in the Shouf area on the west slopes of Mt. Lebanon, Figure 2. These shocks killed 136 people and rendered 5500 houses uninhabitable in 55 communities. The earthquake was felt with an intensity greater than II (MSK) within a radius of about 160 km (Shalem, 1956; Plassard, 1956; Anon., 1958).

Thus, on a comparative basis, the 1202 earthquake had a magnitude greater than 7.3. If we use the relationship derived to predict surface wave magnitudes from macroseismic data of Balkan earthquakes, i.e.

$$M_{SC} = 0.40 + 0.46\,(I_i) + 2.8 \times 10^{-4}\,(R_i) + 1.8 \log(R_i), \tag{1}$$

where R_i is the average radius of isoseismals of intensity I_i, For $R_3 = 1200$ km we find that the magnitude of the 1202 earthquake should be $M_S = 7.6$. This attenuation law is based on macroseismic data of the Balkan region compiled by Shebalin and Kárník (1974) and it has been derived using the procedure employed by Ambraseys (1985b) for Northwest Europe. Assuming that the rupture length associated with this event is defined approximately by the length of the zone of maximum intensity shown in Figure 2, i.e. about 200 km, the relative co-seismic displacement for an earthquake of magnitude 7.6 may be estimated from

$$M_S = 1.1 + 0.4 \log(\mathrm{L}^{1.58}\,\mathrm{D}^2), \tag{2}$$

which yields a relative displacement of D = 2.5 m, most probably wholly strike-slip in nature. Equation (2) is derived from rupture dimensions and dislocation of Eastern Mediterranean and Middle Eastern events (Ambraseys and Melville, 1982). L and D are the length of the fault break and relative displacement, in centimeters; M_S is the corresponding surface wave magnitude. This is all the more interesting in that there is no record of faulting in this earthquake, which occurred on land, far enough from the Syrian coast not to be directly associated with the reported seismic sea-wave east of Cyprus. At any rate, it is unlikely that such a wave would be generated by a mechanism lacking in a substantial dip-slip component. Although the details of damage in Cyprus are not very satisfactorily recorded, the earthquake must have been as damaging in the island as along northern parts of the Syrian coast but not as destructive as further inland. This would enlarge the area of intensive high shaking and reduce the emphasis on an offshore epicentral region that reports of the sea-wave might imply.

The occurrence of a seismic sea-wave between Cyprus and the Syrian coast may be explained through the generation of a large-scale subaqueous slide from the continental margin of Syria, triggered by the earthquake. North of Acre the continental shelf narrows to a few kilometers and off the coast of Lebanon the continental slope steepens from near Acre northwards to an average slope of 10° (Boulos, 1962). Under these circumstances, the principal cause of a seismic sea-wave is submarine slumping. The whole of that coast is certainly prone to slumping because of evaporites in the sedimentary section (Garfunkel, *et al.* 1979).

In conclusion, although it is clear that the damage reported in contemporary sources was due to the main shock on 20 May 1202, there remains the possibility of cumulative damage, particularly from a belated, large magnitude aftershock, perhaps four or five months later, centered in the southern portion of the epicentral region. The main shock was catastrophic, with an on-land meizoseismal area and source dimension of about 200 km. The magnitude of the 1202 earthquake was in excess of 7.5 and it was associated with a protracted sequence of aftershocks and a damaging sea-wave that affected the coasts of Syria and Cyprus.

ACKNOWLEDGEMENTS

The authors would like to thank Dr. Peter Jackson of Keele University for some helpful suggestions and Dr. James Jackson of Cambridge University for his comments.

REFERENCES

1. Arabic Sources:

'Abd al-Latif, Kitab al-ifada, trans. Silvestre de Sacy, Relation de l'Egypte, Paris, 1810.
Abu 'l-Fada'il, Tarikh Mansuri, facs. ed. P. A. Gryaznevitch, Moscow, 1960.
Abu 'l-Fida, Tarikh al-mukhtasar fi akhbar al-bashar, vol. iii, ed. Cairo, 1907.
Abu Shama, Dhail 'ala al-raudatain, ed. al-Zujari and al-Hasani, Cairo, 1947.
al-Dhahabi, Kitab al-'ibar fi khabar man ghabara, vol. iv, ed. S. Munajjid, Kuwait, 1963.
Ibn al-Athir, al-Kamil fi 'l-tarikh, vol. xii, ed. Tornberg, Leiden, 1853.
Ibn al-Furat, Tarikh al-duwal wa 'l-muluk, vol. iv/2, ed. Hassan al-Shamma, Basra, 1969.
Ibn Jubair, The travels of Ibn Jubayr, ed. W. Wright, 2nd ed. J. M. de Goeje, Gibb Memorial Series vol. 5, London, 1907.
Ibn Mankali, al-ahkam al-mulukiyya, fol. 37, quoted in Taher (1979, p. 125).
Ibn Wasil, Mufarrij al-kurub fi akhbar bani ayyub, vol. iii, ed. M. Shayyal, Cairo, 1962.
Sibt b. al-Jauzi, Mir'at al-zaman, vol. viii, printed ed. Hyderabad, 1951.
al-Suyuti, Kashf al-salsala 'an wasf al-zalzala, ed. and trans. S. Nejjar, Fez, 1971 and Rabat, 1974.
Yaqut al-Hamawi, Mu'jam al-buldan, ed. Wustenfeld, 4 vols., Leipzig, 1866-1871.

2. Other Primary Sources:

Salimbene de Adam, Cronica, Mon. German. Hist. Ss. vol. 32.
Francesco Amadi, Chroniques d'Amadi et de Strambaldi, vol. 1, Chronique d'Amadi, ed. Mas Latrie, Coll. de doc. inedit. sur l'hist. de France, Paris, 1891.
Annales de Terre Sainte, ed. G. Raynaud and R. Röhricht, Archives de l'Orient latin, vol. 2b, 1884.
Annales 5689, Annales Terrae Sanctae, Bibl. Nationale de Paris, Fonds Latin no. 5689.
Robert of Auxerre, Chronicon, Mon. German. Hist. Ss. vol. 26; also ed. Bouquet, Rec. Hist. Gaule et de la France, vol. 18.
Bar Hebraeus [Abu 'l-Faraj], Chronography, trans. E. A. W. Budge, London, 1932.
Barletta Ms. in Kohler (1900-1901).
Ralph of Coggeshall, Chronicon Anglicanum, ed. J. Stevenson, Rolls Ser. Vol. 66.
Les Gestes des Chyprois, Rec. Hist. Croisades. Arm., vol. 2.
Chronique de Terre Sainte, 1131-1224, in Gestes des Chiprois, ed. G. Raynaud, Publ. Soc. de l'Orient Lat., vol. 5, Paris, 1887, (repr. Osnabruck, 1968).
Geoffrey of Donjon, Letter in Mayer (1972).
Ernoul, Chronique d'Ernoul et de Bernard le Trésorier, ed. Mas Latrie, Paris, 1871.
L'Estoire d'Eracle, Rec. Hist. Croisades. Occ., vol. 2.
The book of the wanderings of Felix Fabri [1480-83], vol. 2, tr. A. Stewart, Palestine Pilgrims' Text Soc., vol. 9, London, 1893, (repro. AMS, New York, 1971).
Hethum of Gor'igos, Table chronologique, Rec. Hist. Croisades. Arm., vol. 1.
Hethum Patmic, Chronicle, in Hakobyan (1956, p. 61).
Albert Milioli, Cronica Imperatorum, Mon. German. Hist. Ss., vol. 31.
William of Nangis, Chronicon, Rec. Hist. Gaule et de la France, vol. 20.

Nicetas Choniates Acominatos, Historia chronicon, ed. I. Bekker, Corp. Script. Hist. Byz., Bonn, 1835.

Wilbrandus of Oldenborg, Travels, ed. J. C. M. Laurent, Peregrinatores medii aevi quatuor, p. 161-191, 2nd ed., Leipzig, 1873.

Philip of Plessis, Letter in Mayer (1972).

Marino Sanuto, the elder, Liber secretorum (Secrets for the Crusaders) or Gesta Dei per Francos, book III/xi, chapter 1, ed. Hannover, 1611.

3. Other Works:

Abe, K. (1987). Magnitude determination from Milne undamped seismographs (this volume).

Alsan, E., L. Tezucan, and M. Bäth (1975). An earthquake catalogue for Turkey, 1913-1970, *Publ. Seism. Inst. no. 7-75*, Uppsala.

Alsinawi, S. and H. Ghalib (1975). Historical seismicity of Iraq, *Bull. Seism. Soc. Am.*, **65**, 541-547.

Ambraseys, N. (1961). On the seismicity of South-West Asia: data from a XV-century Arabic manuscript, *Rev. Etude Calamités*, **37**, p. 18-30.

Ambraseys, N. (1962). A note on the chronology of Willis' list of earthquakes, *Bull. Seism. Soc. Am.*, **52**, 77-80.

Ambraseys, N. (1968). Early earthquakes in North-Central Iran, *Bull. Seism. Soc. Am.*, **58**, 485-96.

Ambraseys, N. (1974). The historical seismicity of North-Central Iran, *Geol. Survey Iran*, publ. 29, 47-96.

Ambraseys, N. (1985a). The seismicity of western Scandinavia, *J. Earthq. Eng. Struct. Dyn.*, **13**, 361-400.

Ambraseys, N. (1985b). Intensity-attenuation and magnitude-intensity relationships for northwest European earthquakes, *J. Earthq. Eng. Struct. Dyn.*, **13**, 733-778.

Ambraseys, N. and R. Adams (1986). Seismicity of the Sudan, *Bull. Seism. Soc. Am.*, **56** (in press).

Ambraseys, N. and J. Jackson (1985). Long-term seismicity of Britain, *Earthquake Engineering in Britain*, 49-65, London.

Ambraseys, N. and C. Melville (1982). *A history of Persian Earthquakes*, Cambridge.

Ambraseys, N. and C. Melville (1983). Seismicity of Yemen, *Nature*, **303**, 321-323.

Ambraseys, N., *et al.* (1983). Notes on historical seismicity, *Bull. Seism. Soc. Am.*, **73**, 1917-1920.

Anonymous (1958). Le recent tremblement de terre de Litani, *Service des Renseignments, Publ. d'Urbanisme*, Beirut.

Archives Nationales (1759). Serie Bl-Correspondence Consulaire/AE.B1.1120 Tripoli 4 fev. 1760 and B1.88 Aleppo 11 Dec 1759.

Arvanitakis, G. (1903). Essai sur le climat de Jerusalem, *Bull. Inst. Egypt.* **4**, 178-183, Alexandria.

Batman, K. (1581). The Doome warning all men to the Judgement, Newbery: London.

Ben-Menahem, A. (1979). Earthquake catalogue for the Middle East (92 B.C. to 1980 A.D.), *Bol. Geof. Teor. ed Applic.*, **21**.

Berchem, M. van and E. Fatio (1914). Voyage en Syrie, *MIFAO*, **37**, Cairo.

Berloty, R. (1927). Sur le tremblement de terre de Palestine 11 juillet 1927, *Annales Obs. Ksara, Section Seism.*, 62-93, Zahle.

Berryat, J. (1761). Liste chronologique des eruptions de volcans, etc., *Coll. Acad.*, **6**, 488-676, Paris.

Beuther, T. (1601). Compendium terrae motuum, Strassburg.

Blankenhorn, M. (1914). Morphologische uebersicht; Erdbeben Syrien etc., Handbuch der Region. Geol. vol. 5 and Zeitschr. deutsch. Palaest.-Vereins, vol. 28/i (1905), 206-219.

Boulos, I. (1962). Cartes de reconnaissance des côtes du Liban, 1:150,000, Service Topographique, Beirut.

Cahen, C. (1940). La Syrie du Nord à l'époque des Croisades et la principauté franque d'Antioche, Paris.

Coronelli, P. (1693). De'tremuoti accaduti dal diluvio, etc., in *Epitome Cosmografica*, 286-324. Venice.
Critikos, N. (1928). Le tremblement de terre de la mer de Crete du 26 juin 1926, *Annales Obs. Natl. Athens*, **10**, 39-53.
Deschamps, P. (1934-1977), Les châteaux des Croisés en Terre Sainte, 3 vols. in 5, Bibl. Arch. et Hist. vols. 19 (1934), 34 (1939) and 90 (1977), Paris.
Dressdnische Gelehrte Anzeigen (1756). Nachrichten von Erdbeben, No. 2-40, Dresden.
Dussaud, R. (1927). Topographie historique de la Syrie antique et mediévale, *Bibl. Archeol. Hist.*, **4**, Paris.
Elisséeff, N. (1967). Nur al-Din, Inst. Fr. Damas, 3 vols., p. 924.
ENEA-ENEL (1981). Contributo alla caratterizzazione della Sismicita del territorio Italiano, Udine.
Enlart, C. (1928). *Les monuments des Croisés dans le Royaume de Jerusalem*, vol. 2, Paris.
Galanopoulos, A. (1953). On the intermediate earthquakes in Greece, *Bull. Seism. Soc. Am.*, **43**, 159-175.
Garfunkel Z., A. Arad, and G. Almagor (1979). The Palmahim disturbance and its regional setting, *Bull. Geol. Survey Israel*, **72**, 39-40, Jerusalem.
Gassmann, F. (1926). Jahresbericht 1926 des Erdbebendies der Schweizerischen Meteorologischen Zentralanstalt, *Annal. Schweiz. Met. Zentral.*, p. 4, Fig. 6.
Gutenberg, B. and C. Richter (1965). *Seismicity of the earth.* Hafner.
Hakobyan, V. A. (1956). Manr Zhamanakagrut'yunner XIII-XVIII (Armenian short chronicles, 13th - 18th centuries), vol. 2, Erivan.
Hoff, K. von (1840). Chronik der Erdbeben, etc. Gesch. Ueberlief. nachgew. naturl. Veraender. Erdoberfl. vol. 4, Gotha.
Huot, J. (1837). Nouveau cours élémentaire de géologie, vol. 1, p. 108-118, E. Roret: Paris.
Kallner-Amiran, D. (1951-1952). A revised earthquake catalogue of Palestine, *Israel Explor. Jnl.*, **1**, 223-246; **2**, 48-65.
Kárník, V. (1968). *Seismicity of the European area*, Academia, Prague.
Kohler, Ch. (1900-1901). Un rituel et un breviaire du Saint-Sepulchre de Jerusalem, *Rev. de l'Orient lat.*, **8**, 383-469.
Le Strange, G. (1890). *Palestine under the Moslems*, London.
Makropoulos, K. (1978). The statistics of large earthquake magnitudes and an evaluation of Greek seismicity, *Ph.D. thesis*, University of Edinburgh.
Mallet, R. (1852). Third report on the facts of earthquake phenomena, *British Ass. Adv. Sci.*, 1-176, London.
Manetti, G. (1457). De terraemotu, trans. C. Scopelliti, p. 102, ENEA, Rome.
Margottini, C. (1982). Osservazioni su alcuni grandi terremoti con epicentro in oriente; campo macrosismico in *Italia. Publ. Comm. Nazion. energ. Nucleare*, Rome.
Mayer, H. E. (1972). Two unpublished letters on the Syrian earthquake of 1202, Medieval and Middle Eastern Studies in honor of Aziz Suryal Atiya, ed. S. A. Hanna, 295-310, Leiden.
Mayer, H. E. (1984). *The Crusades*, Oxford University Press.
Melville, C. (1983). Seismicity of the British Isles and the North Sea, Rept. SRC, London Centre for Marine Technology, vol. 1, Imperial College, London.
Melville, C. (1984). Sismicité historique de la mer rouge septentrionale, in Tremblements de terre: histoire et archéologie, p. 95-107, APDCA, Valbonne.
Melville, C. (1985). The geography and intensity of earthquakes in Britain; the 18th century, *Earthquake Engineering in Britain*, 7-22 London.
Mihailovič, J. (1928). Annuaire seismique phenomènes macroseismiques, *Ann. Obs. Belgrade*, Ser. A fasc. 4.
Muir Wood, R. G. Woo, and H. Bungum (1987). The history of earthquakes in the northern North Sea (this volume).
Muralt, E. von (1871). Essai de chronographie byzantine, 1057-1452, vol. 1, Basle.
Perrey, A. (1850). Sur les tremblements de terre ressentis dans la Peninsule Turco-Hellenique et en Syrie *Memoires Couronn. Acad. R. Sci. de Belgique*, **23**, Brussels.
Plassard, J. (1956). Note sur le tremblement de terre du 16 mars 1956, *Publ. Obs. Seism. Ksara*, Zahle.

Plassard, J. and B. Kagoj (1968). Catalogue des seismes ressentis au Liban, *Annales-Mem. Obs. Ksara*, **4**, no. 1.

Poirier, J. and M. Taher (1980). Historical seismicity of the Near and Middle East, etc., *Bull. Seism. Soc. Am.*, **70**, 2185-2201.

Postpichl, D. (1985). ed., Atlas of isoseismal maps of Italian earthquakes, Progetto Finalizzato Geodinamica, Bologna.

Rey, G. (1871). Etude sur les monuments de l'architecture militaire des Croisés en Syrie et dans l'Ile de Chypre, Doc. inédits sur l'hist. de France, Paris.

Röhricht, R. (1898). Geschichte des Königreichs Jerusalem, Innsbruck.

Runciman, Sir S. (1971). *A history of the Crusades*, 3 vols., Harmondsworth.

Setton, K. M. (1969). ed., *A history of the Crusades*, vols. 1 and 2, 2nd ed., Univ. of Wisconsin Press, Madison and London.

Seyfart, J. (1756). *Algemeine Geschichte der Erdbeben*, Frankfurt.

Shalem, N. (1956). Isoseismal map of the 16 March earthquake, *Geomorph. Dept. Rept. Geol. Surv. Israel*, Jerusalem.

Shebalin, N. and Kárník, V. (1974). Catalogue of earthquakes in the Balkan region, UNESCO Survey of Seismicity of the Balkan Region, 2 vols. and Atlas, Skopje.

Sieberg, A. (1932a). Erdbebengeographie, in B. Gutenberg, *Handbuch der Geophysik*, vol. 4, Berlin

Sieberg, A. (1932b). Erdbeben und Bruchschollenbau im Ostlichen Mittelmeergebeit, Denk. d. Medizin.-Naturwiss. Ges. zu Jena, vol. xviii, no. 2, Jena.

Sprenger, A. (1843). As-Soyuti's work on earthquakes, *J. R. Asiatic Soc. Bengal*, **12**, 741-749.

Taher, M. A. (1979). Corpus des textes arabes rélatifs aux tremblements de terre de la conquête arabe au xii H./xviii J.C., *Doctoral thesis*, Sorbonne, 2 vols., Paris.

Tholozan, J. (1879). Sur les tremblements de terre qui ont eu lieu en Orient, etc., *Comp. Rend. Acad. Sci.*, **88**, 1063-1066.

Vered, M. and H. Striem (1976). A macroseismic study of the July 11, 1927 earthquake, Publ. IA-LD-1-107, Israel Atomic Energy Commission.

Vigouroux, F. (1928). Dictionnaire de la Bible, vol. 4/ii, 2031, Letouzey-Ane.

Vogt, J. (1984a). Revision de deux seismes majeurs de la région d'Aix-la-Chapelle/Verviers/Liège ressentis en France - 1504, 1692, in *Tremblements de terre: histoire et archéologie*, 9-21, APDCA, Valbonne.

Vogt, J. (1984b). Problèmes de sismicité historique, etc. in *Seismic activity in western Europe: with particular consideration to the Liège earthquake of November 8, 1983*, P. J. Melchior, ed., 205-214, NATO ASI ser. C144, Reidel, Dordrecht.

Willis, B. (1929). Earthquakes in the Holy Land, *Bull. Seism. Soc. Am.*, **18**, 73-103.

Wyss, M. and M. Baer (1981). Earthquake hazard in the Hellenic Arc, *Earthquake Prediction - Maurice Ewing Series*, no. 4, 153-172, Publ. Amer. Geophys. Union.

Preliminary Evaluation of the Large Caracas Earthquake of October 29, 1900, by Means of Historical Seismograms

G. E. Fiedler B.
Chief, Seismological Institute CAR (retired)
Apartado postal 5407 Carmelitas, Caracas 1010-A, Venezuela

1. Introduction

On the 29th of October, 1900, at about 9:11 GMT, which is equal to 4:41 a.m. former Venezuelan legal time or 5:11 a.m. actual Venezuelan legal time (HLV, since Jan. 1, 1965), Caracas and its surroundings were hit by an earthquake. The maximum intensity of this earthquake can be estimated to be 9 on the Mercalli-Cancani-Sieberg-Scale (MCS). The observed duration of the ground motion in Caracas was up to 50 seconds. About 100 victims were counted and great destruction occurred in Caracas, La Guaira, Macuto, and the rest of the so called Litoral Guaireño, and also in Guarenas, Guatire, Higuerote and other places. If we include intensity III (MCS), this earthquake was felt over the whole country. Specific macroseismic information can be found in the following publications: Centeno Grau (1900, 1940), Fiedler (1961, 1968, 1980), Jakubovicz and Larotta (1974), Sievers (1905), Troconis de Veracoechea (1979) and others.

Richter (1958) published the following epicentral ISC-data for this earthquake: Date = October 29, 1900; Origin time = 9:11 GMT; Epicentral coordinates = 11.0° N, 66.0° W; Magnitude = 8.4. This epicenter is located at a point some 65 km west of Tortuga Island or 115 km northeast of Caracas, on a physiographically very important zone, formed of platforms and smaller oceanic basins.

As far as the author of this paper knows, no seismographic records were shown or interpretations besides that of C. F. Richter used in the many publications written about this event. Therefore, an effort was made to find such seismograms and derived seismographic interpretations. In most cases, world wars and other circumstances have resulted in the destruction of photographic or smoked paper records of that event, but two good quality seismograms in the form of copies were obtained: one from Pamplemousses Observatory, Mauritius Island, built in the year 1898, and another from Kew Observatory, England. These two records are shown in Figures 1 and 2. Seismographic record readings were published, as far as could be found out, by: Schütt (1900), Hamburg, Germany (seismic service opened in the year 1898); Laibach, Yugoslavia (1897), Pola and Casamicciola, Italy; Anonymous (1900), Toronto (1897) and Victoria, Canada; Anonymous (1900), Vacoas, Mauritius (Pamplemousses, 1898); Anonymous (1908), Port of Spain, Trinidad (1900); ISC-data and seismogram, Kew, England (1898); Omori (1903), Tokyo, Japan; Kortazzi (1903), Nicolajew, USSR; Mazelle (1901), Triest, Italy.

This event is included in the "Important Earthquake of the Northern Andes" (Gutenberg, 1929, p. 973), and its isoseismal map is also shown in Gutenberg (1929, p. 975). However, Gutenberg (1929) mistook this earthquake to have occurred in 1910 instead of 1900. After the 1900 event, the next Caracas earthquake occurred on December 12, 1915, which is similar to the 1900 event but without damage and with an intensity of about 5 (MCS) in the region of Caracas, Guarenas, Guatire, Higuerote, Macuto, and La Guaira.

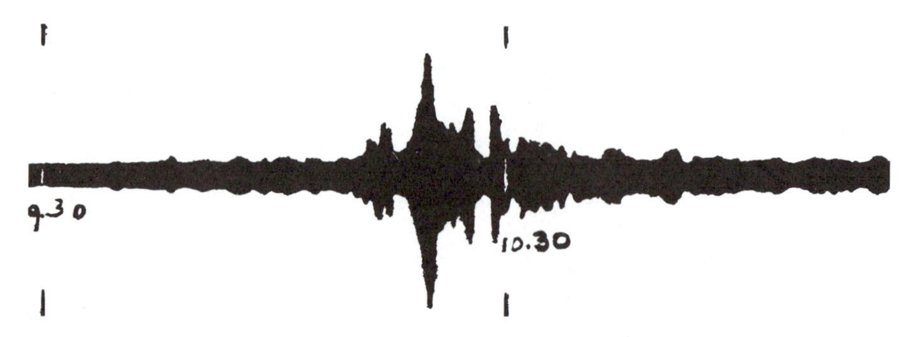

Figure 1. The Caracas earthquake of October 29, 1900, recorded by a Milne horizontal seismograph, installed in the Pamplemousses Observatory, Mauritius Island, at an epicentral distance of about 13,900 km. A photomounting, made by G. Fiedler was necessary, due to problems with the copy. The period of the pendulum was 20.4 sec, damping ratio 1.04, static magnification = 11, mass = 310 gm, the photomounting amplification is 1.17:1 (enlarged). Directionality must be checked.

Figure 2. The Caracas earthquake of October 29, 1900, H = 9:11:00 GMT, recorded by the Milne horizontal seismometer, E-W component, at the Kew Observatory, England (scale 1:1).

2. Interpretation of Seismograms of the Caracas Earthquake of 1900

2.1 Instrumental Constants

The E-W component of the Milne seismograph of the Kew Observatory, following Reid (1910) and Wood (1921), had the intrinsic constants: $T_O = 18$ to 19 sec, static magnification $V = 6.1$, length of the indicator $J = 520$ meters, damping ratio $\epsilon = 1.114$ (in this case and after Reid (1910) is the amplitude ratio of a maximum to the next following minimum, i.e., over one half period or π), friction damping zero (photographic recording), angular displacement 1 mm = 0.55″, mass $M = 255$ gm, distance of the center of oscillation from the axis of rotation $L = 15.6$ cm.

From $J = 520$ meters and $V = 6.1$, we obtain by the relations $V = J/L$ and $T_o = 2\pi\sqrt{L/g}$ the equivalent pendulum length $L = 85.2$ meters and the natural period $T_o = 18.5$ sec, as given by Reid (1910) and Wood (1921). Both published also that for angular displacement, 1 mm = 0.55″. But it is $J/(360° \times 3600''/2\pi) = 2.52$ mm/″ or 1 mm trace amplitude corresponds to ground tilt of 0.39″, unless any unmentioned unknown is still involved.

With $\epsilon = 1.114$ we get a damping factor of only $\varsigma = 0.034$. For short period waves, the dynamic magnification will be similar to the static magnification. For surface waves of period around 18.5 sec, the resonant magnification will be $\tilde{V} = V/(2\varsigma) = 89.7$, and for 20 sec period, $\tilde{V} = 33$.

2.2 Seismogram Reading and Origin Time

From the record (Figure 2) the classification of the recorded seismic waves is the following: eP = 9h 21m 30s GMT, ePPP = 25m 01s, eiPS = 30m 48s, iSS = 34m 16s, eL = 42m 18s, MR = 49m 00s – 52m 00s (T 20 sec). These arrival times yield an epicentral distance from Kew of 67.6° or 7500 km, and an origin time of 9h 10m 35s GMT or 5h 10m 35s a.m. HLV. A correction of the origin time of ±10 sec is possible due to slightly different measurements of the arrival times.

In the seismogram of Mauritius Island (Figure 1), the body waves are not clear enough, but from the photomounting we obtain: ePKP = 9h 30m 00s GMT, ePP = 31m 42s approximately, iS = 39m 18s, iSS = 47m 18s, e(SSS) = 54m 00s, ML = 72.4, MR = 78m 00s – 81m 00s (T 20 sec). From this, the epicentral distance of the Caracas earthquake from Mauritius Island is about 124.3° or 13,800 km, and the origin time is 9h 11m 00s GMT.

By courtesy of Dr. J. B. Shepherd, Seismic Research Unit, University of West Indies, Trinidad, the author obtained a Bulletin (Anonymous 1908) with information on the Milne seismograph records at St. Clair (10°40′ N, 61°31′ W). For this station, to obtain GMT, one has to add to the local time 4 hours, 6 minutes and 2.5 seconds. The arrival time for the Caracas earthquake is given as 5h 6m local time (light shock) and from this an arrival time of 9h 12m 2.5s results. As Trinidad is 580 km east of the macroseismic epicenter, which after Fiedler (1961, 1968) has the coordinates 66.8° W, 10.9° N, we may subtract a travel time of 75.3 sec, and obtain for the origin time 9h 10m 47.2s. Because of the large magnitude of the earthquake, the first onset on the seismogram of Trinidad should have been the P_n-wave with a velocity around 7.7 km/sec.

With the epicenter as given by Richter (1958) as 66° W, 11° N, the epicentral distance to Trinidad is 500 km, the travel time of a P_n-wave is 64.9 sec and the origin time becomes 9h 10m 57.6s.

By courtesy of Mr. W. Person and J. Minsh, National Earthquake Information Service, USGS, Golden, CO, USA, the published arrival times at the stations TNT, VIC, KEW, CMS, POL, TRI, LJU, NIC, MRI, and TOK were used for computation of the epicenter. Depending on what data were kept or dropped, the discrepancy was too large and corrections must be found and made. Nevertheless the obtained origin times varied in most cases between 9h 10m and 9h 11m.

2.3 Magnitude of the Caracas Earthquake from 1900

The Milne seismograph (E-W component) of Kew Observatory recorded both the Caracas earthquake of 1900 and the San Francisco earthquake of April 18, 1906. The epicentral distance from San Francisco to Kew (after Reid, 1910) is about 77.63° or 8,620 km, and from the Caracas epicenter to Kew 67.0° or 7,437 km. The origin time of the San Francisco earthquake, Figure 3, is given by Richter (1958) as 13:12 GMT.

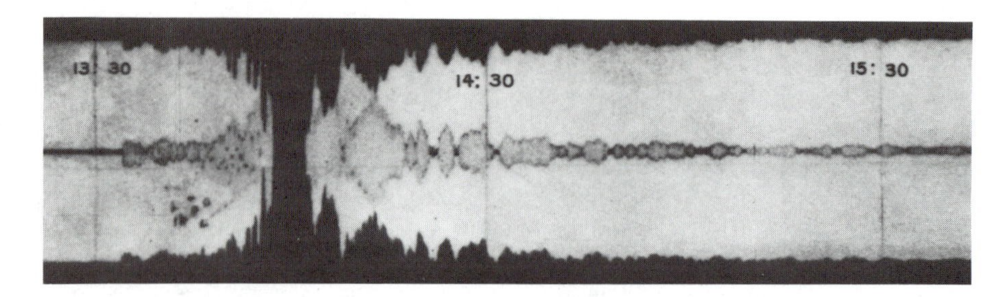

Figure 3. The San Francisco earthquake of April 18, 1906, H = 13:12:00 GMT, recorded by the same Milne horizontal seismometer as in Figure 2, E-W component, at the Kew Observatory, England. Copy taken from Lawson *et al.* (1908). Scale 1:1.

If one compares the seismograms shown in Figures 2 and 3, one finds that the amplitudes of the P-waves of the Caracas earthquake recorded in Kew at 9h 21.5m are larger and clearer than that of the San Francisco event recorded in Kew at 13h 24.5m and 13h 25.6m, even if the azimuthal effect of the orientation of the instrument vs. the direction of propagation of waves is considered. The same is valid for the S-waves, having the San Francisco record a clearer and sharper onset and only where the Rayleigh waves with periods from 30 to 20 sec appear, the trace for the San Francisco event is much larger and strongly over-recorded for four to six minutes.

From Figure 2 we get a zero-to-peak amplitude for the first visible P-arrival of 0.1 mm and this yields with a Q-value of 6.7 and the relation $m_b = Q_v + \log_{10}(A \times 10^3/2vT)$, $m_b = 7.4$ to 7.5, including the correction for azimuth of this east-west seismograph component and assuming a period of 3 sec for the P-wave, recorded by the relative long period Milne seismograph. This result from a horizontal component is questionable, but one should not forget that, for example, for the recent Caracas earthquake of July 29, 1967, the USCGS/NEIS, based on worldwide modern instrumental data, published m_b magnitude values between 4.5 and 7.2, with a mean of 5.6.

Richter (1958) published a magnitude of 8.4 for the Caracas earthquake of 1900. Using information from B. Gutenberg's worksheets, Kanamori and Abe (1979) obtained $M_S = 8.2$, and corrected $M_S = 8.1$. Abe and Noguchi (1983) recomputed magnitudes of historical earthquakes and obtained for the Caracas event $M_S = 7.7$. But they applied the abnormally high damping ratio of 1.55 to get the "average response curve of Milne seismographs" shown on page 6 of their paper.

From the mentioned damping ratio $\epsilon = 1.55$, we define the logarithmic decrement of damping as $D = \log_{10} v = 0.190$, and with this we get the damping constant μ^2 by the relations from B. Galitzin and E. Wiechert (see Sieberg, 1923):

$$\mu^2 = \frac{\pi^2}{\pi^2 + (D/\log_{10} e)^2} = 0.981.$$

From this we get the damping number $h = \sqrt{1 - \mu^2} = 0.137$. This is equal to the damping factor ς, which is used in this paper, because we can write $\ln \epsilon = 0.438$ is the logarithmic decrement "d" for half period and

$$\varsigma = \frac{d/\pi}{\sqrt{1+(d/\pi)^2}} = 0.137$$

or 13.7 % of the critical damping. With this and the static gain of 6, the resonant gain $= 6/(2\varsigma) = 21.5$, so that the dynamic gain used by Abe and Noguchi (1983) is just 3.6 times larger than the static one, and this appears not to be a representative result for most Milne seismometers which are not equipped with a device for viscous damping.

Kanamori and Abe (1979) made important tests of a reconstructed Milne seismograph, the free period of which is 15 sec, and $h = 0.05$ or 5 % of the critical damping. The problem with the dynamic response is discussed in that paper.

From the Kew seismogram (Figure 2), we obtain a zero-to-peak trace amplitude of about 8 mm for the maximum of the 20 sec Rayleigh waves recorded between 9h 49m to 9h 52m. Taking into consideration of the fact that theoretically the vertical component of a pure Rayleigh wave has an amplitude 1.4 times larger than the horizontal one, and that a cosine has to be applied to the trace amplitude, projecting it into the plane of the true oscillation of ground particles, and putting the result into the relation:

$$M_S = \log A + 1.66 \log \Delta^\circ + 2.0$$

where A is the ground amplitude measured in microns and Δ° is the epicentral distance in degrees, we obtain with a dynamic gain of 33 and the ground motion of 400 μ, $M_S = 7.6$ which is about equivalent to $m_b = 7.1$.

The seismogram of the Caracas earthquake recorded on Mauritius Island (Figure 1) is difficult to evaluate because of too many unknown constants and other information. If the seismogram represents the E-W component, the effective gain for 20 sec Rayleigh waves is about 200, and since the maximum amplitude is 19 mm on the seismogram, we obtain $M_S = 7.5$.

Magnitudes as a function of the duration of the earthquake record τ (in minutes) were calculated by Kondorskaya and Shebalin (1977) by the relation (depth less then 70 km):

$$M_S \approx M_\tau = 2.4 \log \bar{\tau} + 1.6$$

with the empirical result of $\bar{\tau} = 3\tau$ for Milne seismographs with an effective gain of 30. As the factor of τ decreases if the gain increases, we may use for the most responsible gain of $\widetilde{V} = 36$, a factor 2.5 for $\log \bar{\tau}$, and with a duration of about 150 minutes, we get $M_S = M_\tau = 7.7$, but also higher values could be obtained as long as "duration" is not well defined or adjusted to a certain level of ground motion.

3. Summary of Results

For the Caracas earthquake of October 29, 1900, the preliminary results of this paper are summarized together with a few data from other authors for comparison.

Origin time (GMT)	Epicenter Lat. (°N)	Epicenter Lon. (°W)	Depth (km)	M_S	References
	11.4	66.1			Centeno Grau (1940)
9h 11m 00s	11.0	66.0		(8.4)*	Richter (1958) - ISC
9h 11m 49s	10.9	66.8	50	7.4	Fiedler (1961)
9h 10m 35s				7.6	Kew record (this paper)
9h 11m 00s				(7.5)	Mauritius record (this paper)
9h 10m 47s					Trinidad data (this paper)
9h 10m 57s					Trinidad data & Richter
				8.1#	Kanamori and Abe (1979)
				7.7#	Abe and Noguchi (1983)

The average M_S without the * entry is 7.66 ± 0.24. The average M_S without the * and # entries is 7.55 ± 0.11.

Summarizing and forming mean values, we obtain the following final focal data for the 1900 earthquake: Origin time H = 09h 10m 50s GMT or 05h 10m 50s HLV; Epicenter Lat. 10.9° N, Long. 66.3° W, focal depth about 50 km; Magnitude $M_S = 7.6$, $m_b = 7.1$.

4. Conclusion

These preliminary results show that historical seismograms are needed for studying historical earthquakes, especially for determining magnitudes and energy releases which are important for seismic risk studies. In the specific case of this large Venezuelan earthquake, a possible repetition may cause much more loss of life and of property damage than in the years 1900 and 1967. To help such studies, a few original or reconstructed historical seismographs should be operating at observatories where historical seismograms of such instruments still exist in order to compare recent earthquakes with historical events.

ACKNOWLEDGMENTS

I am indebted to Mr. I. Dunputh, Meteorological Services, Vacoas, Mauritius, and to Dr. P. W. Burton, Institute of Geological Sciences, Edinburgh for copies of the Milne seismograms of the Caracas earthquake; to Mr. W. Person and his staff at USGS, Golden, CO for trying to get an epicenter location, and to the reviewers of comments and suggestions of the manuscript.

REFERENCES

Abe, K. and S. Noguchi (1983). Revision of magnitudes of large shallow earthquakes, 1878-1912, *Phys. Earth Planet. Interiors*, **33**, 1-11.

Anonymous (1900). Meteorological observations made at the R.A. Observatory Mauritius (1900). Plate II & Table I, page LV, Mauritius.

Anonymous (1900). *Circular No. 3*, Brit. Assoc. Science.

Anonymous (1908). Seismograph records, *Bull. Dept. Agriculture*, **VIII**, 64-67, Governm. Print. Office, Trinidad.

Centeno Grau, M. (1900). El Terremoto Del 29 de Octubre, *Linterna Magica, año IX, No. 3*, Caracas.

Centeno Grau, M. (1940). Estudios Sismologicos, *Monografia Litogr. Comercio*, Caracas (Reprinted by Cartografia Nacional, Caracas, 1969), 365 pp.

Fiedler, G. E. (1961). Areas Afectadas por Terremotos en Venezuela, *Mem. III. Congr. Geol. Venez.*, 1791-1810, MEM, Caracas.

Fiedler, G. E. (1968). Estudio Sismologico de la Region de Caracas con Relacion al Terremoto del 29 de Julio de 1967, *Bol. Inst. Material. Model. Estructural.*, Univ. Central Venezuela, **23-24**, 127-222.

Fiedler, G. E. (1980). Una Contribucion a la Microregionalizacion Sismica en la Region de Caracas, *Proyecto CONICIT*, 1-42, Caracas.

Gutenberg, B. (1929). *Handbuch der Geophysik*, **IV**, 973-975.

Jakubovicz, E. and H. Larotta (1974). El Terremoto del 29 de Octubre de 1900, *Bol. Inst. Material. Model. Estructural.*, Univ. Central Venezuela, **11**, 23-62.

Kanamori, H. and K. Abe (1979). Reevaluation of the turn-of-the-century seismicity peak, *J. Geophys. Res.*, **84**, 6131-6139.

Kondorskaya, N. V. and N. V. Shebalin, (1977). *A New Earthquake Catalog of the U.S.S.R.*, Moscow.

Kortazzi, J. (1903). Les perturbations du pendule horizontal a Nicolajew en 1900, *Beiträge Geophysik*, Band 5.

Lawson, A. C. *et al.* (1908). *Atlas of Maps and Seismograms - California Earthquake of April 18, 1906*, Carnegie Inst., Washington, D.C.

Mazelle, E. (1901). Erdbebenstoerungen zu Triest 1900, *Akademie der Wissensch*, Band 5, p. 41.

Omori, F. (1903). Observations of earthquakes at Hitotsubashi (Tokyo), *Earthquake Inv. Comm.*, **13**, 1, 143.

Reid, H. F. (1910). *The California Earthquake of April 18, 1906, Part 2. Instrumental Records of the Earthquake*, 59-192, Carnegie Inst., Washington, D.C.

Richter, C. F. (1958). *Elementary Seismology*, Freeman and Co., San Francisco, 710 pp.

Schütt, R. (1900). Mittheilungen der horizontalpendel station Hamburg, Oktober 1900. *Berichte der Sternwarte Hamburg Hohenfelde Bol.*, **1**, p. 2.

Sieberg, A. (1923). *Geologische, Physikalische und Angewandte Erdbebenkunde*, Monogr., Verlag G. Fischer, Jena, 572 pp.

Sievers, W. (1905). Das Erdbeben in Venezuela vom 29. Oktober 1900., *Festschrift, Geogr. Vereinigung, Bonn, Germany*, 35-50.

Troconis de Veracoechea, E. (1979). *La Tenencia de la Tierra en el Litoral Central de Venezuela*, Monogr., Univ. Simon Bolivar, Caracas, 185 pp.

Wood, H. O. (1921). A list of seismologic stations of the world, *Bull. Nat. Res. Council (USA)*, **2**, Part 7, No. 15, 397-538.

Evaluation of Damage and Source Parameters of the Málaga Earthquake of 9 October 1680

D. Muñoz and A. Udias
Cátedra de Geofisica
Universidad Complutense de Madrid, Madrid, Spain

ABSTRACT

On the 9th of October 1680, an earthquake caused widespread destruction in the city of Málaga and surrounding region in south Spain. A comprehensive search for contemporary documentation was made and the information used for a detailed analysis of the damage and characteristics of this earthquake. Intensities were evaluated at 38 localities and an isoseismal map has been drawn. Maximum intensity reached VIII–IX in the MSK scale and covers an area of approximately 20 km radius toward the west of Málaga. Damage in the city of Málaga has been studied in full detail, where about 20% of the buildings were totally destroyed and 60% suffered heavy damage, half of them had to be abandoned. Low quality of construction contributed to the damage. Total number of casualties was not high, approximately 60 dead and 150 injured. Radius of perceptibility was fairly large, the shock was felt in Madrid 400 km away with intensity IV. Intensity attenuation and focal parameters have been derived from the intensity distribution. This study shows that contemporary documents must be studied with care in their historical and cultural background to avoid overrating in the evaluation of intensities.

1. Introduction

On the 9th of October 1680, an earthquake occurred in southern Spain producing heavy damage to the city of Málaga and nearby towns and villages. This shock, one of the most important to occur in southern Spain since 1500, is listed in all Spanish catalogues (concerning catalogues of earthquakes in Spain, see Muñoz and Udias, 1982). However, oldest catalogues show some confusion regarding the date. Zahn (1696) and Moreira de Mendonça (1758) date it in November, while Perrey (1847) and Mallet (1858) list three events in September 1679, August and October 1680, which are obviously the same shock. In the recent catalogues, maximum intensity assigned to this shock is: Milne (1912) II (Milne scale), Sánchez Navarro-Neumann (1920) X (Forel-Mercalli scale), Munuera (1963) VIII (Mercalli scale) and IX (Forel Scale). The modern catalogue (Mezcua and Martinez Solares, 1983) gives the following entry: 9 October 1680, 7h, 4° 24′ W, 36° 30′ N, I = IX (MSK scale), south of Málaga. The epicenter is located offshore, but as will be shown, there is not sufficient evidence for this.

To understand the importance of this earthquake and the information contained in the contemporary documents, the historical and demographic situation of Spain and in particular of Andalucia must be taken into account (Ambraseys *et al.*, 1984). The year of the earthquake, 1680, corresponds to the reign of Carlos II (1661-1700) who occupied the throne of Spain after the death of his father Felipe IV in 1665.

His rule represents the darkest period of the Austrian dynasty in Spain. Frequent periods of famine and widespread plagues in 1647-52 and 1676-85, together with emigration to America resulted in a decrease of the population between 1590 and 1717 from 8,485,000 to 7,500,000 in all Spain. The city of Málaga, an active harbour had at that time an estimated population of about 18,000, while that of the province was about 95,400. Estimations of the population of the area, town by town can be obtained from the different surveys made in 1594, 1646, 1694 and 1768 (Anonymous, 1829).

To evaluate the extent of the damage and to determine the focal parameters of this earthquake information has been procured from historical sources. This information provides knowledge of the effects of this earthquake on persons, buildings and the terrain about the epicenter. For this purpose a comprehensive search for historical sources and their careful evaluation has been made. From the obtained macroseismic data, an intensity map has been drawn. Focal parameters such as maximum intensity, macroseismic epicentral location, magnitude, and seismic moment have been estimated from the intensity data.

2. Historical Sources

An extensive search for documents relative to this earthquake was made in libraries and archives. The result is a collection of written contemporary documents of different types. The first type includes eleven published accounts describing the damage experienced in Málaga, Sevilla, Córdoba, Jaen, Antequera, Madrid, and Valladolid. These can be classified under the heading of "Relaciones" (Relations) a word that appears in most of the titles. Although, they are written for the general public and, in general, their style is embellished with frequent figures of speech and exaggerations, some of these documents contain very detailed information, specifying names of villages, of the damage to houses, churches and other buildings and the number of casualties. A second type of sources are ecclesiastic and civil "Memoriales y Actas" (Memorials and Acts) found in cathedrals, churches and city halls, most of them in manuscript form; because of their official character, this type of document contains the most reliable information. To find these documents the archives of cities and towns of Andalucia have been consulted. A third source of information is contemporary publications outside Spain in which this earthquake is mentioned with some detail. The most extensive account is given in the Gazette de France n°98 (1680), which also relates unusual storms and floods in Spain in the same year. R. Hooke's account of the earthquake is practically a translation of the French text. Shorter relations are included in German and Dutch texts. Although all these accounts are based on Spanish sources, especially the "Relaciones" and do not add any new information, they are of interest since they indicate the importance of this event. Secondary sources have been also consulted.

A list of the most important contemporary documents divided in the three mentioned categories and secondary sources is given in the Appendix I. Not all documents are independent firsthand accounts and there is considerable dependence of some documents on others. This must be taken carefully into consideration at the time of assessing the intensities. For example a document printed in Sevilla will give firsthand data from that city and copy from other documents the damage of other cities and towns.

3. Evaluation of Damage and Parameters

The information collected from all sources has been used in the estimation of intensities at 38 sites. The fact that this earthquake coincided with one period of pestilence (1676-1685), in a certain way, has been an advantage in the evaluation of intensities. In many localities where no appreciable damage was produced, fear of the pestilence made the impression of the shock fall into a second place and the earthquake was not even mentioned in contemporary documents. It is probable that in the absence of the plague, more attention would have been given to the earthquake and its effects magnified in the accounts. This could have led to overestimation of intensities in many locations. Since documents, usually reflect only serious damage, they are only found in the region near the epicenter. Felt reports which correspond to lower intensities are given only for important cities such as Madrid, Sevilla, Valladolid, etc. A list of the towns with estimated intensities is given in Table 1. The intensity map resulting from these data is shown in Figure 1. Dashed lines are used for regions with insufficient data. After much consideration maximum intensity has been estimated at VIII–IX (MSK). This estimation is lower than some previously given intensities for this earthquake, as mentioned above.

The zone of maximum intensity covers an area of approximately 25 km radius, centered to the west of Málaga. This zone is well documented from the information of Málaga and surrounding towns; we have estimated that intensity nowhere exceeded VIII–IX (MSK). For example in Alhaurin el Grande, the church was totally ruined, 122 houses collapsed, 53 houses were rendered uninhabitable and the rest were heavily damaged, and 7 persons died. Although the percentage of destroyed houses was large, construction was of very low quality everywhere, most of type A according to the MSK scale. For this reason it is not definite that intensity IX was clearly reached. In Málaga, where churches were more solidly built, only two of them suffered substantial damage (Santos Mártires and San Juan). Areas with intensities VI and VII are also well documented, since the shock was strongly felt and caused sufficient damage to be recorded in contemporary documents. One must be careful in contrasting more than one document to avoid gross errors. For instance, in one account published in Sevilla, it is said that the cathedral of Jaen suffered damage beyond repair, however documents from Jaen itself, such as the Acts of the

Table 1. List of Towns With Estimated Intensities

Intensity	Locality
VIII-IX	Alhaurin el Grande, Alhaurin de la Toree, Alozaina, Benalmadena, Cartama, Coin, Málaga, Mijas, Pizarra
VIII	Alora, Competa
VII	Antequera, Ecija, Osuna
VI-VII	Lucena, Ronda, Vélez-Málaga
VI	Alcala del rio, Andujar, Carmona, Cazalla de la Sierra, Cordoba, Granada, Jaen, Morón de la Frontera, Sevilla
V-VI	San Juan de Azfalrache (Aznalfarache)
V	Jerez de la Frontera, Sanlúcar la Mayor
IV	Cádiz, Madrid, Puerto de Santamaria, Sanlúcar de Barrameda, Toledo
III	Almeria, Valladolid
Not Felt	Alcala la Real, Motril

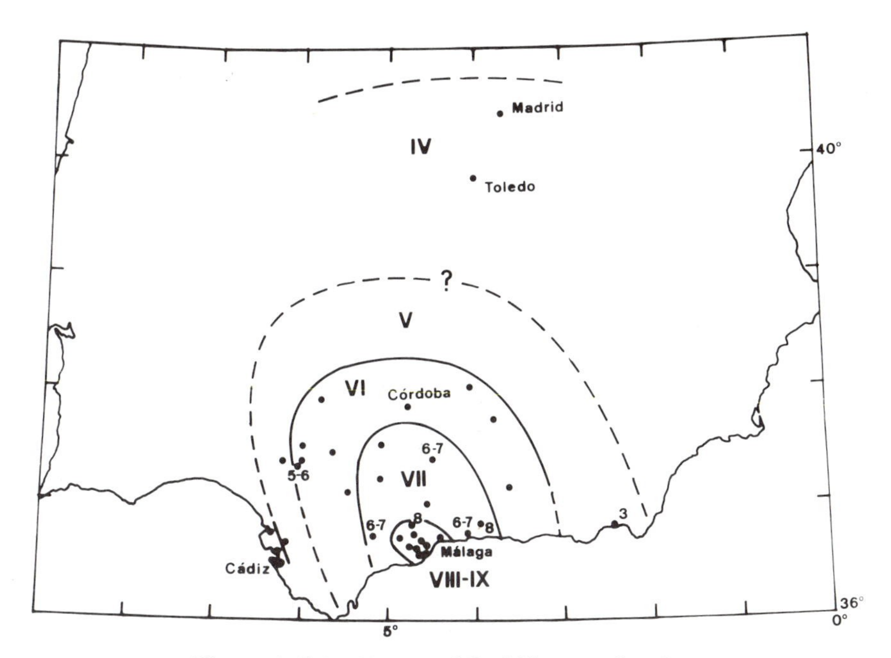

Figure 1. Intensity map of the Málaga earthquake.

Chapter of the Cathedral and of the City Hall, do not mention any damage; in the same way there is no notice of any damage to the cathedral by Galera (1977) in his book on architecture of Jaen. This critical assessment of the liability of the different accounts is very necessary in order to obtain correct values of intensity. This lack of information and of sufficient critical judgement is reflected in the intensity map presented by Gentil and Justo (1985) for this earthquake, where some values are overestimated and the shape of curves distorted.

In Madrid the shock was strongly felt causing the bells to ring and many people to flee from their houses (intensity IV). The shock was also felt further north in Valladolid, giving a radius of perceptibility of about 700 km. However, the total number of casualties was not great, 60 dead and 150 injured. Of the dead, 42 correspond within the city of Málaga.

The area of greater damage, shown in Figure 2, is located in the Betics, a zone of complex tectonics due mainly to superposition of nappes. During the Neogene, several interior depressions were formed (Bousquet *et al.*, 1978), one of them that of the Guadalhorce River to the west of Málaga. These depressions were later filled with Neogene and Quaternary deposits and formed autonomous structural elements. The Guadalhorce basin is surrounded by the Alpujarride complex, formed by Paleozoic and Triassic materials, and the Malaguide complex, formed by non-metamorphic Paleozoic materials. It is well known that the epicenter of an earthquake may not coincide with the area of maximum intensity. The instrumental epicenter is the projection on the horizontal plane of the initial point of rupture, while location of the maximum damage depends greatly on the geometry of the fracture, local geology and terrain conditions. However, it is a common practice to assign also epicenter

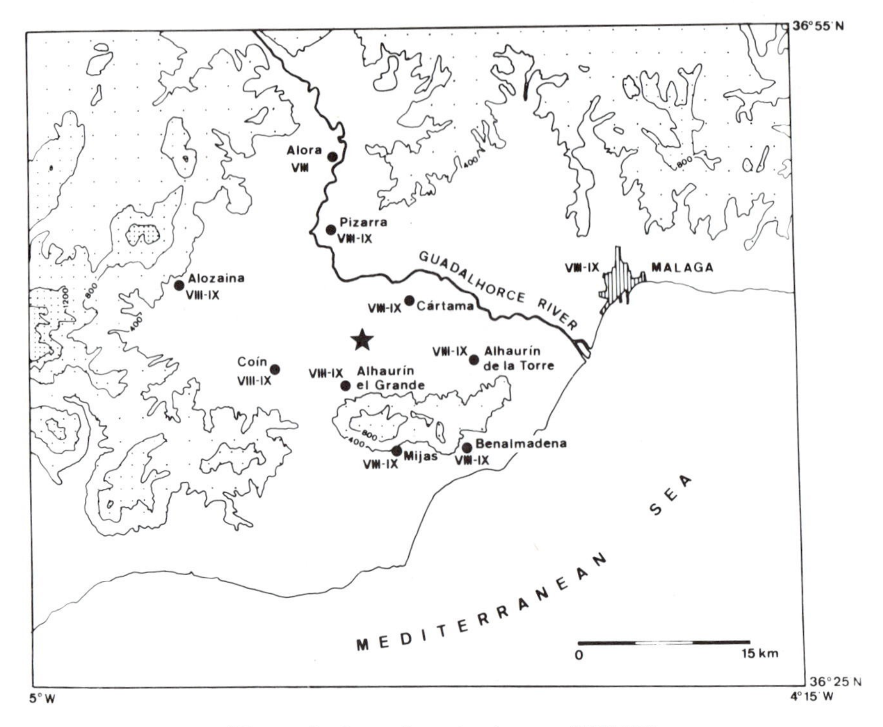

Figure 2. Area of greater damage (VIII-IX).

coordinates to historical earthquakes. These are based on the macroseismic data and are located at the center of the maximum intensity area. In the case of the Málaga earthquake, the isoseismal lines are not closed and it is possible that the epicenter was situated offshore. However, besides the absence of direct evidence of tsunami effects, there are reports in Málaga that the shock came from the west, and the area of maximum damage is located at the Guadalhorce basin west of Málaga where there exist deep geological faults oriented approximately N 110° E. The orientation of these faults agrees with the elongation of the maximum intensity area. These arguments do not completely exclude the location of the epicenter at sea, but better support a location inland to the west of Málaga. From the map shown in Figure 2, the estimated coordinates are 36.7° N, 4.7° W.

Origin time cannot be given with better precision than the hour. Accounts give 7 a.m. and 7:15 a.m. local time and since there is no knowledge of the state of clocks, this may be taken as the best approximation.

From the intensity map the attenuation coefficient and depth has been calculated from the relation

$$I_o - I = a \log\left(\sqrt{\Delta^2 + h^2}\,/h\right). \tag{1}$$

Two directions of maximum and minimum attenuation, corresponding to west and north respectively have been taken. From them, an attenuation coefficient of $a = 6$

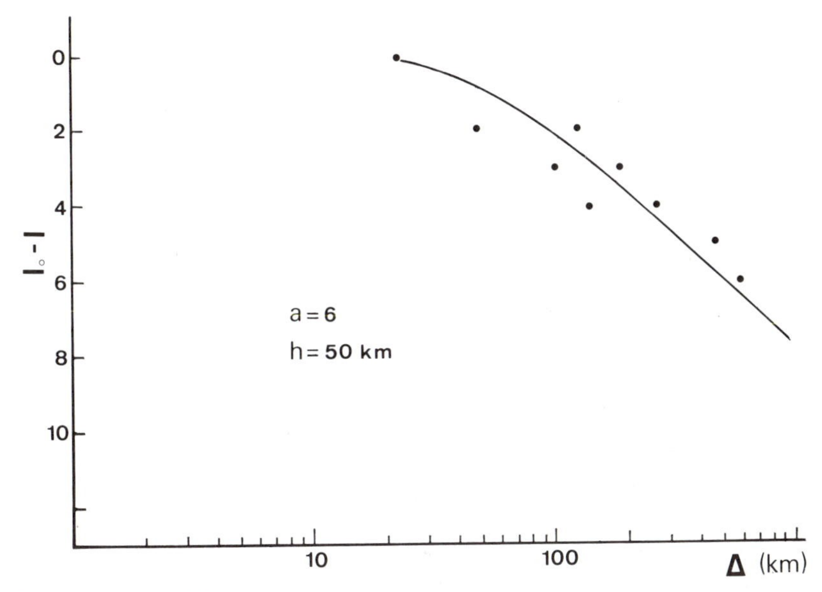

Figure 3. Plot of $I_o - I$ versus Δ (km). Observational values are plotted as dots, and the curve is based on Equation (1).

and an average depth of 50 km are obtained (Figure 3). It must be noticed that this method gives pairs of values for attenuation and depth, that are not independent. The relatively high attenuation value agrees with values obtained for other earthquakes in this zone (Muñoz *et al.*, 1974). The average value for the depth, locates the hypocenter under the crust which in this region has a value of approximately 25 km (Ansorge and Banda, 1980). This depth of the focus is not unusual in this region since in the seismic catalogue (Mezcua and Martinez Solares, 1983) there are several earthquakes with depths between 30 km and 100 km. The values obtained for attenuation coefficient and depth are not very precise because the intensity zones are not well defined, due to the low density of points, except for the highest grades.

From the maximum intensity and depth, a value of magnitude may be estimated using the formula of Sponheuer (1960)

$$M = 0.661\, I_o + 1.7 \log h - 1.4. \tag{2}$$

The value obtained is between 6.8 and 7.4 using $h = 50$ km and taking $h = 30$ km, 6.4 and 7.1 (lower and upper limits correspond to $I_o =$ VIII or IX). Another expression that relates magnitude and intensity is

$$M = 1 + \frac{2}{3} I_o. \tag{3}$$

This relation gives a magnitude between 6.3 and 7.

Seismic moment can be estimated from this value of magnitude according to Brune (1968) giving a value between 10^{18} Nm and 10^{19} Nm. This parameter can also be estimated directly from the area of intensity VI, according to the results given by Herrman *et al.* (1978) for earthquakes in California and Central United States. Although the attenuation of Spanish earthquakes is more similar to that of California than of Central United States, a mean value of both places has been taken. For the radius of intensity VI we have taken 135 km, a mean between the maximum 175 km and the minimum 100 km. Using this value, the area obtained is 1.2×10^5 km^2 and resulting seismic moment

$$M_o = 1.25 \times 10^{19}\,\text{Nm}. \tag{4}$$

A comparison of this earthquake with that occurred between Granada and Málaga the 25th of December of 1884 with maximum intensity IX (Udias and Muñoz, 1979) is of interest. Although the maximum intensity seems to have been higher in 1884, the area of intensity VI is one order of magnitude smaller. Also, the 1680 earthquake was felt with greater intensity in Madrid, even though its epicenter was more distant. The extent of the felt area and the fact that no aftershocks are mentioned anywhere seem to substantiate the fact that this was a deeper shock.

Evaluation of damage in the city of Málaga has been made from all existing documents, but in first place we base our analysis on the Act of the Notary of the Episcopal Audience of Málaga, the 15th of October of 1680, where the number of houses, churches and convents that suffered damage is specified, as well as the kind of damage affecting them. This information is presented on a contemporary map in Figure 4. Location of churches and convents has been identified in the map and the extent of damage is shown by the black sections of the circles. Numbers in the map situating the buildings correspond to those of Table 2. Four convents were destroyed and two had to be abandoned after the earthquake. The city of Málaga was divided into four parishes. The Act gives the total number of houses in each parish, and the number houses that suffered total destruction, heavy damage, etc. This permits a calculation for each parish of the percentages of the damage suffered by the buildings. These percentages are represented in the large circles of Figure 4 for each individual parish; complete destruction is shown in black, heavy damage in narrow lines, appreciable damage in wide lines and light damage in dots. Heaviest damage is located in the parishes of San Juan and Santos Mártires with 39 persons dead, 87 injured, and total destruction of nearly 25% of all buildings and appreciable damage of between 50% and 65%. In the parish of San Juan, the district of Percheles, located on the alluvial deposit of the Guadalmedina River, was almost totally ruined. In Table 2, a summary of the damage suffered by each of the four parishes of the city of Málaga is presented. Of a total of 4320 houses, 795 were destroyed, 355 suffered serious damage, 2385 considerable damage and 635 moderate damage. This means that practically all houses were somewhat affected by the earthquake. Personal casualties are given as 42 dead and 110 injured. All these numbers are based on the Notarial Act mentioned above.

In Figure 5, a present day map of the city of Málaga is presented, where the outline of the limits of the 1600 city has been drawn. It can be seen that major development has taken place to the west, precisely where damage from this earthquake was most intense. As has been mentioned before this zone is built on recent sedimentary terrain of the Guadalmedina and Guadalhorce Rivers. This part of the city can be expected to suffer heaviest damage in the case of a future earthquake similar to that of 1680. In that event it is not sure what will be the behaviour of the modern structures, most of them reinforced concrete.

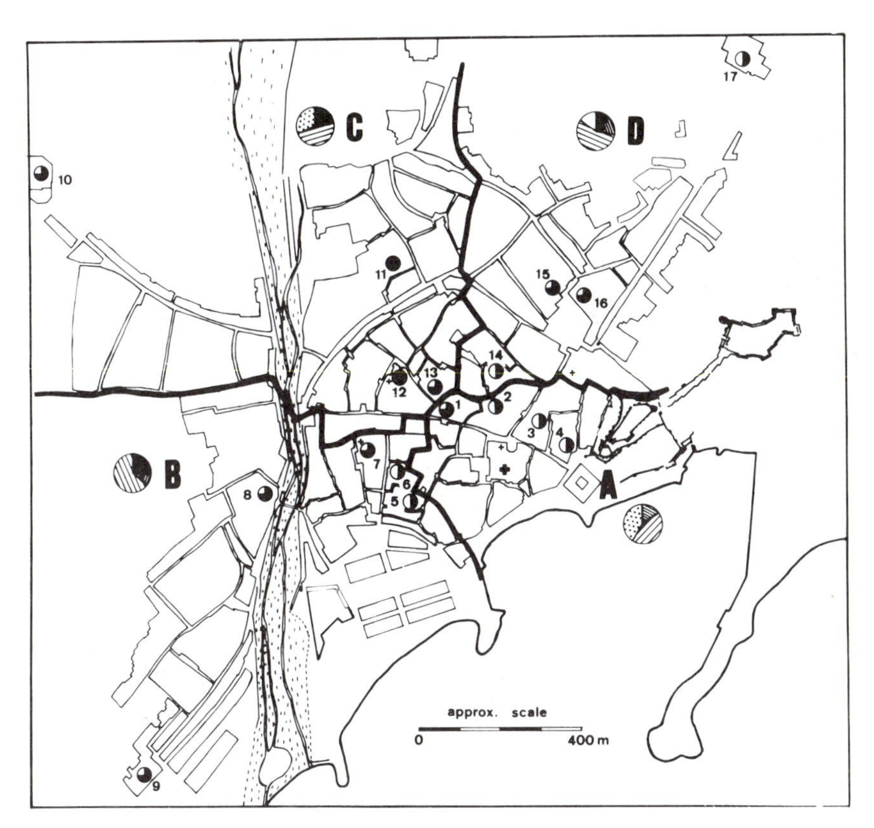

Figure 4. Map of Málaga with damage sites identified.

4. Conclusions

Compilation and critical analysis of historical documents concerning the earthquake of 9 October 1680, has led to the assessment of how it was felt and how much damage was produced in more than 38 localities in Spain. From the intensity data the following parameters have been determined: macroseismic epicenter coordinates 36.7° N, 4.7° W; focal depth about 50 km; maximum intensity VIII-IX (MSK); estimated magnitude 7 and seismic moment 10^{19} Nm; attenuation coefficient 6.

Though not the most destructive one, this earthquake is probably the one with largest magnitude in southern Spain since 1500. If this analysis proves to be correct, seismic risk in southern Spain may have to be revised since it is now biased toward higher hazard in the Granada area.

The analysis of historical sources has shown that documents must be judged very critically, establishing which ones provide first hand information and which are the literary genera used in each case. Also knowledge of the socioeconomical situation, demographic conditions, building characteristics, etc. is necessary to a

Table 2. Damage by Parishes in the City of Málaga

A. Sagrario Parish. 2 dead.

Total houses 390: heavy damage 57 (15%), appreciable damage 173 (44%), light damage 160 (41%).

1. Carmelitas Descalzas. Heavy damage.
2. Santa Clara. Appreciable damage.
3. San Agustin. Appreciable damage.
4. Recoletas Bernardas. Appreciable damage.

— Religiosas Agustinas. Totally destroyed, abandoned after earthquake.

B. San Juan Parish. 24 dead, 61 injured.

Total houses 1211: complete destruction 310 (25%), heavy damage 113 (10%), appreciable damage 788 (65%).

5. Trinitarios Descalzos. Appreciable damage.
6. Colegio Clérigos Menores. Appreciable damage.
7. Iglesia de San Juan. Heavy damage.
8. Santo Domingo. Heavy damage.
9. Carmelitas Descalzos. Heavy damage.

C. Santos Mártires Parish. 15 dead, 26 injured.

Total houses 1642: complete destruction and heavy damage 379 (23%), appreciable damage 788 (48%), light damage 475 (29%).

10. Trinitarios. Heavy damage.
11. San Francisco. Destroyed and uninhabitable.
12. Iglesia de los Santos Mártires. Destroyed.
13. Dominicas. Heavy damage.

— Encarnación. Totally destroyed, abandoned after earthquake.

D. Santiago Parish. 1 dead, 23 injured.

Total houses 1077: complete destruction 106 (10%), heavy damage 185 (17%), appreciable damage 636 (59%).

14. San Bernardo. Appreciable damage.
15. Nª Sra de la Merced. Heavy damage.
16. Nª Sra de la Paz. Heavy damage.
17. San Francisco de Paula. Appreciable damage.

correct assessment of the damage. Lack of these considerations will lead to gross errors in intensity estimations with consequent effects on the evaluation of seismic risk. Finally, studies of historical events are of great importance in the evaluation of seismic hazard, especially in regions where large earthquakes are not frequent.

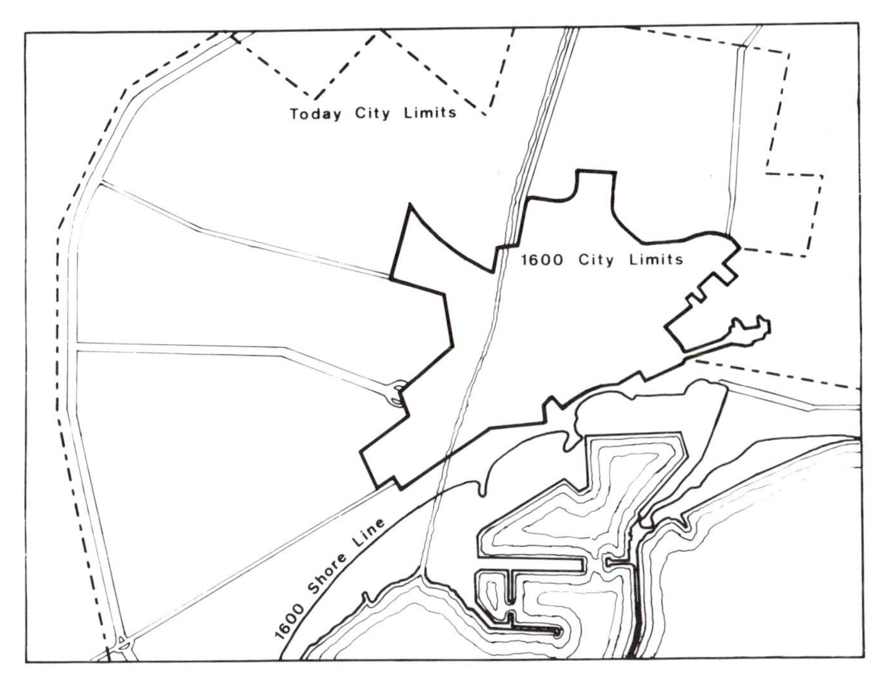

Figure 5. Present day map of the sity of Málaga showing the outline of the limits of the city in 1600.

ACKNOWLEDGEMENTS

The authors wish to thank all persons who contributed in the seach of historical documents and especially V. Vogt., M. de la Torre, J. M. López Marinas, C. Juan Lovera and E. Gómez Martinez. Also they appreciate the anonymous reviewers for their critical and useful commentaries. Contribution No. 263, Cátedra de Geofisica, Universidad Complutense, Madrid (Spain).

REFERENCES

Anonymous (1829). Censo de población de las Provincias y Partidos de la Corona de Castilla en el siglo XVI. Con varios Apéndices para completar la del resto de la Peninsula en el mismo siglo, y formar juicio comparativo con la del anterior y siguidente, según resulta de los Libros y Registros que se custodian en el Real Archivo de Simancas. De orden del Rey N. S., Madrid en la Imprenta Real.

Ansorge, J. and E. Banda (1980). Seismic experiments in the arc of Gibraltar crossing the Ronda peridotita Complex A, *Abstract Geophys. Un. Meeting, Toronto.*

Bousquet, J. C., C. Montenat, and H. Philip (1978). La evolución tectónica reciente de las Cordilleras Béticas orientales *Geodinámica de la Cordillera Bética y del Mar de Alborán,* Universidad de Granada, 59-76.

Brune, J. N. (1968). Seismic moment, seismicity and rate of slip along major fault zone, *J. Geophys. Res.,* **73**, 777-784.

Galera Andreau, P. A. (1977). Arquitectura de los siglos XVII y XVIII en Jaen, Caja General de Ahorros y Monte de Piedad de Granada.

Gentil, P. and J. L. Justo (1985). Mapa de isosistas del terremoto de Málaga de 1680, *Revista de Geofisica*, **41**, 65-70.

Herrmann, R. B., Ch. Shiang-Ho, and O. W. Nuttli (1978). Archeoseismology applied to the New Madrid earthquakes of 1811 to 1812, *Bull. Seism. Soc. Am.*, **68**, 1751-1759.

Mallet, R., and J. W. Mallet (1858). The earthquake catalogue of the British Association, with the discussion, curves, and maps, etc., Transaction of the British Association for the Advancement of Science, London.

Mezcua J. and J. M. Martinez Solares (1983). Sismicidad del área Iberto-Mogrebi, Presidencia del Gobierno, Instituto Geográfico Nacional, Madrid.

Milne, J. (1912). A catalogue of destructive earthquake A.D. 7 to A.D. 1899, *Rep. Bri. Asso. London*, 1911-1912, 549-740.

Moreira de Mendonça, J. J. (1758). Historia universal dos Terremotos que tem havido no mundo, de que ha noticia, desde a sua creacao atè o seculo presente. Con huma Narracam individual do Terremoto do primero de Novembro de 1755 e noticia verdadera dos seus effeitos em Lisboa e todo Portugal, Algarves, e mais partes de Europa, Africa, e America, aonde se estendeu: e huma Dissertacao Phisica sobre as causas geraes dos Terremotos, seus effeitos, differences e prognosticos: e as particulares do ultimo. Lisboa, Na. Offic. de Antonio Vicente da Silva.

Munuera, J. M. (1963). Datos básicos para un estudio de sismicidad en el área de la Peninsula Ibérica, *Memorias del Instituto Geográfico Catastral*, Tomo XXXII, Madrid.

Muñoz, D., A. López Arroyo, and J. Mezcua (1974). Curvas medias de variación de la intensidad sismica con la distancia epicentral, *Memorias I^a Asamblea Nacional de Geodesia y Geofisica*, 327-339, I.G.N., Madrid.

Muñoz, D. and A. Udias (1982). Historical development of Spain's catalogs of earthquakes, *Bull. Seism. Soc. Am.*, **72**, 1039-1042.

Perrey, A. (1847). Sur les tremblements de terre de Peninsule Iberique, *Annales des Sciences Physiques et naturelles d'agriculture et d'industrie*, Tomo X, 461-514.

Sánchez Navarro-Neumann, M. M. (1920). Bosquejo sismico de la Peninsula Ibérica, *La Estación Sismo-lógica y el Observatorio Astronómico de Cartuja*, Granada.

Sponheuer, W. (1960). *Methoden zur herdtiefenbestimung in der makroseismik*, Akademik-Verlag, Berlin.

Udias, A. and D. Muñoz (1979). The Andalusian earthquake of 25 December 1884, *Tectonophysics*, **53**, 291-299.

Zahn, J. (1696). Speculae Physico-Mathematico-Historicae Notabilium & Mirabilium Sciendorum, Scutinium IV Disquisitio I, (Geo-scopica), Norimbergae.

APPENDIX I

A. Relations

- Relacion verdadera de la lastimosa Destruicion, que padeciò la Ciudad de Málaga, por el espantoso Terremoto que sucediò el Miercoles 9. de Octubre deste presente año de 1680. Biblioteca Nacional V. E. 69-4 and 69-71. Madrid. Archivo Histórico Nacional. Madrid.

- Segunda relacion del horrible Temblor de Tierra que padeciò la Ciudad de Málaga el Miercoles 9. de Octubre deste año de 1680. Refierense las circunstancias que faltaron a la Primera, assi de lo sucedido en dicha Ciudad, como en todos los Lugares de sus Contornos. Archivo Histórico Nacional. Madrid.

- Relacion sucinta de los sucedido en la ciudad de Málaga con el terremoto, y temblor de tierra, Miercoles 9, de Octubre deste presente año de 1680. Biblioteca Universitaria de Granada. Fondo Montenegro, A-31-126 Leg. 17bis.

- Temblor de Tierra en Málaga este año de 1680. Biblioteca Universitaria de Granada. Fondo Montenegro, A-31-126 Leg. 16.

- Temblor de Tierra en Málaga este año de 1680. Archivo de la Catedral de Granada, 8-504-4. Biblioteca Nacional, Ms. 18654-82. Madrid.
- Relacion verdadera en que da quenta de la Ruina que â Causado, el Temblor de Tierra en la Ciudad de Málaga, y Lugares de su Comarca, y assi mismo lo que causo en Madrid, sucedido el dia nueve de Octubre, este presente año de 1680. Impreso en Sevilla por Juan Francisco de Blas. Biblioteca del Ayuntamiento de Málaga, 12856-114.
- Tercera relacion, en que se da quenta de las ultimas Noticias de las Tempestades sucedidas en el pasado mes de Setiembre deste presente año de 1680. Y assimismo se refiere el espantoso Temblor de Tierra que sobrevino à la Coronada Villa de Madrid, Corte Augusta de nuestro Monarca Carlos Segundo (que Dios guarde) el Miercoles nueve de Ocyubre de dicho año, en punto de las siete de la mañana. Archivo Historico Nacional, Madrid.
- Relacion verdadera, que da cuenta del espantoso temblor de tierra, que en la muy Noble, y muy Leal Ciudad de Sevilla sucedio el dia Miercoles 9, de Octubre deste año de 1680. Impreso en Sevilla por Juan Cabecas, 1680. Biblioteca Universitaria de Granada. Fondo Montenegro A-31-126 Leg. 17.
- Segunda, y verdadera relacion, en la qual se da cuenta de los estragos, y ruinas que hizo el temblor de tierra el dia 9. de este presente mes de Octubre, en las Ciudades, Villas, y lugares de estos Reynos. Impreso en Sevilla por Juan Cabecas, 1680. B.C.C. ms 82-5-21, ff 146 and 146 bis. In "Memorias de Sevilla (1600-1678)". Edición introductión y notas de F. Morales Padrón. Monte de Piedad y Caja de Ahorros de Cordoba, 1981.
- Relacion de el Terremoto, que el dia nueve de Octubre de mil y seiscientos y ochenta padeció la Ciudad de Cordoba, y de las demonstraciones que en accion de gracias hizieron los Illustrisimos Cabildos de la Santa Iglesia Cathedral, y de la Ciudad. Impreso en Cordoba por el Licenciado Antonio de Cea, y Paniagua Presbytero, 1680. Archivo Museo de Artes Populares J. Diaz Escovar y Urbano. Málaga.
- Relacion verdadera, en que se refiere lo sucedido el Miercoles nueve de Octubre deste presente año de 1680. con el espantoso Temblor de Tierra, que generalmente se padeció à las siete de la mañana en estos Reynos, y especialmente en Cordova, Valladolid, Iaen, Antequera, y otras partes, con otras Noticias que vera el Curioso. Biblioteca Nacional, Sala Varios. Madrid.

B. Memorials and Acts

- Archivo Municipal de Málaga; Actas Capitulares 1680. Jueves 10, Sabado 12 Octubre.
- Archivo Catedral de Málaga; Actas Capitulares 1680.
- Archivo Municipal de Sevilla; Actas Capitulares 1680. Miercoles 23 Octubre.
- Archivo Catedral de Sevilla; Actes Capitulares 1680. Miercoles 9, Viernes 11, Lunes 21, Viernes 25 Octubre.
- Archivo Histórico Municipal de Cordoba; Actas Capitulares 1680. Viernes 11 Octubre.
- Archivo Catedral de Cordoba; Actas Capitulares 1680, Miercoles 9, Viernes 11, Martes 15 Octubre.

- Archivo Municipal de Jaen; Actas Capitulares 1680. Miercoles 9, Viernes 11 Octubre.
- Archivo Histórico Catedral de Jaen; Actas Capitulares 1680. Miercoles 9 por la mañana, Miercoles 9 por la tarde Octubre.
- Archivo Parroquial de San Bartolome (Jaen); Libro de Bautismos de 1676 a 1730 de la Parroquia de San Lorenzo Folio n° 18.
- Archivo Municipal de Almeria; Actas Capitulares 1680. Martes 22 Octubre, Martes 5 Noviembre.
- Archivo Histórico Municipal de Andujar; Actas Capitulares 1681. Viernes 24 Enero.
- Certificacion: Manuel Fernando de Velasco, Notario oficial mayor desta Audiencia Episcopal de Malaga que al presente exerco el Oficio de Notario mayor della. Biblioteca Universitaria de Granada. Fondo Montenegro A-31-126 Leg. 16.
- "SEÑOR". Memorial al Rey pidiendo exenciones tributarias, elevado por la Santa Iglesia Catedral de Malaga. Archivo Museo de Artes Populares J. Diaz Escovar y Urbano. Málaga.
- Breve relacion de la mision apostolica que hizo el Colegio de San Pablo de la Comp[a] de Jesus de la Ciudad de Granada con occasion del horrible temblor de tierra sucedido en dicha Ciudad miercoles 9 de Oct[e] a las 7 de la mañana este año de 1680 y de su causa y effectos. Biblioteca Universitaria de Granada. Fondo Montenegro A-31-126 Leg. 15.

C. Publications Outside Spain

Gazette de France N. 98, pp. 609-620, 1680. Tremblement de Terre, arrive le 9[e] Octobre dans toute l'Espagne: Avec les particularitez des dommage causez par les tempestes & les inondations dans le mesme Royaume.

R. Hook. V. Discourses of Earthquakes, their Cause and Effects, and Histories of several; to which are annext, Physical Explications of several of the Fables in Ovid's Metamorphes, very different from other Mythologick Interpreters. In the Posthumous Works of Robert Hook, M.D.S.R.S. Geom. Prof. Gresh. Containing his Cartesian Lectures, and other Discourses, read at the Meetings of the Illustrious Royal Society.

G. H. Petri. Onder-Aadze Storm-klok Door de Almachtige en alleen Wonder-doende hand Gods, in een onverwachte en verschrikkelyke Aardbeeving, By Dirk en Hendrik Bruyn, Amsterdam 1692.

Anonymous. Christelijke Aanmerkingen Op de Sware Aardbeevinge, By Willem Jansz, Utregt 1692.

Anonymous. Unglücks Chronica, vieler grausahmer and erschrecklicher Erdbeben, By Thomas von Wiering, Hamburg 1963.

D. Secondary Sources

C. Garcia de la Leña (1792). Conversaciones históricas Malagueñas. Biblioteca Museo Diocesano. Málaga.

F. de Botella y Hornos (1885). Los Terremotos de Málaga y Granada. Bol. Soc. Geográfica de Madrid. T. XVIII.

J. Diaz de Escovar y Urbano (1885). Los Terremotos de Málaga. Datos históricos coleccionados por D. y U. Acrhivo Museo de Artes Populares J. Diaz Escovar y Urbano. Málaga.

J. Estrada Sagalerva (1971). Efemérides Malagueñas. Biblioteca Canovas del Castillo. Málaga.

J. B. Olaechea Labayen (1980). La sismicidad en la capital de España y su región, con referencia especial al terremoto de 1755. Artes Gráficas Municipales. Madrid.

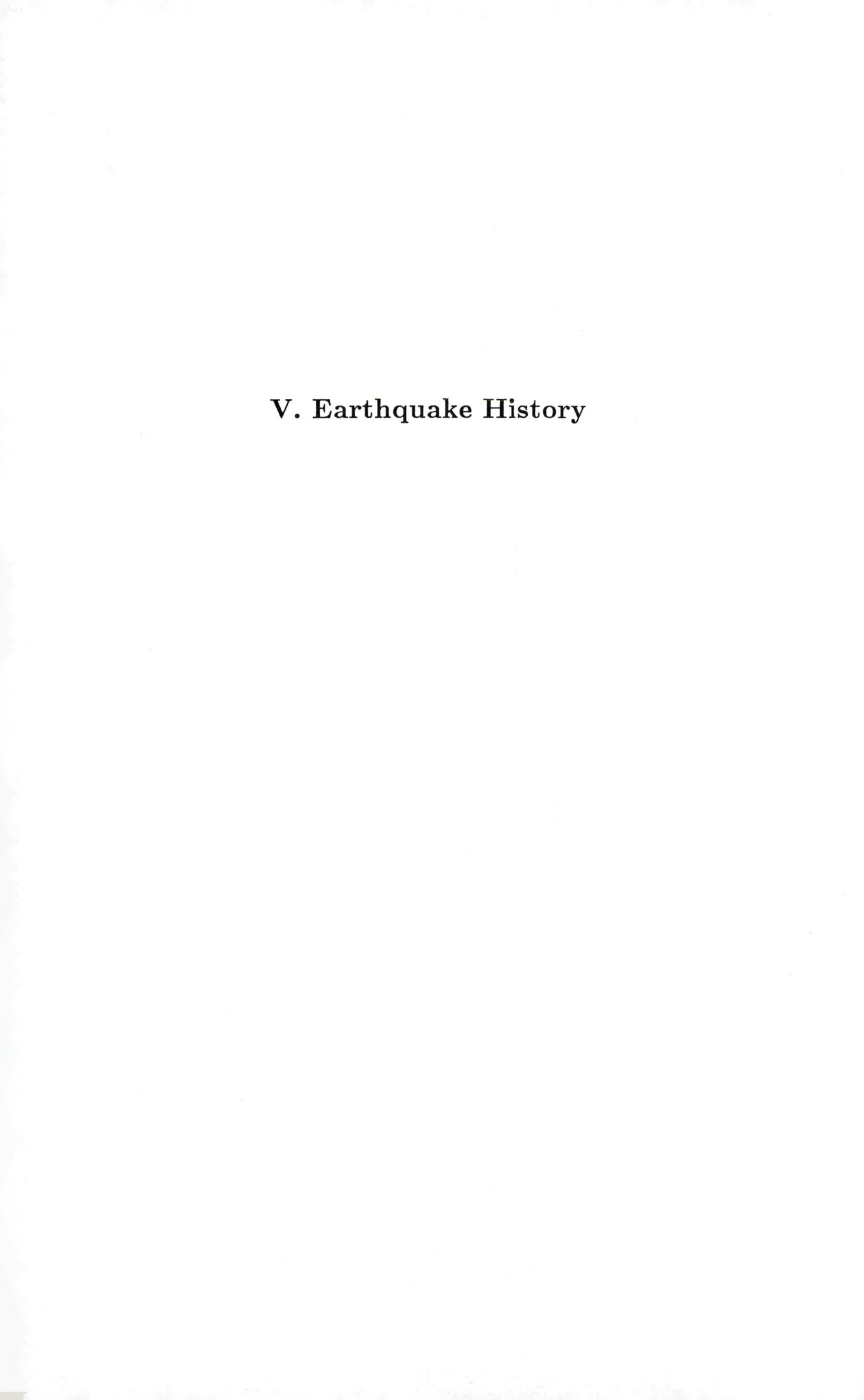

V. Earthquake History

Studies of Earthquakes on the Basis of Historical Seismograms in Belgium

M. De Becker and T. Camelbeeck
Centre de Géophysique Interne, Royal Observatory of Belgium
3, Avenue Circulaire, 1180 Brussels, Belgium

ABSTRACT

The history of the seismological stations in Belgium is told, with a special emphasis on the description of the oldest instruments. The first Belgian station was set up in the Royal Observatory in Brussels in 1899 and was equipped with a Rebeur-Ehlert seismometer. This station was later equipped with two Wiechert penduli (horizontal and vertical components) as well as with three Galitzin seismometers.

In 1984, the "Centre de Géophysique Interne" started to review the seismicity of Belgium using historical documents and seismograms. The brief study of two main earthquakes that occurred during this century in Belgium show the usefulness of the analysis of historical seismograms to better assess earthquake hazards, especially in regions with quite low seismicity such as Belgium.

The first seismological station in Belgium was set up in the grounds of the Royal Observatory at Uccle, a suburb of Brussels, at the beginning of 1899. The geophysics cellar, originally the private property of E. Solvay, a well-known patron of science, was offered to the government in 1904. The first earthquake was recorded on 21 March 1899. In 1903, stations also were set up in a quarry of Quenast and in the mines at Frameries, in the hope of recording and predicting explosions of fire-damp, which were claiming many victims. No seismograms from either station have survived.

The instruments at all three stations were Rebeur-Ehlert pendulums made by the firm of Bosch in Strassburg. Somville (1907) gives a full description of these instruments. They consist of three identical horizontal pendulums oriented in directions 120° apart, and enclosed in a heavy brass box provided with levelling screws, and covered with a glass plate. Each pendulum carried a mirror, which enabled its movements to be recorded photographically. The total weight of the instrument was 68.3 kg.

The pendulums were regularly calibrated, and operated at a period of about 10 sec. With the axis placed horizontally (as in determining a reduced pendulum length) the period was 0.64 sec. A static magnification of about 160 was obtained. Because of the slow speed of the recording drum, times cannot be read reliably to better than about 8 sec. The Royal Observatory still has the instrument and some of its records.

In June 1906, two "Strassburg type" heavy horizontal pendulums, with a mass of 100 kg, were installed at Uccle. These "tronometers" oriented N-S and E-W, were much less sensitive than the Rebeur-Ehlert instruments, and their periods proved to be very unstable, varying between 10 and 25 sec (Somville, loc. cit.). On 18 May 1907, the E-W component from the Rebeur-Ehlert instrument was provided with damping, obtained by means of a small glass fibre trailing in oil.

At the beginning of 1909, a 1000 kg Wiechert astatic pendulum, made by Spindler and Hoyer of Göttingen, was added together with a clock that was changed daily by telephone to the Royal Observatory time-service. A timing accuracy of about 1 sec was obtained. The first vertical instrument, a 1300 kg Wiechert pendulum by the same makers, was added in May 1910. The equipment was further augmented the following year by the addition of two classical Galitzin horizontal instruments, made by Masing in St. Petersburg (Galitzin, 1911).

In May 1912, the Bosch tronometers were given to the Ministry for the Colonies and installed at Elisabethville, Congo (now Lubumbashi, Zaire).

Instrumental constants were regularly determined, and are reported in the published monthly bulletins, so that the seismograms retain their full use. The Wiecherts proved to be very sensitive to temperature changes in the vault, but the periods of the Galitzins remained very stable, at just under 25 sec. In June 1913, the clock was electrically synchronized with the Royal Observatory time-service, and timing accuracy to 0.1 sec became possible.

Two further additions to the instrumentation were made. In April 1930, a vertical Galitzin-Wilip with a period of about 8 sec was installed. Its sensitivity was to be improved and its period lengthened to 10 sec in 1936, when the original steel springs were replaced by Elinvar. Three Sprengnether instruments obtained in the early 1950's are still in use.

In 1984, the "Centre de Géophysique Interne" began a systematic study of Belgian seismicity, which involved an inspection of all the available seismograms of shocks before 1950, including those made abroad. Table 1 shows preliminary results for the two best-documented earthquakes, a magnitude $M_S = 5$ shock in the Brabant massif on 11 June 1938, and the earthquake of 3 April 1949. The first of these was the largest in Belgium since 1800, and the second, a magnitude smaller, is representative of shocks in the Hainaut province, where it caused considerable damage. The earthquake of 11 June 1938 is similar in magnitude to the biggest historical events in this region, as shown in Figure 1. The correct assessment of the seismic hazard in Belgium needs especially a detailed study of those main historical events. Concerning the 11 June 1938 earthquake, original seismograms or copies were collected from 19 stations (Figure 2). Epicentres have been determined using well-determined P_n arrival times, at 19 stations at distances from 140 to 800 km in the first case, and 10 stations from 140 to 500 km in the second case (Table 2). A P_n velocity of 8.1 km/sec was adopted as an appropriate average value for the region. Adopted focal depths were derived from macroseismic data as Uccle, the nearest station, was more than 50 km from the epicentre.

Table 1. Characteristics of the Two Earthquakes Studied

	Date	Origin time	Lat. °N	Long. °E	Depth (km)	M_S	m_{bLg}
1	1938 Jun 11	10h 57m 33s	50.78	3.59	15-25	5.32±0.17(10)	5.59 ±0.19 (11)
2	1949 Apr 03	12h 33m 41s	50.45	4.11	shallow	4.27±0.31(05)	4.59 ±0.25 (07)

The numbers in parentheses indicate the number of stations used for the computation of these parameters.

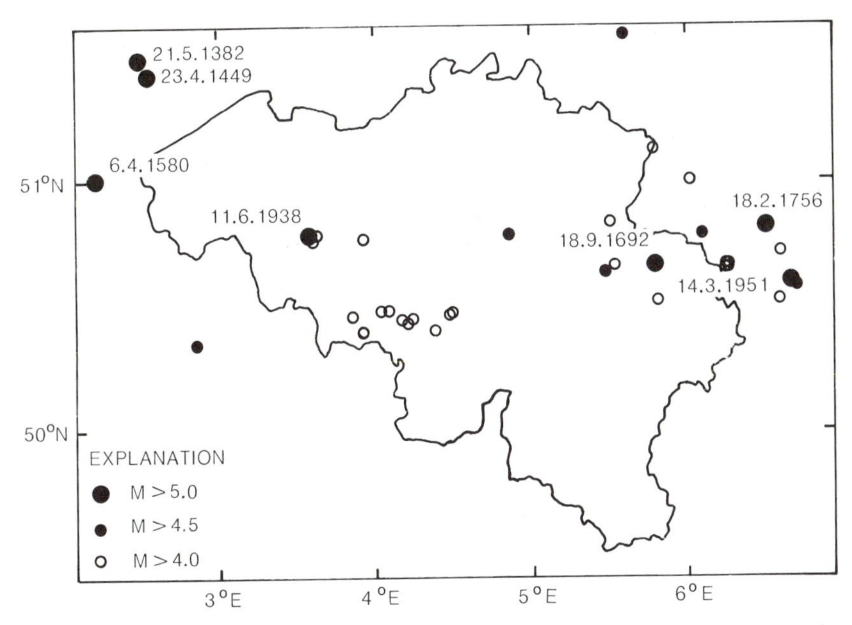

Figure 1. Main earthquakes felt in Belgium since 1380.

Figure 2. Stations from which original seismograms or copies were received for the June 11, 1938 earthquake.

Table 2. Stations and Parameters Used for Location of the Earthquake of April 3, 1949 at 12h 33m

Stations	Δ (km)	P_N	O-C residues (sec)
HEE	140.9	12h 34m 6.0s	−0.10
PAR	216.7	12h 34m 15.0s	−0.45
STR	308.8	12h 34m 30.0s	+3.17
BAS	411.7	12h 34m 39.2s	−0.33
STU	412.3	12h 34m 38.5s	−1.10
NEU	437.5	12h 34m 42.1s	−0.61
JRS	468.4	12h 34m 47.0s	+0.48
ZUR	474.4	12h 34m 46.9s	−0.37
RAV	499.7	12h 34m 50.0s	−0.40
CHU	566.2	12h 34m 58.9s	+0.30

Magnitudes were determined in two ways: 1) M_S from the Prague formula (Kárník, 1969); and 2) m_{bLg} from the maximum amplitude of the vertical component of the Lg-waves (Camelbeeck, 1985), or from the vector sums of the maximum amplitudes of the Lg-waves recorded on the horizontal components when no data are available from the vertical component (in this case the amplitude value is divided by a factor 2.5).

$$M_S = \log(A/T)_{\max} + \sigma(\Delta) + \Delta M_S \quad (1)$$

$$m_{bLg} = 2.60 + 0.833 \log_{10}(\Delta/10) + 0.0011\,\Delta + \log_{10} A(\Delta) \quad (2)$$

The values of the function $\sigma(\Delta)$ given in Table 3 are derived from Kárník (1969). Period and amplitude data for the earthquake of 11 June 1938 are summarized in Table 4. Table 5 shows the limits of distance and period within which the Prague formula holds.

It is interesting to compare the values of magnitude obtained from different stations. The scatter of both M_S and m_{bLg} values from the mean and the variation of M_S with epicentral distance are plotted in Figure 3. Considering the date at which the records were made, the scatter is gratifyingly small.

Figure 4 shows the maximum ground displacements for Lg-waves of one second period as a function of epicentral distance for both events, together with curves for shocks of different magnitude. There does not appear to be any systematic dependence on epicentral distance, suggesting that historical seismograms can be made to produce consistent results for studies of this kind, and therefore have a special value for the study of regions with low seismicity.

Table 3. $\sigma(\Delta)$ for LgH

$\Delta°$	0.5	1.0	1.5	2.0	2.5	3.0	3.5	4.0	5.0	6.0	8.0	9.0
$\sigma(\Delta)$	2.2	2.7	3.0	3.3	3.6	3.8	4.0	4.1	4.4	4.7	4.80	4.86

Table 4. Period and Amplitude Data for the Earthquake of 11 June 1938

Station	Dist. (km)	Instr.	A_V (μm)	T_V (sec)	A_E (μm)	T_E (sec)	A_N (μm)	T_N (sec)	ΔM_S
Paris	233				170	1.1	70	1.1	0.0
Bochum	266				52	1.3	49	1.3	0.0
Kew	282	WA			74	*	73	*	−0.1
Strasbourg	388	W	53	3.5	80	2.8	98	3.2	−0.1
Karlsruhe	400		42	1.7	72	2.5	27	2.5	−0.2
Göttingen	453	W			12.5	2	13	2	+0.1
Stuttgart	461	G			22.5	3	21	4	+0.1
Bâle	463	Q	14.5	1.6	19.5	1.6	40.5	1.9	
Zürich	526	Q			13.5	*	11.5	1.7	−0.1
Hambourg	533	W	8.3	3.6	18	4.4	24	3.6	−0.3
Jena	562	W	8.5	2.1			10	2.5	+0.2
Edinburgh	721	M			12.8	4.0			
Vienne	966	W			17	3.3	21	3.3	−0.1

Instruments: WA = Wood-Anderson; W = Wiechert; G = Galitzin; Q = Quervain-Picard; M = Milne-Shaw. The symbol (*) indicates that the period of ground motion is unreadable. (A_V, T_V), (A_N, T_N), (A_E, T_E) are the amplitudes and periods read on the vertical, north-south and east-west components of the seismograms.

Table 5. $T(\Delta)$ for LgH

$\Delta°$	1-2	3	4	5-6	7-8	9-12	13-19	20-30	30-50	50-80	80-120
T (sec)	1	1.5	2	3	6	8	10	12	15	18	20

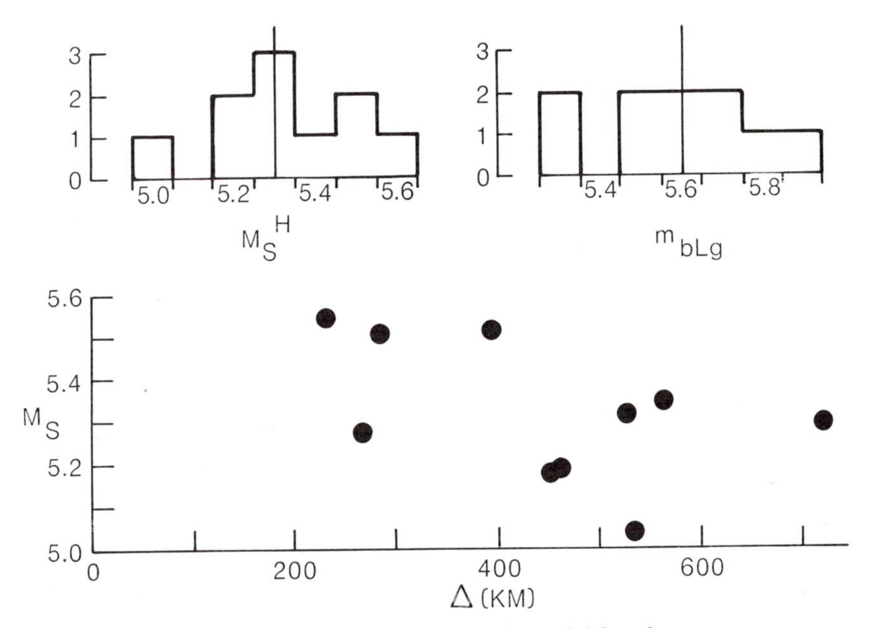

Figure 3. Dispersion of the m_bLg- and M_S-values.

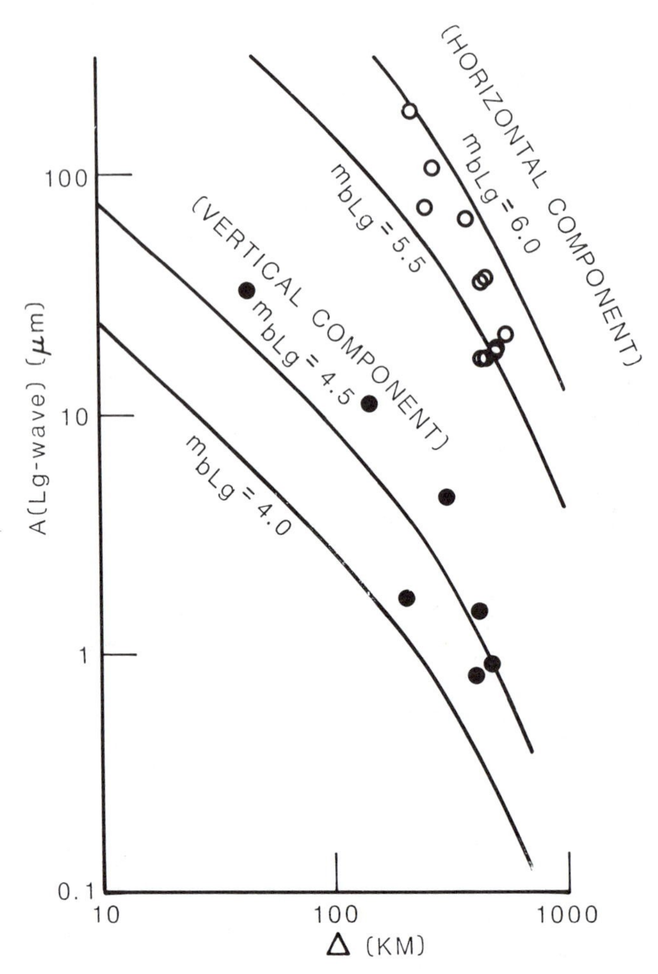

Figure 4. Maximal ground displacement of the 1 sec Lg-wave as function of the epicentral distance.

REFERENCES

Ambraseys, N. N. (1985). Magnitude assessment of Northwestern European earthquakes, *J. Earthquake Eng. Struct. Dyn.*, **13**, 307-320.

Camelbeeck, T. (1985). Some notes concerning the seismicity in Belgium, in *Seismic Activity in Western Europe*, D. Reidel Publ. Co., 448 pp.

Galitzin, B. (1911). Ueber ein neues aperiodisches Horizontalpendel mit galvanometrischer Fernregistrierung, *Imperial Academy of Sciences*, t 4., liv. 1.

Kárník, V. (1969). *Seismicity of the European Area*, D. Reidel Publ. Co., 364 pp.

Somville (1907). Annales de l'Observatoire Royal de Belgique.

ADDENDUM

Table 6, which follows, supersedes Table 4. It includes additional amplitude-period data for the Belgian earthquake of 11 June 1938 supplied by Prof. N. Ambraseys. Where Lg-phases are shown, M_S values were calculated using the calibration curves derived for Northwestern Europe (Ambraseys, 1985). Where Lm is shown, magnitudes were estimated from the Prague formula.

The average value of M_S for the Belgian earthquake of 11 June 1938 in Table 3 from 10 stations is 5.3 ($\pm$ 0.17). For the data in Table 6, with station corrections derived by Kárník, we find from 17 station estimates, 5.0 ($\pm$ 0.27).

Table 6. Period and Amplitude Data for the Earthquake of 11 June 1938

Station	Dist. (km)	Instr.	P	A_V (μm)	T_V (sec)	A_E (μm)	T_E (sec)	A_N (μm)	T_N (sec)	M
Paris	233	w	Lg			170.0	1.0	80.0	1.0	5.52
Paris	233	w	–			170.0	1.1	70	1.1	
Bochum	266	P17	–			52.0	1.3	49	1.3	5.14
Kew	282	WA	–			74.0	*	73	*	5.42
Strasbourg	388	w	–	53.0	3.5	80.0	2.8	98	3.2	
Strasbourg	388	G	Lg			90.0	3.0	90.0	3.5	5.20
Karlsruhe	400	M	–	42.0	1.7	72.0	2.5	27	2.5	5.08
Göttingen	453	w				12.5	2	13	2.0	
		w	Pm			0.3	0.3	0.5	0.3	
		w	Lg			13.8	1.0	13.5	1.0	4.96
		W	Lg			12.5	1.0	13.8	1.2	
		w'	Lg					15.8	2.9	
		w'		15.5	3.5					
Stuttgart	461	G				22.5	3	21	4.0	4.59
Bâle	463	Q		14.5	1.6	19.5	1.6	40.5	1.9	5.05
Neuchatel	488	Q	Lg					11.4	1.8	4.64
		Q		3.8	0.8					
Zürich	526	Q				13.5	*	11.5	1.7	4.72
Hamburg	533	w		8.3	3.6	18.0	4.4	24	3.6	
Hamburg		w	Lm	7.0	4.5					
		w	Lm			16.0	5.0	22.0	4.0	5.20
Jena	562	w		8.5	2.1			10	2.5	
		w	Sm			2.0	3.0	2.0	1.0	
		w	Lg			15.0	2.0	2.0	2.0	4.68
		W	Lg					10.0	2.0	
Eger	622	w	Lm			20.0	6.0	20.0	6.0	5.22
Münich	644	w	Lm			21.0	5.5	17.5	3.7	5.34
Edinburgh	721	m				12.8	4.0			4.66
Copenhagen	799	G	Lg					13.0	3.0	4.96
Vienna	966	w				17.0	3.3	21	3.3	
	966	w	Lm					32.1	4.2	
Ogyalla	1099	M	Sm					4.0	17.0	
Pulkovo	1944	G	Lm					1.0	10.0	4.46

Instruments: w = normal Wiechert (100 kg); W = big Wiechert (15 ton); w' = very light Wiechert (10 kg); P17 = 17 ton pendulum; M = Mainka; W.A. = Wood-Anderson; G = Galitzin; Q = Quervain-Picard; m = Milne-Shaw. The symbol (*) indicates that the period of groundmotion is unreadable. (A_V, T_V), (A_N, T_N), (A_E, T_E) are the amplitudes and periods read on the vertical, North-South and East-West components of the seismograms.

Documenting New Zealand Earthquakes

G. A. Eiby
Seismological Observatory
P.O. Box 1320, Wellington, New Zealand

ABSTRACT

The record of even large earthquakes in New Zealand is likely to be seriously incomplete until organized colonization began in 1840, but accounts of over 300 earlier shocks exist. Colonists soon recognized the practical importance of earthquakes, and in 1868 the reporting of felt earthquakes was included in the duties of an organized network of climatological observers. This reporting still continues in a greatly expanded way.

The archives of the Seismological Observatory, Wellington, contain seismograms dating from 1900, but the earliest instruments were more effective in recording teleseisms than in recording local shocks. The present network of short-period instruments was started in 1930, and stored seismograms now cover about 140 square metres of floor-space.

Several earthquake catalogues have been prepared, the most comprehensive being a list of 24,000 shocks stored on magnetic tape. Fully referenced historical material forms the basis of a "Descriptive Catalogue" of which two sections, up to the year 1854, are published. A great deal of later information exists in manuscript, but is incomplete.

Origins and magnitudes based upon instrumental data recorded before 1964 are being recalculated using improved knowledge of crustal structure, wave velocities, and propagation characteristics, and are being published together with the associated felt reports and other comment. Several monographs dealing with the more important events revealed by these studies have appeared.

1. Introduction

New Zealand has an earthquake of magnitude 7 or more about once a decade (Figure 1). The shocks have been widely distributed, and only one has resulted in more than a score of deaths. Nevertheless, some eighty per cent of the population (which totals a little over three million) is urban, and recognizes that the evaluation of seismic risk is a practical problem of some urgency.

In comparison with the countries of Europe and the ancient civilizations of Asia our history is brief, but we are fortunate in the detail and completeness of our seismic record. Until about 1000 A.D., when the first Polynesian voyagers arrived, the country was completely without human inhabitants; and although Maori oral tradition records an important earthquake in about 1460, it is not until 1773, when Captain Furneaux of the "Adventure", James Cook's companion on his second voyage, experienced a moderate shock in the Marlborough Sounds, that we have a written description giving the date, place, and time of a New Zealand earthquake.

During the next three quarters of a century, the explorers were joined by sealers and whalers, and mission stations and trading-posts appeared. Unfortunately, most of them were in the far north, the least seismic part of the country, but the letters,

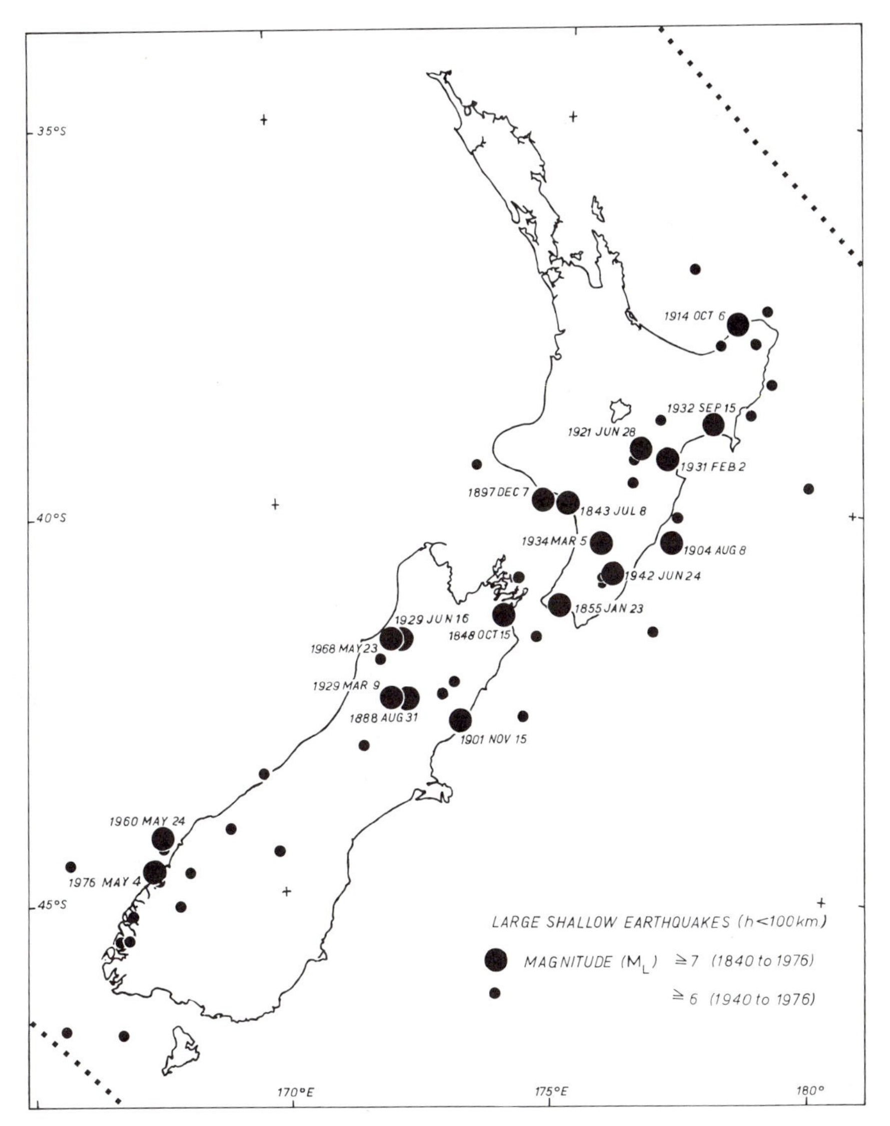

Figure 1. Distribution of shallow New Zealand earthquakes since European settlement began.

diaries, and log-books surviving from this period are well seasoned with references to earthquakes. In 1840 systematic colonization began. Coastal towns grew rapidly, and settlers moved inland in search of farming country. Many of the early colonists were highly literate. Newspapers were started, descriptions of the fauna, flora, and geology of the country were published, and gentleman amateurs of science began keeping records of the weather, of signs in the heavens, and of tremblings of the Earth.

From a seismic point of view the earliest years of the colony were unusually active. Within a few months of their arrival Wellingtonians experienced a strong earthquake and the sequence of aftershocks that followed. In 1843 the newly-built houses of Wanganui were badly damaged, and in 1848 Wellington also suffered damage from shocks centered across Cook Strait in Marlborough. New Plymouth chimneys fell in 1853, and two years later a magnitude 8 shock in southern Wairarapa was felt throughout the whole country. Few Wellington buildings escaped damage, and parts of the northern shore of Cook Strait were lifted several metres. It is no surprise to find these events well documented, and that even the occurrence of small shocks long continued to attract the notice of the press.

In 1868, government climatological observers were appointed, and their duties included the reporting of felt earthquakes. Annual summaries were published, and the possibility of making recording instruments was discussed. Milne sent details of his work in Tokyo, and in 1884 a horizontal pendulum based on his ideas began recording in Wellington. (Young, 1984; Young *et al.*, 1984). None of its records are known to have survived, but descriptions of its indications are to be found in the press. In 1898, the Government bought two Milne pendulums, which were set up in Wellington in 1900 and in Christchurch in 1901. Data were published by the Seismological Committee of the British Association for the Advancement of Science, and this is usually considered to be the beginning of instrumental seismology in New Zealand.

The first two decades of the twentieth century mark a period of confusion during which it is impossible to be sure that records are complete. Leading seismological personalities died or retired, patterns of government administration changed, and the disruptions of the first World War included delays in publication. In 1921, however, the first of the "New Zealand Seismological Reports" appeared, and these have continued under the direction of what has now become the Seismological Observatory, Wellington, surviving a number of confusing changes in the Observatory's name and its administrative control. The New Zealand Department of Scientific and Industrial Research has a statutory duty to carry out research "that is in the national interest", and the attention the Observatory can give to such matters as compiling historical catalogues or conserving existing records is consequently limited by political priorities.

2. Earthquake Catalogues

The earliest attempts to make comprehensive lists of past New Zealand earthquakes were prompted by a meeting of the Australasian Association for the Advancement of Science in Christchurch in 1891. The first (Hogben, 1891) lists 775 shocks between 1848 and 1890, and is based upon reports found in leading newspapers, in lists of climatological observations published annually since 1868, and in a few minor sources. The other (Hector, 1891) summarises the climatological reports from 1868 to 1890 and appends some early listings of felt earthquakes. Neither provides any index of severity beyond the selection of a few shocks for more extended comment. These remain the major published lists of pre-instrumental shocks, but there is much unpublished material in the Observatory's files. R. C. Hayes's "Catalogue of New Zealand Earthquakes up to and including the Year 1855" has been superseded by the published sections of my "Descriptive Catalogue" (Eiby, 1968a, 1973). Barnett's "Catalogue of Non-Instrumental Earthquake

Reports", which ends in 1899, is unfinished. It consists of a number of overlapping files, some in manuscript and some typed with manuscript additions and emendations. The contents are in the main drawn from Hogben and Hector, but there are also some personal comments, some estimates of radii of perceptibility, and references to papers published in the "Transactions of the New Zealand Institute".

Bastings (1935) published a list of 69 shocks between 1835 and 1934 that he regarded as "destructive". It was intended to alert New Zealanders to earthquake dangers, and suffers from its tenacity of purpose. A single report of fallen plaster was sufficient to secure inclusion; at least one event never occurred; and many details are doubtful or incorrect. The work of Hayes (1953), who selected sixteen shocks as "major" and assigned them magnitudes on the basis of the radii of perceptibility and other details of the felt effects, is of more lasting value. He shows insufficient appreciation of the effect of time of day and distribution of population, but it remains a useful contribution to discussions of risk.

The defects of these catalogues are largely the result of factors beyond the control of their compilers, who not only had more pressing responsibilities, but lacked the facilities now provided by the Alexander Turnbull and General Assembly Libraries, and by National Archives. Not only have there been great improvements in the catalogues of most of our major libraries, but the many transcripts that now exist provide paths through the thickets of impenetrable handwriting that so often surround historic documents.

The "Descriptive Catalogue" already mentioned is based on a systematic exploration of all relevant published material and on documents in archives and private hands. Two sections have already been published. The first discusses eighty events before 1846, and the second a further three hundred between 1846 and 1854. Shocks (and doubtful or incorrect reports) are listed chronologically, with an estimate of the probable magnitude and epicentre, a brief description including more extended quotations from unpublished material, isoseismal maps in suitable cases, and full references to published work and the location of unpublished material.

It is of course impossible to accumulate the data for a study of this kind chronologically. Anything likely to be relevant is at once copied and filed for later assessment. At the present time I have three drawers of 20 × 12.5 cm cards listing minor events, and 14 foolscap drawers containing xeroxed or handwritten transcripts of more detailed accounts. The process of assessment and correlation will naturally reduce its bulk to a more manageable size. A further section of the work, bringing it up to 1870, awaits only completion of a full study of the earthquake of 1855, and assessment of the accounts of a major shock in Hawke's Bay in 1863.

Much of the value of projects like the "Descriptive Catalogue" lies in their comprehensive nature. They cannot be completed quickly; nor is it permissible to begin by concentrating upon major events. Much of New Zealand is sparsely populated, and even today earthquakes of magnitude 5 can occur without there being a single report that they were felt. When there is a strong probability that no one at all was in the epicentral region, the only certain way to be sure that all major events have been identified is to correlate the reports of all shocks, including those of slight or moderate intensity. To yield to the temptation to concentrate upon shocks already considered to be "major" or "interesting" is not only a scientifically reprehensible way of treating data, but increases the risk that excuses of cost will be used to justify abandoning the study of the "minor" earthquakes, and that important information will be overlooked for a long time.

Immediate practical problems nevertheless call for an interim listing of earthquakes that have been considered large, or are of some other special interest. In an attempt to supply this need an "Annotated List" of about 170 shocks between 1460 and 1965 was produced (Eiby, 1968b). It contains appropriate warnings that it is a compilation of traditional wisdom and not a considered assessment, and must therefore be used with caution. Some curious conclusions have nevertheless been based upon it, justifying continued insistence upon the orderly and comprehensive approach used for the "Descriptive Catalogue".

A less fallible aid is a file of epicentre and magnitude data for about 24,000 shocks stored on magnetic tape (Smith, 1976). It includes all instrumentally-determined epicentres, and the shocks that appear in the "Descriptive Catalogue" and the "Annotated List", and is amended and extended to include new material every six months. From it, lists of shocks within selected limits of time, space, or magnitude can readily be prepared. It will eventually become a complete catalogue of New Zealand earthquakes from which any desired sub-set can be obtained, and whose limits of completeness can be assessed. Its principal deficiency at the present time is the lack of fully assessed data for the second half of last century and the first four decades of this one.

3. Revision of Instrumental Data

The routine operations of the Observatory are now summarized in annual volumes known as "New Zealand Seismological Reports". They contain lists of epicentres, focal depths and magnitudes, and the station readings on which they are based, together with summaries of the felt reports received, isoseismal maps of the larger shocks, station constants, and some descriptive matter. Since 1964 a computer has been used, and the typesetting of the whole volume is now carried out by computer. Before 1964, graphical methods were in use, but similar data have been published since about 1950. An effective recording network has covered that part of the country lying north of Banks Peninsula since 1942, and there are some instrumentally-determined epicentres of fair quality dating from as early as 1930.

In much of the pre-computer period, little was known of the complex crustal structure of the country, and of the appropriate wave-velocities. Epicentre calculations were based upon Japanese and European tables, and only in the last few years has a method of calculating local magnitudes that takes local peculiarities of propagation into account been introduced.

Two programmes of revision of this material are proposed. All epicentres and magnitudes found between 1950 and 1964 are to be re-calculated, but priority is being given to carrying the work backwards from 1950 as far as the available data will permit. The results will appear, together with new epicentre and isoseismal maps and previously unpublished felt information and discussion, in special volumes of the "Seismological Reports". Two volumes, covering the years 1948-1950 and 1945-1947 have already appeared, and calculations for 1943 and 1944 are nearly complete.

An important matter found to need urgent attention was the accuracy of adopted station positions, particularly in the case of long-established stations. Although they satisfied the needs of individual seismologists and the international agencies concerned with teleseisms when they were adopted, and were accurate enough for local earthquake studies while there were still large uncertainties in crustal structure and wave velocities, they did not match the timing accuracies it was possible to obtain from the records, and limited the value of the revisions. With some difficulty, the positions of all stations that have operated in New Zealand have now been established to a second of arc. Once the actual position of the instrument was known, reference to large-scale Lands and Survey Department maps yielded accurate coordinates. The operators of several stations that had been closed for many years were found to have died, but it proved possible to find children or other members of the family who could remember where the instruments had stood, and could in some cases even describe them in sufficient detail to establish the orientation of horizontal components. Seismologists in any doubt whether the coordinates of their stations are accurately known in terms of a uniform survey are urged to attend to the matter without delay.

4. Major Events

The dangers that can arise from concentrating the study of past earthquakes upon the events that are known or believed to be major have already been stressed. Nevertheless, closer study of particular events is not only historically valuable in itself, but the necessary preliminary to such projects as the "Descriptive Catalogue" and the revised "Seismological Reports". Neither is intended to contain exhaustive accounts of individual earthquakes, but abstracts of incompletely assessed material are of small worth.

There are few connected accounts of New Zealand earthquakes in the pre-instrumental period, and the farther our work takes us into the past, the less systematic we find the collection of data to have been. The sources are nevertheless voluminous, and the Observatory files contain information about quite recent earthquakes that has never been correlated and assessed.

During the years of depression and the second World War, and for some time afterwards, the resources of the Observatory did not allow the staff to travel except for instrument maintenance, but a great deal of information about the larger earthquakes was collected and filed. At this period New Zealand was well served by local newspapers, whose reporters provided detailed accounts of the effects they observed, and recorded verbatim their interviews with people affected. The Observatory also elicited reports from police and postal officials, Ministry of Works engineers, harbour authorities, and other informed people in the epicentral area. Even now witnesses can be found, and building repairs made at the time preserve evidence of damage. Detailed assessment of this material forms a substantial part of the work of revision for the "Seismological Reports".

The problems are well exemplified in my work on the Marlborough earthquakes of 1848 (Eiby, 1980), the magnitude 8 Wairarapa earthquake of 1855, and two tsunami-generating shocks in 1946 (Eiby, 1982a, 1982b). The existence of a fresh-looking fault scarp, and the belief that an elderly lady interviewed by a noted field geologist at the end of last century had been an eye-witness of the breakage led geologists to believe that the 1848 earthquakes had been centered in the Awatere Valley, and Richter (1957) devotes a lengthy section of his well-known book to the "Awatere Earthquake". Historical records established that the lady did not reach the valley until more than five years later, and that the valley was uninhabited at the time of the earthquake. Non-technical documents included clear descriptions of fresh breakage on a parallel fault in the Wairau Valley thirty kilometres to the north-west; but subsidences in this valley that were asserted to have been the result of this shock, and had been "explained" in two papers by an internationally celebrated geomorphologist are clearly shown by well-dated watercolours to have existed beforehand. Mistaken traditions of this kind, especially when accepted by well-known technical authorities, have serious consequences for seismic zoning, and provide a strong practical argument for rigorously conducted historical studies.

Similar confusion and mis-interpretation surrounds the 1855 Wairarapa earthquake. With the exception of a half-legendary event in pre-European times, it is New Zealand's largest known earthquake, and occupies an important place in seismological history because of the accounts of it to be found in later editions of Sir Charles Lyell's "Principles of Geology", one of the canonical scriptures of Victorian science, and one of the earliest discussions of the relationship between earthquakes and geological faulting.

Lyell's information came from three New Zealanders he met in England. Two of them had some geological knowledge, and the third was a military engineer who had already submitted an official report on the event to his superiors. When the shock occurred he was supervising coastal road-works within a few kilometres of the probable epicentre. He describes the extent and amount of coastal uplift, but makes no mention of any kind of faulting. Of Lyell's other informants, one had made his only visit to the district before the earthquake, and the other had moved from it to the epicentral region of the 1848 earthquake, of which he left accounts, before 1847. It was recognized by Ongley (1943) that Lyell had confused the two events, but not that the man on whose authority an improbably great length of fresh faulting has been generally accepted could not himself have seen it, even supposing it to have taken place at that time.

The study of historic earthquakes cannot be confined to the examination of written evidence. It must be accompanied by field-work; but it is important not to assign historic dates to geological occurrences when written evidence is lacking. It is also important to remember that statements in quite recent documents may not mean what they would mean had they been written today. Usage and context change very quickly. When studying the 1855 shock, I found it helpful to follow about 150 kilometres of the trace of the West Wairarapa Fault on foot and to visit all the places mentioned in contemporary accounts. I soon found that many names appearing on modern maps are not in the places to which they were applied in the past. It is essential to consult maps of the same period as the historical records being studied.

The two tsunami-generating shocks in 1946 (Eiby, 1982a, 1982b) occurred at a time when the Observatory was without the resources to undertake field work. Every effort was made to collect information from eye-witnesses, but there was no opportunity to make a proper assessment. The absence of a full study was realized only in the course of revising the instrumental data. New Zealand tsunamis of local origin are very rare. Since the associated earthquakes were not large, it was assumed that the height of the wave had been greatly exaggerated, but it proved to be established by photographs of damage to structures on beach and river terraces, and reports of seaweed caught in electric power-lines. The seismograms showed the earthquakes to have been unusual, and the wave is now attributed to the explosive expulsion of mud from a ruptured diapiric fold, a phenomenon that has several times been observed at inland places nearby.

5. Historic Seismograms

All seismograms are historic, though some are more historic than others. Much of the information embodied in New Zealand seismograms has already been extracted and published in a form that is more convenient for most seismologists than a copy of the record would be; but every generation of seismologists poses new questions that only direct examination of the records can answer.

The oldest seismograms that survive in the archives of the Seismological Observatory in Wellington come from the Milne instruments installed at Wellington in 1900 and Christchurch in 1901. A few of them carry important records of large New Zealand shocks, but most show only teleseisms. In 1923, the Wellington Milne instrument was replaced by a Milne-Shaw. This greatly improved the recording of local shocks, but the emphasis on teleseisms remained.

The magnitude 7.9 Buller earthquake that occurred in 1929 was the stimulus that obtained for New Zealand seismologists the instruments that they needed if they were to contribute effectively to the solution of local problems. It was felt over the whole country, and caused 17 deaths – an unprecedently large number, and unacceptable to public opinion. The following year Wood-Anderson seismographs were installed at Wellington, Christchurch, and New Plymouth, and by 1935 nine Jaggar shock-recorders were operating in different parts of the country. Although the Jaggars had no absolute timing, they produced some records of sufficient quality to be used in early studies of crustal structure. Improvements in teleseismic recording followed the installation of a set of classical Galitzins at Christchurch, and a vertical component Galitzin-Wilip at Wellington.

In terms of relevance to the continuing research programmes of the Observatory, the records made from 1930 onwards are the most important part of the archive. They should enable a substantially complete catalogue of shocks above about magnitude $4\frac{1}{2}$ or 5 to be compiled for the central part of the country. This will certainly involve a re-examination of the original records of most of the earthquakes recorded before 1942.

In 1942 further Wood-Anderson seismographs were added to the network, gradually replacing the Jaggars. At first, every seismic arrival on every record was meticulously measured, whether it could lead to an epicentre determination or not. By 1950, this was no longer possible, and rules were evolved to eliminate the work on shocks of less than magnitude 4, unless they had been reported felt. This threshold has recently been lowered by a quarter of a magnitude, and coverage now extends to the whole country.

So far, it has been the policy to keep all records made at New Zealand stations. A few have been lost through a variety of mishaps, mainly those of important teleseisms lent to overseas researchers and never returned. There are two main archives, each with a floor area of about 90 square metres. They are reasonably secure from the hazards of fire and flood, but neither has such refinements as temperature and humidity control, and some of the older records are becoming brittle. Deficiencies in photographic processing need attention, and cataloguing and shelving could be improved.

It is not easy to decide what fraction of the resources of a government research institution should be devoted to conservation, but there is no doubt that the criterion it uses to assign its priorities for deploying money and available staff must be importance to the study of New Zealand earthquakes. Few seismologists concerned with near-earthquake studies consider photographic copies satisfactory. Preserving original records and improving their safety and accessibility will inevitably come first, and not all the resources that might be wished for are to be had. Any request from abroad for extensive photographic copying of teleseisms can be considered only in the light of its importance in the context of New Zealand's international obligations. Substantial grants are already made to the International Seismological Centre and to analogous organizations serving other sciences, and a request would not necessarily be turned down; but if photographic copying of records is to be justified to the New Zealand taxpayer it must be shown not only to be a legitimate activity, but the best possible use of whatever allocation of severely limited funds and scientific manpower might become available.

REFERENCES

Bastings, L. (1935). Destructive earthquakes in New Zealand, 1835-1934, *N. Z. J. Sci. Technol.*, **17**, 406-411.

Eiby, G. A. (1968a). A descriptive catalogue of New Zealand earthquakes, Part I. Shocks felt before the end of 1845, *N. Z. J. Geol. Geophys.*, **11**, 16-40.

Eiby, G. A. (1968b). An annotated list of New Zealand earthquakes, 1460-1965, *N. Z. J. Geol. Geophys.*, **11**, 630-647.

Eiby, G. A. (1973). A descriptive catalogue of New Zealand earthquakes, Part II. Shocks felt from 1846 to 1854, *N. Z. J. Geol. Geophys.*, **16**, 857-907.

Eiby, G. A. (1980). The Marlborough earthquakes of 1848, *DSIR Bulletin 225*, Wellington, 82pp.

Eiby, G. A. (1982a). Earthquakes and tsunamis in a region of diapiric folding, *Tectonophys.*, **85**, T1-T8.

Eiby, G. A. (1982b). Two New Zealand tsunamis, *J. Roy. Soc. N. Z.*, **12**, 337-351.

Hayes, R. C. (1953). Some aspects of earthquake activity in the New Zealand region, *Proc. 7th Pacif. Sci. Congr.*, **2**, 629-636.

Hector, J. (1891). Report of the committee ... upon seismological phenomena in Australasia, *Rept. 3rd Mtg. Austral. Assoc. Adv. Sci.*, Christchurch, **3**, 504-532.

Hogben, G. (1891). The earthquakes of New Zealand, *Rept. 3rd Mtg. Austral. Assoc. Adv. Sci.*, Christchurch, **3**, 37-57.

Lyell, C. (1875). *Principles of Geology*, John Murray, London, 12th ed., **2**, 82-89.

Ongley, M. (1943). Surface trace of the 1855 earthquake, *Trans. Roy. Soc. N. Z.*, **73**, 84-89.

Richter, C. F. (1957). *Elementary Seismology*, W. H. Freeman & Co. San Francisco, 768 pp.

Smith, W. D. (1976). A computer file of New Zealand earthquakes, *N. Z. J. Geol. Geophys.*, **19**, 393-394.

Young, R. M. (1984). Early New Zealand seismographs, *Bull. Roy. Soc. N. Z.*, **21**, 113-118.

Young, R. M., J. H. Ansell, and M. A. Hurst (1984). New Zealand's first seismograph: the Hector seismograph, 1884-1902, *J. Roy. Soc. N. Z.*, **14**, 159-173.

Historical Earthquakes and the Seismograms in Taiwan

Pao Hua Lee
Central Weather Bureau
Taipei, China

ABSTRACT

Although the most active seismicity in the Taiwan area is on the east of the island, earthquake risk is greater on the west part in Chaiyi-Tainan region. Other areas of high earthquake risk are on east Taiwan in the Hwalien-Ilan and Taitung-Lanyu regions.

The Central Weather Bureau has established a remote recording system consisting of 14 seismological stations. This system is linked to a VAX 11/750 computer, which is used exclusively for data processing and data studies.

The accumulated seismograms are going to be microfilmed for the historical seismogram archive. On completion, a catalogue will be published and seismological data exchange with other countries will be encouraged.

1. Introduction

The purpose of this paper is to describe the Seismological Observatory of the Central Weather Bureau (CWB) and the general status of seismological activities from 1897 to 1985 in Taiwan area.

Instrumental observations in the Taiwan area began on December 1, 1897, when a Gray-Milne seismograph was installed at Taipei. By 1945, 16 seismological stations had been opened in Taiwan. But in March 1945 seismic observations were halted because all supplies were cut off by the air raids of World War II. On January 1, 1946, after the war ended, seismological activities were restarted, and today there are 17 seismic stations in Taiwan.

2. The Seismological Observatory, CWB

The Seismological Observatory of CWB conducts seismological research and furnished seismicity information to the general public of the Taiwan area. To meet these responsibilities, the Observatory maintains a network of recording stations on and around the island.

In 1985, the Seismological Observatory operates 17 stations (see Figure 1 for geographic locations). Before 1982, most seismographs of these stations – except Anpu, a World Standard Seismograph Network Station – were mechanical instruments, such as Wiechert, Omori, and other horizontal portable seismographs. In 1982, all seismographs were upgraded and a remote recording system consisting of 14 stations was completed.

The upgraded seismographs consist of three types of instruments: short-period seismometers, long-period seismometers, and strong-motion accelerographs. Each of the 14 remote stations contains the necessary sensors, signal conditioners, microprocessors, digital printers, communication interfaces, and analog recorders. The

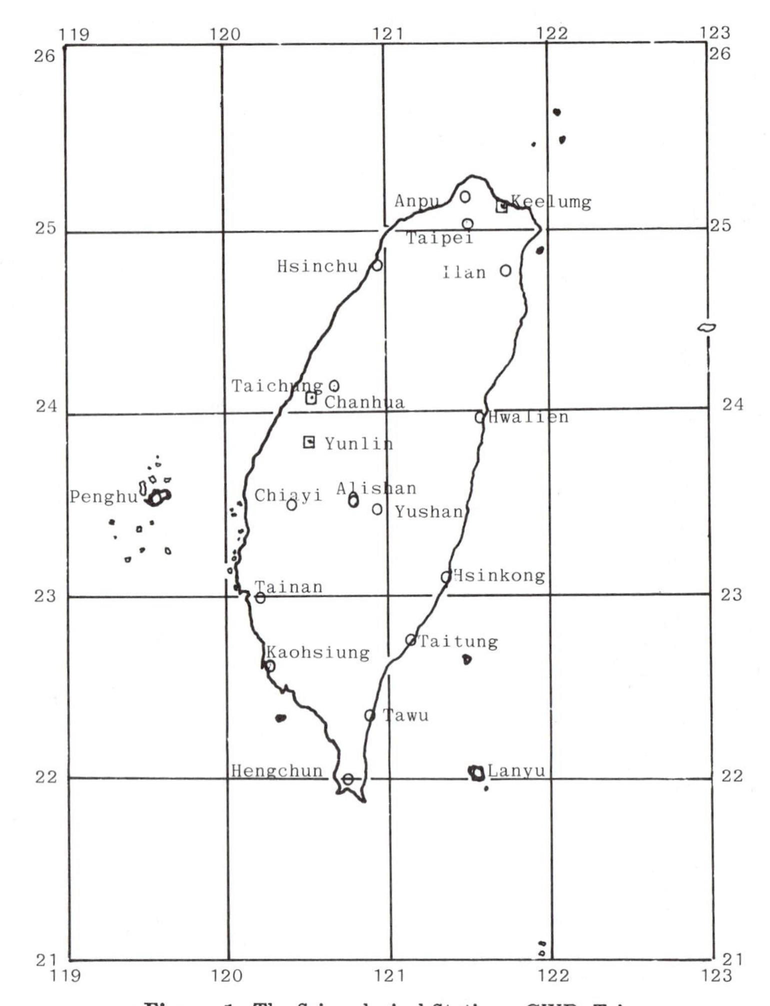

Figure 1. The Seismological Stations, CWB, Taiwan.

heart of the remote stations is the microprocessor, which is based on a microcorder that detects the earthquakes, then sends the origin time and other information to the digital printer and records the data on a cassette tape. The cassettes are later sent to the central station for playback onto computer tapes for archival and future use. Our interest is mainly in the larger earthquakes ($M \geq 4.0$), and therefore this system is designed to record and analyze earthquakes in the magnitude range 4.0 to 9.0.

For data processing, communications equipment has been established to permit the remote stations to link automatically with the central computer – a VAX 11/750. The stations transmit earthquake information that permits online analysis of location and magnitude. Offline generation of archival data tapes will be performed in the future using this computer.

3. Earthquakes

Taiwan is located at the center of the western part of the circum-Pacific seismic zone. Therefore, earthquakes are frequent and sometimes strong and destructive. Although Taiwan lies in the active circum-Pacific seismic zone, seismicity patterns show that three smaller zones exist in the area (see Figure 2):

(1) Western Seismic Zone: It starts south of Taipei and extends southward to Tainan. The zone is about 80 km in width and runs nearly parallel to the island axis. Compared to other two zones, earthquakes here are infrequent, but aftershocks are generally frequent and cause damage over large areas. Focal depths are shallow (about 10 km), and severe crustal movements sometimes occur.

(2) Eastern Seismic Zone: It extends from northeast of Ilan on a line south by southwest to Taitung, and on to Luzon, Philippine Islands. It forms an arc toward the Pacific, extends about 130 km in width, and is parallel to the island axis. In this zone, earthquakes are frequent, and their hypocenters are deeper than those of the western zone.

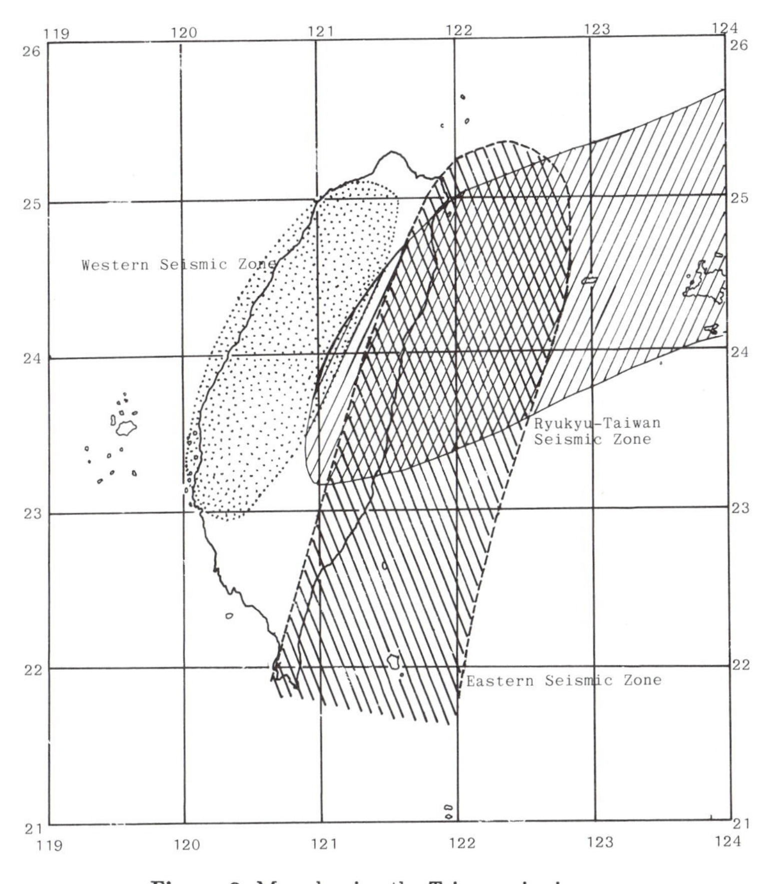

Figure 2. Map showing the Taiwan seismic zones.

(3) Ryukyu-Taiwan Seismic Zone: This zone extends from Ryukyu Island southwestward to Hwalien and Iland. Focal depths in this zone vary from shallow up to as much as 300 km.

Further, we have studied the distribution and energy release [calculated by using the Gutenberg-Richter formula: $\log E = 11.8 + 1.5M$; E = Energy release (erg); M = Earthquake magnitude] of major earthquakes, from 1935 to 1984 having magnitudes $M \geq 5.0$. To accomplish this, we have divided the area into 72 equal grids, extending from 21° N to 25.5° N, and 119° E to 123° E (see Figure 3). According to Figure 3, earthquake occurrence west of Taiwan is low; the percentage in most grids is less than 1.0%, except in the region between 23.0° N to 24.5° N and 120.5° E to 121.0° E, where the percentage ranges between 1.0% and 2.0%.

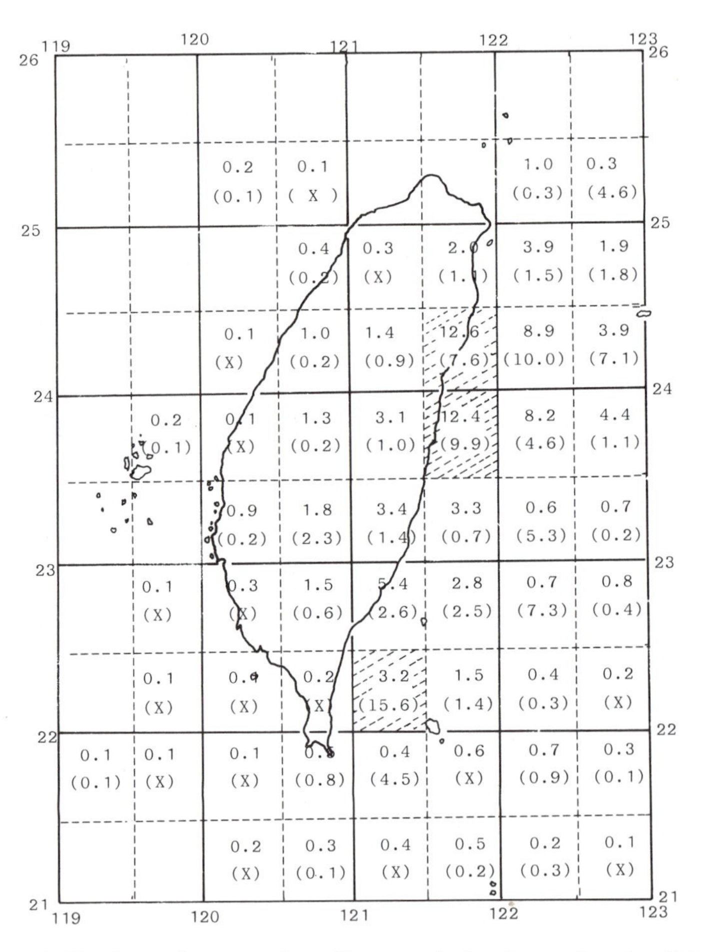

Figure 3. Number and energy release (in percent) of major earthquakes ($M \geq 5.0$) in the Taiwan area, from 1935 to 1984. In each grid the upper numerals represent percent of earthquake occurrence and the lower numerals in parentheses represent the percent of energy release of the earthquakes. X represents energy release of less than 0.1%.

In the east area of the island, between 23.5° N to 24.5° N and 121.5° E to 123.0° E, the percentage of earthquake occurrence is high – 58.2%. This means that over half of the earthquakes in the Taiwan region occur in this area of the island – especially within 23.5° N to 24.5° N and 121.5° E to 122.0° E, the most active seismic region in Taiwan.

The highest percentage of energy release, 15.6%, occurs within 22.0° N to 22.5° N and 121.0° E to 121.5° N. Northeast of Hwalien (between 24.0° N to 24.5° N and 122.0° E to 122.5° E) and southeast of Hwalien (between 23.5° N to 24.0° N and 121.5° E to 122.0° E) the percentages are 10.0% and 9.9%, respectively. In the west part of the island, however, the percentage of energy release in almost all grids is less than 1.0%. The largest percentage, 2.3%, is found in the area between 23.0° N to 23.5° N and 120.5° E to 121.0° E.

Figure 3 also shows that the earthquakes in Taiwan are distributed from northeast to southwest, and that the most active seismic area is on east Taiwan between Hwalien and Ilan. Further, even though the percentage of earthquake occurrence is not high in the area between Taitung and Lanyu (22.0° N to 22.5° N and 121.0° E to 121.5° E), that area has the highest percentage of energy release in Taiwan, which means that large-magnitude earthquakes occur there frequently.

Historically, 12 disastrous earthquakes have occurred in this region between 1644 and 1897 (see Table 1). Since 1897, 73 disastrous earthquakes have occurred in Taiwan area according to records at CWB (see Table 2). The most severe shocks, occurring on April 21, 1935 in the Hsinchun-Taichung area (magnitude 6.6), killed 3,276 people and injured 12,053. The most recent destructive earthquake was a magnitude 6.3 in Tainan on January 18, 1964. This strong shock killed 106 people, injured 650, and collapsed more than 10,502 houses.

Table 1. Catalog of Disastrous Earthquakes before Instrumental Observation, 1644 to 1896

Date	Districts damaged	Remarks
July 30, 1644	Tainan	Crack and tilt of castle wall
Nov. 1, 1720	Tainan	Houses destroyed, number of lives lost
Jan. 29, 1736	Tainan, Chiayi	Many people died
Dec. 11, 1776	Chiayi	Houses destroyed, number of lives lost
Aug. 7, 1792	Chiayi, Chanhua	More than 100 people killed
1816	Ilan	Number of houses destroyed
June 27, 1839	Chiayi	Houses destroyed, landslide
Dec. 3, 1848	Yunlin, Chanhua	Houses destroyed
June 5, 1862	Tainan, Chiayi, Chanhua	Houses destroyed, many people killed
Dec. 18, 1867	Keelung	Number of houses were washed away by tsunami and several hundred people drowned
Apr. 19, 1881	Taipei	Houses destroyed, some people killed and wounded
Apr. 22, 1892	Tainan	Number of houses destroyed

Table 2. Major Earthquakes in Taiwan, 1897 to 1984

Year	Date	Time* hr:mn	Lat °N	Long °E	h km	M_L	Remarks
1901	Jun 7	08:05	24.7	121.8	—	6.0	Ilan earthquake, 1 house destroyed, 57 houses partially destroyed
1904	Apr 24	14:39	23.5	120.5	—	6.1	Chaiyi earthquake, 3 people killed, 10 wounded; 66 houses destroyed, 152 partially destroyed, 688 damaged
1904	Nov 6	04:25	23.5	120.3	—	6.3	Chaiyi earthquake, 145 people killed, 158 wounded; 661 houses destroyed, 1112 partially destroyed, 2067 damaged; fissure
1905	Aug 28	00:22	24.2	121.7	—	5.6	Hwalien earthquake, 1 house partially destroyed, 8 damaged
1906	Mar 17	06:42	23.6	120.5	sh	7.1	Chaiyi earthquake, 1258 people killed, 2385 wounded; 6769 houses destroyed, 10,585 damaged; fault, fissure
1906	Mar 26	11:29	23.7	120.5	—	5.0	Chaiyi earthquake, 1 person killed, 5 wounded; 29 houses destroyed, 43 partially destroyed, 486 damaged
1906	Apr 4	20:42	—	—	—	—	Chaiyi earthquake, 5 houses destroyed
1906	Apr 7	00:52	23.4	120.4	—	5.5	Tainan earthquake, 1 person killed, 6 people wounded; 63 houses destroyed, 96 partially destroyed, 187 damaged; fault
1906	Apr 14	03:18	23.4	120.4	20	6.6	Tainan earthquake, 15 people killed, 84 wounded; 1794 houses destroyed, 2116 partially destroyed, 7921 damaged; fissure, fault
1906	May 4	—	—	—	—	—	Tainan earthquake, 3 houses destroyed
1908	Jan 11	11:35	23.7	121.4	sh	7.3	Hsingkong-Hwalien earthquake, 2 people killed; 3 houses destroyed, 1 partially destroyed, 4 damaged; fissure, fault
1909	Apr 15	03:54	25.0	121.5	80	7.3	Taipei earthquake, 9 people killed, 51 wounded; 122 houses destroyed, 252 partially destroyed, 798 damaged
1909	May 23	06:44	24.0	120.9	—	5.6	Taichung earthquake, 6 people wounded; 10 houses destroyed, 32 partially destroyed
1909	Nov 21	15:36	24.4	121.8	sh	7.3	Hwalien earthquake, 4 people wounded; 14 houses destroyed, 25 partially destroyed, 14 damaged
1910	Apr 12	08:22	25.1	122.9	200	8.3	North-east Taiwan earthquake, 13 houses destroyed, 2 partially destroyed, 57 damaged
1913	Jan 8	06:50	24.0	121.6	—	6.4	Hwalien earthquake, 2 houses destroyed; fissure
1916	Aug 28	15:27	23.7	120.9	—	6.4	Nantou earthquake, 16 people killed, 159 wounded; 614 houses destroyed, 954 partially destroyed, 3931 damaged
1916	Nov 15	06:31	24.1	120.7	—	5.7	Taichung earthquake, 1 person killed, 20 wounded; 97 houses destroyed, 200 partially destroyed, 772 damaged

(*continued*)

Table 2. (*Continued*)

Year	Date	Time* hr:mn	Lat °N	Long °E	h km	M_L	Remarks
1917	Jan 5	00:55	23.9	120.9	—	5.8	Nantou earthquake, 54 people killed, 85 wounded; 130 houses destroyed, 230 partially destroyed, 395 damaged
1917	Jan 7	02:08	23.9	120.9	—	5.6	Nantou earthquake, 21 people wounded; 187 houses destroyed, 221 partially destroyed, 277 damaged
1918	Mar 27	11:52	24.6	121.9	—	6.2	Suao earthquake, 3 people wounded, 6 houses damaged
1920	Jun 5	12:22	24.0	122.0	sh	8.3	Hwalien earthquake, 5 people killed, 20 wounded; 273 houses destroyed, 277 partially destroyed, 980 damaged
1922	Sep 2	03:16	24.6	122.2	sh	7.6	Northern district earthquake, 5 people killed, 7 wounded; 14 houses destroyed, 22 partially destroyed, 139 damaged
1922	Sep 15	03:32	24.6	122.3	—	7.2	Northern district earthquake, 5 people wounded; 24 houses destroyed, 24 partially destroyed, 365 damaged
1922	Sep 17	06:44	23.9	122.5	—	6.0	Hwalien earthquake, 1 person wounded; 6 houses destroyed, 2 partially destroyed, 195 damaged
1922	Oct 15	07:47	24.6	122.3	—	5.9	Suao earthquake, 6 people killed, 2 wounded; 14 houses damaged
1922	Dec 2	11:46	24.6	122.0	—	6.0	Suao earthquake, 1 person killed, 2 wounded; 1 house destroyed, 33 damaged
1922	Dec 13	19:26	24.6	122.1	—	5.5	Suao earthquake, 1 person wounded; 13 houses damaged
1923	Feb 28	18:12	24.6	122.0	—	—	Suao earthquake, 1 house destroyed
1923	Mar 5	08:10	24.5	121.8	—	—	Suao earthquake, 1 house destroyed
1923	May 4	18:41	23.3	120.3	—	5.7	Tainan county earthquake, 1 house destroyed
1923	Sep 29	14:51	22.8	121.1	—	5.5	Taitung earthquake, 1 person wounded; 1 house destroyed, 5 partially destroyed, 75 damaged
1925	Jun 14	13:38	24.1	121.8	—	5.6	Hwalien earthquake, 1 person wounded; 339 houses damaged
1927	Aug 25	02:09	23.3	120.3	—	6.5	Tainan county earthquake, 11 people killed, 63 wounded; 214 houses destroyed, 225 partially destroyed, 984 damaged
1930	Dec 8	14:20	23.3	120.4	—	6.1	Tainan county earthquake, 4 people killed, 25 wounded; 49 houses destroyed, 277 partially destroyed, 172 damaged; fissure
1930	Dec 22	08:08	23.3	120.4	—	6.5	Tainan county earthquake, 14 people wounded; 121 houses destroyed, 424 partially destroyed, 2295 damaged
1931	Jan 24	23:02	23.4	120.4	—	5.6	Tainan-Chaiyi earthquake, 698 houses damaged
1933	May 4	07:30	24.2	121.5	—	—	Hwalien earthquake, 1 person killed

(*continued*)

Table 2. (*Continued*)

Year	Date	Time* hr:mn	Lat °N	Long °E	h km	M_L	Remarks
1934	Aug 11	16:18	24.8	121.8	—	6.5	Ilan earthquake, 3 people wounded; 7 houses destroyed, 11 partially destroyed
1935	Apr 21	06:02	24.3	120.8	5	6.6	Hsinchun-Taichung earthquake, 3276 people killed, 12,053 wounded; 17,907 houses destroyed, 11,405 partially destroyed, 25,376 damaged; fault, landslide
1935	May 5	07:02	24.5	120.8	sh	6.0	Hsinchun-Taichung earthquake, 38 people wounded; 28 houses destroyed, 98 partially destroyed, 473 damaged
1935	May 30	03:43	24.1	120.8	sh	5.5	Hsinchun-Taichung earthquake, 2 houses destroyed, 24 partially destroyed
1935	Jun 7	10:51	24.2	120.5	sh	5.7	Hsinchun-Taichung earthquake, 2 people wounded; 5 houses destroyed, 16 partially destroyed, 174 damaged
1935	Jul 17	00:19	24.6	120.7	30	6.3	Hsinchin-Taichung earthquake, 44 people killed, 391 wounded; 1734 houses destroyed, 1850 partially destroyed, 4037 damaged
1935	Sep 4	09:38	22.4	121.2	20	6.8	Taitung earthquake, 114 houses damaged
1936	Aug 22	14:51	22.0	121.2	50	7.3	Hengchun earthquake, 3 people wounded
1939	Nov 7	11:53	24.4	120.8	sh	5.8	Hsinchun-Taichung earthquake, 4 houses destroyed, 20 damaged
1941	Dec 17	03:19	23.4	120.5	10	6.9	Chiayi earthquake, 358 people killed, 753 wounded; 4320 houses destroyed, 6910 partially destroyed, 4176 damaged; landslide
1943	Oct 23	00:01	23.8	121.5	5	6.2	Hwalien earthquake, 1 person killed, 1 wounded; 1 house destroyed, 148 damaged; landslide
1943	Nov 3	00:51	24.0	121.8	sh	5.6	Hwalien earthquake, 87 houses damaged
1943	Nov 24	05:51	24.0	121.7	sh	5.7	Hwalien earthquake, 479 houses damaged; landslide
1943	Dec 2	13:09	22.5	121.5	40	6.1	Taitung earthquake, 3 people killed, 11 wounded; 139 houses destroyed, 55 partially destroyed, 229 damaged; fault
1944	Feb 6	01:20	23.8	121.4	5	6.4	Hwalien earthquake, 2 houses destroyed, 8 partially destroyed, 380 damaged; landslide
1946	Dec 5	06:47	23.1	120.3	sh	5.8	Tainan earthquake, 74 people killed, 482 wounded; 1954 houses destroyed, 2084 partially destroyed; landslide
1951	Oct 22	11:29	24.1	121.8	20	7.0	Hwalien earthquake, 68 people killed, 856 wounded; 2382 houses destroyed; fissure, landslide
1951	Nov 25	02:47	23.0	120.9	5	6.5	Taitung earthquake, 17 people killed, 326 wounded; 1016 houses destroyed, 582 partially destroyed; fault, landslide

(*continued*)

Table 2. (*Continued*)

Year	Date	Time* hr:mn	Lat °N	Long °E	h km	M_L	Remarks
1955	Apr 4	19:12	21.8	120.9	5	6.7	Hengchun earthquake, 7 people wounded; 22 houses destroyed, 30 partially destroyed, 141 damaged
1959	Apr 27	04:41	25.0	122.6	80	7.2	East Taiwan earthquake, 1 person killed; 9 houses destroyed, 4 damaged
1959	Aug 15	16:57	21.7	121.3	20	7.2	Hengchun earthquake, 16 people killed, 63 wounded, 789 houses destroyed, 752 partially destroyed
1959	Aug 17	16:25	23.3	121.3	20	5.5	Tawu earthquake, 6 houses destroyed
1959	Aug 18	08:34	22.1	121.7	15	6.1	Hengchun earthquake, 32 houses destroyed, 5 partially destroyed
1959	Sep 25	10:37	22.1	121.2	10	6.5	Hengchun earthquake, 3 people wounded; 3 houses destroyed, 28 partially destroyed, 37 damaged
1963	Feb 13	16:50	24.4	122.1	47	7.2	Ilan earthquake, 15 people killed, 3 wounded; 6 houses destroyed, 6 partially destroyed
1963	Mar 4	21:38	24.6	121.8	5	6.4	Ilan earthquake, 1 person killed
1963	Mar 10	10:53	24.5	121.8	5	6.0	Ilan earthquake, 3 houses partially destroyed
1964	Jan 18	20:04	23.2	120.6	20	6.3	Chaiyi-Tainan earthquake, 106 people killed, 650 wounded; 10,502 houses destroyed, 25,818 partially destroyed; landslide
1964	Feb 17	13:50	23.1	120.6	10	5.3	Chaiyi-Tainan earthquake, 3 people wounded; 422 houses destroyed, 4203 partially destroyed
1965	May 18	01:19	22.5	121.3	21	6.5	Tawu earthquake. 1 person wounded; 21 houses destroyed, 70 partially destroyed
1967	Oct 25	08:59	24.5	122.2	65	6.1	Ilan earthquake, 2 people killed, 2 wounded; 23 houses destroyed, 27 partially destroyed
1972	Jan 25	10:07	22.5	122.3	70	7.3	Taitung earthquake, 1 person killed, 1 wounded; 2 houses destroyed, 3 partially destroyed, 1 damaged
1972	Apr 24	17:57	23.5	121.5	3	6.9	Hwalien earthquake, 5 people killed, 17 wounded; 50 houses destroyed, 98 partially destroyed
1978	Dec 23	19:23	23.3	122.1	35	6.8	Hsinkong earthquake, 2 people killed, 3 wounded; 1 house partially destroyed, 1 damaged
1982	Jan 23	22:11	24.0	121.6	15	6.5	Hwalien earthquake, 1 person killed, a few houses damaged

* = Origin Time (120°E); h = Depth; sh = shallow.

Although earthquakes are frequent in this area, tsunamis caused either by local or distant earthquakes seldom occur. Historically, we found only one tsunami – on December 18, 1867 – that caused damage. In "The Island of Formosa, Past and Present", James W. Davidson (1903) wrote:

> "... an earthquake, the most severe ever experienced on the island since the first day of the Dutch, ... It occurred on the 18th of December 1867, and the vicinity of the town of Kelung sustained the greatest damage, ... Foreigners in the Custom service reported that at Kelung about 15 shocks were felt during the day, but that it was the first movements that caused the damage. In 15 seconds after the first perceptible shock the damage was done, and the town of Kelung was in ruins. The force of the earthquake may be judged when it is noted that the water of Kelung Harbor ran out, leaving the bottom of the bay exposed. Fortunately, there were no foreign vessels present, but the Chinese junks that were there, large and small, were in one second left dry on the bottom and in another, were dashed into the town with fearful speed to work havoc on the remaining houses left near the shore. Multitudes of fish were thrown upon the shore and were promptly gathered by the populace ... The number of lives lost is unknown. It is doubtful if a count was made, but probably several hundred perished ..." [Note: 'Kelung' is now spelled 'Keelung'.]

Another tsunami occurred in 1917. On east coast of Taiwan, its wave height was reported 3.7 meters, but there was no damage description.

4. Historical Seismograms

The Central Weather Bureau has retained about 350,000 seismographic records since 1897. The oldest records in existence, though not complete, are of 1929 coming from Wiechert seismographs. Seismograms earlier than 1929 were all lost.

By 1987, all seismograms in the Central Weather Bureau will be photographically reproduced in microfilm for archives. Now, there are 20,000 records of Taipei seismographs from 1956 to 1982 which have been put into microfiche already. The size of the above mentioned microfiche is 10 × 15 cm containing 6 images that can be enlarged 22 times for reading. Each microfiche includes the information such as time, station name, type of seismograph and its reduction rate. All these records are arranged by station, year, and type of seismograph in archives.

For user's easy access, an index will be provided and when all seismograms have been microfilmed, a catalog will be accordingly made. For international data exchange, we will prepare a duplicate copy set to supply World Data Center A.

5. Discussion and Conclusions

Since 1897, 73 significant disastrous earthquakes have occurred in Taiwan area according to records at CWB. Among them are 19 major earthquakes, those having magnitudes of 7.0 or larger. Forty-four of the 73 earthquakes occurred on land, and 29 occurred on west Taiwan (see Figure 4). Although most of the earthquakes have occurred in the east region of Taiwan, a mountainous area with few inhabitants, earthquakes in the west region have been more damaging to life and property. Therefore, earthquake risk is higher in the west than in the east. Earthquake risk is especially high in the Chaiyi-Tainan area, and moderately high in the areas of Hwalien-Ilan and Taitung-Lanyu.

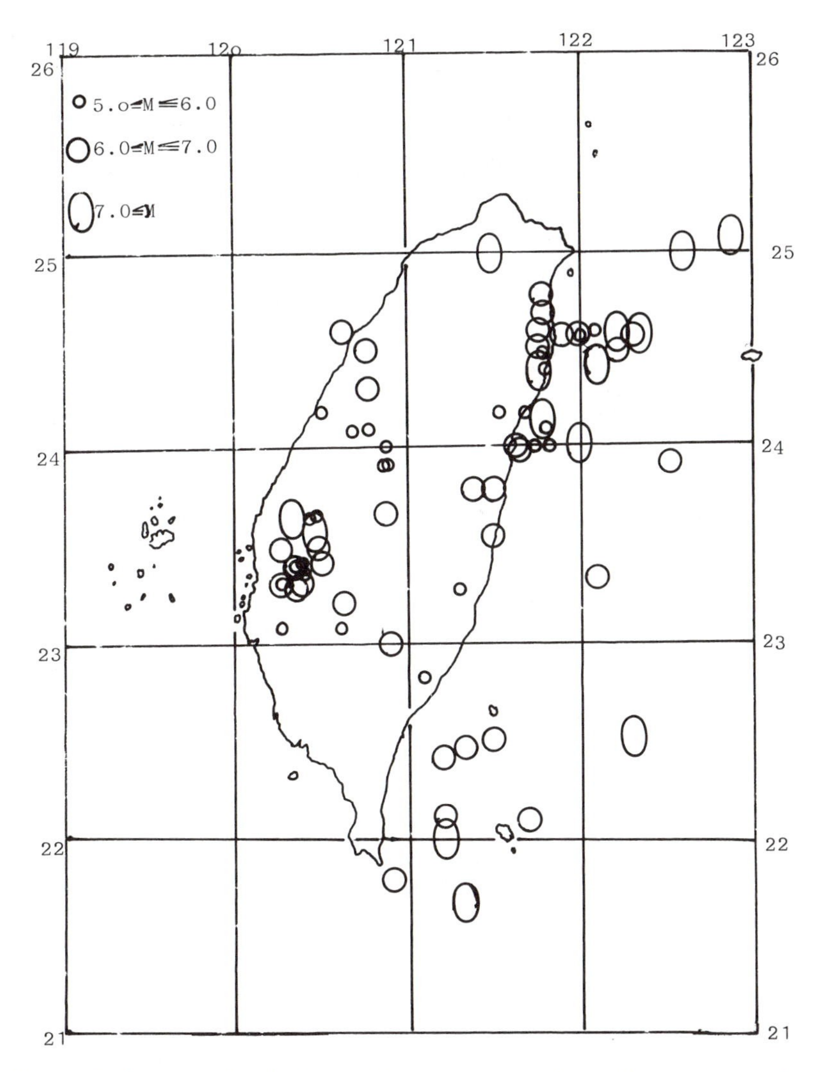

Figure 4. Epicenter distribution of disastrous earthquakes in Taiwan, 1897 to 1984.

The CWB has established a remote seismic recording system, which uses a VAX 11/750 computer for data processing and studies. This system was set up to enhance the scientific study of earthquakes in Taiwan, and to furnish the public with accurate seismic information to reduce the loss of life and property from severe earthquakes. Both the quality and quantity of earthquake data will be therefore improved.

The historical seismograms are going to be microfilmed for archiving. From 1956 to 1982, 20,000 records microfilming of the Taipei seismographs have been completed in June 1985. When completed, CWB will publish a catalog of records available and will exchange seismic data with other countries.

REFERENCES

Central Weather Bureau, Taiwan (1954-1984). *Seismological Bulletins*, **1-31**.

Central Weather Bureau, Taiwan: Taiwan Disastrous Earthquake Record, Unpublished Manuscript.

Davidson, James W. (1903). *The Island of Formosa. Past and Present*, McMillan & Co., London & New York, 187 pp.

Hsu, M. T. (1980). Earthquake Catalogs in Taiwan from 1644 to 1979, Earthquake Engineering Research Center, National Taiwan University, 6 pp.

Lee, P. H. (1980). Earthquake, Monsoon Publication Co., 46-47.

Lee, P. H. and Y. L. Liu (1983). The Strong Earthquake Recorders of 1900 to 1972, Reanalysis and Occurrence Time, Spatial Distributions Study in the Taiwan Area, Chinese National Science Council, 4-5.

Historical Earthquakes of Thailand, Burma, and Indochina

S. Prachuab
Meteorological Department, Bangkok, Thailand

ABSTRACT

Historical earthquake data from 642 BC to 1983 has been analyzed and reported in this paper in order to understand the seismicity and associated phenomena. From quantitative seismicity analysis, Thailand, Burma and Indochina have been divided into 12 seismic zones corresponding to the seismotectonic structure and geologic features in this region.

Using the well-known "magnitude-frequency" relation of Gutenberg and Richter and a modification of this relation by Peter Welkner, the values of a and b are investigated for the period 1900 to 1983. The values of a and b obtained for the region are 6.654 and 0.75 respectively, which are typical of moderately active seismic zones. The recurrence periods for earthquakes of magnitude 7.0 and 7.5 are calculated as 3 and 8 years, respectively.

1. Introduction

A study of historical earthquakes in Thailand, Burma and Indochina provides not only a basis for statistical analysis of earthquakes but also for the earthquake hazard studies in this area. The data used for analysis are taken from the report *Seismicity Data of Thailand and Adjacent Areas* (Nutalaya and Sodsri, 1983). This earthquake report contains two main catalogs. The first catalog contains the historical earthquake data for the period 642 BC to 1900, accumulated from the historical texts, annuals, stone inscriptions and astrological documents, and from the Geological Survey of India Records and Memoirs. The second catalog contains the instrumental earthquake data for the period 1900 to 1983, accumulated from various international seismological centers (eg. USGS, ISC, MOS) and from the Thai Meteorological Department. The seismicity map of the region between 5°N to 25°N latitude and 90°E to 110°E longitude for the period 642 BC to 1983 is shown in Figure 1.

Nutalaya and Sodsri (1984) divided Thailand, Burma and Indochina into 12 seismic source zones, using quantitative seismicity analysis (Figure 2). Table 1 lists the zone name, the number of seismic events, and the maximum magnitude event for each seismic source zone. These seismic source zones were identified by plotting the historical epicenters along with the geological and tectonic information. The size and shape of each source zone reflects the geographic distribution of seismicity together with the regional tectonic style. Only zone C is characterized by seismicity associated with individual active faults as documented by historical records.

Magnitude-frequency analysis has been determined for the entire region for the period 1900 to 1983. The objective of this analysis is to investigate the earthquakes that occurred during this century, the results of which indicate the status of historical seismicity in the region.

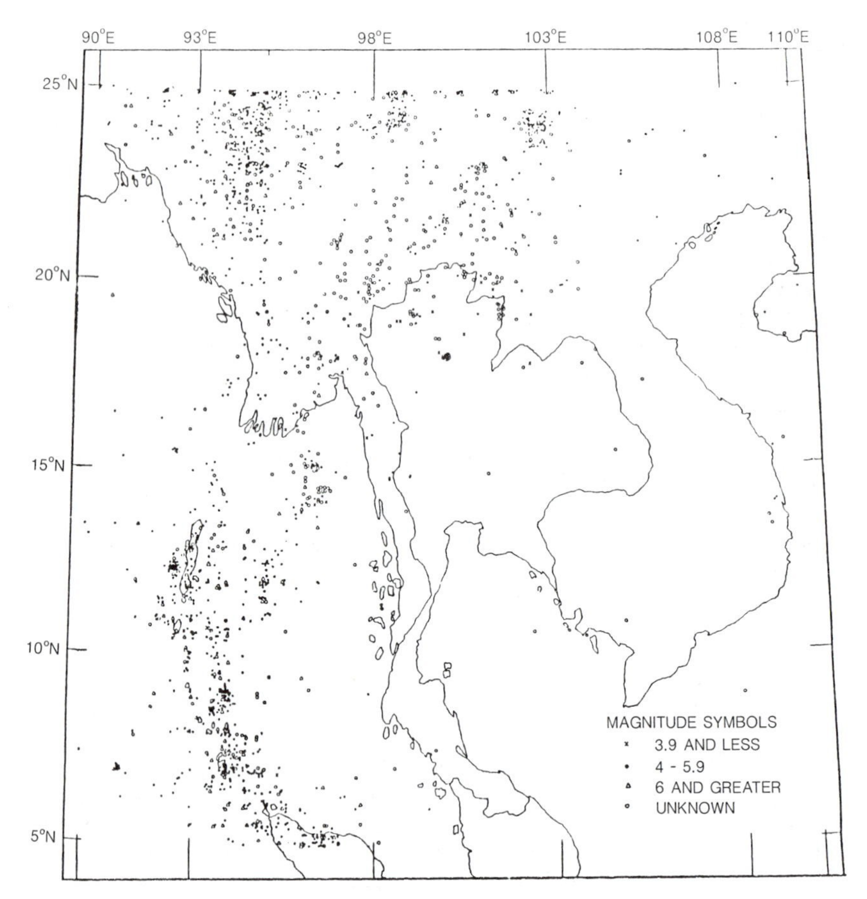

Figure 1. Seismicity map of Burma, Thailand, Indochina, and Andaman – Nichobar Islands.

2. Historical Earthquakes and Tectonics

Table 2 lists the date, location, and description of major historical earthquakes in Thailand, Burma and Indochina for the period 642 BC to 1900. The details of earthquake damage are reported only from seismic zones A, B, C, D, I and J.

Earthquakes having intensity X to XI occurred in zone B (West-Central Burma Basin) and zone C (East Central Burma Basin) during August 1858 and March 1839, respectively. Intensity maps of these two events are shown in Figure 3(a, b). The active tectonics in zone B is due to the subduction zone below the Indoburman Ranges. In zone C, the "Sagaing fault" is clearly defined as an earthquake source. The Sagaing fault runs in the NS direction near the eastern margin of the eastern trough of the Central Basin, and extends over 800 miles from northern Kachin State to the Gulf of Mataban, as shown in Figure 4.

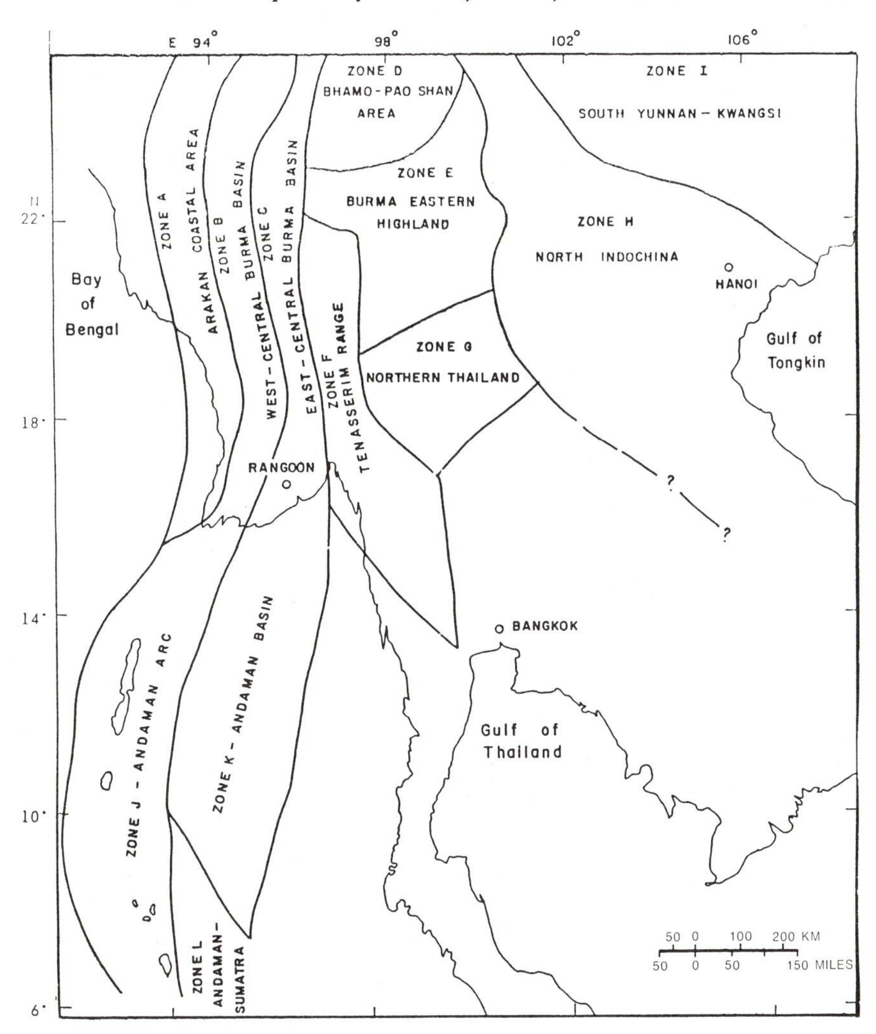

Figure 2. Seismic source zone of Burma – Thailand – Indochina.

In zone I (South Yunnan-Kwangsi), the most destructive earthquake was the Qiongzhou earthquake of July 1605, having maximum intensity XI (M.M. scale) at Hainan Island. The main seismic zone is located in South Yunnan Province, north of the Red River fault between the Red River fault and the area where the left-lateral Kang Ting fault seems to terminate near the Red River fault, as shown in Figure 5.

Figure 6 shows the location and date of historical earthquakes during the period 1900 to 1983 having magnitude greater than 7. The greatest earthquake (magnitude 8.1) occurred during June 1941, and was located in the NW of the Andaman Islands (zone J). The active tectonics in the Andaman Sea is due to the underthrusting of the Indian Plate.

Table 1. 12 Seismic Source Zones of Burma, Thailand and Indochina Areas

Zone	Name of zone	Number of events	Maximum intensity	Maximum magnitude
A	Arakan Area	134	X-XI	5.75 PAS
B	West-Central Burma Basin	225	XI	7.40 PAS
C	East-Central Burma Basin	94	XI	7.50 PAS
D	Bhama-Hao Shan Area	74	VIII	6.50 PAS
E	Burma Eastern Highland	101	—	7.30 PAS
F	Tenasserim Range	160*	—	7.90 PAS
G	Northern Thailand	50	VII	4.70 m_b**
H	North Indochina	150	VIII	6.75 PAS
I	South Yunnan-Kwangsi	49	XI	7.70 m_b
J	Andaman Arc	329	—	8.10 PAS
K	Andaman Basin	224	—	6.50 PAS
L	Andaman-Sumatra	107	—	6.30 UPP

* including aftershocks; ** 1977-1983 only.

Table 2. Major Historical Earthquakes from 642 BC to 1900

Zone A: Arakan Area

Date	Location	Brief description
1761	Arakan	Very strong, the coast sections of land rose as much as 7 m.
1762 Apr 02	All over Bengal Bay	Most severe, water in tanks and river rose, openings in earth, fountain of sand and mud, coast of Arakan elevated. I_{MAX} = X to XI.
1822 Apr 03	Bengal	District shocks.
1833 Jun 26	Kyaupyu	Violent earthquake, flames issued to a height of several hundred feet.
1843 Feb 06	Kyaukpyu	Eruption of mud volcanoes.
1848 Oct 30	Sandoway	Violent shock.
1858 Aug 24	Thayetmyo and Prome	Most severe near location given, but severe also at Akyab and apparently False Island, situated SE of Cheduba Is. disappeared following the earthquake.
1881 Feb 27	Bengal	Eruption of a volcano on Cheduba accompanied by flames and trembling of the earth.
1881 Dec 31	Burmese coast	Believed to have originated in the Bay og Bengal, affected large portion of India, Bengal, and Andaman–Nicobar Is. Followed by mud volcano in Ramri Is. and broad, massive flames of fire from Cheduba volcano.
1882 Dec 31	Arakan	Earthquake shock followed by volcanic eruptions on Cheduba Is.

Zone B: West–Central Burma Basin

Date	Location	Brief description
422 BC	Mt. Popa	A great earthquake roared through central Burma
1858	Thayetmyo and Prome	Felt as far as Moulmein in the south, houses collapsed, pagodas toppled. Course of river reversed, I = X to XI (M.M. scale).

(continued)

Table 2. (Continued)

Zone C: East–Central Burma Basin

Date	Location	Brief description
868	Pegu	Shwemawdaw Pagoda fell.
875	Pegu	Shwemawdaw Pagoda fell.
1429	Ava	Fire-stopping enclosure walls fell.
1467	Ava	Pagoda, solid, hollow, and brick monasteries destroyed.
1482	Sagaing	Hsinmyashin Pagoda destroyed.
1485	Sagaing	Pagodas destroyed.
1501	Ava	Pagodas fell.
1564	Pegu	Pagodas fell.
1567	Pegu	Pagoda fell.
1582	Pegu	Pagoda fell.
1588	Pegu	Pagodas, etc. fell.
1590	Pegu	Great incumbent Buddha destroyed.
1620	Ava	Ground broken into pieces.
1629	Rangoon	Umbrella on top of Pagoda toppled.
1664	Rangoon	Pagoda seriously damaged, top portion toppled.
1699	Ava	Pagodas destroyed.
1714	Ava	Pagodas fell. Water from the river gushed into the city.
1739	Pegu	Top portion of Pagodas toppled.
1757	Pegu	Pagoda damaged.
1768	Pegu	Pagoda fell.
1776	Ava	Pagoda fell.
1830	Pegu	Umbrella of the Pagoda fell.
1838	Ava	Severe damage to buildings, earth-collapsed.
1838	Sagaing,Mingun	Pagoda destroyed.
1839	Ava	Severest damages, not a Pagoda seen standing intact. Every building in town destroyed. Felt for thousands of miles, I = XI.
1888	Pegu	Pagoda collapsed.

Zone D: Bhamo–Pao Shan Area

Date	Location	Brief description
1577	SW Yunnan (25 N, 98.6 E)	170 people killed, I = VIII.
1906	SW Yunnan (24.6 N, 98.6 E)	Two people killed, Magnitude = 5.3.

Zone I: South Yunnan–Kwangsi

Date	Location	Brief description
1499	Yunnan (25 N, 103 E)	Extremely damaged, 20,000 people killed.
1500	Yunnan (24.5 N, 103 E)	Many people killed, I = IX.
1560	Yunnan (24.2 N, 102.7 E)	10 people killed.

(*continued*)

Table 2. (Continued)

Zone I: South Yunnan–Kwangsi

Date	Location	Brief description
1571	Yunnan (24.1 N, 102.7 E)	10 people killed, I = VIII.
1588	Yunnan (24.0 N, 102.8 E)	Some people killed, I = VIII.
1605	Hainan Island	Severe damage, land collapsed on a large scale, I = XI.
1606	Yunnan (23.6 N, 102.8 E)	1,000 people killed, I = IX.
1680	Yunnan (25.0 N, 101.5 E)	2,700 people killed, severely damaged, I = IX.
1750	Yunnan (24.7 N, 102.9 E)	37 people killed, I = VII.
1755	Yunnan (24.7 N, 102.2 E)	270 people killed, I = VIII.
1755	Yunnan (23.8 N, 102.7 E)	70 people killed, I = VIII.
1761	Yunnan (24.4 N, 102.5 E)	120 people killed, I = VIII.
1761	Yunnan (24.4 N, 102.5 E)	50 people killed, I = VII.
1762	Yunnan (24.3 N, 102.8 E)	1,000 people killed, I = VIII.
1789	Yunnan (24.2 N, 102.8 E)	1,000 people killed, I = IX.
1799	Yunnan (23.8 N, 102.4 E)	2,000 people killed, I = IX.
1814	Yunnan (23.7 N, 102.5 E)	200 people killed, I = VIII.
1884	Yunnan (23.0 N, 101.1 E)	17 people killed.
1887	Yunnan (23.7 N, 102.5 E)	2,000 people killed, I = IX.
1909	Yunnan (24.4 N, 103.0 E)	19 people killed, I = VIII.

Zone J: Andaman Arc

Date	Location	Brief description
1846	Nicobar	Shock produced great landslips.
1847	Nicobar	Great earthquake with numerous aftershocks.
1881	W. Andaman Island	Damage in Andaman and Nicobar Is. Waves were formed and rolled against the coast for several hours. Damage to building at Port Blair.

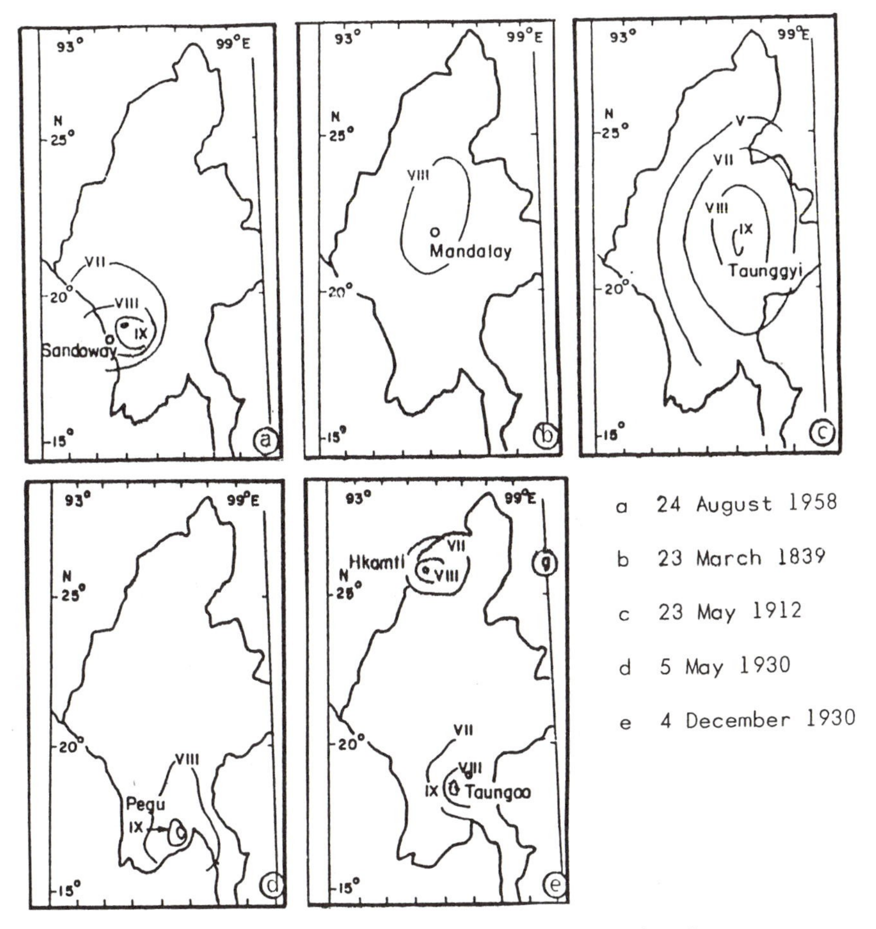

Figure 3. Intensity maps of great historical earthquakes.

In zone F (Tenasserim Range), the Taungyi earthquake of May 1912, having a magnitude about 8 and maximum intensity IX (R.F. scale), was felt throughout Burma and part of Thailand. The intensity map of the Taungyi earthquake is shown in Figure 3(c). Major faults in this area are the Taungyi fault, Pan Laung fault, Moei-Uthai Thani fault zone, Si Sawat fault and Three Pagoda fault (Figure 7).

In zone C, the Pyu earthquake of May 1930 and the Pegu earthquake of December 1930, each having magnitude 7.3 and maximum intensity IX (M.M. scale), caused intensive damages throughout Burma and Thailand. The intensity maps of these two events are shown in Figure 3 (d,e).

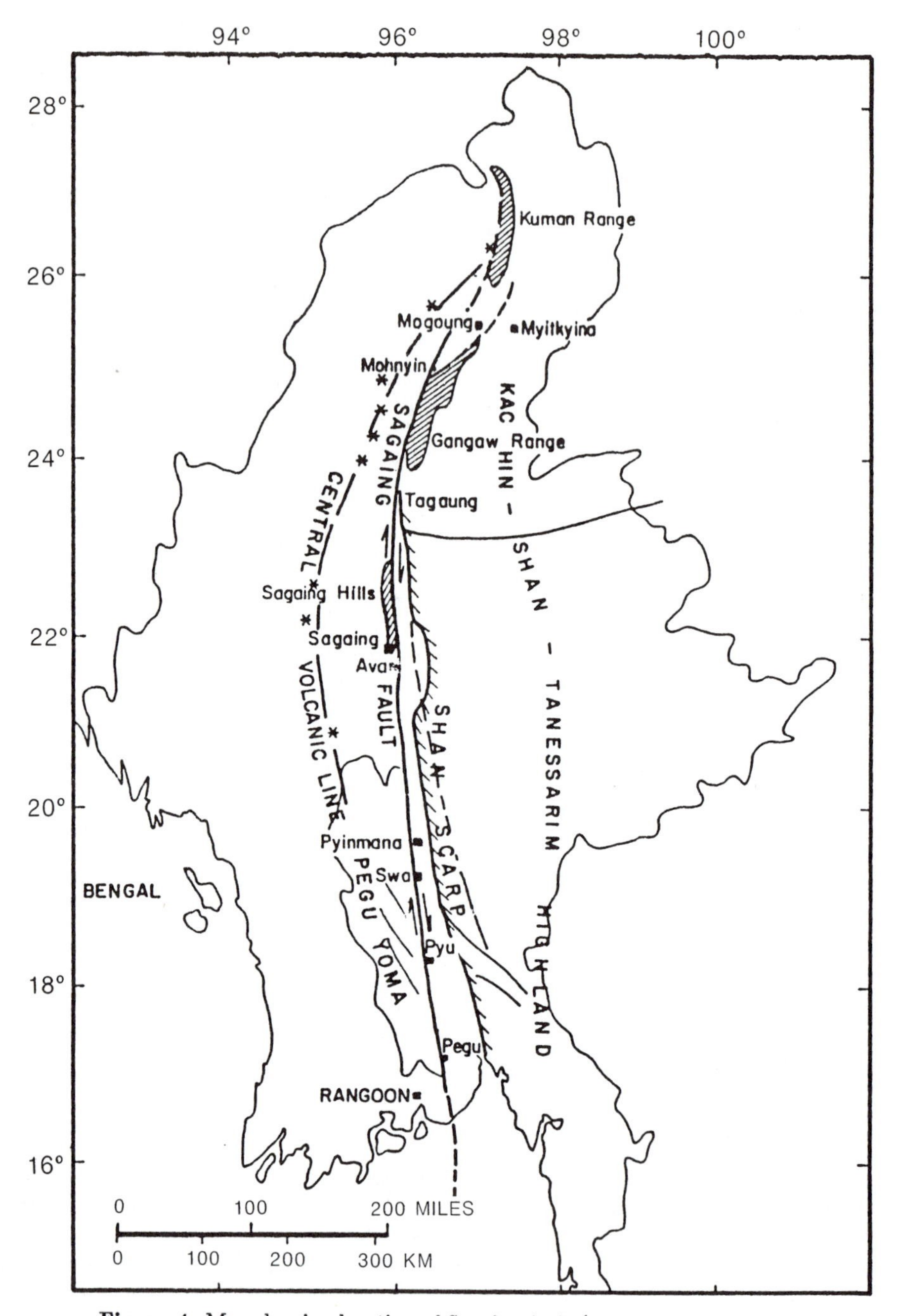

Figure 4. Map showing location of Sagaing fault (after Win Swe, 1980).

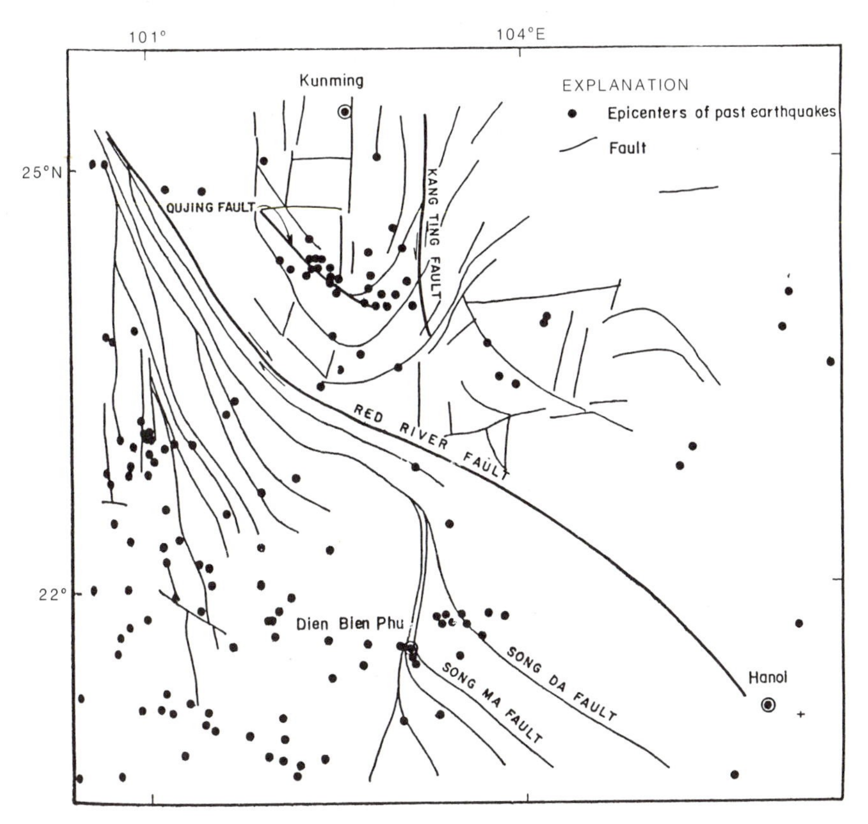

Figure 5. Seismic and tectonic structures of Yunnan Province, China and North Vietnam. Fault information compiled from Fontaine and Workman (1978) and Tectonic Map of China (1975, in Chinese).

3. Magnitude-frequency Analysis

The seismic data used for this magnitude-frequency analysis is taken from various international seismological institutes such as USGS, ISC, and MOS, and from the Thai Meteorological Department. The area of study is bounded by the coordinates 5°N to 25°N latitude and 90°E to 110°E longitude; the time period of seismic data is 1900 to 1983.

The temporal variation of earthquake occurrence is shown in Figure 8. After the year 1960, the total number of earthquakes increased due to installation of the World-wide Standardized Seismograph Network (WWSSN). The lack of data prior to the WWSSN era is due mainly to the absence of sufficient seismic instruments. Most of the earthquake observations prior to the instrumentation era were based on human perceptions, therefore earthquakes having insufficient strength to be felt by an observer would not be recorded.

Figure 9 shows the magnitude-frequency distribution. The range of magnitude value used was $m_b > 4.0$. Figure 9 shows that earthquakes having magnitude 4.5

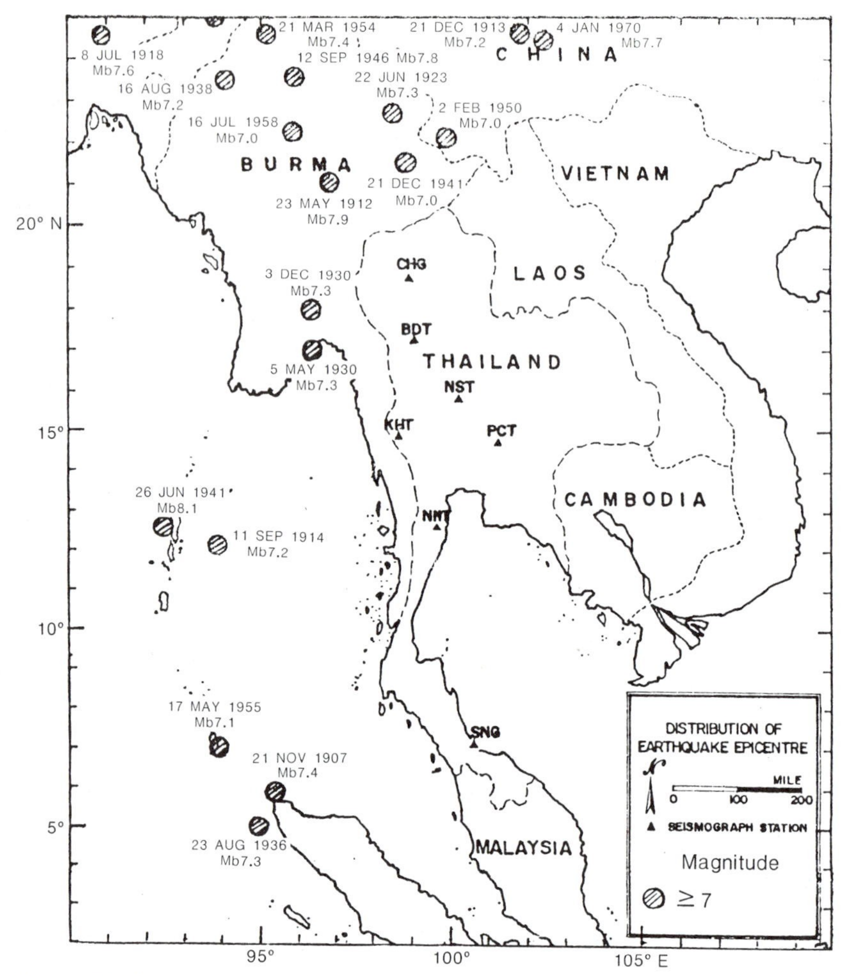

Figure 6. Historical earthquakes of magnitude ≥ 7.0 during the period 1900 to 1983.

are the most prevalent. It is presumed that there are more earthquakes having magnitude ≤ 4.5, that would be detected only by a larger seismic network.

An empirical relationship has been established by Gutenberg and Richter (1954). This relation usually is called the magnitude-frequency relation and is expressed as

$$\log N = a - bM, \tag{1}$$

where N is the number of earthquakes having magnitude M, and a and b are constants depending on the period of observation and the tectonic region.

Using the equation given by Utsu (1967):

$$b = 0.4343/\ \overline{M} - M_o, \tag{2}$$

where $\overline{M}$ is the mean which is $\sum M_i / \sum i$, and M_o is the lowest magnitude.

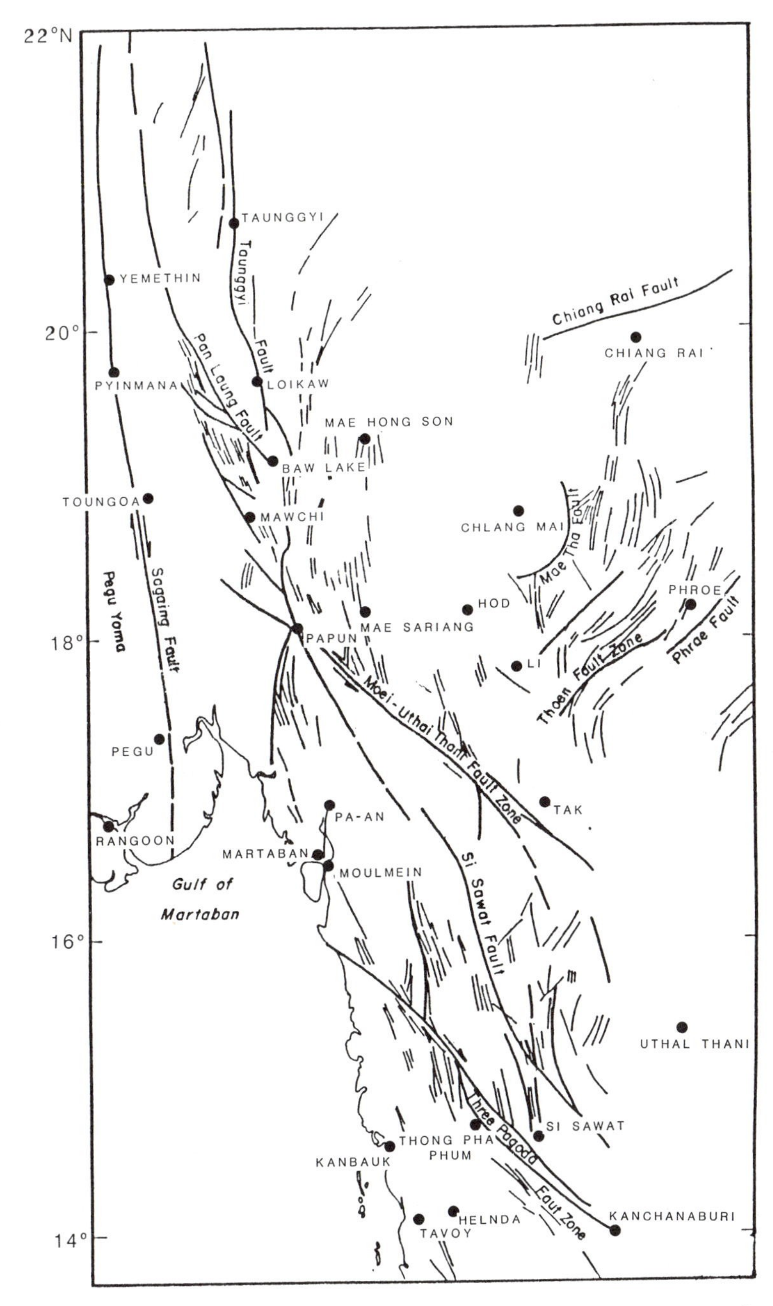

Figure 7. Major faults in the Tenasserim Ranges and Northern Thailand (modified from Bender, 1983).

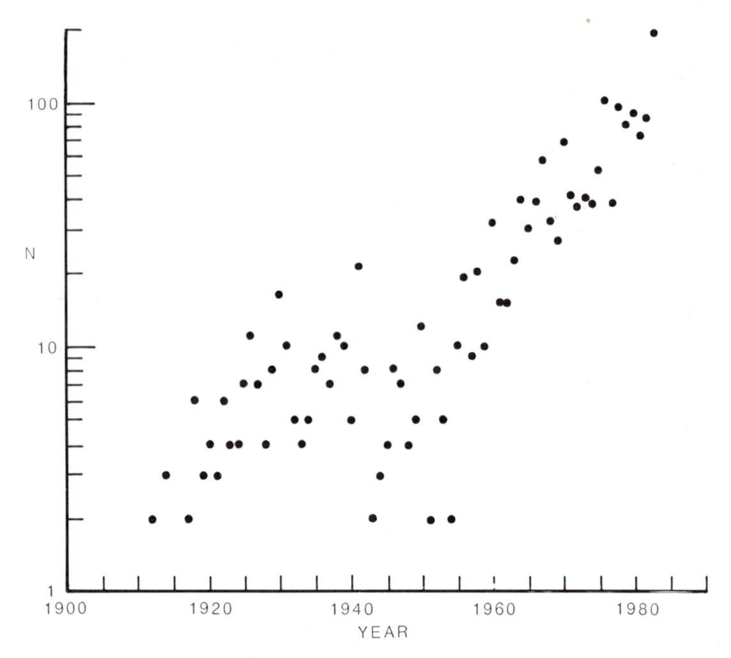

Figure 8. Time variation of earthquake occurrence.

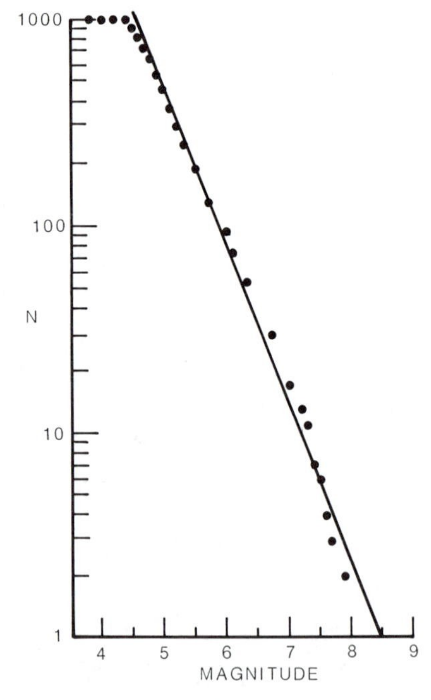

Figure 9. Magnitude-frequency distribution.

By using b-value, a-value can be determined by using the formula of Welkner (1965) as follows:

$$a = \log N_{(M \geq 5.0)} + \log(b \ln 10) + 5.0\, b \tag{3}$$

where $N_{(M \geq 5.0)}$ is the number of earthquakes having magnitude 5.0 or more.

The annual number of earthquakes can be calculated theoretically by dividing N by the period of observation, T_o. Putting

$$a_1 = a - \log T_o, \tag{4}$$

the annual frequency of earthquake is equal to

$$N(M) = 10^{a_1 - bM}, \tag{5}$$

the mean return period is equal to

$$T = \frac{1}{N(M)} \text{ years.} \tag{6}$$

4. Results and Conclusion

From the magnitude-frequency analysis, the values of a and b are 6.654 and 0.75 respectively. This means that there are about 54,000 earthquakes having magnitude greater than zero and ten earthquakes having magnitude greater than 5.0 occurring annually in this region on the average. The value of b obtained in this paper agrees well with S. Hattorri's (1974) result.

Table 3 lists the mean return period (per month) and the frequency (per year) for earthquakes in the magnitude range 5.0 to 7.5. Figure 10 shows the relation between magnitude and the mean return period. It can be seen that an earthquake having magnitude 5.0 may occur approximately every month for this area, but for the earthquakes having magnitude 7.0 and 7.5, the recurrence periods are 3 and 8 years respectively.

In conclusion, the above analysis of historical earthquake data indicates that the region encompassing Thailand, Burma and Indochina is one of the seismically active zones of the world.

Table 3. Mean Return Period and Frequency of Earthquakes for Magnitude 5.0 to 7.5

Magnitude (M_b)	Return period (month)	Frequency (per year)
5.0	1.2	9.659
5.5	2.9	4.073
6.0	7.0	1.718
6.5	16.6	0.724
7.0	39.3	0.305
7.5	93.0	0.129

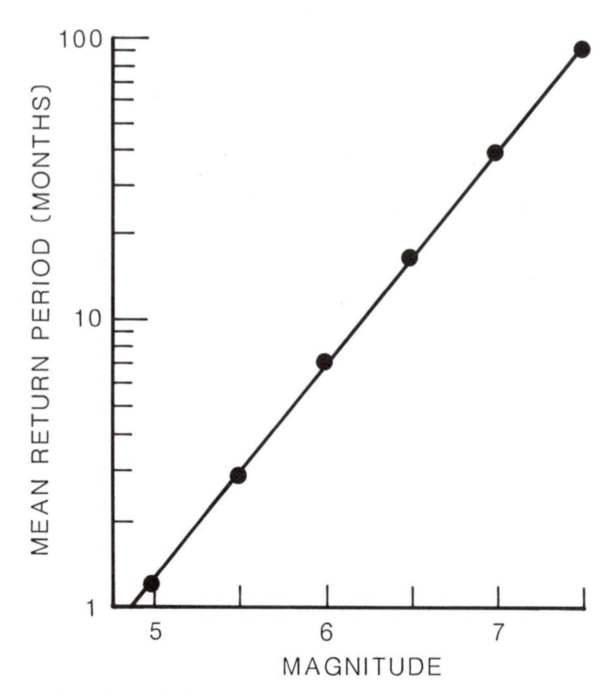

Figure 10. The relation between magnitude and mean return period.

REFERENCES

Bender (1983). *Geology of Burma*, Gebruder Borntrayer, Berlin, 293 pp.

Fontaine, H. and D. R. Workman (1978). Review of the geology and mineral resources of Kamphuchea, Laos and Vietnam, Reg. Conf. Geol. Miner. Resour. Southeast Asia Proc., **3**, 539-603.

Gutenberg, B. and C. F. Richter (1954). *Seismicity of the Earth and Related Phenomena*, Princeton Univ. Press, 310 pp.

Hattori, S (1974). Regional distribution of *b*-value in the world, *Int. Inst. Seismol. Earthquake Eng. Bull.*, **12**, 39-57.

Nutalaya, P. and S. Sodsri (1983). *Seismicity Data of Thailand and Adjacent Areas*, Asian Inst. Tech., Bangkok, Thailand, A1-A5, B1-B109, C1-C5.

Nutalaya, P. and S. Sodsri (1984). *Seismic Source Zones and Earthquakes Recurrence Relationship of the Burma-Thailand-Indochina Area*, Asian Inst. Tech., Bangkok, Thailand, 1-46.

Utsu, T. (1967). Some problems of the frequency distribution of earthquakes in respect to magnitude (I), *Geophys. Bull. Hokkaido Univ.*, **18**, 53-69.

Welkner, P. M. (1965). Statistical analysis of earthquakes occurrence in Japan, *Int. Inst. Seismol. Earthq. Eng. Bull.*, **2**, 1-27.

Win Swe (1980). A major Strike-Slip Fault in Burma, **1**, 63-72 (presented at The Fifth Burma Research Congress, 1970, Revised and title changed, 1980).

Earthquake History of California

T. R. Toppozada, C. R. Real, and D. L. Parke
California Department of Conservation
Division of Mines and Geology
630 Bercut Drive, Sacramento, CA. 95814, USA

1. Introduction

The history of earthquake occurrences is basic to assessing and mitigating earthquake hazards. Statewide seismographic coverage in California was established in 1932. Before 1932, the record of earthquakes comes from the effects reported by the population. Interpretation of the pre-instrumental record provides the approximate location and sizes of earthquakes. Because the pre-instrumental record covers a longer period than the instrumental record, it includes a major proportion of the large earthquakes, which are rare events.

The earthquake history shows what the strongest earthquake effects have been in various parts of the state. Critical structures such as nuclear power plants, dams, and hospitals are then designed to withstand at least these historical effects. The earthquake history also suggests the location of seismically active fault zones. If further geological study of these faults shows that they were active during the Holocene (last 10,000 years), they are zoned so that structures are not built on active fault traces.

The rate of occurrence of earthquakes in various fault zones, when combined with geologic and geodetic rates of deformation, helps to define the recurrence interval of destructive earthquakes in these zones. The fault zones that are most likely to rupture in the near future are then selected for the development of emergency response plans, and for the concentration of earthquake prediction monitoring efforts.

2. Completeness of the Record

The written history of damaging California earthquakes begins in 1800, when the first earthquake damage was reported at Mission San Juan Bautista. From that time on, earthquakes were reported only if they were damaging (Modified Mercalli intensity VII or greater) at any of the missions, which were located along the coast in central and southern California. Thus, the record is probably complete back to 1800 only for earthquakes of about magnitude 7 ($M7$) or greater that occurred within 70 km of the coast south of San Francisco.

Newspaper coverage came to California with the 1849 gold rush. Newspapers reported earthquake felt effects as well as damage effect. During the 1850's, 1860's, and 1870's, newspaper coverage was concentrated around San Francisco Bay and in the gold mining area east of Sacramento, with limited coverage along the coast. Thus, from 1850 the record is probably complete for earthquakes of about M 6 or greater around San Francisco Bay and in the gold country. From 1850-1880, in other areas of California the record is probably complete only for $M7$ or larger earthquakes. After 1880, with the increase of newspaper coverage, the statewide record of earthquakes is probably complete for events of $M6$ or greater. Starting in the early 1900's, with the advent of statewide seismographic recording, the record is complete for $M6$ or greater earthquakes.

3. Method of Analysis

Catalogs of felt earthquakes were prepared by Townley and Allen (1939) through 1927, and by the U.S. Department of Commerce (*U.S. Earthquakes*) starting in 1928. Since 1950, *U.S. Earthquakes* has also included isoseismal maps showing the areal distribution of the reported intensities. For events prior to 1950, we prepared isoseismal maps for 140 damaging earthquakes dating back to 1812 (Toppozada *et al.*, 1981, 1982). In addition to using the intensity information available from the catalogs, we also used information from original sources such as newspapers. Some 14,000 newspaper issues were searched, and about a quarter had earthquake reports. Obtaining information from the original sources is preferable because it avoids any previous misinterpretations.

Damaging earthquakes were identified from the existing catalogs, and newspapers were searched for the surrounding area and the days following the earthquake. The reported effects were interpreted in terms of the Modified Mercalli (MM) intensity scale and isoseismal maps were drawn showing the localities reporting the earthquake and the intensity at each locality. The resulting isoseismal maps were used to determine the approximate epicentral locations and magnitudes of the earthquakes. The epicentral location was estimated to fall at or near the center of the zone of highest intensity and to be near any locations reporting foreshocks or aftershocks. The magnitude was estimated by comparing the size of the areas shaken at various levels of intensity to those of modern earthquakes in the same region that have known instrumental magnitudes (Toppozada, 1975). For this comparison to be most meaningful, the isoseismal maps of the instrumental and pre-instrumental earthquakes should be prepared in the same way. The interpretation of reported effects in terms of the intensity scale can vary, and it is important that a given value of intensity be based consistently on the same reported effects. For example, people's reactions and ground failure are less reliable indicators of intensity than are permanent effects on buildings and their contents.

4. Seismicity, $M > 6$

Some 70 earthquakes of $M > 6$ are known to have occurred since 1800 in California and within 50 km of the state borders. These occurrences are listed in Table 1 and are numbered sequentially. The locations of these earthquakes are shown in Figures 1, 2, and 3 and are identified by the numbers used in Table 1. The areas damaged (MM intensity VII or greater) are outlined. For earthquakes not damaging a definable area, the epicentral location is plotted as a square.

The largest earthquakes of about $M8$ or greater occurred in 1857, 1872, and 1906. The only one of these to clearly affect the seismicity before it occurred was the 1906 San Francisco earthquake. Any increase in local seismicity preceding the 1857 and 1872 earthquakes would have been difficult to detect because they occurred in remote and sparsely populated areas. In the decade or two before the 1906 earthquake, there was a remarkable increase of damaging earthquakes within about 50 km of the zone that was to rupture. That level of seismicity is the highest for the Bay area, particularly the north Bay, during the century and a half of record.

For pre-1900 earthquakes, the uncertainty of the epicentral location is indicated under "Quality" in Table 1. Quality "A" indicates either an epicentral uncertainty less than 25 km, or that surface faulting was identified. An example of the latter

Table 1. California Earthquakes, $M > 6$

No.	Date (Greenwich)	Lat.°N	Long.°W	Quality	I	M
01.	1800/11/22	33.0	117.3	C	VII	6.5
02.	1812/12/08	33.7	117.9	C	VII	7
03.	1812/12/21	34.2	119.9	C	VII	7
04.	1836/06/10	37.8	122.2	A	VIII	6.8
05.	1838/06	37.6	122.4	A	VIII	7
06.	1852/11/29	32.5	115.0	C	IX	6.5
07.	1857/01/09	35.3	119.8	A	IX+	8
08.	1858/11/26	37.5	121.9	B	VII	6.1
09.	1860/03/15	39.5	119.5	C	VI+	6.3
10.	1865/10/08	37.3	121.9	A	IX	6.3
11.	1868/10/21	37.7	122.1	A	IX+	6.8
12.	1869/12/27	39.4	119.7	B	VII	6.1
13.	1872/03/26 10:30	36.7	118.1	A	IX+	8
14.	1872/03/26 14:06	36.9	118.2	C	V	6.5
15.	1872/04/03	37.0	118.2	C	V	6.1
16.	1872/04/11	37.5	118.5	C	IX	6.6
17.	1873/11/23	42.0	124.0	C	VIII	6.7
18.	1885/04/12	36.4	121.0	C	VII	6.2
19.	1887/06/03	39.2	119.8	B	VIII	6.3
20.	1890/02/09	33.4	116.3	C	VI+	6.3
21.	1892/02/24	32.7	116.3	B	IX	6.7
22.	1892/04/19	38.4	122.0	A	IX	6.4
23.	1892/04/21	38.5	121.9	A	IX	6.2
24.	1892/05/28	33.2	116.2	C	VI	6.3
25.	1897/06/20	37.0	121.5	A	VIII	6.2
26.	1898/03/31	38.2	122.4	A	IX	6.2
27.	1898/04/15	39.2	123.8	B	IX	6.4
28.	1899/07/22	34.3	117.5	B	VIII	6.5
29.	1899/12/25	33.8	117.0	A	IX	6.6
30.	1901/03/03	36.0	120.5		VIII	6+
31.	1906/04/18	37.7	122.5		IX+	8
32.	1906/04/19	32.9	115.5		VIII	6+
33.	1908/11/04	36?	117?		VII	$6\frac{1}{2}$
34.	1909/10/28	40.5	124.2		VIII	6+
35.	1911/07/01	37.25	121.75		VII	6.2
36.	1914/04/24	39.5	119.8		VIII	6.4
37.	1915/06/23	32.8	115.5		VIII	6.3
38.	1918/04/21	33.8	117.0		IX	6.8
39.	1922/03/10	35.75	120.25		VIII	$6\frac{1}{2}$
40.	1923/01/22	40.5	124.5		VIII	7.2
41.	1923/07/23	34.0	117.3		VIII	6.3
42.	1925/06/29	34.3	119.8		IX	6.3
43.	1926/10/22	36.5	122.2		VII	6.1
44.	1927/11/04	34.7	120.8		VIII	7
45.	1932/06/06	40.75	124.5		VIII	6.4
46.	1933/03/11	33.62	117.97		IX	6.3
47.	1934/01/30	38.28	118.36		IX	6.3
48.	1940/05/19	32.73	115.5		IX	6.7
49.	1941/10/03	40.4	124.8		VII	6.4
50.	1942/10/21	32.97	116.0		VI+	6.5
51.	1946/03/15	35.73	118.06		VIII	6.3
52.	1947/04/10	34.98	116.55		VII	6.2

continued

Table 1. *Continued*

No.	Date (Greenwich)	Lat.°N	Long.°W	Quality	I	M
53.	1948/12/04	33.93	116.38		VII	6.5
54.	1952/07/21	35.0	119.0		XI	7.2 & 6.4
55.	1952/07/23	35.37	118.6		V	6.1
56.	1952/07/29	35.4	118.75		VII	6.1
57.	1952/11/22	35.73	121.2		VII	6+
58.	1954/03/19	33.3	116.2		VI	6.2
59.	1954/12/21	40.8	123.9		VII	6.5
60.	1966/09/12	39.42	120.15		VII	6+
61.	1968/04/09	33.2	116.13		VII	6.4
62.	1971/02/09	34.4	118.4		XI	6.4
63.	1979/10/15	32.63	115.32		IX	6.4
64.	1980/05/25	37.6	118.83		VII	6.3 & 6.4
65.	1980/05/27	37.48	118.82		VI	6.3
66.	1980/11/08	41.0	124.64		VII	6.9
67.	1983/05/02	36.2	120.3		VIII	6.4
68.	1984/04/24	37.32	121.7		VII	6.2

Notes:

"No." is also Identifying Number on the maps.

Quality of pre-1900 epicenters: A = fault rupture identified, or uncertainty less than 25 km; B = uncertainty up to 50 km; C = uncertainty up to 100 km.

Quality of post-1900 epicenters is generally "A" by the above criteria.

I = Maximum reported Modified Mercalli intensity.

case is the great 1872 earthquake (event 13), for which 100 km of surface faulting was reported on the Sierra Nevada fault system. Even though the epicenter could have been anywhere on this rupture, the quality is A because faulting was identified.

Figure 1 shows the areas damaged by the 11 earthquakes that occurred from 1800 to 1868. Earthquakes 1, 2, and 3 were damaging at the Missions on the southern California coast, and earthquakes 4 and 5 were damaging around San Francisco Bay. Event 6 was damaging near Fort Yuma. As stated previously, before the 1849 gold rush, records of earthquake damage were restricted to the coast southward from San Francisco. During the 1850's, newspaper coverage spread northeastward from San Francisco into the gold country east of Sacramento, providing reports for event 9. Event 7 (area from Agnew and Sieh, 1978) was a great earthquake on the San Andreas fault having no obvious foreshocks or aftershocks, although two possible foreshocks of $M \leq 6$ occurred in the preceding two hours (Sieh, 1978). Events of $M \leq 6.5$ may not have been identified in central California in 1857. By contrast, the increase in seismicity in San Francisco Bay area that started in 1858 and culminated in 1868 is well substantiated (events 8, 10, and 11).

Figure 2 shows the areas damaged by earthquakes that occurred from 1869 to 1906. Two great earthquakes occurred in this period. The 1872 Sierra Nevada earthquake had identifiable aftershocks that progressed northward (events 13, 14, 15, 16). No obvious foreshocks were identified, although events smaller than $M6.5$ could have been missed in eastern California. The 1906 San Francisco earthquake was preceded by a remarkable increase in damaging earthquakes starting in 1892 (events 22, 23, 25, 26, 27). No significantly damaging aftershocks of the 1906 event were noted.

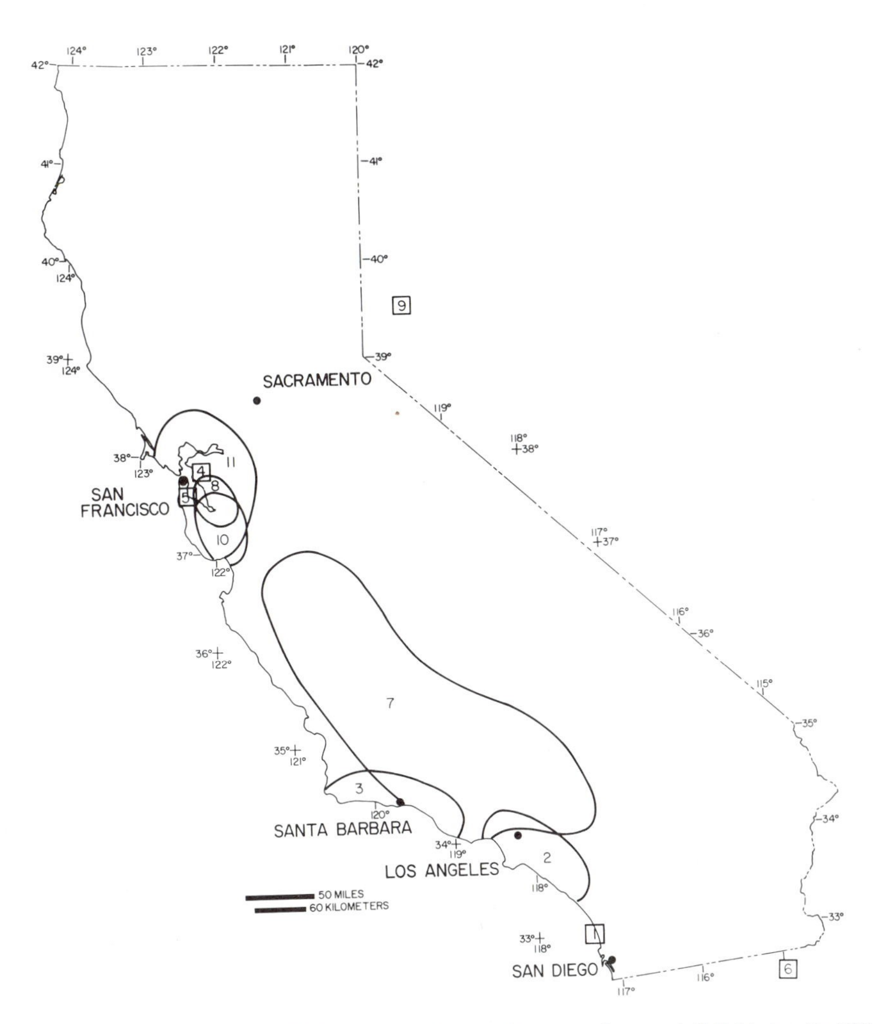

Figure 1. Earthquakes of $M > 6$, 1800-1868. The areas damaged (MM intensity VII or greater) are outlined. Earthquakes not damaging a definable area are indicated by a square at the epicenter. The earthquakes are numbered chronologically according to Table 1.

Figure 3 shows the location of earthquakes that have occurred since 1907. This map shows less earthquake damage around the San Francisco Bay area than during the preceding 50 years. This suggests that the 1906 earthquake relieved the stresses in the Bay area, and that the stresses have not yet built up to the level of the 1860's. Throughout the state, seismicity has been lower after 1906 than before. The largest California event since 1906 was the $M7.2$ earthquake of 1952. The most destructive events since 1906 occurred in 1933 (event 46, $M6.3$) and in 1971 (event 62, $M6.4$) in the densely populated Los Angeles area.

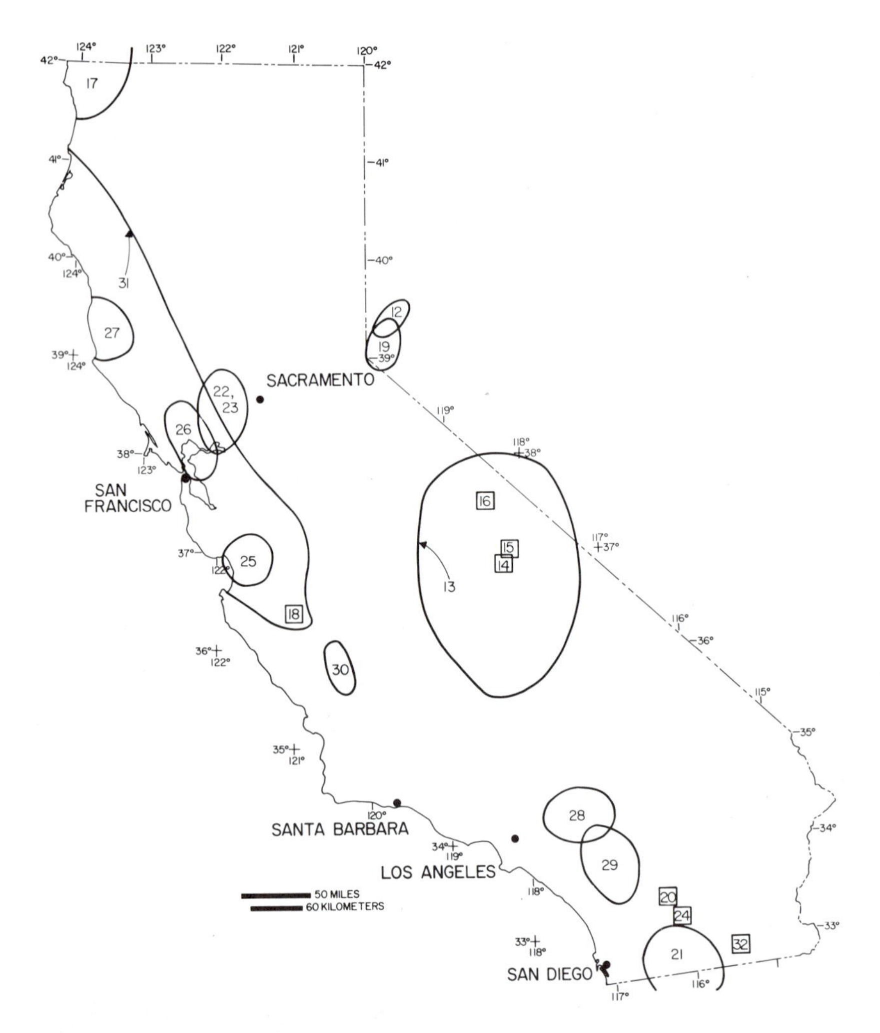

Figure 2. Earthquakes of $M > 6$, 1869-1906. The areas damaged (MM intensity VII or greater) are outlined. Earthquakes not damaging a definable area are indicated by a square at the epicenter. The earthquakes are numbered chronologically according to Table 1.

Even with today's population distribution, earthquake damage is a poor indicator of $M > 6$ seismicity. There remain unpopulated areas where such earthquakes cause no damage. For example the $M6.4$ and $M6.3$ Mammoth Lakes earthquakes of 25 May 1980 (event 64) damaged a very small area in the Sierra Nevada. Also, the $M6.2$ Borrego earthquake of 1954 (event 58) occurred in the desert, and caused no significant damage. Comparing Figure 3 with Figures 1 and 2, illustrates that the modern record is more complete for earthquakes that were not significantly

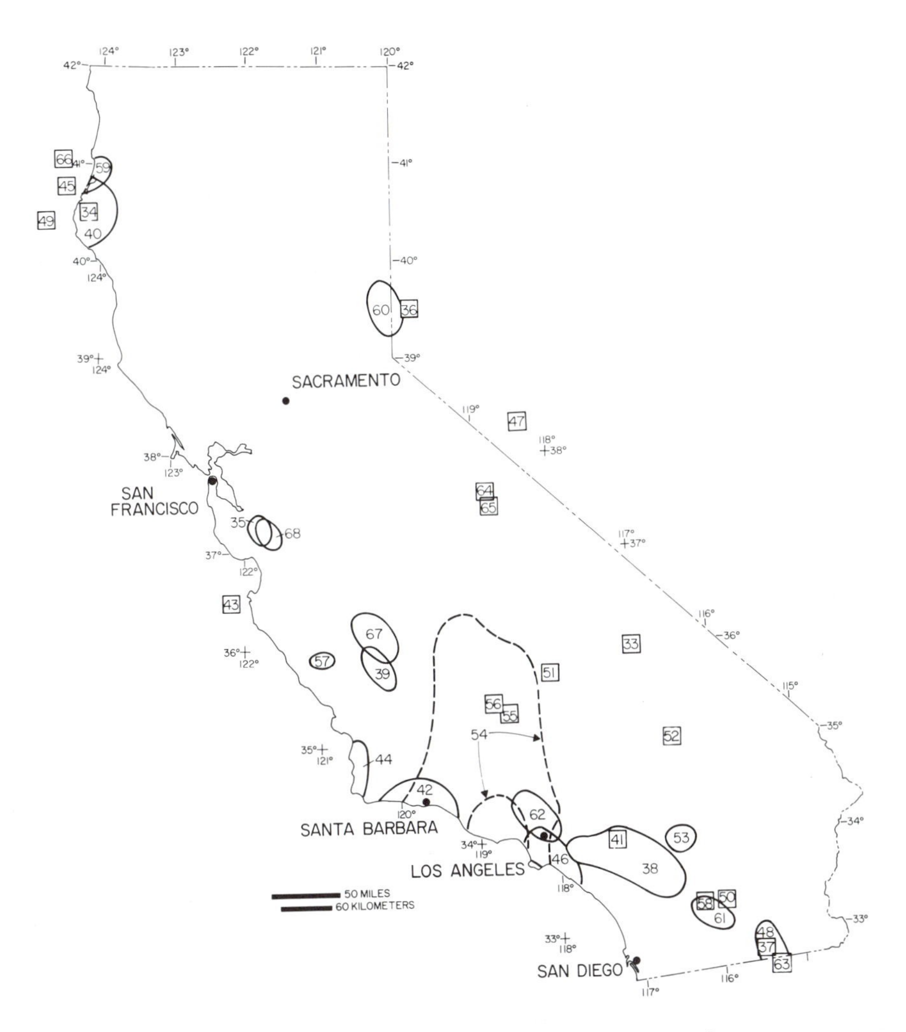

Figure 3. Earthquakes of $M > 6$, 1907-1984. The areas damaged (MM intensity VII or greater) are outlined. Earthquakes not damaging a definable area are indicated by a square at the epicenter. The earthquakes are numbered chronologically according to Table 1.

damaging over large areas. Before 1907 such events were incompletely reported, but large earthquakes which damaged large areas occurred more frequently than after 1907.

By superimposing Figures 1-3, the outline of all the areas damaged is shown in Figure 4. This shows that since 1800 the seismic hazard has been highest near the San Andreas fault zone and the Sierra Nevada fault zone.

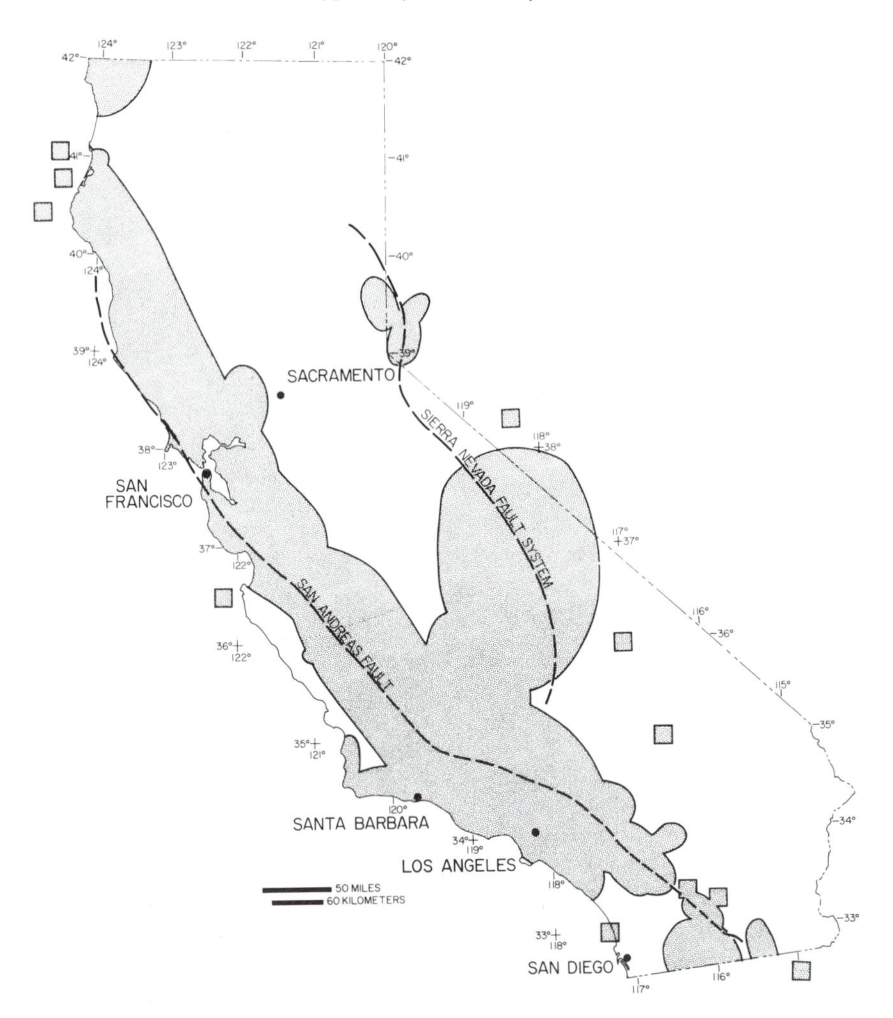

Figure 4. Outline of areas damaged by $M > 6$ earthquakes, 1800-1984. Earthquakes that did not damage a definable area are indicated by a square at the epicenter.

ACKNOWLEDGEMENTS

This paper was reviewed by J. H. Bennett, L. Mualchin, and R. W. Sherburne, and was typed by C. L. Pingree. The suggestions by J. H. Bennett are greatly appreciated.

REFERENCES

Agnew, D. C. and K. E. Sieh (1978). A documentary study of the felt effects of the great California earthquake of 1857, *Bull. Seism. Soc. Am.*, **68**, 1717-1730.

Sieh, K. E. (1978). Central California foreshocks of the great 1857 earthquake, *Bull. Seism. Soc. Am.*, **68**,1731-1749.

Toppozada, T. R. (1975). Earthquake magnitude as a function of intensity data in California and Western Nevada, *Bull. Seism. Soc. Am.*, **65**, 1223-1238.

Toppozada, T. R., C. R. Real, and D. L. Parke (1981). Preparation of isoseismal maps and summaries of reported effects for pre-1900 California earthquakes, *Calif. Div. Mines Geol. Open-File Rept1-11 SAC.*

Toppozada, T. R. and D. L. Parke (1982). Areas damaged by California earthquakes 1900-1949, *Calif. Div. Mines Geol. Open-File Rept. 82-17 SAC.*

Townley, M. D. and M. W. Allen (1939), Descriptive catalog of earthquakes of the Pacific coast of the United States, 1769 to 1928, *Bull. Seism. Soc. Am.*, **29**, 297.

U.S. Earthquakes, Annual publication of the U.S. Department of Commerce, starting in 1928.

Study of Historical Earthquakes in Japan (2)

Tatsuo Usami
Faculty of Engineering
Shinshu University, Nagano, Japan

ABSTRACT

In Japan, a project on historical earthquakes (i.e., the collection and publication of old materials describing historical earthquakes) is accelerated recently. The purpose of this article is to introduce the project and some of the scientific works based on these materials and to show that historical documents can aid modern scientific studies, such as seismotectonics and long-range prediction of large earthquakes.

We define "historical earthquake" as before 1873. Studies on historical earthquakes carried out before 1970 are summarized in Section 1. These studies begin with collection and publication of historical materials relating to old earthquakes. The first and second phases of the project are introduced in Section 2, and the status of the third phase being carried out by the present author and collaborators is given in Section 4.

Among studies based on these historical materials, the following three topics were carried out by the present author:
(1) Change of seismicity of destructive earthquakes in Japan is documented in Section 3. Materials for destructive earthquakes are more reliable and more uniform than for non-destructive earthquakes.
(2) The 1858 Omati earthquake is studied employing recently collected materials (Section 5). Other historical earthquakes occurring in valleys connecting Itoigawa-Matumoto-Iida-Toyokawa cities are also considered. It is shown that the damaged area of each earthquake contacts each other, covering valleys with almost no gaps.
(3) The 1793 Kansei earthquake is studied in Section 5. Comparison with recent earthquakes suggests that the Kansei event is quite similar to the 1896 off-Sanriku event of reverse type. Comparison of this result with other large earthquakes off the Sanriku coast of Japan, provides a clue for the occurrence time of the next off-Sanriku earthquake of reverse type.

1. Introduction

Although this is the second paper of the same title, the introductory parts of the first paper (Usami, 1979) are reproduced. In this article, we consider earthquakes which took place in and near Japan, that is, Hokkaido including the southern Kurile, Honshu, Shikoku, Kyushu and Ryukyu Islands (Figure 1). In this article, the word "historical" refers to the time period before 1873. The year 1873 was chosen as the year of division for the following reasons:

(1) In this year, instrumental observations of earthquakes started.
(2) At the end of this year, the Gregorian calendar replaced the Japanese lunar calendar. On 9 November of the fifth year of the Meiji era, 1 January 1873, the Gregorian calendar was first put into general use.

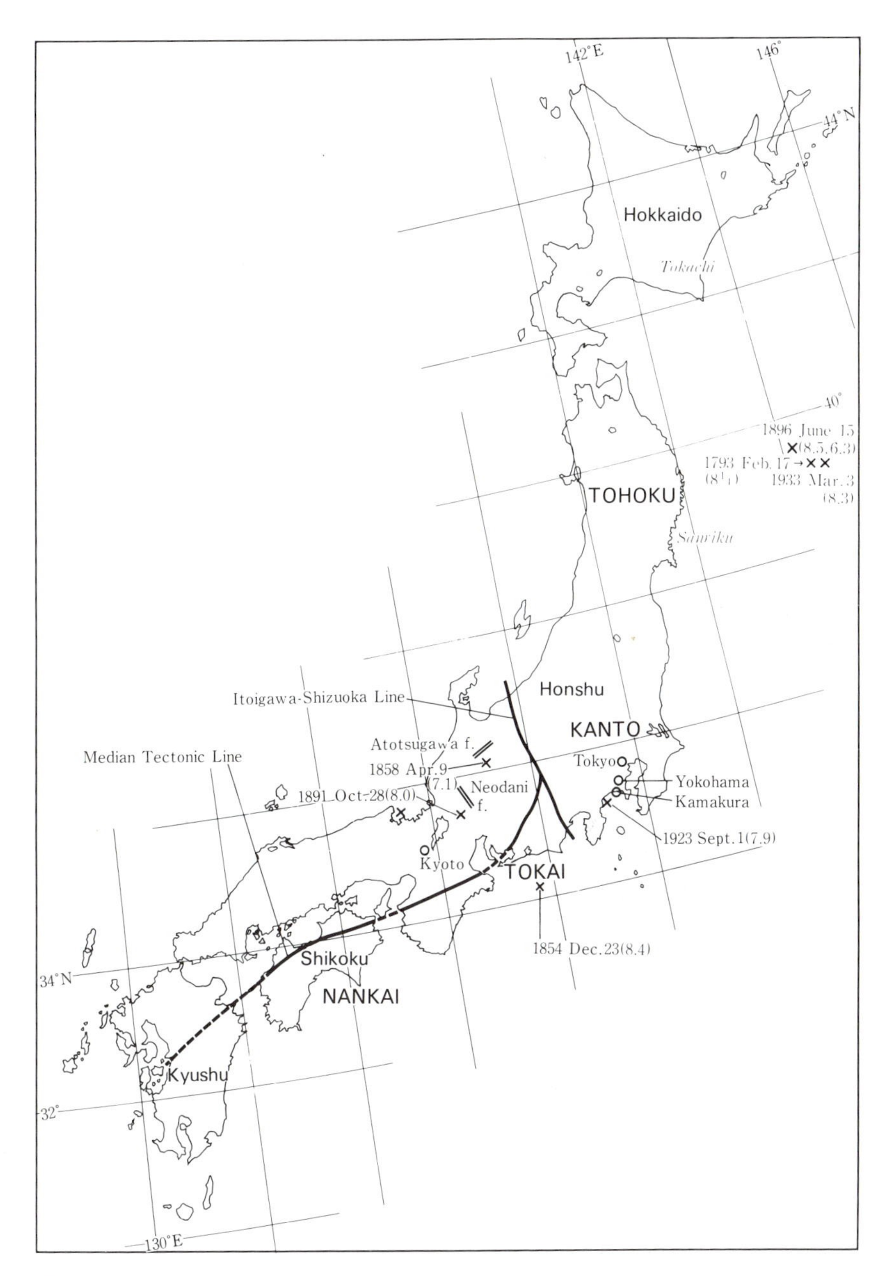

Figure 1. Map of Japan. Small gothic letters indicate the name of the island; large gothic letters indicate the district name; open circles – city name; cross – earthquake epicenter; double lines – fault. Numbers in parentheses indicate earthquake magnitude.

(3) Documents published after 1 January 1873 are more readily available than those before 1873 and are printed in type which is more readable to young Japanese than those of historical period.
(4) The collection of documents before 1873 is incomplete and many documents have remained untouched. Old documents were written with carbon ink and brush pen and young Japanese have difficulty in reading them. Reading the old documents is a hard and time consuming job and that couldn't have been done without special training.

2. Brief Sketch of Historical Earthquake Studies before 1970

During the historical period, when a disastrous earthquake took place, older earthquakes were usually referred to and listed for comparison. Considering this kind of work as a study of historical earthquakes, the history of such study can be traced back to before the 10th century. Although such non-comprehensive lists of big events could not be said to be scientific, the records include basic data for modern scientific study of historical earthquakes. Employing old records, including lists of earthquakes as stated above, I. Hattori, Esq. made an extensive study on historical earthquakes in Japan (Hattori, 1878). This probably was the first work of its kind in Japan. Thirty-four old documents were employed and 149 destructive earthquakes were found after the 5th century.

Naumann (1878) published a paper on Japanese earthquakes and volcanic eruptions. On 22 February 1880, a moderate earthquake took place near Yokohama ($M = 5.9$). Foreign professors who had been invited to teach modern science and technology and had lived near Tokyo were intrigued by the event. On 26 April of the same year, the Seismological Society of Japan (which is the first seismological society in the world) was established through the efforts of foreign scholars.

John Milne, one of the invited professors, published a study on historical earthquakes in Japan (Milne, 1881). He used sixty-four old records (including thirty-four used by Hattori) and listed 366 earthquakes.

From about this time, short introductions concerning specific historical earthquakes were published in various kind of bulletins. On 28 October 1891, the Nobi earthquake ($M = 8.0$) took place in central Japan. Due to the impact of this event the Imperial Earthquake Investigation Committee was established on 25 June 1892, in order to make investigations for preventing disasters due to earthquakes. One of the eighteen purposes of the Committee set forth at the start was the "investigation of earthquakes in historical ages, namely, the compilation of earthquake history." Tayama was appointed as a part-time member of the Committee on 13 July 1893 and was entrusted to execute the compilation of old earthquakes. His first work (Tayama, 1899) is a table containing 1,896 earthquakes (including aftershocks) from the years 416 to 1864. The date, locality, magnitude (not instrumental), and references are tabulated for the earthquakes. His second work (Tayama, 1904), perhaps his last work on historical earthquakes, consists of two volumes, totalling 1,201 pages. They include earthquakes from the years 416 to 1865. This is the first comprehensive and scientific work in the field of historical earthquakes and marked an epoch in the history of the study of historical earthquakes in Japan.

On 1 September 1923, the great Kanto earthquake ($M = 7.9$) hit the southern part of the Kanto district including the Tokyo and Yokohama area. After the event, on 13 November 1925 the Imperial Earthquake Investigation Committee was replaced by the Earthquake Research Institute in order to promote more intensive earthquake studies. In 1928, K. Suyehiro, the first director of the Earthquake Research Institute, asked K. Musha to revise and enlarge the work by Tayama. After thirteen years of efforts, Musha (1942) published the first volume of the revised edition of *Historical Data on Japanese Earthquakes*. The second and third volumes were published in 1943 (Musha, 1942-1943). These three volumes were mimeographed. The fourth volume (Musha, 1951) was published in 1951 in type. The four volumes amount to 4,000 pages in all, and include 6,000 earthquakes (excluding aftershocks) from the years 416 to 1867. The revised edition by Musha is too voluminous to read through. He picked up large disastrous earthquakes, gave appropriate summaries for each event, and arranged them in a report (Musha, 1950-1953). This report was also mimeographed. Since then the importance of such a chronological table has been acknowledged among seismologists. The first seismologist who recognized the importance of this table was the late Dr. H. Kawasumi. He estimated magnitude and epicenter location for destructive historical earthquakes and his efforts resulted in two works: (1) sixty-nine year periodicity of strong earthquakes in southern Kanto district (Kawasumi, 1963); and (2) the so-called Kawasumi's map (Kawasumi, 1952) showing maximum acceleration anticipated in every 75, 100, and 200 years.

3. Short Summary of the Study of Historical Change of Seismicity

The present author (Usami, 1977) made a descriptive table of destructive earthquakes in Japan. The number of such earthquakes is 693, during the period 416 to 1973. Of these, 458 earthquakes occurred in historical time. Figure 2 shows the number of destructive earthquakes in ten-year intervals for Kyoto, Kamakura, and Tokyo. Kamakura became the capital of Japan at the beginning of the 13th century and Tokyo at the beginning of the 17th century. Figure 2 shows that disastrous earthquakes in the vicinity of the capital increase suddenly at the time of the opening of a new government.

Comparison of disastrous earthquakes according to old and new materials is shown in Table 1. Here, "old" means materials collected by Tayama and Musha. It should be noted that new materials are four times more voluminous than old ones, and the more we search, the more new materials will be found. There seems no limit for finding additional materials. Total number of disastrous earthquakes tabulated from old and new materials is 282 and 350 respectively. Table 1, shows that the addition of new materials changes the history of earthquakes and opens a new viewpoint relevant to modern seismology.

4. Recent Work on the Collection of Old Documents

Since the publication of *Historical Data on Japanese Earthquakes* by Musha in 1951, no extensive collection of old documents has been done by any Japanese until 1971, when the present author began collecting them. Since then, interest in historical earthquakes has increased among seismologists, especially in relation to the long-range prediction of earthquakes and mitigation of earthquake disasters. At the end of 1976, by the strong desire of the Ministry of Education, the Historiographical Institute of Tokyo University agreed to cooperate with the Earthquake Research Institute in the collection and reading of old documents. Historians know about old

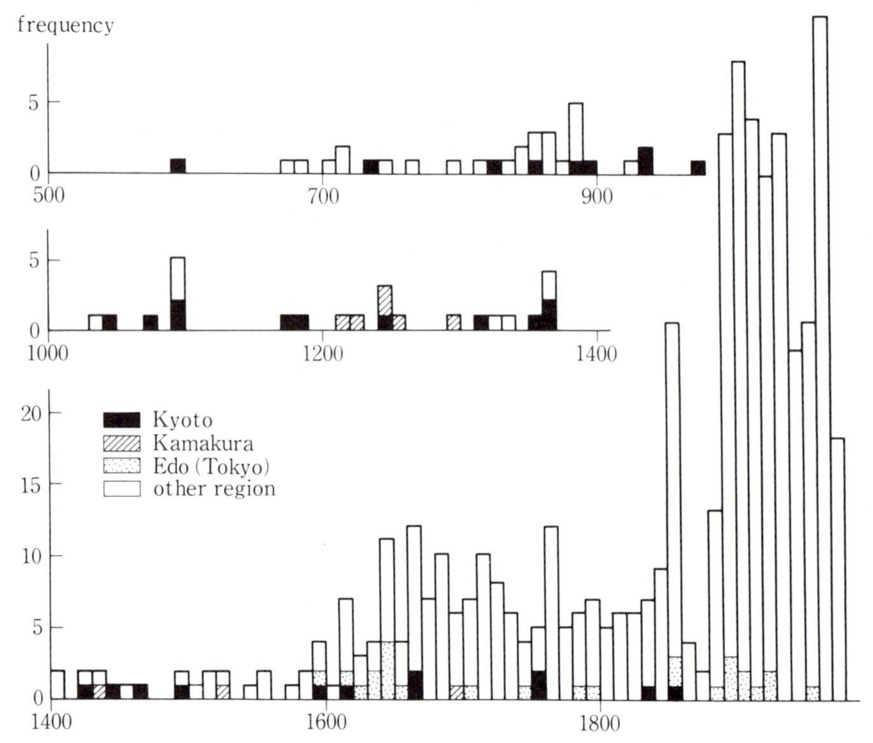

Figure 2. Number of disastrous earthquakes in every ten years.

Table 1. Number of Disastrous Earthquakes in Japan Obtained from Old and New Materials

Judgement due to		Number of earthquakes (before 1872)
Old materials	New materials	
D	D	261
D		21
	D	89

D – acknowledged as disastrous earthquakes; blank – not acknowledged as disastrous earthquakes

documents and advise seismologists where to go and what to look for. Through this cooperation, the collection of old documents has been accelerated. The materials thus collected are arranged in chronological order and have been published annually beginning in 1981 (Earthquake Research Institute, 1981-1987). The published part amounts to 11,000 pages, just three quarters of the materials collected to date. On the average, about 5,000 sheets of photographs and/or copies of old documents are being collected annually.

5. Basic Study on Historical Earthquakes Due to New Materials

On 9 April 1858, the so-called Hi-etsu earthquake ($M = 7.0$) took place in central Japan. As shown in the first paper (Usami, 1979), this earthquake was caused by the activity of the Atotsugawa-fault which trends ENE-WSW. Near the epicenter, in the mountainous areas on the west foot of Mt. Tateyama (3,015 m), there occurred many landslides and mudflows which blocked the flow of rivers at various places. One of the natural dams gave way suddenly at about 11 a.m. of 23 April 1858 and a flood containing mud, rock and big trees ran down to the Toyama plain, causing loss of lives and damaging houses and rice paddies. At about 8 a.m. of the same day, an earthquake took place near Matsushiro, about 50 km east of Mt. Tateyama, causing slight damages. Owing to the 50 km distance, this event has not been considered as the cause of failure of the natural dam. New materials indicate that the earthquake took place near Omati-city, about 20 km east of Mt. Tateyama, and caused medium damage to villages near Omati-city. Moreover, the occurrence time of this event is about 11 a.m. of 23 April 1858. This event will be called the Omati earthquake. The similar occurrence time of the Omati earthquake and the natural dam failure suggests to us that the Omati event may have triggered the dam failure.

The damaged area of the Omati earthquake is shown in Figure 4, by the hatched area indicated by the occurrence year. The intersection of the elongated part of Atotsugawa-fault and the Itoigawa-Shizuoka tectonic line running N-S near Omati-city is found within the damaged area of the Omati earthquake. This fact may have some geophysical meaning which is a theme for future study.

Stimulated by this study, damaged areas due to medium-size earthquakes along the Itoigawa-Shizuoka line and along the median tectonic line south of Matsumoto city are plotted in Figures 3 and 4. Figure 3 is drawn employing old materials and Figure 4 is drawn by adding new materials. Two earthquakes in the year 715 took place within two days and their damaged areas are shown by rough estimation. From Figure 3, we can not deduce definite conclusions about the seismic activity along the considered tectonic lines. However, Figure 4 shows that in the northern portion, medium-size earthquakes move from north to south, and in the southern portion, from south to north, and they meet near Omati and Matsumoto, that is, the junction of the Itoigawa-Shizuoka line and the median tectonic line.

In the northern portion, damaged areas contact each other, indicating that in these 300 years, seismic energy in this area has been released. On the other hand, for the southern portion there still remain intermittent areas where no seismic activity was recorded during historical time. However, it is not certain that earthquakes have not occurred in the intermittent areas, because the old documents are not completely collected.

From fault studies in Japan, it is said that inland faults give rise to earthquakes every several hundred to several thousand years. It is not clear whether or not these considerations and the serial occurrence of earthquakes in Figure 4 in these 1000 years are concordant.

The Kansei earthquake of 17 February 1793 (7 January of the 5th year of Kansei era) occurred off the Pacific coast of northeastern Japan. Characteristic features of this earthquake according to old materials are:

(1) The few documents available have descriptions at several coastal places between 39° and 40° N.

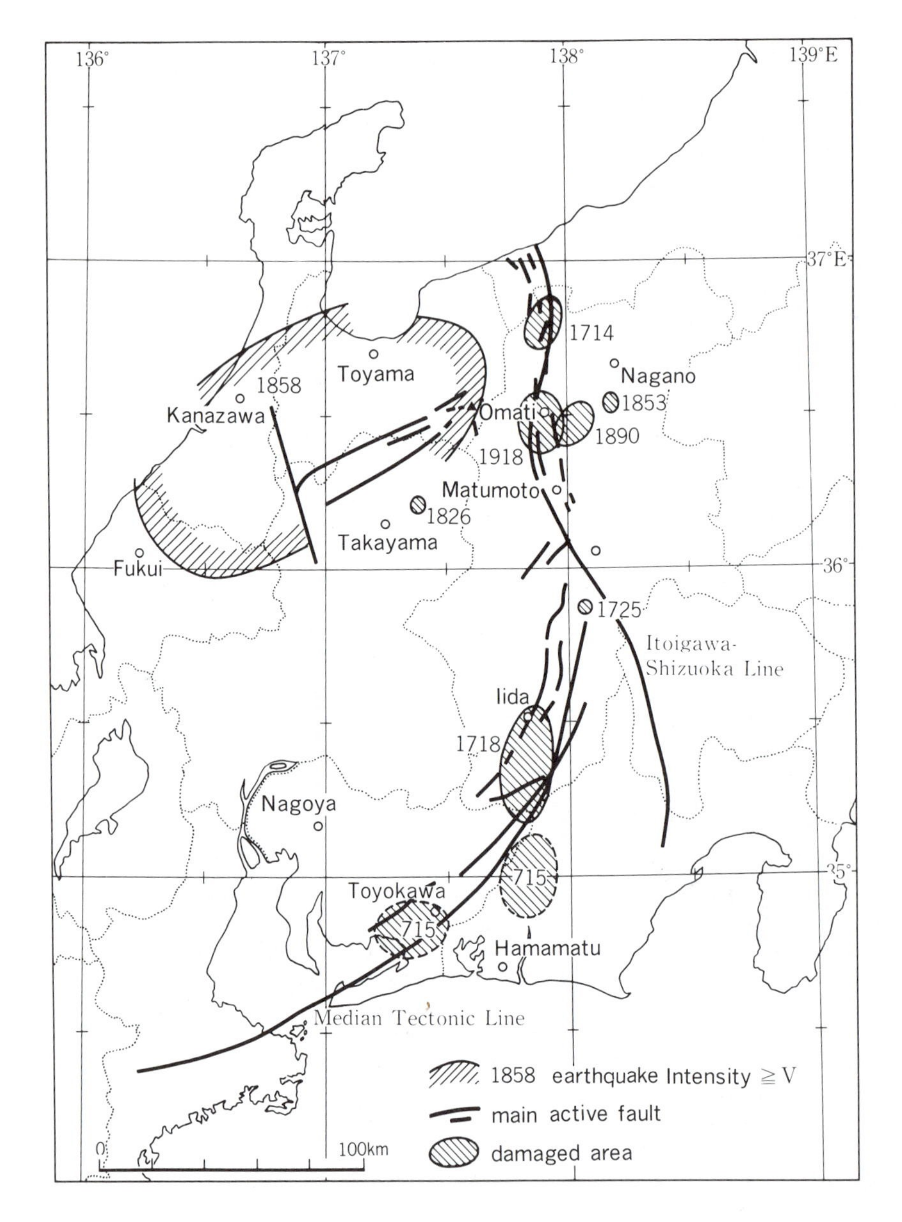

Figure 3. Damaged areas of historical earthquakes along 138° E line based on old materials.

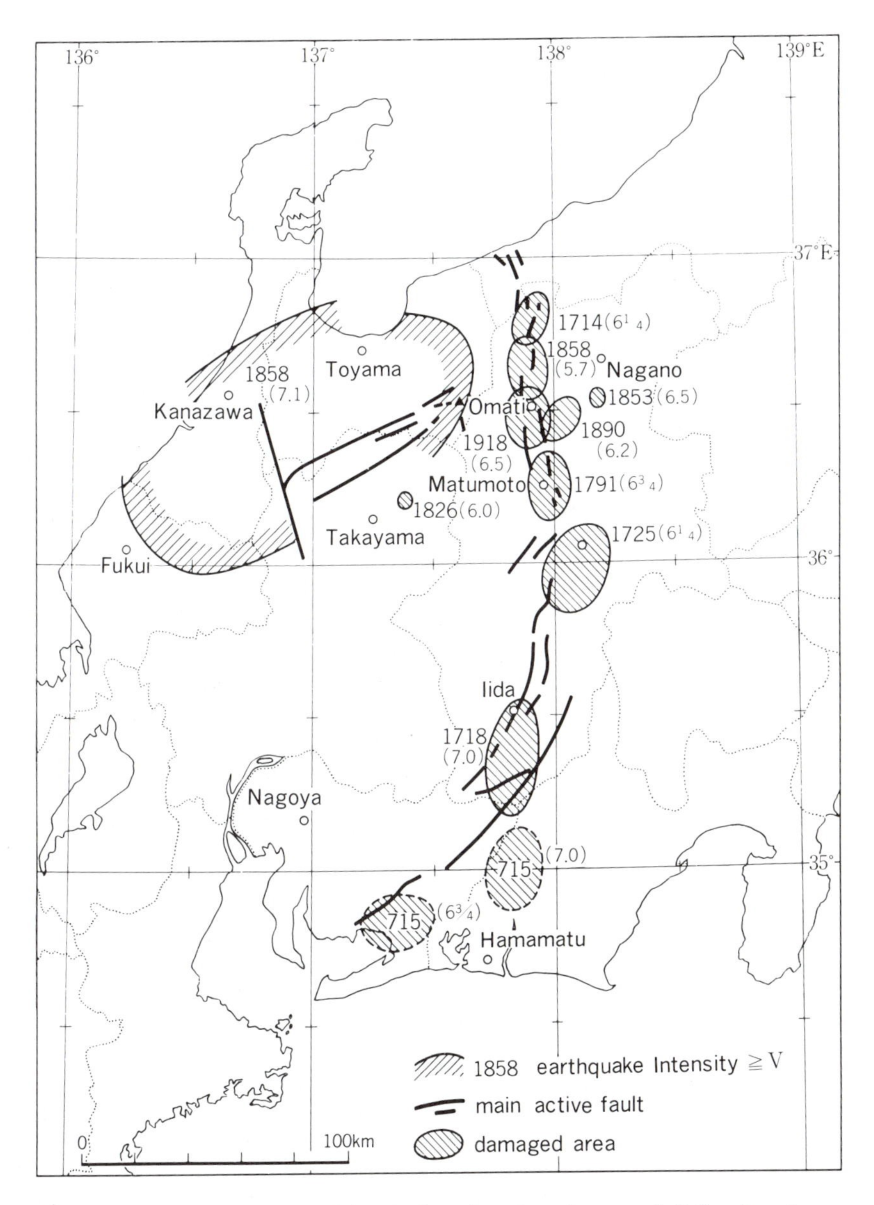

Figure 4. Damaged area of historical earthquakes along 138° E line based on new materials. Numbers in parentheses indicate earthquake magnitude.

(2) A tsunami attacked the Pacific coast between 39° and 40° N and caused several hundred deaths and several hundred houses were washed away.
(3) Occurrence time is about 10 a.m.
(4) Beginning noon of the same day, swarm earthquakes occurred in Tokyo and continued until 23 February. The total number of felt earthquakes was more than 70. No damages were reported.

New materials reveal the following facts:

(1) Intensity (Japanese scale) distribution is shown in Figure 5. For comparison, isoseismals are drawn in Figure 6 for some other earthquakes which took place off the Sanriku coast. The area of intensity $\geq$ 4 resembles those in Figure 5, however the area of intensity $\geq$ 5 is smaller than the Kansei earthquake.
(2) Aftershock series observed at various places is shown in Figure 5. They show that the number of aftershocks decrease abruptly one week after the main shock. At some places, aftershocks occur more frequently than in Tokyo. The daily aftershock number observed at Sendai is as follows:

17 February	54-55
18 February	30
19 February	17
20 February	7-8
21 February	5-6

No modern earthquake in Table 2 gives a persuasive explanation for the number of aftershocks at Tokyo due to the Kansei event. Table 3 lists two large off-Sanriku earthquakes in 1896 and 1933. The number of aftershocks for the 1896 event is greater than for the 1933 event. Except for Tokyo, aftershock number of the Kansei earthquake could be explained by considering that it gave rise to a few more aftershocks than the 1896 event. The intensity distribution of the Kansei earthquake could be reasonably explained if we make the magnitude somewhat larger than the 1896 event.

The distribution of areas hit by tsunami and the tsunami height estimated from damages resemble those of 1896 and 1933 earthquakes. The 1896 earthquake is due to reverse faulting, and the 1933 event is due to normal fault movement.

Let us try to determine the fault type of the Kansei earthquake employing the initial motion of the tsunami wave observed at Nakamura, i.e., the "pull" motion. However, we do not know the initial motion at Nakamura for the 1896 and 1933 earthquakes. The nearest point where the initial motion is known is Ayukawa, about 100 km north of Nakamura. At Ayukawa, the initial motion of tsunami wave is "pull" for 1896 earthquake and "push" for 1933 earthquake. So, for the time being, it may be possible to consider that Kansei earthquake is reverse fault type. In Table 3, we compare the Kansei earthquake with some typical types of earthquakes occurring off the coast of northeastern Japan. From Table 3, the type of earthquake which most resembles the Kansei earthquake is the 1896 off-Sanriku earthquake of reverse type.

Historical large earthquakes of the types 1 to 3 (as defined in Table 3) are arranged in Table 4. From plate tectonics theory, it is said that reverse fault movements repeat every one hundred years or so, but there is no definite repeat interval for the normal fault movement. If the Kansei earthquake is of reverse type, we can see

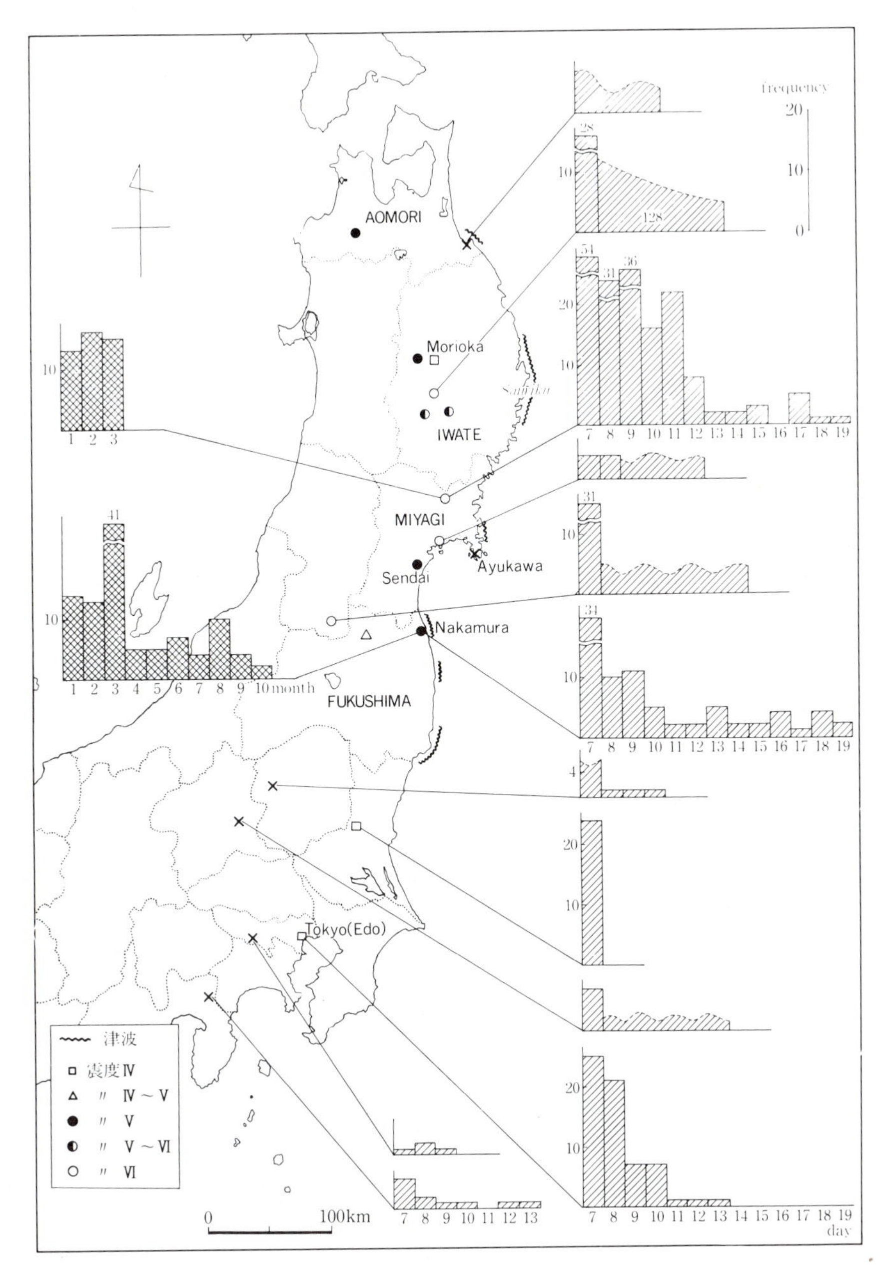

Figure 5. Intensity at various places and aftershock series of the 1793 Kansei earthquake. On the right, daily aftershock number is shown from January 7th (Japanese lunar calendar) to 19th. On the left, monthly aftershock number from January to October is shown. For January, those after 20th are shown.

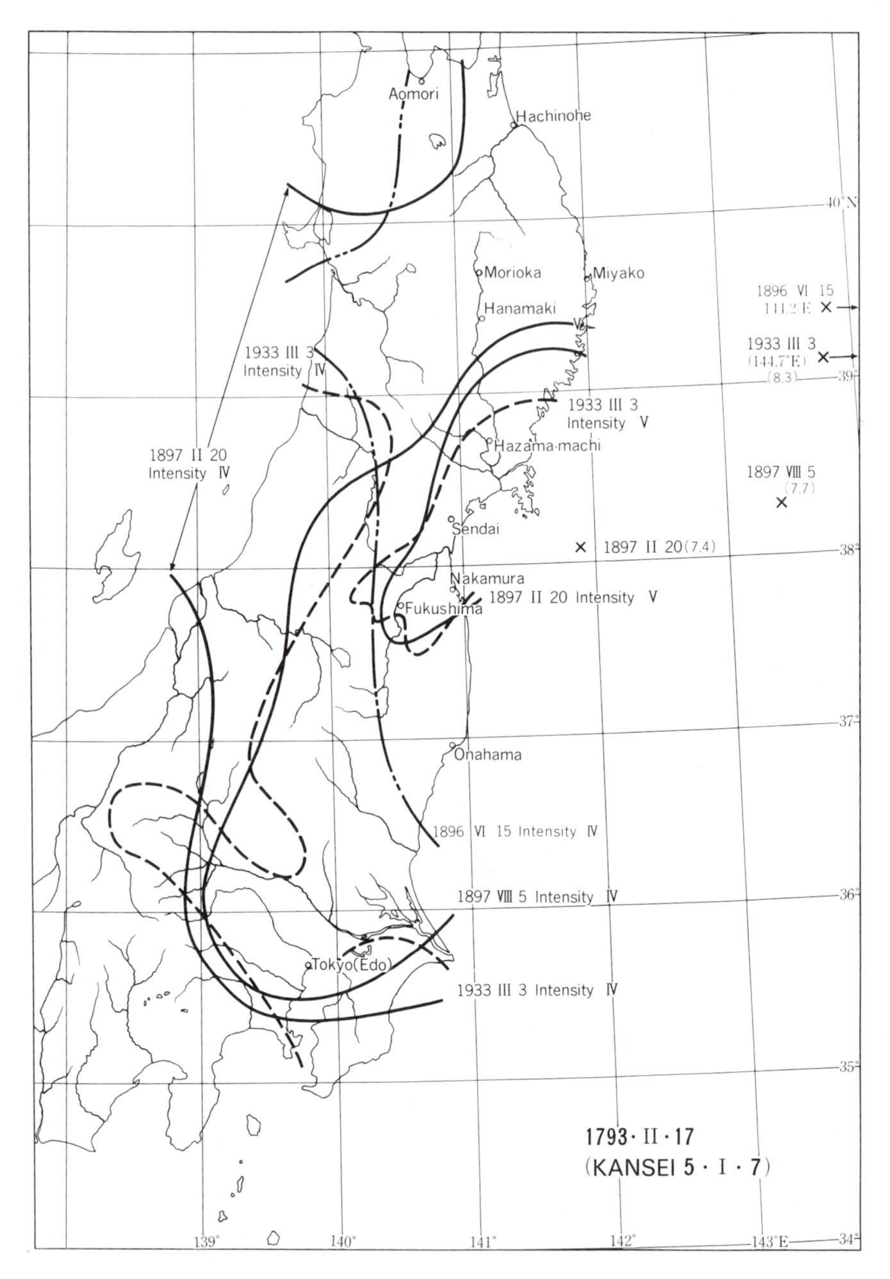

Figure 6. Isoseismals for four earthquakes occurring off the coast of northeastern Japan. Comparison should be tried with intensity in Figure 5. Numbers in parentheses indicate earthquake magnitude. The 1896 event was formerly considered to be magnitude 8.5, but was recently revised to 6.8.

Table 2. Comparison of Large Earthquakes Off the Coast of Northeastern Japan

Type	Date year.mo.da	λ (°E)	φ (°N)	M	Number of felt aftershocks: Tokyo (a.)	Tokyo (b.)	Ha	Mo	Mi	S	F	O	N	H	Hn	A	Rem.
2	1896.06.15	144	$39\frac{1}{2}$	6.8	(1)	0			62							95	(1)
4	1897.02.20	141.9	38.1	7.4	0	(2)											
4	1897.08.05	143.3	38.1	7.6	2+(1)	1+(3)											
4	1915.11.01	142.9	38.3	7.5	1+(1)	(4)											(2)
3	1933.03.03	144.7	39.1	8.3	1	0		29	17	3	14	3					(3)
5	1938.05.23	141.4	36.7	7.1	1	3	6	5	12	5	21	23					(4)
5	1938.11.05	141.7	37.1	7.7	12	8	16	26	29	77	164	91					(5)
	1793.02.17	142.4	38.3	7.1	≥ 70	?							69	170	184		(6)

(a.) = within 7 days after the main shock; (b.) = 8 to 30 days after the main shock.
Ha = Hachinohe; Mo = Morioka; Mi = Miyako; S = Sendai; F = Fukushima; O = Onahama; N = Nakamura; H = Hazama-machi; Hn = Hanamaki; A = Aomori.
Parentheses in columns (a.) and (b.) indicate epicenter is not certain, but considered as aftershock by the present author.
Remarks: (1) = felt aftershock in June; (2) = felt earthquake in November; (3) = all felt earthquakes in March; (4) = all felt earthquakes in May and June; (5) = felt earthquake in November; (6) = felt earthquake within one week after the main shock.

Table 3. Comparison Between 1793 Kansei Earthquake and Other Typical Earthquakes Occurring off the Coast of Northeastern Japan

Type	Typical earthquake referred	Tsunami	Distribution of intensity	Number of aftershocks: Edo	General
1	Off Tokachi earthquake (1677, 1763, 1856)	X	X	X	○
2	Off Sanriku earthquake (reverse type, 1896)	○	◐*	X	○
3	Off Sanriku earthquake (normal type, 1933)	○	○	X	X
4	Off Miyagi Pref. earthquake (1897-1, 1897-2, 1915)	X	○	X	X
5	Off Fukushima Pref. earthquake (1938)	X	◐	X	○

Type = classification due to epicentral area and fault movement. Edo = former name for Tokyo. Similarity to 1793 Kansei earthquake: ○ = good; ◐ = moderate; X = bad. (*) = isoseismal is similar, but the value of intensity is different.

from Table 4 that reverse type earthquakes take place 30-40 years after a type 1 event, therefore the next reverse fault type earthquake may take place circa the year 2000. The interval between earthquakes is longer when the energy release by the previous earthquake is large, and is average (about 100 years), when the energy release by the previous one is average. So, the prediction of the year for the next off-Sanriku earthquake of reverse type depends upon the estimation of the released energy by the 1896 earthquake.

Table 4. Occurrence Series of Large Earthquakes of Types 1 to 3

Type	1 Tokachi	2 Sanriku (reverse)	3 Sanriku (normal)
1968	○		
1933			○
1896		○	
1856	○		
1793		○	
1763	○		
1677	○		
1611		○* (spanning 2 and 3)	

(*) – isoseismal is similar, but the value of intensity is different.

REFERENCES

Earthquake Research Institute (1981-1987). New collection of materials for the history of Japanese earthquakes, vols. 1-5, with 8 supplementary vols.

Hattori, I., Esq. (1878). Destructive earthquakes in Japan, *Trans. Asiatic Soc. Japan*, **6**, 249-275.

Kawasumi, H. (1952). Distribution of earthquake danger in Japan (in Japanese), *Shigen data book*, **6**, 1-14.

Kawasumi, H. (1963). On the expectation of earthquake intensity at Kamakura (in Japanese), Saigai Kagaku Kenkyukai, 1-24.

Milne, J. (1881). Notes on the great earthquakes of Japan, *Trans. Seim. Soc. Japan*, **3**, 65-102.

Musha, K. (1942-1943). *Historical Data on Japanese Earthquakes* (in Japanese), revised ed., **1-3**, Shinsai Yobo Hyogikai.

Musha, K. (1951). *Historical Data of Japanese Earthquakes* (in Japanese), Mainichi Press.

Musha, K. (1950-1953). *Catalogue of Great Earthquakes in and near Japan* (in Japanese), Shinsai Yobo Kyokai.

Naumann, E. (1878). Ueber Erdbeben und Vulcanausbrueche in Japan, *Mit. Deutsch. Gesellschaft f. Nat. und Voelkerkunde Ostasiens*, **2**, 163-216.

Tayama, M. (1899). Catalogue of historical data on Japanese earthquakes (in Japanese), *Rep. Imp. Earthq. Invest. Comm.*, **26**, 3-112.

Tayama, M. (1904). Historical data on Japanese earthquakes (in Japanese), *Rep. Imp. Earthquake Invest. Comm.*, **46**, A1-606, B1-595.

Usami, T. (1977). *Descriptive Table of Disastrous Earthquakes in Japan* (in Japanese), Univ. of Tokyo Press.

Usami, T. (1979). Study of Historical Earthquakes in Japan, *Bull. Earthq. Res. Inst.*, **54**, 399-439.

Studies of Philippine Historical Earthquakes

R. G. Valenzuela and L. C. Garcia
Philippine Atmospheric, Geophysical and Astronomical Services Administration
Quezon City, Philippines

1. Introduction

Historical records from the year 1589 to the present show that the Philippines frequently experienced earthquakes of all intensities. This is to be expected since the Philippine archipelago lies at the collision junction between the Philippine Sea Plate and the China Sea Plate.

A certain degree of development in science and culture was attained by the Filipinos long before foreign civilizations reached the shores. However, early contact with Chinese traders did not seem to include any technology transfer on earthquake studies (the Chinese started seismological studies as early as the 2nd Century B.C.).

The earliest record of a Philippine earthquake dates back to 13 July 1589, more than half a century after the arrival of the Spaniards. The Spaniards built churches and other civil structures, mostly mortared stone blocks and bricks. In those days, earthquake occurrences and effects were recorded, especially the violent and destructive ones, for the purpose of reporting to authorities for repair requisitions for damaged structures.

In 1865, formal seismological observation and studies were started in the Philippines with the establishment of the Manila Observatory by the Jesuits. Father Federico Faura, S.J., was the founder. In the observatory reports of 1865, there was mention of the construction of a horizontal seismoscope and a vertical seismometer. In a report issued by the Spanish government on an earthquake of 1851, there was also mention of a pendulum in operation at a Southern Luzon town.

2. Macroseismic Data

Many compilations of historical earthquakes are extant, but it was only in 1982 that a more concerted effort was made to collect macroseismic data. This was in conjunction with the implementation of the "Earthquake Hazard Mitigation Program in Southeast Asia" under the auspices of the Southeast Asia Association of Seismology and Earthquake Engineering (SEASEE) in cooperation with the Office of Foreign Disaster Assistance of the U.S. Agency of International Development (OFDA/USAID) and the U.S. Geological Survey (USGS).

The richest source of historical earthquake information was the "Catalogue of Philippine Earthquakes, 1589-1899" by Repetti (1946). A group of researchers from the Philippine National Library compiled a volume of data, most of which were translations from Spanish to English. Materials from American libraries were obtained from the USGS. Information from newspapers, journals, official and private communications and technical papers also form part of the catalogue sources.

A catalogue of Philippine earthquakes in four volumes was subsequently produced. Each volume represents a division of the history of seismological service in the Philippines.

The first volume contains a listing of seismic events from 1589 to 1864. This period covers the years when there was yet no organized seismological service in the archipelago. Earthquake records were in various formats, but nevertheless, informative and useful.

The second volume contains data from 1865 to 1900. The year 1865 marked the beginning of seismology in the Philippines. During this period and up to 1906, a local intensity scale of I–VI was used. It is notable that in 1884, the Manila Observatory was subsidized by the Spanish government as the Central Weather Bureau of the Philippines, which included seismological work.

The third volume covers events from 1901 to 1942. In 1901, the Manila Observatory, while continuing to function as the Weather Bureau was re-organized under the American Regime. From this year onward, more accurate observations were made with the addition of better instruments. Reports were classified into instrumental and non-instrumental bulletins published periodically. The original Rossi-Forel Intensity Scale replaced the local intensity scale. Repetti revised the Rossi-Forel to a scale of I to IX in 1934. As a consequence of the Japanese invasion, the Manila Observatory ceased its function as a government institution in 1942, but continued its operation by the Jesuits.

The fourth volume covers events from 1948 to 1983. From the restoration of the Weather Bureau after the war, information on only one earthquake was found for 1948. Instrumental records in other countries between 1942 and 1947, however, showed earthquakes occurred during this period.

The present compilation of earthquake data cannot be considered complete. There are possibilities of finding data in Spain, e.g. in the National Library at Madrid and the Archivo de Indios (Archive of the Indies) in Sevilla, and in other countries which have had relations, in one way or another, with the Philippines.

3. Destructive Earthquakes

From the list of events of the nearly 400 years of written seismic history, a catalogue of 63 destructive earthquakes was prepared. This includes records which give very long narratives of damages on natural and man-made structures.

Forty of these destructive events are listed in Table 1. Each event has a Modified Mercalli intensity re-assessed macroseismic descriptions. Figure 1 shows the epicenters of these listed events. Many of these epicenters were determined only from the isoseismal maps. Others were obtained from the work of Lomnitz (1974).

4. Preparation of Isoseismal Maps

Earthquake intensities in given localities were evaluated using the 1956 version of the Modified Mercalli Intensity Scale for the purpose of drawing isoseismal maps. Maso (1895) gave the original local scale in use prior to the Rossi-Forel Intensity Scale. Su (1985) suggested the correlation of the adapted Rossi-Forel Scale and the Modified Mercalli Scale. This is shown in Table 2.

In the process of evaluating intensities and preparation of isoseismal maps, some difficulties were encountered such as: (a) Inapplicability of the descriptions to local structures – both Rossi-Forel and Modified Mercalli scales were developed in foreign countries where the engineering structures are different to conditions in the Philippines. (b) Assessing earthquakes based only on reported intensities – the levels of

Table 1. Destructive Philippine Earthquakes, 1589-1983

	Date	Time (LST)	Lat. °N	Long. °E	Magnitude M_S	I_o (MMI)
1.	1599 Jun 21	10:00 a.m.	14.60	121.00		VIII
2.	1619 Nov 30	12:00 p.m.	18.17	121.60		X
3.	1743 Jan 12	5:00-6:00 p.m.	14.00	121.60		X
4.	1787 Jul 13	6:45 a.m.	10.70	122.55		X
5.	1796 Nov 05	2:00 p.m.	16.05	120.30		X
6.	1852 Sep 16	6:30 p.m.	13.95	120.40		IX
7.	1863 Jun 03	7:20 p.m.	14.63	121.04		X
8.	1869 Aug 16	3:00 p.m.	12.17	123.69		IX
9.	1869 Oct 01	11:15 a.m.	14.82	120.82		IX
10.	1873 Nov 14	5:30 p.m.	13.11	122.98		VIII
11.	1880 Jul 18	12:40 p.m.	16.00	121.85		X
12.	1885 Jul 23	10:45 a.m.	8.43	123.60		X
13.	1889 May 26	2:23 a.m.	13.59	121.19		VIII
14.	1892 Mar 16	9:01 p.m.	16.06	120.42		IX
15.	1893 Jun 21	3:30 p.m.	6.88	125.83		X
16.	1897 Sep 21	1:15 p.m.	7.11	122.11	8.7 *	IX
17.	1897 Oct 19	7:52 p.m.	12.40	125.00	8.1 *	IX
18.	1902 Aug 21	7:17 p.m.	8.10	124.25		X
19.	1907 Nov 24	9:59 p.m.	13.10	123.40		X
20.	1911 Jul 12	12:09 p.m.	9.00	126.00	7.7 *	X
21.	1913 Mar 14	4:47 p.m.	4.50	126.50	7.9 (PAS)	IX
22.	1917 Jan 31	12:02 p.m.	5.60	124.80		IX
23.	1918 Aug 15	8:20 p.m.	5.50	123.00	8.3 *	X
24.	1924 Apr 15	12:22 a.m.	6.50	126.50	8.3 *	IX
25.	1924 Aug 30	11:07 a.m.	8.50	126.50	7.3 (PAS)	IX
26.	1925 Nov 13	8:16 p.m.	13.00	125.00	7.3 (PAS)	VIII
27.	1929 Jun 13	5:26 p.m.	8.50	127.00	7.2 (PAS)	X
28.	1931 Mar 19	2:26 p.m.	18.30	120.20	6.9 (PAS)	VIII
29.	1937 Aug 20	7:59 p.m.	14.20	122.10	7.5 *	VIII
30.	1948 Jan 25	1:46 a.m.	10.90	122.10	8.3 *	IX
31.	1954 Jul 02	10:46 a.m.	13.00	124.00	6.75 (PAS)	IX
32.	1955 Apr 01	2:17 a.m.	8.00	124.00	7.5 (PAS)	X
33.	1968 Aug 02	4:19 a.m.	16.50	122.30	7.3 *	IX
34.	1970 Apr 07	1:34 p.m.	15.80	121.70	7.3 (NEIS)	IX
35.	1973 Mar 17	4:31 p.m.	13.41	122.87	7.0 (NEIS)	XI
36.	1976 Aug 17	12:11 a.m.	7.30	123.60	7.9 (NEIS)	X
37.	1977 Mar 19	5:43 a.m.	16.70	122.31	7.0 (NEIS)	VIII
38.	1981 Nov 22	11:06 p.m.	18.71	120.65	6.7 (NEIS)	VIII
39.	1982 Jan 11	2:11 p.m.	14.00	124.50	7.1 (NEIS)	VIII
40.	1983 Aug 17	8:18 p.m.	18.33	120.87	6.5 (NEIS)	VIII

* Lomnitz (1974).

accuracy of the assessment of many of the earthquakes may not have been uniform considering the variance in the degree of subjectivity different observers have in evaluating intensities of earthquakes. (c) Limitation of the spatial distribution of earthquake effects observations – in many cases, reports are confined to a region of one intensity level. For cases like these, isoseismal maps were not drawn.

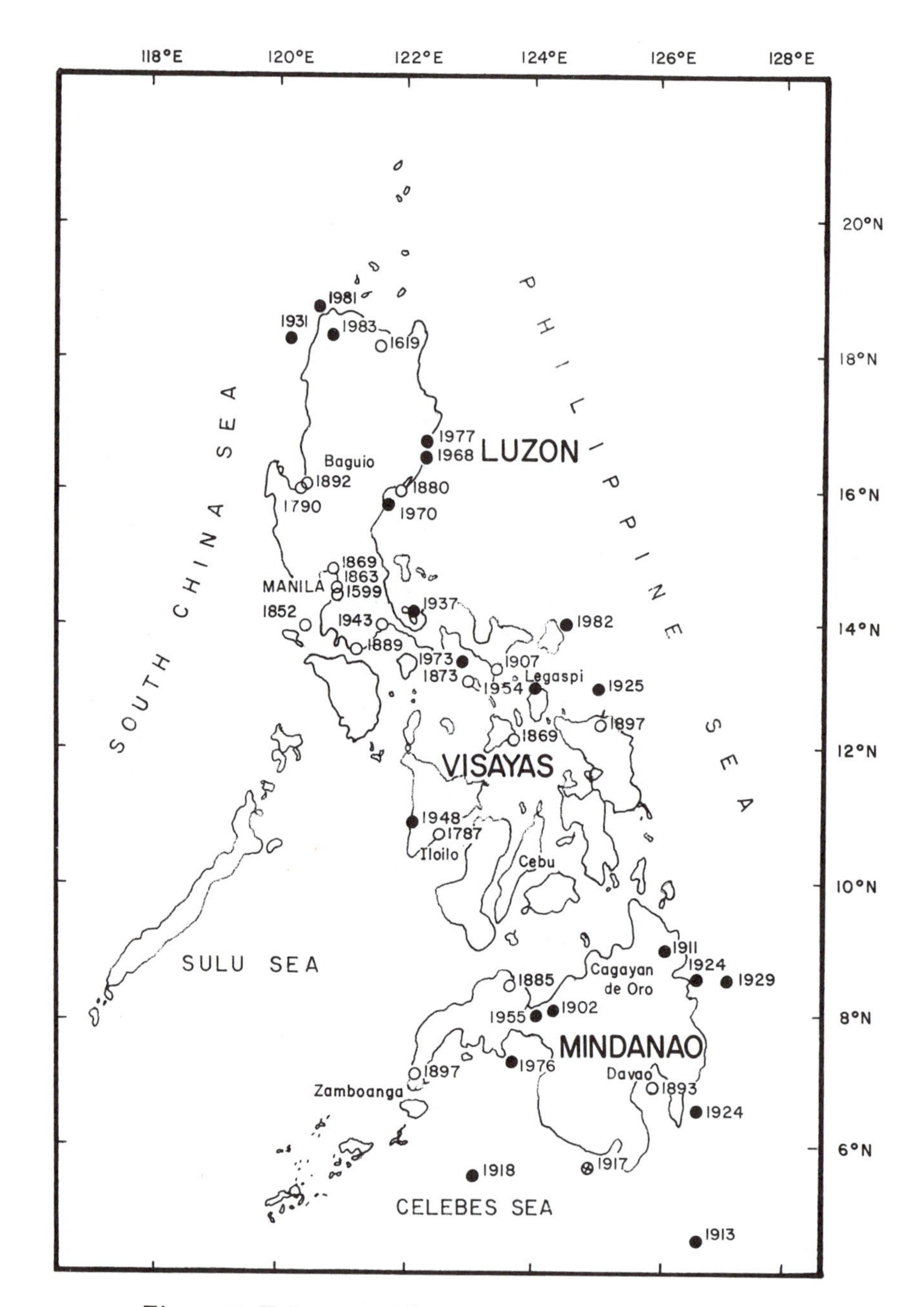

Figure 1. Epicenters of destructive Philippine earthquakes.

Table 2. Comparison of Intensity Scales

	Maso	R-F (original)	R-F (adapted)	MM
I	Perceptible	2–3	1	2
II	Light	4	2	3
III	Normal	5	3	4
IV	Strong	6–7	4	5
V	Violent	8	5	6
VI	Destructive	9–10	6	7–8 (lower)
			7	8 (upper) –9 (lower)
			8	9 (upper) –10
			9	11–12

5. Intensity Attenuation Data

Isoseismal maps for 35 earthquakes were used for calculating attenuation functions. An equation derived by Su (1980) for the Philippine region in the form:

$$I(R) - I_o = a - bR - c \log R,$$

where $I(R)$ = intensity as a function of hypocentral distance, I_o = intensity at the epicenter, and R = hypocentral distance in kilometers, was used in obtaining attenuation curves. In this equation, the coefficient b is related to the absorption factor while the coefficient c is associated with the geometric spreading factor (Howell and Schultz, 1975).

Inasmuch as earthquakes in the Philippines are either inland or offshore earthquakes, two methods of determining I_o were employed. For the first category, I_o's were taken to be the maximum observed intensity while for the second category, I_o's were extrapolated values from the individual intensity decay curves.

A regression analysis was performed on the data from the 35 earthquakes. Table 3 gives the list of events and their respective a, b, and c values. Results were then correlated with the seismic source zones delineated by Su (1985) as shown in Figure 2. Events were grouped according to zones as follows:

Zone	No. of events per zone
1	8
2	8
3	1
4	12
5	0
6	0
7	6

The values of a, b, and c for all zones, except zones 5 and 6 where there are no sample events, are listed in Table 4.

In this study, three events were screened out due to either very large or very small attenuation rates deviating greatly from the average values for their respective groups. Further analysis of these earthquake events needs to be done. Figure 3 shows the attenuation curves for the different zones.

Table 3. Earthquakes Analyzed with their Respective a, b, and c Values

	Date	Lat. °N	Long °E	I_o (MMI)	a	b	c
1.	1852 Sep 16	13.95	120.40	IX	5.531552	-0.013249	-3.192130
2.	1880 Jul 18	16.00	121.85	IX	4.574782	-0.008053	-2.187242
3.	1892 Mar 16	16.06	120.42	IX	1.566331	-0.010209	-0.795276
4.	1911 Jul 12	9.00	126.00	X	3.377046	-0.008302	-2.622990
5.	1913 Mar 14	4.50	126.50	IX	3.935529	-0.004738	-2.030975
6.	1931 Mar 19	18.30	120.20	VIII	2.528296	-0.003874	-1.429308
7.	1932 Aug 24	16.50	120.50	VII	3.643132	-0.006104	-2.417826
8.	1937 Aug 20	14.20	122.10	VIII	4.476718	-0.039064	-2.945239
9.	1949 Dec 29	17.00	121.63	IX	4.651885	-0.006746	-3.765001
10.	1950 Jan 03	17.60	121.10	VII	2.786550	-0.503386	-1.278303
11.	1951 Jun 01	18.00	119.00	VII	2.333964	-0.011687	-1.050067
12.	1952 Mar 19	9.40	126.10	VII	7.412699	-0.000100	-3.671602
13.	1954 Jul 02	13.00	124.00	IX	1.721129	-0.003003	-2.306835
14.	1955 Mar 31	8.00	124.00	X	6.625746	-0.018409	-1.385440
15.	1955 Apr 10	7.80	124.10	VIII	5.549073	-0.000355	-3.156348
16.	1956 Oct 28	13.94	122.98	VII	3.454500	-0.036860	-2.446632
17.	1957 Jun 12	17.88	120.24	VIII	2.508133	-0.010871	-1.297766
18.	1959 Jul 19	15.50	120.33	VII	1.068560	-0.006645	-0.898640
19.	1968 Aug 01	16.50	122.30	IX	6.530390	-0.008943	-3.000000
20.	1970 Jan 10	6.80	126.70	VII	9.273395	-0.000254	-4.051376
21.	1970 Apr 07	15.80	121.70	IX	3.033319	-0.018431	-1.277658
22.	1971 Jul 04	15.60	121.85	VI	7.754283	-0.004342	-2.677510
23.	1972 Apr 26	13.40	120.30	VII	5.073628	-0.012358	-2.005167
24.	1972 May 22	16.10	122.30	VII	6.639731	-0.008229	-2.488747
25.	1973 Mar 17	13.41	122.87	XI	3.448681	-0.054480	-3.332591
26.	1974 Feb 19	13.98	122.17	VII	3.534985	-0.002128	-2.184250
27.	1975 Oct 31	12.47	126.01	VII	6.122996	-0.000319	-3.179876
28.	1976 Aug 16	7.30	123.60	X	10.902350	-0.000050	-5.725272
29.	1976 Nov 11	8.36	126.58	VIII	5.757759	-0.009318	-1.907273
30.	1977 Mar 19	16.70	122.31	VIII	6.120612	-0.012222	-2.754127
31.	1980 Mar 31	16.04	121.88	VII	1.837912	-0.012474	-0.500000
32.	1980 Oct 26	12.04	126.33	VII	2.049114	-0.014935	-0.910102
33.	1981 Nov 22	18.71	120.65	VIII	1.524204	-0.214798	-0.613689
34.	1982 Jan 11	14.00	124.50	VI	3.119324	-0.009309	-1.100000
35.	1983 Aug 17	18.33	120.87	VIII	1.951468	-0.006583	-1.232840

Table 4. Values of a, b, and c for Seismic Source Zones

Zone	a	b	c
1	6.391013	-0.008066	-3.045896
2	3.036227	-0.009080	-1.714270
3	3.448681	-0.054480	-3.332591
4	3.686528	-0.016540	-1.893100
7	4.958582	-0.085856	-2.590441

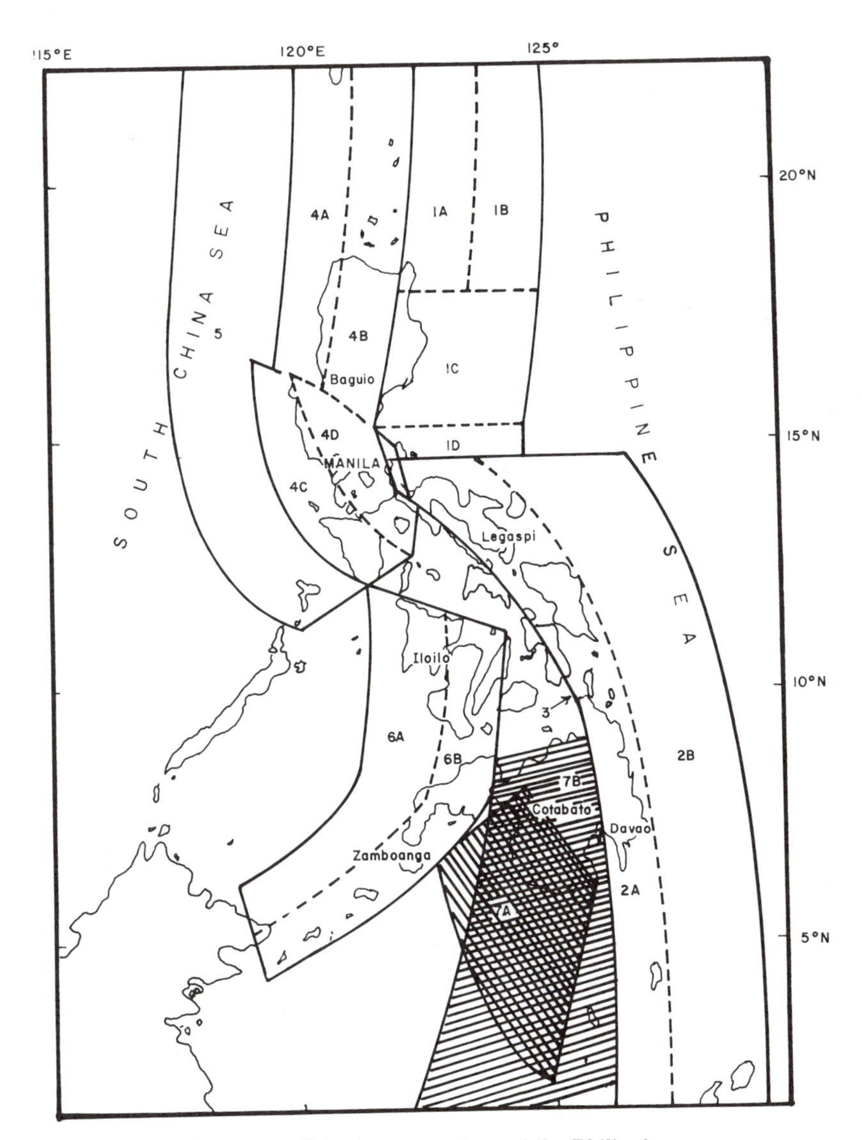

Figure 2. Seismic source zones of the Philippines.

6. Conclusions

It can be seen from Table 4 that Zone 3 has the highest c-value (-3.33259). This could be attributed to the fact that Zone 3 comprises the Philippine Fault. Zones 2 and 4 have similar c-values. Zone 2 is characterized by the West Luzon Trench. Zones 1 and 7, likewise have c-values fairly close to each other. Zone 1 is the East Luzon Trough while Zone 7 is a complicated double-micro-lithosphere, one on top of the other.

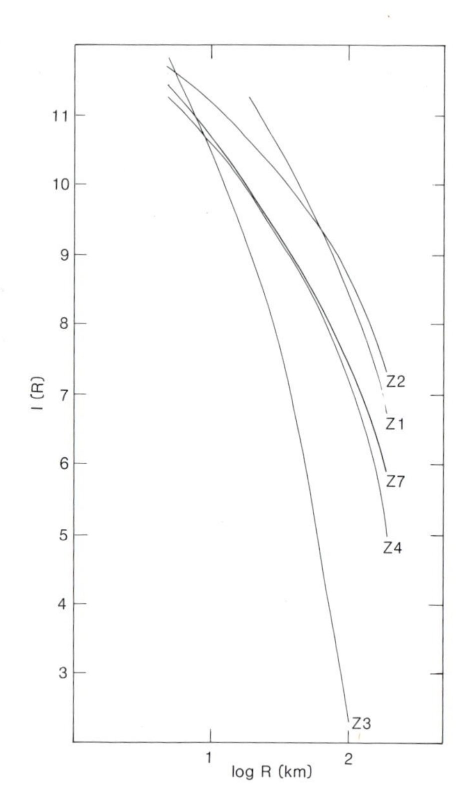

Figure 3. Attenuation curves for the different zones.

REFERENCES

Howell, B. F. and T. R. Schultz (1975). Attenuation of modified Mercalli intensity with distance from epicenter, *Bull. Seism. Soc. Am.*, **65**, 651-665.

International Seismological Centre (1964-1976). Monthly Bulletin of Epicenters.

Lomnitz, C. (1974). *Global Tectonics and Earthquake Risks*, Development in Geotectonics 5, Elsevier, The Netherlands, 231 p.

National Earthquake Information Service (1983). Preliminary Determination of Epicenters.

PAGASA compilation (1985). Catalogue of Philippine earthquakes, 1589-1983, Vols. I–IV, (unpublished).

PAGASA compilation (1985). Catalogue of Philippine earthquake epicenters, 1975 September – 1983 May (1985), (unpublished).

Repetti, W. C., S. J. (1946). Catalogue of Philippine earthquake, 1589-1899, *Bull. Seism. Soc. Am.*, **36**, 133-322.

Repetti, W. C., S. J. (1945). Seventy five years of seismology in the Manila Observatory (Speech delivered in 1945).

Su, S. S. (1980). Attenuation of intensity with epicentral distance in the Philippines, *Bull. Seism. Soc. Am.*, **70**, 1287-1291.

Su, S. S. (1985). Seismic source zones of the Philippines, 1985 (unpublished).

Valenzuela, R. G. and L. C. Garcia (1985). Epicentral estimates of pre-1900 Philippine earthquakes, 1985 (unpublished).

The History of Earthquakes in the Northern North Sea

R. Muir Wood and G. Woo
Principia Mechanica Ltd.
50 Vineyard Path, East Sheen, London SW14 8ET, UK

H. Bungum
NTNF/NORSAR
P. O. Box 51, N-2007 Kjeller, Norway

ABSTRACT

The northern North Sea is a region of moderate seismicity in which considerable offshore investments have taken place in the past twenty years. Whereas earthquake catalogues in the past were prepared out of academic interest only, the need has arisen for a comprehensive revised catalogue which can be used in the assessment of seismic hazard. In this paper, a review is given of historical sources of information on earthquakes in the region, emphasizing the incompleteness and national bias of early catalogues. This is followed by a description of the procedure used for defining intensities and contouring isoseismals, and locating epicenters and assigning magnitudes to historical events. A new catalogue of felt earthquakes in the northern North Sea is presented, and it is shown that the pattern of historical seismicity is consistent with that of recently recorded high quality instrumental data for the past five years. The paper concludes with a brief reference to regional seismotectonics.

1. Introduction

The northern North Sea is a continental shelf area rich in oil and gas fields and with the greatest density of economic investment of any offshore region on earth. In order to comprehend the seismic hazard in this area, it is necessary to utilize land-based seismological observations. Located on the northern margin of Europe, this region has been relatively poorly served by instrumental seismic monitoring right up to the late 1970's. Networks of seismological stations were then installed on the British side in Scotland, the Shetland islands, and on the sea-bottom close to the Beryl Field, and on the Scandinavian side in southern Norway and southern Sweden. Subsequent improvements included new stations in western Norway, plans for ocean-bottom seismometers close to the Statfjord and the Oseberg Fields, and a new network in northern Norway. In the 1980's the seismicity of the northern North Sea will therefore be well monitored, and with steady improvements.

Instrumental seismology was introduced in northern Europe around 1900, then experienced a decline during the 1930's, and a gradual build-up in the 1960's. During most of these periods near-shore and onland events around the northern North Sea are more poorly covered instrumentally than macroseismically, in terms of detection thresholds as well as location accuracy. Only for events located at great distance from land are the instrumentally determined solutions better than those from macroseismic observations. The macroseismic record is therefore of great importance in determining the geographical distribution and return periods of significant earthquakes within this region.

In this paper we report on studies that have led to the construction of a new northern North Sea seismicity catalogue. In order to create such a catalogue independent of the old partial and national catalogues it has been necessary where possible to return to the primary earthquake information. The original data have been interpreted according to a uniform intensity scale and mapped according to a standard method, and the work also includes a reevaluation of earthquake magnitudes. This process of reconstruction has been undertaken recently for the region of Great Britain, the North Sea and Scandinavia over several years (Principia, 1982, 1985; Woo and Muir Wood, 1984) and has been continued through a number of special studies for individual sites in the northern North Sea (NORSAR *et al.*, 1984; NGI *et al.*, 1985; NORSAR *et al.*, 1985). Historical earthquake studies for the same areas have also been published by Neilson (1979) and by Ambraseys (1985a,b).

2. Review of Historical Sources

The existence and survival of records of earthquakes from the peripheral countries of the North Sea reflects the history of settlement and learning in the northwest European region. This history is itself in part the product of geological factors. The fertile countries around the southern North Sea have supported large agricultural populations that have nucleated cities and centers of documentation and learning, while to the north the rocky glaciated lands have sustained only scattered communities that throughout much of their history have remained as effective colonies of dominant southern neighbors. Thus it is only from around the beginning of the 19th century, in both northern Scotland and western Norway, that earthquakes were regularly chronicled. This was done through the establishment of newspapers and from the presence of local natural historians and scientific societies. However, in both Scotland and Norway the random reporting of earthquakes goes back two centuries earlier, but these reports are hard to use because there rarely was more than one chronicler, reporting only that the earthquake was felt at his place of residence. In western Norway it is only for earthquakes located in and around the Skagerrak, reported also in Denmark and Sweden, that it is possible to draw even crude maps of events prior to 1800. This is in marked contrast to England, Belgium, France and Germany where it is possible to map significant earthquakes back as early as 1200 AD.

Throughout the historical record there is a marked asymmetry in the reporting of earthquakes to either side of the northern North Sea. While many earthquakes have been reported from the coast of western Norway alone, all earthquakes felt along the coast of eastern Scotland either have an epicenter in mainland Scotland or they have also been reported from Norway. Throughout the historical period eastern Scotland has been more developed than western Norway, has accumulated more detailed accounts of earthquakes in local newspapers and has been more highly populated. Thus, this variation in reported earthquakes represents a genuine and important variation in seismicity levels to either side of the northern North Sea, which has been amply confirmed through recent seismic monitoring. Apart from a few events reported by lighthouse keepers in Shetland in the late 19th century, the seismicity between those islands and Norway also shows a similar imbalance in that earthquakes chiefly have occurred on the Norwegian side of the intervening North Sea basin.

Detailed earthquake documentation in mainland Scotland was inspired by a series of prolonged seismic swarms beginning in the late 18th century, and by a damaging shock at Inverness in 1816. This detailed reporting was kept up throughout the whole of the 19th century, but with no North Sea earthquakes to be reported. On Shetland, after a large earthquake offshore Kristiansund was felt at the lighthouse on Unst, lighthouse keepers were encouraged to report shocks. By 1909, some ten events had been noted in this way. In Norway the occurrence of two widely felt earthquakes in 1834 (8/17 and 9/3) encouraged Keilhau (1836) to attempt the first catalogue of Norwegian earthquakes. This catalogue which was compiled from very few sources, and most notably appropriated Swedish earthquakes from the catalogues of Gissler, was published in the mid-18th century in the Proceedings of the Swedish Academy. In 1838 the French earthquake historian Perrey compiled a list of Norwegian earthquakes absorbing Keilhau's catalogue and adding a few new events reported by secondary sources in Germany and France. An earthquake of 1866 was widely reported and merited a scientific account in the journal *Naturen* (Reusch, 1887). However, it was not until after a widely felt earthquake on October 25, 1886, at the culmination of the scientific interest in macroseismic reports of earthquakes in the 1880's, that Hans Reusch sought reports of earthquakes from the public. The response to Reusch's initiative was extraordinary: it was as though a vast resource of observations had previously been left untapped. These reports showed that parts of western Norway are remarkable for suffering numerous earthquakes, although never any of sufficient size to cause appreciable damage. The lighthouse keeper at Ytteroen, an island located close to the northwest corner of western Norway, mentioned that "since the island was first inhabited it had been like a rolling wagon", and within this first year's catalogue there were five earthquakes noted at the island.

Reusch's enterprise survived, and after passing into the hands of various colleagues, it became institutionalized by the Bergen Museum, later to form part of the University of Bergen. On the receipt of a report of an earthquake, questionnaires were sent to a large number of "centers" (police stations, post offices, etc.) scattered across the country in order to gain the community's responses to the earthquake. While in many other parts of the globe the interest in earthquakes dwindled from 1920 through to 1960, in Norway the comparative isolation, the institutionalization and the reasonably steady supply of several earthquakes each year has allowed the enterprise to survive uninterrupted through to the present day.

While Bergen Museum's data collection has continued, the form in which the observations have been presented has changed quite markedly, and this has important implications for how the information can be reinterpreted. Reusch's catalogue simply reported descriptions of earthquakes from many different localities, but from 1888 (Thommassen, 1888) these descriptions were supplemented with an intensity assignment according to the 1882 Rossi-Forel scale. The early reports also recorded and analyzed information on the azimuth of the perceived shock wave. A number of the larger earthquakes from the early period merited separate reports, and for several years the quantity of macroseismic information extended to more than 100 pages. In 1912 the intensity assignments were switched to the Mercalli-Cancani scale, and from 1913 Kolderup introduced a significant change in the reporting whereby detailed descriptions of felt effects were abandoned for the unambiguous and historically "opaque" listing of locations and intensities. From 1938 even the individual intensity assignments were dropped as the events were indicated solely with a map marked with a felt area boundary or isoseismals. The number of small events reported in this period also showed a marked decline. However, the data gathering, the posting of directed questionnaires, continued, and when in 1983 Sellevoll, Almaas and Kijko finally caught up with the backlog of data from 1953-1975 they provided maps with individual intensity points noted, also marked with the instrumentally located epicenters.

3. Analysis of Macroseismic Data

In order to derive important seismological parameters from the record of historical earthquakes, it is important that the data have been processed and mapped according to a uniform methodology. Intensity is a compound measure of the amplitude, spectral content and period of earthquake waves at a surface location. Intensity at a given site varies according to the properties of the geology, ground and resonant frequency of the building in which an observer is located. Thus assessments of intensity for different observers in the same village or town may vary by up to one intensity grade. The intensity scale utilized here is the standard European MSK scale, with some specific adaptations made for the circumstances of Norwegian wooden houses.

Intensity assessment cannot be divorced from the means by which the information was gathered. For example, solicited questionnaires placed in newspapers will tend to gain extreme value, rather than average observations, since correspondants will tend to be those with the most interesting effects to report. The subsequent isoseismals will therefore also be of extreme value, including all the outermost locations where a given intensity grade was perceived. In contrast, the intensities within this present study have been plotted as modal community intensities using grid averages, taking into consideration the range of perceptions, including also information as to where the earthquake was not felt. The individual intensity points have then been contoured, and wherever possible the individual intensity assignments have been made from the primary data. However, where only intensity maps have been published without individual intensity assignments (as between 1937 and 1953), these have been recontoured according to a comparison of both intensity scales and the original method of contouring. An example of an important relocalization is given in Figure 1.

The collection of isoseismal maps for felt earthquakes in the North Sea region forms the basis for a quantitative analysis of macroseismic data. From historical observations of earthquakes, the essential seismological parameters of event location and magnitude can be estimated even though no instrumental records may exist. The fact that this is possible is due mainly to the availability of 20th century macroseismic and instrumental data which allow earlier events to be quantified via a correlation procedure.

The need for a correlation procedure to determine the epicenters of felt earthquakes is a consequence of the number of important offshore events in the region. For such events, the land intensity distribution is often inadequate to locate the epicenter geometrically with enough precision. As Figure 2 illustrates, there is a possible solution to this problem if another event can be found with a similar intensity distribution, and which has been instrumentally located. This event, if it

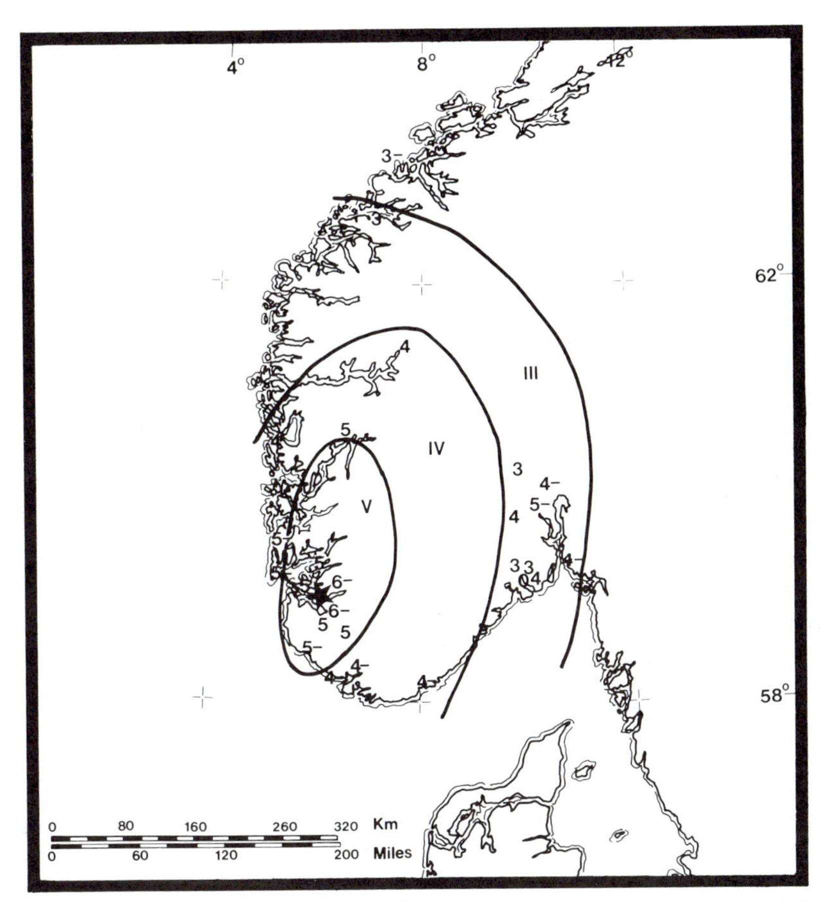

Figure 1. As a result of detailed researches of contemporary reports preserved in local newspapers, the 1865/5/7 earthquake, considered in all previous catalogues as having an epicenter in the North Sea, has been shown to have been located onshore in Rogaland, to the northeast of Stavanger.

can be found, serves as a template to locate the epicenter of the earlier event, for which there is no instrumental record. For an event felt on both sides of the North Sea, geometrical constraints may be sufficient in themselves to locate the epicenter, as demonstrated in Figure 3.

Once an event has been located, its magnitude needs to be determined. In the past the assignment of magnitude values to historical earthquakes in the region has been a major stumbling block. This has been resolved with the establishment of a correlation between surface wave magnitude (M_S) and felt area ($A_{\mathrm{III}}, A_{\mathrm{IV}}$) within the outer isoseismals III and IV. This takes the form (Principia, 1985):

$$M_S = -0.356 + 1.00 \log A_{\mathrm{III}} \qquad (\ \sigma = 0.12) \tag{1}$$

$$M_S = 0.91 + 0.818 \log A_{\mathrm{IV}} \qquad (\ \sigma = 0.13) \tag{2}$$

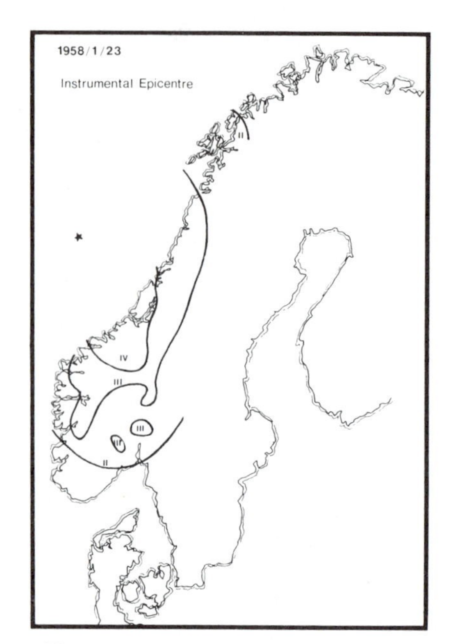

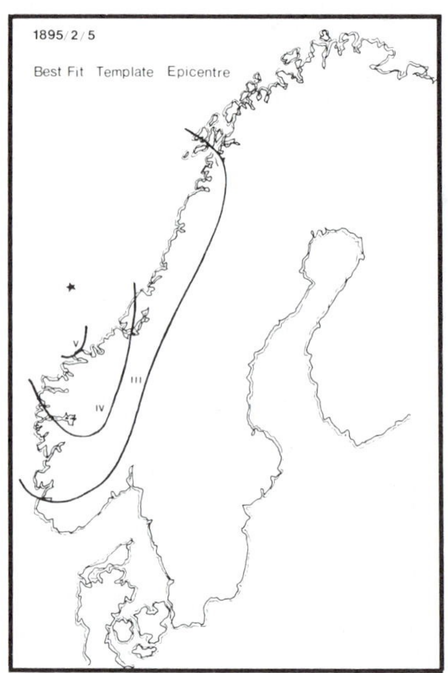

Figure 2. Macroseismic map for the 1895/2/5 earthquake (right) and the 1958/1/23 earthquake (left). The epicenter to the left is instrumentally determined, and the figure illustrates the use of model (or template) events for locating historical earthquakes.

This correlation does not include an explicit depth dependence. The focal depths of North Sea events of engineering significance ($> 4.0\ M_S$) are probably fairly homogeneous, and there is thus little need to introduce a depth dependence. Hypocentral depths in this region are moreover known only with poor precision, and any attempt to include a depth factor would therefore have to rely on a choice of average depths which would lead to very similar results.

The magnitude-felt area correlation given above has several important features: (a) magnitudes can be assigned when either A_{III} or A_{IV} is known, and when both are known a consistency check is possible; (b) the logarithmic dependence on felt area ensures the stability of the correlation against macroseismic error.

The first feature enables magnitudes to be assigned for most documented felt earthquakes, even though during the past centuries the observation threshold has varied substantially in the region. The second feature allows M_S values to be assigned to offshore events for which an estimate of felt area has to be based on an incomplete arc of data. Because of the logarithmic dependence of M_S on felt area, and because of the low regional attenuation which results in comparatively small coefficients for this logarithmic dependence, the error in M_S due to incomplete data is only about ± 0.1 or ± 0.2. In the catalogue shown in Figure 4 and Table 1, the M_S values for felt events are based on the above correlation, except where an instrumental value has been measured.

Because of the varying observational geographical and temporal detection thresholds, the map of felt events (Figure 4) is not a complete picture of the historical seismicity of the region. The same can be said of the ISC instrumental catalogue

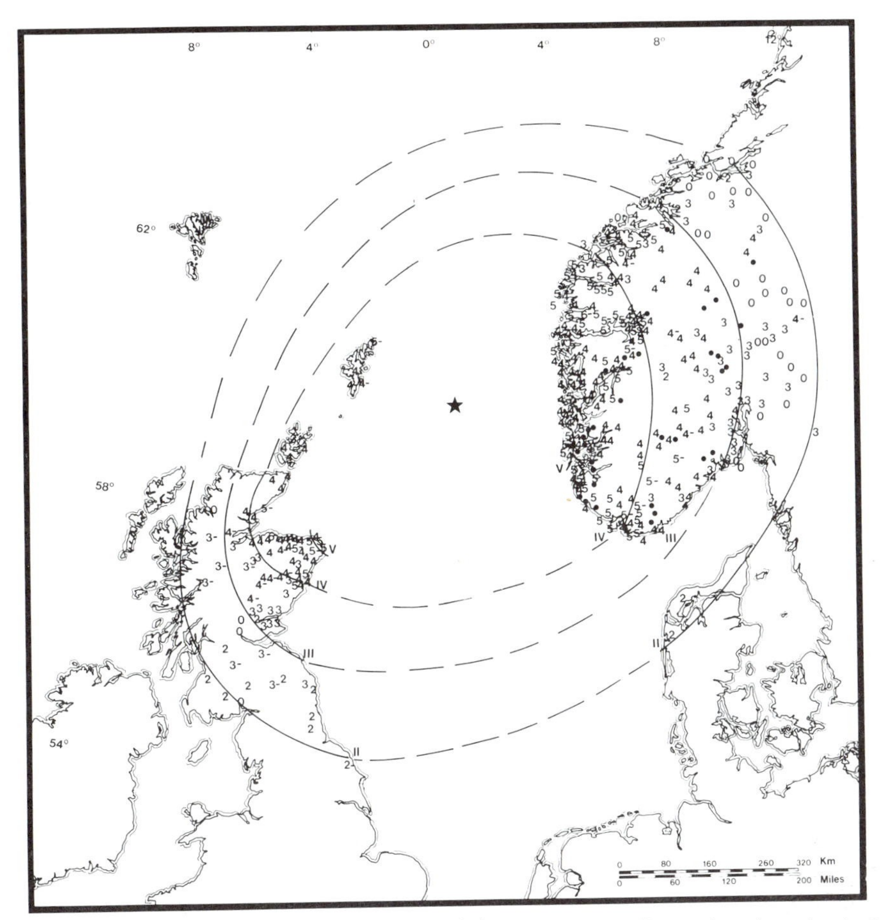

Figure 3. Macroseismic map for the 1927/1/24 earthquake, illustrating the epicentral constraints that can be obtained when an event is felt at opposing coastlines.

for the post-war years, which is limited by the biases and high detection thresholds of early seismological networks. The most accurate picture of the seismicity, albeit for a short period of 5 years, is that provided by the distribution of events above 2.3 M_L between 1980 and 1984 (Bungum *et al.*, 1985). This picture, shown in Figure 5, indicates a pattern of seismicity which is consistent with the macroseismic data plot shown in Figure 4.

The distribution of seismicity in the northern North Sea requires explanation within a regional seismotectonic model. The seismicity of the eastern half of the northern North Sea is delineated to the north, west and south by regions generally lacking seismicity (see Figures 4-5). The western boundary of the seismic province is itself approximately defined by the Viking Graben, the location of the epicenter of the largest observed earthquake of the northern North Sea region on 24 January

Table 1. Felt Earthquakes in the Northern North Sea, with $M_S \geq 4$

Date yr.mo.da	Time hr:mn	Epicentre		I_{Max} MMI	M_S	Log felt area (km^2)					Regional location main affected area
		Lat°N	Lon°E			VI	V	IV	III	II	
1834. 8.17	00:30	61.5	4.1	6.0	4.9						Western Norway
1834. 9. 3	20:00	59.5	7.9	5.0	5.0						Southern Norway
1841. 4. 3	16:00	57.0	8.5	6.0	4.5	3.0	3.7	4.5	4.9		Northwest Denmark
1865. 5. 7	14:20	59.0	6.1	5.0	4.9						Southern Norway
1879. 1. 4	01:30	61.0	2.0	3.0	4.5				4.9		Shetland & Sognefjord
1880. 7.18	12:20	61.0	1.0	4.0	4.0				4.4		Shetland
1880. 8. 4	06:30	63.6	3.9	3.0	4.5				4.8		N. coast Western Norway
1883. 6.13	14:00	61.5	5.7	4.0	4.1				4.6		Aalesund – Bergen
1886. 1.16	04:00	57.2	6.4	4.0	4.5				4.9		South Norway
1886. 9. 5	13:00	61.3	5.0	4.0	4.0				4.4		Nordfjord
1886.10.25	12:15	62.0	6.9	5.0	4.8		4.0	4.7	5.2		Western Norway
1892. 5.15	14:51	61.4	5.1	6.0	5.2		4.1	5.2	5.6		South Norway, Shetlands
1892.11.20	22:30	59.7	5.7	5.0	4.4			4.3	4.8		South west Norway
1895.11.27	03:30	59.4	6.0	4.0	4.0				4.4		Hardangerfjord
1895.12.16	13:45	57.6	7.9	4.0	4.0				4.4		South Norway & Denmark
1896. 1. 7	03:00	61.9	6.4	4.0	4.1				4.5		Western Norway
1896. 1.28	22.30	61.7	3.6	4.0	4.3				4.7		Bergen – Aalesund
1899. 1.31	12:30	60.1	5.5	5.5	4.6		3.5	4.5	5.0		South Norway
1901. 3. 8	06:00	61.8	3.0	4.0	4.2				4.7		Sunnfjord & Nordfjord
1902. 2. 9	03:50	59.5	5.5	5.0	4.1			3.8	4.5		Stavanger – Bergen
1905. 2. 6	18:10	61.5	5.1	5.5	4.5		3.3	4.4	4.9		Bergen – Kristiansand
1906. 6. 3	04:30	57.6	6.2	5.0	4.5			4.4	4.8		South west Norway
1906.11.17	20:30	61.9	6.0	4.0	4.1			4.0	4.5		Sunnmore, W.coast Norway
1906.12.10	17:15	58.0	5.7	4.0	4.1				4.5		South west Norway
1907. 6.29	21:00	60.5	7.8	5.0	4.2			4.0	4.6		Hardanger & Telemark
1911. 8.24	22:50	60.0	5.9	5.0	4.5			4.4	4.9		Western Norway
1912.12. 1	12:00	56.7	7.7	4.0	4.0			3.7	4.4		W. Jutland & Denmark
1913. 7.29	05:00	56.1	8.2	5.0	4.2		3.4	4.1	4.5		West coast of Denmark
1913. 8. 4	08:38	61.3	5.2	5.0	4.9		4.0	4.8	5.3		Western Norway
1918. 4.10	23:26	61.5	5.9	5.0	4.8		3.8	4.7	5.2		Western Norway
1922. 7.13	20:00	61.7	5.8	5.0	4.4		3.3	4.3	4.7		Sunnmore – Bergen
1923. 3.23	02:10	61.4	4.5	5.0	4.4			4.2	4.8		Sognefjord
1923. 5. 5	04:10	62.5	4.8	4.0	4.5			4.4	4.9		North west Norway
1924. 5. 5	07:20	61.8	4.7	4.5	4.1				4.5		Sognefjord & Sunnmore
1927. 1.24	06:18	59.9	1.8	5.0	5.3		5.1	5.5	5.7	5.9	Northern North Sea
1927. 6.15	07:20	61.0	4.7	4.0	4.0				4.4		Nordfjord & Sognefjord
1929. 5.23	19:36	57.2	6.6	4.0	4.5			4.7	5.1		Skagerrak
1929. 5.30	00:31	57.3	6.4	4.0	4.3			4.5	4.8		Skagerrak
1935.10. 6	08:45	62.0	3.9	4.0	4.3			4.2	4.7		Sunnmore – Nordhordland
1938. 3.11	17:10	61.6	4.1	4.5	4.9		4.0	4.8	5.3		Western Norway
1939.10. 9	11:10	59.3	8.4	5.0	4.1			4.0	4.5		South Norway
1941. 1.27	03:21	61.0	5.0	4.5	4.1			4.0	4.5		Hordaland & Sognefjord
1942.11.26	04:10	59.9	6.2	5.0	4.5		3.3	4.3	4.9		South Norway
1943. 8.29	06:35	58.9	5.9	4.5	4.5			4.4			Rogaland
1953. 1.29	20:36	61.1	4.8	4.5	4.2			4.1	4.6		Nordfjord – Bergen
1954. 7. 7	00:25	59.7	4.9	5.0	4.9		4.2	4.8	5.3		Western Norway
1955. 6. 3	11:40	61.9	4.1	5.0	4.8			4.7	5.2		Western Norway
1957. 7. 8	12:45	62.0	4.0	3.0	4.0				4.4	4.7	Nordfjord
1957.11.17	16:19	57.7	8.9	4.5	4.1			4.0	4.5		Southern Norway
1958. 3.20	14:47	57.2	7.0	3.0	4.0				4.4		Southernmost Norway
1958. 8. 6	17:16	59.6	5.8	5.5	4.5			4.4	4.9		Western Norway
1960. 1.19	00:05	58.5	6.5	4.0	4.0				4.4		Southernmost Norway
1961. 4. 4	22:43	61.8	1.5	3.5	4.5				4.9		West coast of Norway
1962. 2.21	12:44	61.2	3.4	3.0	4.1				4.5		Bergen & coast of Norway
1964. 7.14	05:34	57.2	7.0	4.0	4.1				4.5		Southern Norway
1967. 8.21	13:42	57.3	4.7	4.0	4.5				4.9		South west Norway
1968. 4.29	21:59	57.9	8.3	4.0	4.1				4.5		Southern Norway
1971. 1.14	02:30	62.1	5.3	5.0	4.0			3.7	4.4		Nordfjord & Inland
1971. 8.20	19:06	61.7	4.7	5.0	4.3			4.2	4.7		Western Norway
1972. 4. 7	20:20	62.2	5.6	5.5	4.0		2.9	3.8	4.4		North west Norway
1975. 4. 3	06:39	59.5	5.2	5.0	4.0			3.7	4.4		Stord Island & Stavanger
1977. 4. 6	20:32	61.0	5.0	5.0	4.0			3.9	4.3		South Norway
1978. 3.20	04:58	62.5	6.6	5.0	4.1			3.9	4.4		Moere
1982. 4.19	11:50	61.7	4.4	4.5	4.0				4.4		Nordfjord
1982.12.15	07:45	62.3	5.4	4.5	4.0			3.7	4.4		Sunnfjord
1983. 3. 8	19:44	59.7	5.6	5.0	4.2		2.9	4.1	4.6		Vestlandet

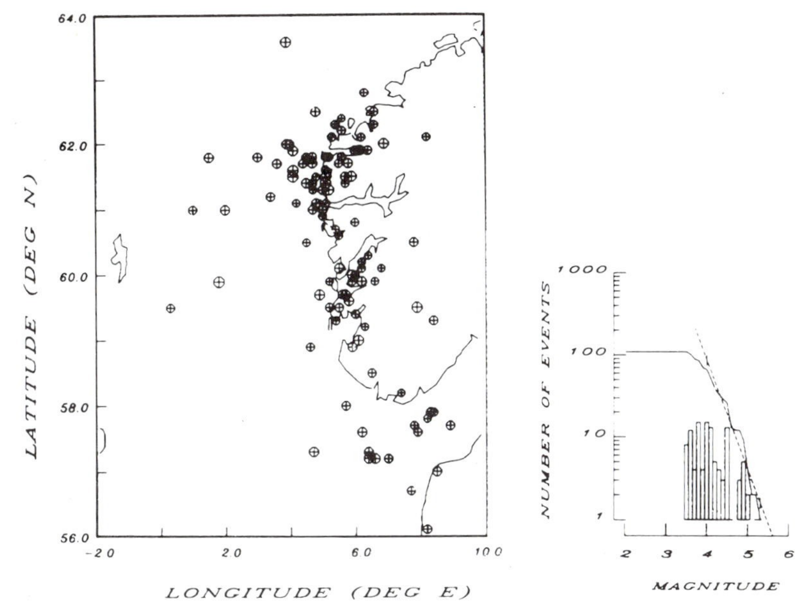

Figure 4. Epicentral distribution of felt and macroseismically reinterpreted earthquakes in the northern North Sea between 1834 and 1983. The map contains 109 earthquakes above magnitude (M_S) 3.5 (cf. Table 1), with magnitude-frequency distribution as shown to the right, where a slope (b-value) of 1.33 is indicated.

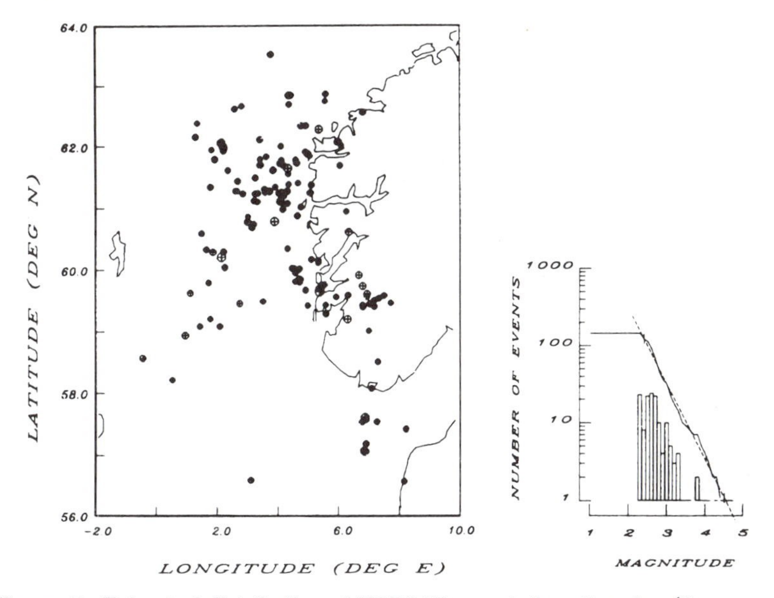

Figure 5. Epicentral distribution of NORSAR-reported earthquakes (Bungum *et al.*, 1985) between 1980 and 1984. The map contains 145 events above magnitude (M_L) 2.3, with magnitude-frequency distribution as shown to the right, where a slope (b-value) of 0.98 is indicated. The good linear fit is an indication of the completeness of the catalogue.

1927, with $M_S = 5.3$ (Figure 3). The large Oslofjord earthquake in 1904 (23 October) has, in comparison, an M_S magnitude of 5.4 when calculated based on the same magnitude-felt area relationship.

From the end of the Cretaceous to the present, which is a period of around 65 million years, the North Sea region has not followed the fundamental pattern of rifting and extensional basin tectonics that characterized the crustal deformation over the previous 200 million years. The most significant tectonics has involved episodes of regional dome uplift associated with density changes in the underlying asthenosphere. These have not been symmetrically disposed about the North Sea in time; the uplift of Scotland and the Shetland platform took place around 60 Ma, while the beginning of the uplift of western Norway can be dated at around 30 Ma (Groth and Muir Wood, 1985). The current seismotectonic regime appears to reflect the continuation of some of the most significant late Tertiary tectonic events through into the present.

REFERENCES

Ambraseys, N. N. (1985a). Magnitude assessment of Northwestern European earthquakes, *J. Earthq. Eng. Struct. Dyn.*, **13**, 307-320.

Ambraseys, N. N. (1985b). The seismicity of western Scandinavia, *J. Earthq. Eng. Struct. Dyn.*, **13**, 361-399.

Bungum, H., J. Havskov, B. Kr. Hokland, and R. Newmark (1985). Contemporary seismicity of northwest Europe, *Annales Geophysicae*, **4B**, 567-576.

Groth, A. and R. Muir Wood (1985). The Tertiary uplift of western Norway, in preparation.

Keilhau, B. M. (1836). Efterretninger om Jordskjælv i Norge, *Mag. for Naturvidenskaperne*, **12**, 83-165.

Kolderup, C. F. (1913). Norges Jordskjælv, *Bergens Museums Aarbok*, **8**, 152 pp.

Neilson, G. (1979). Historical seismicity of the North Sea, in *Energy in the Balance*, Brit. Ass. Adv. Science, Edinburgh, 119-137.

NGI, NORSAR, and Principia (1984). Seismic Hazard Evaluation for the Sleipner Field, Report prepared for Statoil a.s., Stavanger, Norway, 172 pp.

NORSAR, NGI, and Principia (1984). Earthquake Loads for the Troll Field, Report prepared for Norsk Hydro a.s., Oslo, Norway, 260 pp.

NORSAR, NGI, and Principia (1985). Gullfaks C Seismic Hazard Analysis, Report prepared for Statoil a.s., Stavanger, Norway, 185 pp.

Principia Mechanica Ltd (1982). British Earthquakes, Report prepared for CEGB, SSEB and BNFL, London, U.K.

Principia Mechanica Ltd (1985). The Seismicity of the North Sea, Report prepared for the Department of Energy, London, U.K.

Reusch, H. (1887). Jordskjælvet Natten til 25 Oktober 1866, *Naturen*, **7**.

Sellevoll, M. A., J. Almaas, and A. Kijko (1983). A Catalogue of Earthquakes Felt in Norway 1953-1975, Report, Seismological Observatory University of Bergen, Norway.

Thommassen, T. C. (1888). Berichte über die, wesentlich seit 1834, in Norwegian eingetroffenen Erdbeben, Bergens Museums Aarsbertning 1888, 52 pp.

Woo, G. and R. Muir Wood (1984). British seismicity and seismic hazard, *Proceedings, Eighth World Conference on Earthquake Engineering, Volume I*, 39-44, Prentice-Hall Inc., New Jersey, U.S.A.

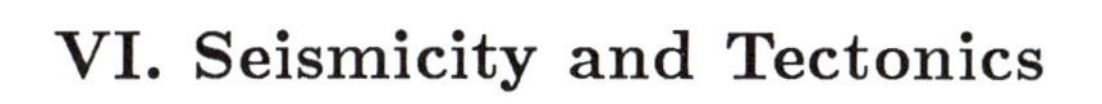

VI. Seismicity and Tectonics

Magnitude–Fault Length Relationships for Earthquakes in the Middle East

N. N. Ambraseys
Dept. of Civil Engineering, Imperial College of Science and Technology
London SW7 2BU, England, UK

An estimate of magnitudes from the dimensions of fault rupture for Eastern Mediterranean and Middle Eastern earthquakes may be made from the major axis solution

$$M_{SC} = 4.63 + 1.43 \log(L), \tag{1}$$

where L is the observed length of faulting in kilometers and M_{SC} is the surface-wave magnitude of the associated event (Ambraseys and Melville 1982, p. 156). It is of interest that as it can be seen from Figure 1, the data for Iran recently presented by Nowroozi (1985) fit Equation (1).

A better estimate of magnitudes from the dimensions of rupture and dislocation for events in the same region may be made from:

$$M_{SC} = 1.1 + 0.4 \log(L^{1.58} R^2), \tag{2}$$

where L and R are the length of the fault break and relative displacement respectively, in centimeters (Ambraseys, 1976). Figure 2 shows the relation between surface-wave magnitude M_S and predicted magnitude M_{SC} from Equation (2) using the fault parameters given by Nowroozi (1985) for Iran. The goodness of fit for this set of data is 0.88. The same figure shows the relation between M_S and the values of M_{SC} predicted from Equation (2) for seismic events in other parts of the world (Bonilla, *et al.*, 1984), with a goodness of fit of 0.74.

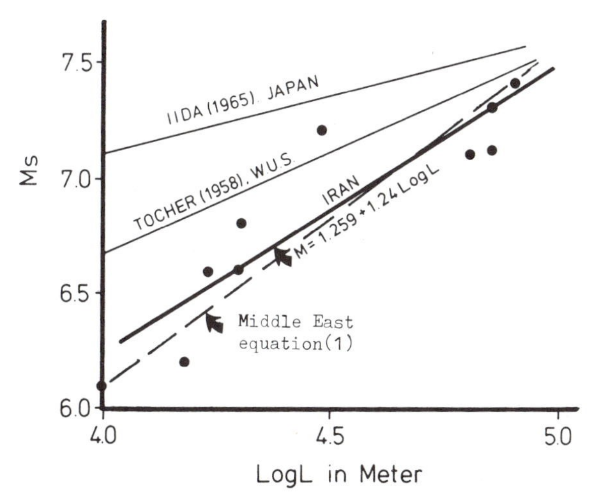

Figure 1. Equation (1) superimposed on data plot for Iran (Figure 3 of Nowroozi, 1985).

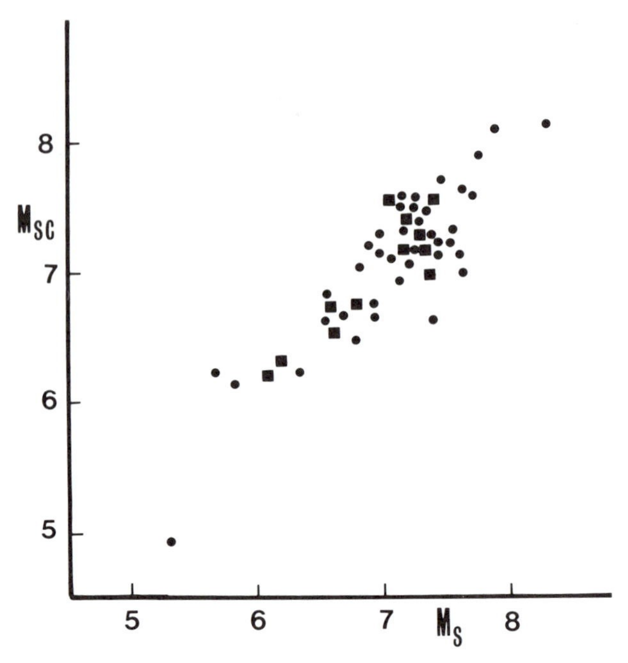

Figure 2. Plot of magnitude M_{SC} predicted from Equation (2) using fault parameters in Bonilla, *et al.* (1984) circles, and Nowroozi (1985) squares, versus corresponding surface wave magnitude M_S reported by these authors. This figure includes events No. 1 and 10 in Nowroozi's Table 1, the latter with the correct M_S-value of 7.4. The same figure excludes event 46 of the Calingiri earthquake, for which $R = 0.1$m and $M_S = 5.2$. The Iranian data give $M_{SC} = 0.53 + 0.94 M_S$ and the world data give $M_{SC} = 0.93 + 0.87 M_S$.

Equations (1) and (2) may be used to estimate the magnitude of historical events in the Alpine region from faults of known or inferred length and mobility. Of these two equations, the highest correlation coefficient is obtained for Equation (2) or for equations of the type $M_{SC} \sim \log(L^m R^n)$.

REFERENCES

Ambraseys, N. (1976). Earthquake epicentres in Iran, *Cento Seminar Proc. on Recent Advanc. Earthq. Hazard Minimization*, 70-80, Tehran.

Ambraseys, N. and C. Melville (1982). *A History of Persian Earthquakes*, Cambridge Univ. Press.

Bonilla M., R. Mark, and J. Lienkaemper (1984). Statistical relations among earthquake magnitude, surface rupture length and surface fault displacement, *Bull. Seism. Soc. Am.*, **74**, 2379-2411.

Nowroozi, A. (1985). Empirical relations between magnitudes and fault parameters for earthquakes in Iran, *Bull. Seism. Soc. Am.*, **75**, 1327-1338.

Applications of Fuzzy Mathematics in Studying Historical Earthquakes and Paleoseismicity

De-yi Feng
Seismological Bureau of Tianjin, Tianjin, China

Ming-zhou Lin
Seismological Bureau of Shanghai, Shanghai, China

Kun-yuan Zhuang
Seismological Bureau of Anhui Province, Hefai, China

ABSTRACT

Some concepts and methods of fuzzy mathematics, such as the membership function, fuzzy multifactorial evaluation, principle of choosing the closest degree of approaching, and others have been applied in the quantitative evaluation of intensity of a historical earthquake and approximate determination of its source mechanism based on historical materials.

A method of fuzzy description of macroscopic criteria for recognition of paleoseismic events is discussed briefly. For illustration of the suggested methods of fuzzy mathematics, the intensity and source mechanism of the 1920 Haiyuan ($M = 8.5$) great earthquake, the intensity of the 1937 Hozhei ($M = 7.0$) earthquake, and some relics of ancient earthquakes appearing along the Xianshuihe fault zone are studied.

1. Introduction

Historical materials about earthquakes and paleoseismicity are very important for the studies of earthquake hazard assessment and earthquake engineering, because instrumental seismic data are available only for the last 100 years. However. these materials often contain non-precise and very fuzzy information. Therefore, in studying historical earthquake and paleoseismicity, we must deal with treatments and analyses of fuzzy data, and some methods of fuzzy mathematics may be applicable.

Since L. A. Zadeh advanced the fuzzy set theory in 1965, fuzzy mathematics has developed rapidly. At present, it attracts attention in various scientific and technical domains (see e.g., Dubois and Prade, 1980). In particular, fuzzy mathematics (fuzzy sets theory) has been applied to various problems in earthquake research, including earthquake engineering, earthquake hazard assessment, and earthquake prediction (Feng *et al.*, 1983*b*; Feng and Liu, 1985).

This paper presents some preliminary results obtained in applying fuzzy mathematics to the study of historical earthquake and paleoseismicity. Our emphasis is on developing some application methods of fuzzy mathematics. Obviously, our suggestions are tentative and require further examination and evaluation.

2. Fuzzy Assessment of Seismic Intensity of Historical Earthquakes

2.1 Method

In the assessment of seismic intensity of a historical earthquake, we may use a method based on the concept of degree of approaching for normal fuzzy sets and by means of fuzzy multifactorial evaluation as described by Feng *et al.* (1984).

The damage conditions of buildings and the fissure conditions of the earth's surface after an earthquake may be taken as macroscopic standards for the assessment of seismic intensity. Let us suppose that the macroscopic standards corresponding to a determinate degree in the seismic intensity scale can be characterized approximately by the normal distribution function as:

$$A = \frac{N_i}{N_o} = \exp\left[-\left(\frac{x_i - a}{b}\right)^2\right], \tag{1}$$

where x_i is the value of i-th sample, a is the average value of x_i, b is the standard deviation, N_i is the number of i-th sample with value x_i, and N_o is their total number.

According to the statistics obtained on the basis of abundant macroscopic observational data, we have compiled the corresponding quantitative comparison tables for each standard, as shown in Tables 1 and 2. Table 1 shows a rough comparison between intensities (from VI to XI degrees) and the building damage standard, because the building damage conditions described in the historical materials often can not be classified in detail. Table 2 compares intensities with the earth fissure standard, and it can be used only for assessing the extremely high intensity in the epicentral region of a strong earthquake.

Let us suppose that the macroscopic standards for different degrees in the seismic intensity scale form a series of normal fuzzy sets A_j, and the evaluated sample set also forms a normal fuzzy set A_o. Then, we can take the corresponding distribution functions (1) as their membership functions with parameters (a_j, b_j) and (a_o, b_o) respectively. For two normal fuzzy sets A_j and A_o, the degree of approaching can be defined as

$$(A_j, A_o) = 1/2\left\{\exp\left[-\left(\frac{a_j - a_o}{b_j + b_o}\right)^2\right] + 1\right\}. \tag{2}$$

According to the principle of choosing the closest approach, the fuzzy sample set A_o is most approaching to the fuzzy model set A_i if we have:

$$r_i = (A_o, A_i) = \max(A_o, A_j), \quad j = 1, 2, 3, \ldots \tag{3}$$

The parameters a_o and b_o for the assessed sample fuzzy set may be obtained from the collected macroscopic data, and the parameters a_j and b_j for different model fuzzy sets may be taken directly from the corresponding quantitative comparison values in Tables 1 and 2.

Then, we may calculate all elements $r_{k,j}$ for different kinds of macroscopic standards k and different degrees j in the seismic intensity scale, and obtain the matrix of degree of approaching $R = (r_{k,j})$ which may be reformulated into a normalized matrix $Q = (q_{k,j})$ by means of averaging and normalization.

Table 1. Rough Comparison between Intensity Degrees from VI to XI and Building Damage Standard

Type	XI		X		IX		VIII		VII		VI	
	a	b	a	b	a	b	a	b	a	b	a	b
I	1	0+	0.991	0.052	0.944	0.065	0.762	0.293	0.400	0.081	0.154	0.040
II	1	0+	0.877	0.192	0.738	0.258	0.560	0.223	0.260	0.119	0.055	0.025
III	0.953	0.122	0.557	0.182	0.296	0.550	0.136	0.083	0.050	0.034	0	0

"Type" refers to building type. Damage conditions are "destroy" and "severe destroy"

Table 2. Quantitative Comparison between Intensities and the Earth Fissure Standard Represented by Its Width

Type of fissures	XII		XI	
	a (cm)	b (cm)	a (cm)	b (cm)
Rock fissure	163	100	50	27
Ground fissure	720	260	140	100

Finally, considering a suitable weight distribution function W for different cases, we may apply the fuzzy multifactorial evaluation on the basis of the matrix Q. The corresponding formula is

$$P = W \circ Q = (P_1, P_2, \ldots, P_m), \tag{4}$$

where the symbol "$\circ$" denotes the compositional operation which is defined by

$$P_j = \max_k(\min(W_k, q_{k,j})), \quad j = 1, 2, \ldots, m. \tag{5}$$

By using the principle of choosing the closest approach, we may obtain an estimated earthquake intensity, i.e., the degree in the seismic intensity scale for a given event. This intensity must correspond to the maximal component of the vector P in equation (4).

2.2 Examples

(1) Intensity of the 1937 Hozhei ($M = 7.0$) Earthquake. According to the historical records (Gu, 1983), about 30% of buildings in the Hozhei region and 20% of buildings in the Donming and Dintau regions were destroyed by this earthquake. Let us suppose that the buildings destroyed in the given regions in 1937 are basically type II buildings, and the conditions of building damage may be considered as "destroy" and "severe destroy". By using Table 1, we may obtain the degree matrices of approaching corresponding to intensities (XI X IX VIII VII VI) as

$$R_1 = (0.5 \quad 0.5 \quad 0.528 \quad 0.628 \quad 0.947 \quad 0.5)$$

for the Hozhei region, and

$$R_2 = (0.5 \quad 0.5 \quad 0.506 \quad 0.537 \quad 0.888 \quad 0.5)$$

for the Donming and Dintau regions together. Their normalized forms are

$$Q_1 = (0.139 \quad 0.139 \quad 0.147 \quad 0.174 \quad 0.263 \quad 0.139), \text{ and}$$

$$Q_2 = (0.146 \quad 0.146 \quad 0.147 \quad 0.157 \quad 0.259 \quad 0.146).$$

From the components of matrices-vectors Q_1 and Q_2, we may conclude that the intensity of the Hozhei earthquake is equal to VII$^+$ in the Hozhei region, and VII in the Donming and Dintau regions. For comparison, Figure 1 shows the isoseismal curves obtained from conventional assessment, and the dashed isoseismal line by the fuzzy sets method.

(2) Intensity of the 1920 Haiyuan ($M = 8.5$) Earthquake. By use of the historical data about the disaster of the Haiyuan earthquake (Seismological Institute of Lanzhou and Seismological Bureau of Ningxia Region, 1980) we may take approximately the average width of the earth fissure in the Haiyuan region as: $a = 250$ cm, $b = 0$ for rock fissure; and $a = 550$ cm, $b = 0$ for ground fissure. By using Table 2, we obtain the degree matrix of approaching as:

$$R = \begin{pmatrix} 0.735 & 0.5 \\ 0.826 & 0.5 \end{pmatrix},$$

where the rows correspond to rock fissure and ground fissure, and the columns correspond to intensity XII and intensity XI. Its normalized form is

$$Q = \begin{pmatrix} 0.60 & 0.40 \\ 0.62 & 0.38 \end{pmatrix}.$$

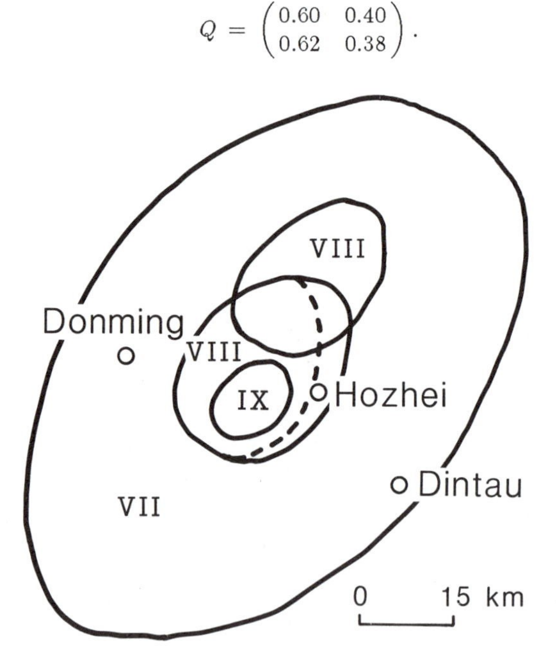

Figure 1. Intensity of Hozhei Earthquake. The dashed line denotes a corrected isoseismal line.

Supposing that the rock fissure and ground fissure have equal weight for evaluating the earthquake intensity, i.e., taking $W = (0.5, 0.5)$, we may obtain the vector for fuzzy multifactorial evaluation as

$$P = W \circ Q = (0.5, 0.4).$$

According to the principle of choosing the closest approach, the intensity in the Haiyuan region may be estimated as XII degree (see Figure 2).

3. Fuzzy Determination of Source Mechanism of Historical Earthquakes

3.1. Fuzzy Description of Historical Phenomena on Earthquake Effects

In the absence of enough seismograms, the source mechanism of a historical earthquake may be determined on the basis of macroscopic phenomena related to the earthquake effects. Among them, the geomorphological and geological effects caused by a historical earthquake often become very important. For example, local upheaval, spouting of water and sand, etc. may be considered as the compressional effects of an earthquake, but ground fissure, collapse, etc. must be related to its rarefactional effects.

Generally, we may consider any set of macroscopic phenomena to be a fuzzy set characterized by some membership function. A direct method for a fuzzy description of earthquake effects is based upon the membership functions of different phenomena. Therefore, its effectiveness depends on the technique of constructing the membership functions.

Here, for a fuzzy description of the compressional and rarefactional effects of an earthquake, we simply construct the corresponding membership functions as:

$$\mu_{compr.} = \mu_{upheaval} \vee \mu_{spouting} \vee \mu_{other\ compr.\ effects},$$

$$\mu_{raref.} = \mu_{fissure} \vee \mu_{collapse} \vee \mu_{other\ raref.\ effects},$$

where the sign $\vee$ denotes the disjunction, i.e., $\mu_1 \vee \mu_2 = \max(\mu_1, \mu_2)$. Then, the character of the mechanical effects of the earthquake at an observational point can be estimated by the total membership function:

$$\mu = \mu_{compr.} \vee \mu_{raref.}.$$

The $\mu_{upheaval}$, $\mu_{spouting}$, $\mu_{fissure}$, $\mu_{collapse}$, etc. must be defined empirically on the basis of abundant macroscopic phenomena. Here we define:

$\mu_{upheaval}$ = 0.7 when the local upheaval appears, or = 0 otherwise;
$\mu_{spouting}$ = 0.9 when the spouting of water and sand occur, or = 0 otherwise;
$\mu_{other\ compr.\ effects}$ = 0.5 if some other compressional effects are discovered, or = 0 otherwise;
$\mu_{fissure}$ = 0.3 when a ground fissure with width $d < 5$ cm appears, or
= 0.5 when $d \approx 5$ cm, or
= 0.7 when $d > 5$ cm, or
= 0 when no remarkable fissure is observed;
$\mu_{collapse}$ = 0.9 when the ground collapses, or = 0 otherwise;
$\mu_{other\ raref.\ effects}$ = 0.5 if some other rarefactional effects are discovered, or = 0 otherwise.

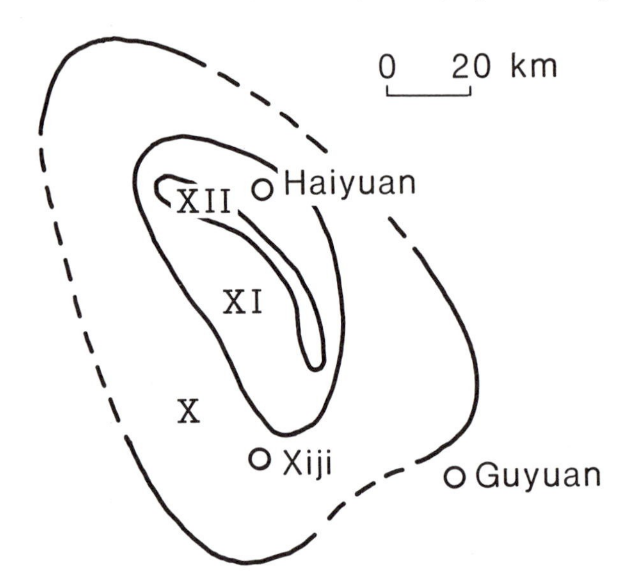

Figure 2. Intensity of Haiyuan earthquake.

By using the fuzzy description method mentioned above, we may determine the characters of the mechanical effects of a historical earthquake at all recording points, and therefore, recognize the compressional points to be denoted by the sign "+", and the rarefactional points to be denoted by the sign "–". As an example, Table 3 shows the data of compressional and rarefactional points for the 1920 Haiyuan ($M = 8.5$) great earthquake. These data may be used to determine its source mechanism.

3.2. Method

Using the data of compressional and rarefactional points obtained from recorded macroscopic phenomena of a historical earthquake, we may determine its source mechanism, namely, the fault plane and the principal compressional and dilatational stress axes by applying the method based on the first motion signs of P waves. However, there may be a range of source mechanisms because of insufficient data points. Consequently, we often obtain several possible models of source mechanism for the same earthquake. For choosing the optimal model, we may apply the method of fuzzy multifactorial evaluation (Feng *et al.*, 1983*a*). This method consists of the following main steps.

First, we must take several factors for fuzzy multifactorial evaluation. In this paper, three factors were used:

(1) Comparison with regional seismotectonics. Suppose we have n possible fault plane solutions for an earthquake, then we can compare them with the main regional geological fault and find the corresponding standard deviations in orientation, $\sigma_1, \sigma_2, \ldots, \sigma_n$. The values $1 - \sigma_i, i = 1, 2, \ldots, n$ may be considered as a factor for fuzzy multifactorial evaluation.

Table 3. Data Used for Determination of Source Mechanism of the 1920 Haiyuan ($M = 8.5$) Earthquake

No.	Place	A_z°	Δ (km)	i_h°	Sign	μ
1	Haiyuan	105.3	66.5	70.9	+	0.9
2	Xiji	136.8	108.5	78.0	−	0.7
3	Guyuan	122.4	143.5	80.9	−	0.3
4	Jingning	150.9	152.25	81.4	−	0.9
5	Tongwei	168.9	168	82	+	0.9
6	Qinan	160.7	215.25	84	−	0.9
7	Tianshui	163.3	245	85	+	0.9
8	Gangu	170.1	224	84	+	0.9
9	Jingchuan	124.4	269.5	53	−	0.7
10	Lingwu	39.4	197.75	83.4	+	0.9
11	Liangdang	157.7	336	53	−	0.7
12	Zhenyuan	118.1	236.25	84.4	−	−
13	Weixian	161.4	343	53	−	0.7
14	Lintan	212.6	264.25	53	+	0.9
15	Huaxian	116.7	503	53	−	0.9
16	Zhaoyi	111.4	518	53	−	0.9
17	Hongdong	91.5	609	53	−	0.9
18	Guide	258.1	316.75	53	+	0.9
19	Xiangfen	95.7	592	53	−	0.9

(2) The ratio between the number of points having consistent stress signs and the total number of points having stress signs. For each possible fault plane solution the ratio

$$S_i = 1 - \left(\sum_{k=1}^{m_i} \mu_k\right) / N_i, \quad i = 1, 2, \ldots, n$$

may be taken as a factor for fuzzy multifactorial evaluation, where μ_k is the membership function of the k-th inconsistent point, N_i is the total number of points, and m_i is the number of inconsistent points for the i-th fault plane solution.

(3) Quarterly distribution of compressional and rarefactional points. For each fault plane solution we may count the percent of points distributed in the j-th quarter

$$k_{ij} = r_{ij}/N_i, \quad i = 1, 2, \ldots, n; j = 1, 2, 3, 4;$$

where r_{ij} denotes the number of points in the j-th quarter for the i-th fault plane solution.

Taking $k'_{ij} = 0.25/k_{ij}$, when $k_{ij} > 0.25$; or $k'_{ij} = k_{ij}/0.25$, when $k_{ij} < 0.25$, the average value $\bar{k}_i = \sum k'_{ij}/4$ may be considered as a factor for fuzzy multifactorial evaluation.

Then, we may obtain a fuzzy evaluation matrix Q as:

$$Q = \begin{pmatrix} (1-\sigma_1) & (1-\sigma_2) & \ldots & (1-\sigma_n) \\ s_1 & s_2 & \ldots & s_n \\ \bar{k}_1 & \bar{k}_2 & \ldots & \bar{k}_n \end{pmatrix}. \quad (6)$$

Taking a suitable weight vector $W = (W_1, W_2, W_3)$ for three factors, we may obtain the fuzzy multifactorial evaluation on the basis of the formula:

$$P = W \bullet Q = (P_1, P_2, \ldots, P_n) \tag{7}$$

where the symbol "•" denotes product of W and Q.

According to the principle of choosing the closest approach, we may obtain the optimal source mechanism of an historical earthquake.

The formulas (4) and (7) represent two possible ways of fuzzy multifactorial evaluation.

3.3. Example

As an example, we have studied the source mechanism of the 1920 Haiyuan ($M = 8.5$) earthquake. The original macroscopic phenomena were taken from the Seismological Institute of Lanzhou and the Seismological Bureau of Ningxia Region (1980). Based on these phenomena, we obtained the data for compressional and rarefactional points (Table 3). By use of the mentioned method, we derived four models of source mechanism (Figure 3). The optimal model of source mechanism is Model 1 in Figure 3. The nodal plane trends N 17° W dipping SW at an angle of 80°, the another nodal plane strikes N 86° E, dipping NW at 39°, the principal compressional stress axis has an orientation of N 72° W with dip angle of 41°, and the principle dilatational stress axis is oriented N 45° E with dip angle of 35°.

4. Fuzzy Description of Macroscopic Criteria for Recognizing Paleoseismic Events

4.1. Method

So far, the data relevant to recognition criteria of paleoseismic events have been accumulating, but all these criteria are often described by very fuzzy languages. We now attempt to describe these criteria by use of membership functions with landfall and landslide as examples.

(1) Landfall. We take the following five fuzzy subsets as the recognition criteria for a paleoearthquake:

$\tilde{U}_1$: In the case of intact rocks such as granite, and massive limestone without mickle interlayer, break phenomena and the trace of river erosion suggest that great collapse occurred.

$\tilde{U}_2$: The size of the collapsed body is great, the valley bottom of the collapsed sphere is very deep, and its depth often equals to 300-500 m.

$\tilde{U}_3$: The collapsed body is displaced very far, its distance could be farther than twelve kilometers.

$\tilde{U}_4$: On the end levee of the collapsed body high-dip-angle steep gradient appeared, it is higher than the background by 50-100 m.

$\tilde{U}_5$: The collapse occurred at a slope whose angle is smaller than 30 degrees.

Then, we may consider the paleoearthquake as a fuzzy set:

$$\tilde{U} = \tilde{U}_1 \cup \tilde{U}_2 \cup \tilde{U}_3 \cup \tilde{U}_4 \cup \tilde{U}_5,$$

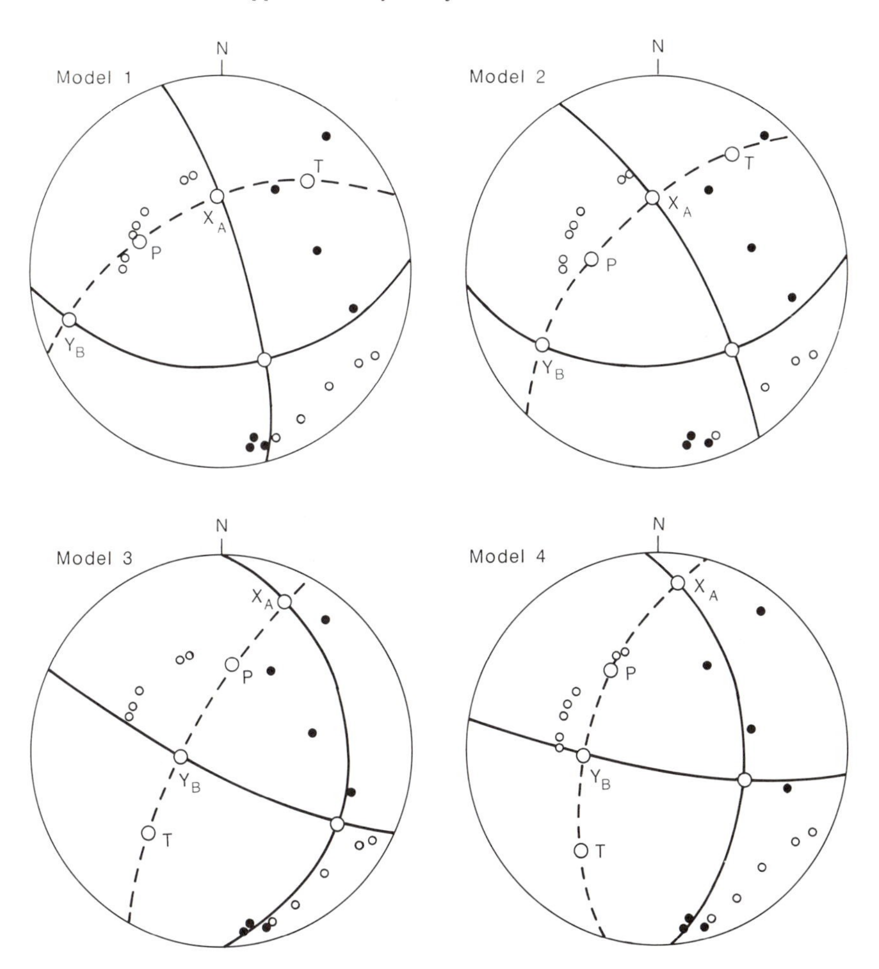

Figure 3. Source mechanism of the 1920 Haiyuan earthquake.

and the membership function is expressed by

$$\mu = \mu_1 \vee \mu_2 \vee \mu_3 \vee \mu_4 \vee \mu_5.$$

The μ_1 may be defined empirically as: (1) $\mu_1 = 1$ for granite and massive rocks in which the clay interlayer, the break, and the trace of river erosion are absent; (2) $\mu_1 = 0.5$ for granite and massive rocks is which the clay interlayer is absent but the trace of river erosion exists; or there is clay interlayer and break, but the trace of river erosion is absent; and (3) $\mu = 0.1$, for granite and massive rocks in which the clay interlayer, the break, and the trace of river erosion exist.

For μ_2, μ_3, and μ_4, the upward semi-normal distribution may be taken as: $\mu(x) = 0$, or $\mu(x) = 1 - \exp[-k(x-a)^2]$. On the basis of empirical data, we have obtained the following preliminary results:

$\mu_2 = 0$ for $0 \leq x \leq 0.3$, or $\mu_2 = 1 - \exp[-69.31(x-0.3)^2]$ for $0.3 < x$.

$\mu_3 = 1 - \exp[-0.011x^2]$.

$\mu_4 = 0$ for $0 \leq x \leq 50$, or $\mu_4 = 1 - \exp[-0.001(x-50)^2]$ for $50 < x$.

For μ_5, the downward semi-normal distribution is taken as $\mu(x) = \exp[-kx^2]$. When $x = 30°$, we took $\mu = 0.5$, so that $k = 2.5$. Then we have $\mu_5 = \exp[-2.5x^2]$. The parameter x in the expressions of μ_2, μ_3, μ_4, and μ_5 is in units of kilometers, meters, kilometers, and radians, respectively.

(2) Landslide. We can take the downward semi-normal distribution as $\mu(x) = \exp[-kx^2]$, where x is the slope angle measured in radians, k is an empirical coefficient depending on earthiness and saturation. For non-saturated loessial mild clay, $k = 3.61$.

4.2. Example of Application

An example was taken from the Seismological Bureau of Shanxi (1982). At the old Yulin grike of Kangding on the Xianshuihe fault zone, where the slope angle is 28°, the accumulation of many collapsed bodies was discovered.

Based on the third criterion, from $\mu_3 = \exp[-2.5x^2]$, we obtain $\mu_3 = 0.547 > 0.5$. As mentioned above, this phenomenon may be recognized as a possible relic of paleoearthquake.

5. Discussions

It should be emphasized that this paper is mainly for describing methods of fuzzy research. It is only a first step that we have done. Obviously, the methods suggested by the authors need to be further examined, tested and developed in the future.

REFERENCES

Dubois, D. and H. Prade (1980). *Fuzzy Sets and Systems: Theory and Application*, Academic Press.

Feng, D. Y. and X. H. Liu (Editors) (1985). *Fuzzy Mathematics in Earthquake Research*, Seismological Press of China.

Feng, D. Y., M. Z. Lin, and others (1983*a*). Application of fuzzy integrated evaluation and close press to the studies of seismic source, *Di Zhen (Earthquake)*, No. 4.

Feng, D. Y., S. B. Lou, and others (1983*b*). *The Methods of Fuzzy Mathematics and Their Applications*, Seismological Press of China.

Feng, D. Y., S. B. Lou, and others (1984). Quantitative evaluation of earthquake intensity based on the fuzzy standards of seismic scale, *Proceedings of Eighth World Conference on Earthquake Engineering*, Vol. II, 819-826, San Francisco.

Gu, G. X. (Editor) (1983). *Catalogue of Earthquakes in China*, Scientific Press.

Seismological Bureau of Sansi (1982), *Papers of Prehistoric Earthquakes and Quaternary Geology*, Scientific Technical Press of Sansi.

Seismological Institute of Lanzhou and Seismological Bureau of Ningxia (1980). *The 1920 Haiyuan Great Earthquake*, Seismological Press of China.

On the Seismicity of the Middle East

Samir Riad
Geology Department
Assiut University, Assiut, Egypt

Herbert Meyers
World Data Center-A for Solid Earth Geophysics
Boulder, CO 80303, USA

Carl Kisslinger
CIRES and Department of Geological Sciences
University of Colorado, Boulder, CO 80303, USA

ABSTRACT

Middle East earthquake data were collected from various files at the World Data Center and from several other publications. Earthquakes of magnitude $m_b \geq 4.5$ or of intensity $I_o \geq$ VI for the period 1900-83 were considered in the present study. The total number of these events was about 8,000, of which 2,400 were reported as suspected duplicates. Duplicates were removed by giving preference to those sources that reevaluated earthquake parameters and to those that included more complete or recent information.

Relations between the reported values of m_b, M_S, M_L, and I_o were obtained and then used to assign calculated values for all events. The M_S/m_b relation for the Middle East was obtained using 405 events, in the form:

$$M_S = 1.604\, m_b - 3.374 \pm 0.49. \tag{1}$$

The range of magnitude used was $4.5 - 7.1$ and $3.2 - 7.4$ for m_b and M_S, respectively. The relation between M_L and I_o was obtained, using 438 events, in the form:

$$M_L = 0.5\, I_o + 1.36 \log h - 0.133 \pm 0.44. \tag{2}$$

Relations between other parameters were also deduced for the Middle East as a whole and for selected areas in the Middle East.

The collected data were used in this way to compile an earthquake catalog of the Middle East, with lower limit of $M_S \geq 3.9$, which corresponds approximately to a value of $m_b \geq 4.5$ and $I_o \geq$ V.

Recurrence-rate relations for the Middle East have been derived and are given in the form:

$$\log N = 7.73 - 1.306\, M_L\,, \tag{3}$$

$$\log N = 8.44 - 0.979\, I_o\,, \tag{4}$$

where N is the cumulative number of events per year of magnitude $\geq M_L$ or intensity $\geq I_o$. Similar relations also were obtained for selected areas.

These relations are used to obtain the mean return period for events of magnitude equal to or larger than a given magnitude M_L. For the Middle East, the following relation may be used:

$$t(M_L) = 10^{1.306M_L - 7.73} \text{ years,} \tag{5}$$

which gives an average return period equal to 25 years for events of $M_L \geq 7.0$.

The value of b given by equation (3) above, $\log N = a + bM_L$ shows that the study area belongs to relatively young orogenic belt zones. Values of b equal to 0.65, 1.4, and 1.8 were found for Pakistan, Turkey, and southern Iran, respectively.

1. Introduction

The Middle East covers an area of more than 13 million square kilometers, including countries geographically belonging to three different continents. It also includes areas that have significant differences in their tectonic evolution – the spreading center and formation of new oceanic crust in the Red Sea, the transcurrent faulting and rifting in the Dead Sea area, the collision and subduction in the eastern Mediterranean, and the subduction and thrusting in western Iran. Therefore, the earthquake activity and seismotectonics of each of these areas definitely will be different.

The major part of the Middle East is comprised of developing countries. Some of them are oil producing, and a major part of their revenues is used in social and economic development. Complete towns and large engineering establishments have been built, and the development is still continuing. Many of the Middle East countries are planning to build nuclear power plants.

Different parts of the Middle East have been affected by destructive earthquakes in historic times. Great property damage and loss of life have been reported in recent times (Ganse and Nelson, 1981; Ambraseys and Melville, 1983; Barazangi and Rouhban, 1983). With the present trend of social and economic development, the socio-economic hazards of earthquakes also increase.

Reliable seismological and seismotectonic studies have not been done in many parts of the Middle East. These kinds of studies are necessary for reliable site studies and design factors for nuclear power plants, as well as other major engineering projects. They also are essential to reduce the socio-economic impact of earthquake hazards.

Study of the seismicity and seismotectonics of an area depends largely on the availability of information and its completeness and reliability. Therefore, an earthquake catalog for the Middle East countries has been prepared and used to compile a seismicity map of the Middle East at a scale 1:8,000,000. The aim of the present paper is to present the main information included in the catalog, and to provide detailed seismicity relations for selected areas in the Middle East.

The prepared catalog includes 47 regions (Flinn and Engdahl, 1965) that cover the Middle East countries (Figure 1). Table 1 shows the geographic locations of these regions. Earthquake data were collected from different files at the World Data Center, Boulder, Colorado, USA, and from other published catalogs (Ganse and Nelson, 1981; Ben-Menahem, 1979; Kárník, 1969) and from individual arti cles (Ambraseys, 1978; Ambraseys and Melville, 1982, 1983; and Barazangi and

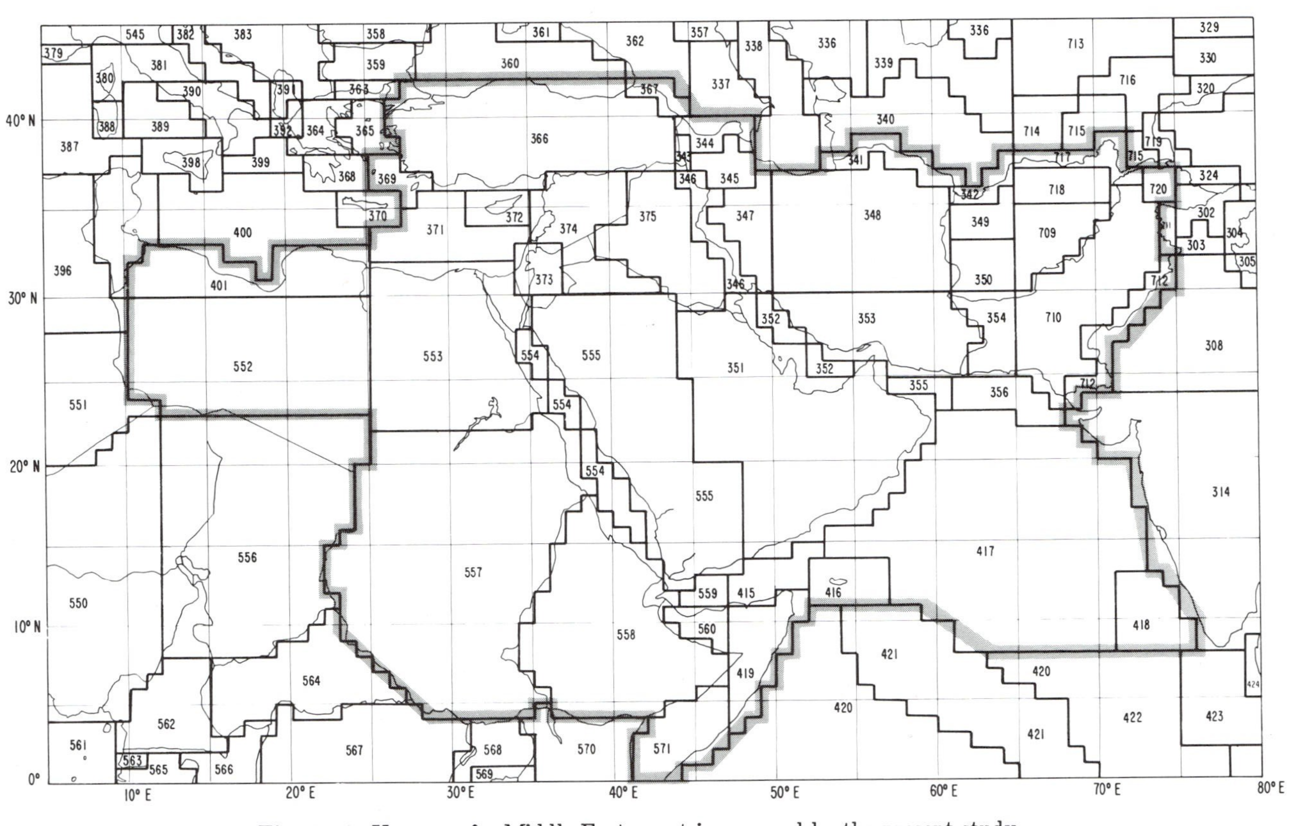

Figure 1. Key map for Middle East countries covered by the present study.

Table 1. Flinn-Engdahl Seismic Regions Included in the Middle East Catalog

No.	Geographical Region	No.	Geographical Region
341	Iran–USSR Border Region	401	Near Coast of Libya
342	Turkmen–Afghanistan Border Region	415	Eastern Gulf of Aden
343	Turkey–Iran Border Region	416	Socotra Region
344	N.W. Iran–USSR Border Region	417	Arabian Sea
345	Northwestern Iran	418	Laccadive Islands Region
346	Iran–Iraq Border Region	419	Northeastern Somalia
347	Western Iran	552	Libya
348	Iran	553	United Arab Republic
349	Northwestern Afghanistan	554	Red Sea
350	Southwestern Afghanistan	555	Western Arabian Peninsula
351	Eastern Arabian Peninsula	556	Central Africa
352	Persian Gulf	557	Sudan
353	Southern Iran	558	Ethiopia
354	Western Pakistan	559	Western Gulf of Aden
355	Gulf of Oman	560	Northwestern Somalia
356	Near Coast of West Pakistan	571	Southern Somalia
366	Turkey	709	Afghanistan
367	Turkey–USSR Border Region	710	Pakistan
369	Dodecanese Islands	711	Southwestern Kashmir
371	Eastern Mediterranean Sea	712	India–Pakistan Border Region
372	Cyprus	717	Afghanistan–USSR Border Region
373	Dead Sea Region	718	Hindu Kush Region
374	Jordan–Syria Region	720	Northwestern Kashmir
375	Iraq		

Rouhban, 1983). The total number of earthquakes collected in the first stage was more than 22,000. This number was reduced to about 8,000 by limiting the investigation to events of magnitude $m_b \geq 4.5$ or of intensity $I_o \geq$ VI. Of these 8,000 events, 2,400 were reported as suspected duplicates, and were removed by reviewing the different sources of data. Preference was given to sources that reevaluated earthquake parameters and to those including more complete or recent information. After removing the duplicates, reported values of the different parameters for the same event (e.g., m_b, M_S, M_L, and I_o) were used to obtain relations between these parameters. Magnitudes given without indications of their type were grouped under catagory M_o. They were found to be approximately equal to M_S, but were not identical. The relations thus obtained were used to assign missing values of all parameters to the events included in the catalog. According to the obtained relations a lower limit of $M_S \geq 3.9$, which is approximately equal to $m_b \geq 4.5$ and $I_o \geq$ V, was considered for the catalog and the accompanying seismicity map for the Middle East. A summary of the earthquakes included in the catalog is given in Table 2.

2. Magnitude-Magnitude Relations

Normally, magnitudes based on surface and body waves are different for the same earthquake. This is because m_b and M_S sample different parts of the characteristic earthquake spectrum. It has been noted by Gutenberg and Richter (1954) that the difference between m_b and M_L is a function of magnitude, and so it follows that the magnitude scales based on surface and body waves are not consistent with each other.

Table 2. Summary of the Middle East Catalog by Year and Magnitude M_S

Year	Mag not given	Less than 3.9	3.9 to 4.4	4.5 to 4.9	5.0 to 5.4	5.5 to 5.9	6.0 to 6.4	6.5 to 6.9	7.0 to 7.4	7.5 to 7.9	8.0 to 8.4	8.5 to 8.9	Totals
1900	0	0	0	0	2	1	0	0	0	0	0	0	3
1901	0	0	4	3	2	1	0	0	0	0	0	0	10
1902	0	0	1	3	2	0	0	2	0	0	0	0	8
1903	0	0	0	2	8	4	2	2	0	0	0	0	18
1904	0	0	3	5	1	7	3	0	0	0	0	0	19
1905	0	0	1	6	5	3	1	2	1	0	0	0	19
1906	0	0	2	1	2	2	1	2	0	0	0	0	10
1907	0	0	5	4	4	0	3	1	1	0	0	0	18
1908	0	0	0	8	6	2	1	2	1	0	0	0	20
1909	1	0	0	2	5	3	0	2	2	1	0	0	16
1910	0	0	1	4	3	1	2	0	0	0	0	0	11
1911	1	0	0	0	1	1	2	2	1	1	0	0	9
1912	0	0	1	4	2	3	3	3	0	1	0	0	17
1913	1	0	0	2	6	1	1	0	0	0	0	0	11
1914	0	0	9	10	2	0	0	3	2	0	0	0	26
1915	0	0	2	0	5	1	1	1	0	0	0	0	10
1916	0	0	4	4	0	0	1	0	1	0	0	0	10
1917	0	0	2	5	2	3	0	0	1	0	0	0	13
1918	0	0	0	5	7	5	1	0	0	0	0	0	18
1919	0	0	6	8	10	4	0	0	1	0	0	0	29
1920	0	0	2	6	7	3	0	0	0	0	0	0	18
1921	0	0	0	3	8	3	0	1	0	0	1	0	16
1922	0	0	0	6	4	1	2	1	0	1	0	0	15
1923	0	0	2	4	6	2	1	3	0	0	0	0	18
1924	0	0	4	13	8	4	3	1	1	0	0	0	34
1925	0	0	13	20	14	9	1	0	0	0	0	0	57
1926	0	0	4	19	17	10	0	1	0	1	2	0	54
1927	0	0	4	4	5	2	6	0	0	0	0	0	21
1928	0	0	10	10	10	4	5	2	0	0	0	0	41
1929	0	0	9	14	10	8	4	0	2	0	0	0	47
1930	0	0	10	21	16	4	0	0	0	1	0	0	52
1931	0	0	7	12	12	3	4	2	2	0	0	0	42
1932	0	0	10	7	18	2	0	0	0	0	0	0	37
1933	0	0	3	8	9	3	2	1	0	0	0	0	26
1934	0	0	3	3	13	3	5	2	1	0	0	0	30
1935	0	0	8	11	12	10	7	2	1	1	0	0	52
1936	1	0	2	14	7	2	0	1	0	0	0	0	27
1937	0	0	5	3	3	1	2	1	1	0	0	0	16
1938	0	0	3	14	8	3	0	2	0	0	0	0	30
1939	0	0	2	11	23	5	1	2	0	1	1	0	46
1940	0	0	7	11	18	12	4	0	0	0	0	0	52
1941	0	0	6	7	14	7	5	0	0	0	0	0	39
1942	1	0	2	5	15	4	3	0	1	1	0	0	32
1943	1	0	0	13	14	0	4	2	1	1	0	0	36
1944	2	0	4	9	15	9	3	1	1	0	0	0	44
1945	0	0	2	9	6	3	1	0	0	0	1	0	22
1946	0	0	1	6	9	3	1	0	0	0	0	0	20
1947	0	0	7	5	6	2	1	1	1	0	0	0	23
1948	0	0	13	12	3	4	1	1	2	0	0	0	41
1949	0	0	8	22	6	0	0	1	0	1	0	0	38
1950	2	0	5	16	3	3	0	0	0	0	0	0	29

(*continued*)

Table 2. (*Continued*)

Year	Mag not given	Less than 3.9	3.9 to 4.4	4.5 to 4.9	5.0 to 5.4	5.5 to 5.9	6.0 to 6.4	6.5 to 6.9	7.0 to 7.4	7.5 to 7.9	8.0 to 8.4	8.5 to 8.9	Totals
1951	0	0	7	14	4	2	1	2	0	0	0	0	30
1952	0	0	11	15	7	4	0	0	0	0	0	0	37
1953	2	0	20	22	11	2	2	2	1	0	0	0	62
1954	0	0	11	19	8	3	0	0	0	0	0	0	41
1955	0	0	7	8	3	3	2	2	0	0	0	0	25
1956	0	0	13	25	11	10	5	1	1	1	0	0	67
1957	1	0	10	31	12	5	2	2	6	0	0	0	69
1958	2	0	11	20	12	9	1	2	0	0	0	0	57
1959	0	0	11	19	14	2	3	1	0	0	0	0	50
1960	0	0	17	20	13	3	5	0	0	0	0	0	58
1961	2	0	18	28	11	9	5	1	1	0	0	0	75
1962	1	0	11	20	6	2	1	0	2	0	0	0	43
1963	0	0	28	30	14	4	2	3	1	0	0	0	82
1964	0	0	45	30	19	8	3	0	1	0	0	0	106
1965	0	0	56	45	16	5	0	0	0	1	0	0	123
1966	0	0	100	70	22	3	3	2	2	0	0	0	202
1967	0	0	88	67	16	8	4	3	1	0	0	0	187
1968	0	0	62	49	27	6	1	1	1	0	0	0	147
1969	0	0	78	65	21	13	4	0	1	0	0	0	182
1970	0	0	91	62	21	4	0	1	1	0	0	0	180
1971	0	0	86	45	17	8	0	0	1	0	0	0	157
1972	0	0	63	49	37	6	4	1	0	0	0	0	160
1973	0	0	53	52	20	2	0	0	0	0	0	0	127
1974	0	0	37	27	15	2	1	0	1	0	0	0	83
1975	0	0	87	68	30	5	2	3	1	0	0	0	196
1976	0	0	75	47	14	3	2	0	1	0	0	0	142
1977	0	0	125	78	21	11	1	1	1	0	0	0	238
1978	0	0	75	31	11	5	3	0	0	1	0	0	126
1979	0	0	87	33	14	4	3	2	1	0	0	0	144
1980	0	0	69	26	11	5	2	0	0	0	0	0	113
1981	0	0	80	23	14	2	0	2	1	0	0	0	122
1982	0	0	71	23	9	2	0	1	0	0	0	0	106
1983	0	0	89	30	13	3	3	1	2	0	0	0	141
Totals	18	0	1889	1585	893	334	153	84	53	14	5	0	5028

Many investigators have established m_b/M_L and m_b/M_S relations for different parts of the world. Some examples are given in Table 3. They show that the magnitude relations are different for various parts of the world.

Using the data available in the Middle East catalog, relations between M_S and m_b, M_S and M_L, M_S and M_o, m_b and M_L were obtained (Table 4). Plots representing some of the obtained relations are given in Figures 2 and 3. These relations were used to assign a value for earthquake parameters not reported in the original sources. Similar relations were also obtained for selected areas in the Middle East (Table 5). They also show wide variations in the values of the constants, which are probably due to the variations in the seismotectonic characteristics of each region. Special attention was given to Turkey, as the reported data provided more information about earthquake parameters for this area.

Table 3. Summary of Magnitude–Magnitude Relations for the Different Parts of the World

Relation	Reference
$m_b = 0.63\ M_S + 2.5$	Gutenberg and Richter, 1956
$m_b = 1.7 + 0.8\ M_L - 0.01\ M_L{}^2$	Gutenberg and Richter, 1956
$m_b = 0.56\ M_S + 2.9$	Bath, 1978
$m_b = 0.48\ M_L + 2.43$	Willis, 1974
$m_b = 4.4 - 0.48\ M_L + 0.11\ M_L{}^2$	Willis, 1974
$M_S = 1.33\ m_b - 1.91$	Willis, 1974
$M_S = 8.33 - 2.59\ m_b + 0.37\ m_b{}^2$	Willis, 1974
$M_L = 1.34\ m_b - 1.71$	Brazee, 1976

Table 4. Magnitude–Magnitude–Intensity Relations for the Middle East

Relation	Standard deviation	No.of events used
$M_S = 1.604\ m_b - 3.374$	0.491	405
	($m_b = 4.3 - 7.1$; $M_S = 3.2 - 7.4$)	
$M_S = 1.009\ M_o - 0.281$	0.303	42
$M_S = 1.069\ M_L - 0.524$	0.480	57
	($M_L = 3.8 - 7.0$; $M_S = 3.5 - 7.1$)	
$m_b = 0.420\ M_S + 3.092$	0.251	405
$m_b = 0.692\ M_L + 1.603$	0.410	107
	($M_L = 3.8 - 7.0$; $m_b = 3.5 - 7.1$)	
$M_L = 0.735\ m_b + 1.226$	0.423	107
$M_L = 0.626\ M_S + 1.995$	0.367	57
$M_L = 0.500\ I_o + 1.361\ \log\ h - 0.133$	0.440	438
$M_L = 0.504\ I_o + 1.580$	0.600	438
$I_o = 1.021\ M_L + 1.754$	0.848	438

3. Magnitude-Intensity Relations

The relation between the instrumental quantity M_L and the macroseismic quantity I_o was determined for the Middle East in general and for Turkey in particular (Table 6). The general form of such a relation is:

$$M_L = a\,I_o + b\log h + c, \tag{1}$$

or in a more simple form

$$M_L = a\,I_o + c, \tag{2}$$

where h is the focal depth.

Relations of this kind have been determined for different parts of the world (Gutenberg and Richter, 1942; Shebalin, 1959; Aivazov, 1961; Kárník, 1969; Brazee, 1976; Ben-Menahem, 1979; and Chiburis, 1981). Values obtained for the constants a, b, and c in the empirical relations show large scattering. However, the value of a is relatively constant between 0.5 and 0.8 (Kárník, 1969).

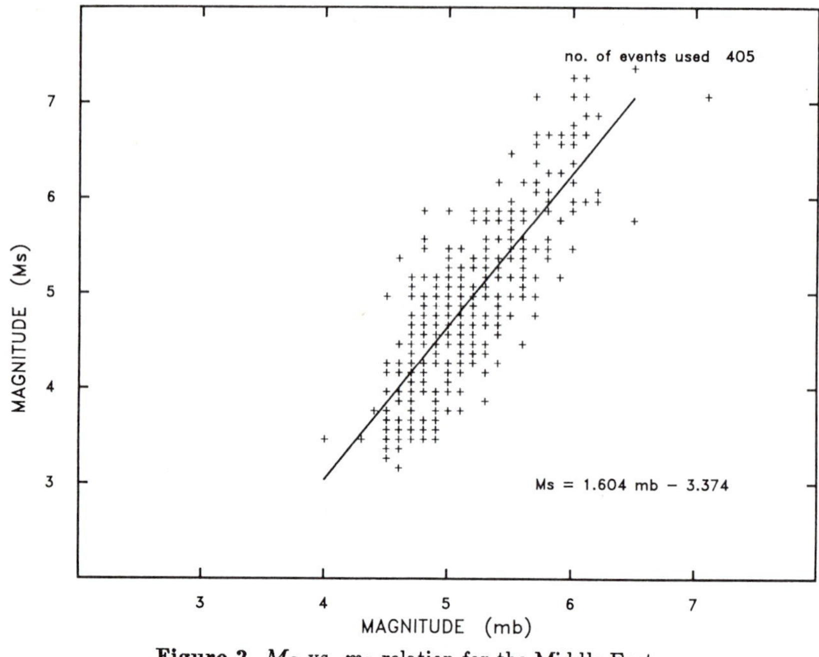

Figure 2. M_S vs. m_b relation for the Middle East.

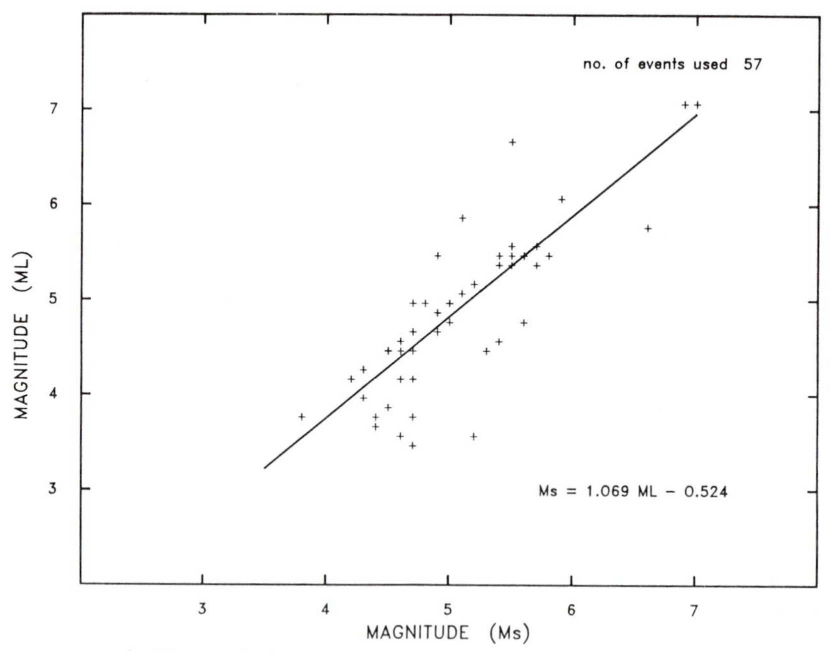

Figure 3. M_S vs. M_L relation for the Middle East.

Table 5. Magnitude-Magnitude Relations for Selected Areas in the Middle East

Region	Relation	Standard deviation	No. of events used	Location
341	$M_S = 1.935\ m_b - 4.981$	0.864	3	Iran-USSR
346	$M_S = 2.333\ m_b - 7.534$	0.655	7	Iran-Iraq
347	$M_S = 2.058\ m_b - 5.948$	0.483	11	W. Iran
348	$M_S = 1.542\ m_b - 2.828$	0.595	39	Iran
353	$M_S = 1.253\ m_b - 1.379$	0.437	40	S. Iran
366	$M_S = 1.420\ m_b - 2.118$	0.489	43	Turkey
415	$M_S = 0.440\ m_b + 2.817$	0.643	15	E. Gul. Adan
417	$M_S = 1.504\ m_b - 2.836$	0.356	16	Arabian Sea
553	$M_S = 2.500\ m_b - 8.380$	0.698	5	Egypt
554	$M_S = 1.891\ m_b - 4.679$	0.450	7	Red Sea
558	$M_S = 1.339\ m_b - 1.862$	0.398	20	Ethiopia

Table 6. Magnitude–Magnitude–Intensity Relations for Turkey

Parameters	Relation	No. of events used	Remarks
m_b, M_S	$M_S = 1.420\ m_b - 2.118$	43	
M_L, M_S	$M_S = 0.792\ M_L + 1.434$	10	
m_b, M_L	$M_L = 1.214\ m_b - 1.088$	11	
I_o, M_L	$M_L = 0.332\ I_o + 3.160$	115	All depths
I_o, M_L	$M_L = 0.325\ I_o + 2.830$	13	$h \leq 10$
I_o, M_L	$M_L = 0.369\ I_o + 2.680$	26	$10 < h \leq 15$
I_o, M_L	$M_L = 0.332\ I_o + 3.100$	25	$15 < h \leq 20$
I_o, M_L	$M_L = 0.446\ I_o + 2.540$	30	$20 < h \leq 30$
I_o, M_L	$M_L = 0.456\ I_o + 2.630$	10	$30 < h \leq 40$
I_o, M_L	$M_L = 0.209\ I_o + 4.300$	11	$h > 40$
I_o, M_L	$M_L = 0.332\ I_o + 1.187\ \mathrm{Log}\ h + 1.665$		$h \leq 100$

The magnitude-intensity relations for the Middle East have been determined using 438 events in the form

$$M_L = 0.5\,I_o + 1.36\log h - 0.133 \pm 0.44 \tag{3}$$

and

$$M_L = 0.5\,I_o + 1.58. \tag{4}$$

Plots showing relation (3) is given in Figure 4.

4. Recurrence Rate Relations

This kind of relation has become a basic approach in statistical seismology. The original form for the magnitude-frequency relation proposed by Gutenberg and Richter (1954) is:

$$\log N = a + b\,(8 - M_L). \tag{5}$$

However, many authors express the magnitude-frequency relation in the form (Kárník, 1969):

$$\log N = a + b\ M_L, \tag{6}$$

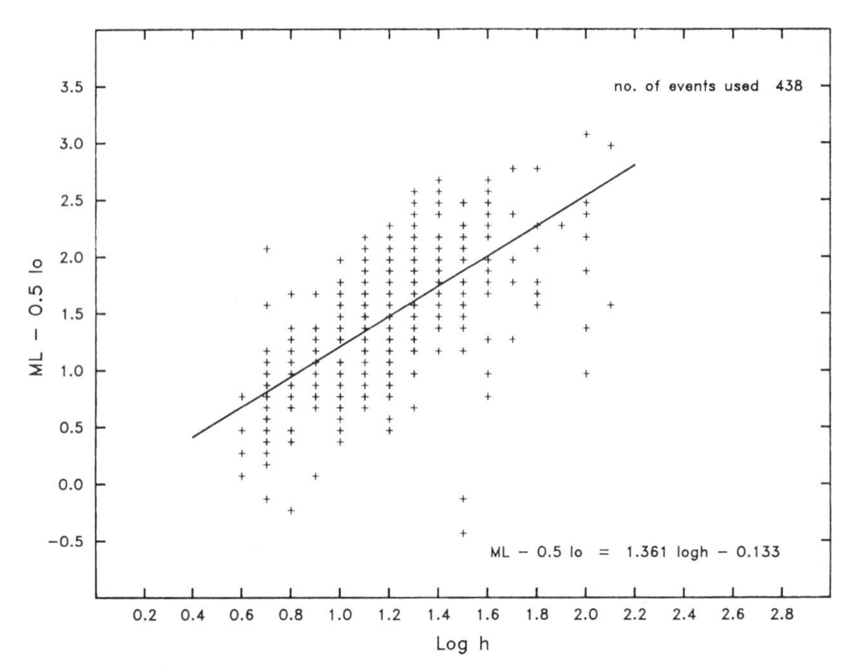

Figure 4. $M_L - I_o$ vs. log h relation for the Middle East.

where N is the cumulative number of earthquakes per year of magnitude $\geq M_L$. The constant a depends on the period of observation, the size of the investigated area and on the level of the seismicity as well. The constant b is assumed to be related to the geologic age of the seismotectonic zone (Miyamura, 1962), as well as depending on other factors such as heterogeneity of source medium and ambient stress level (Mogi, 1962). The value of b has been reported to vary between 0.5 and 1.2 for the USSR, and between 0.9 and 1.2 for the Caucasus (Solov'ev, 1961). On a worldwide scale b was found in the range 0.4 to 1.8 (Kárník, 1969). High values of b (1.0-1.8) have been found in circum-Pacific and Alpine orogenic zones, including island arcs; medium values of b (0.6-0.7) correspond to continental rift zones and platform blocks; low values of b are typical for shields (Miyamura, 1962).

Recurrence-rate relations were determined for the Middle East in general and for selected areas for the period 1900-83 (Table 7). Relations obtained for the Middle East are given in the form:

$$\log N = 7.73 - 1.306\, M_L; \tag{7}$$

$$\log N = 8.44 - 0.979\, I_o, \tag{8}$$

where N is the cumulative number of earthquakes per year of magnitude $\geq M_L$ or intensity $\geq I_o$. Plots representing these relations are shown in Figures 5 and 6. Plots representing the originally reported value of the given parameter (M_L or I_o) are also shown. The misleading results of using incomplete information is obvious in these plots.

Table 7. Recurrence–Rate Relations for the Middle East and for Selected Areas in the Middle East

Parameters	Relation	Standard deviation	No. of events used	Remarks
I_o, N	$\log N = -0.979\ I_o + 8.44$	0.12	4962	Middle East
M_L, N	$\log N = -1.306\ M_L + 7.73$	0.14	5006	Middle East
M_L, N	$\log N = -1.404\ M_L + 7.64$	0.19	1297	Turkey
I_o, N	$\log N = -0.800\ I_o + 6.59$	0.17	1299	Turkey
M_L, N	$\log N = -0.981\ M_L + 5.14$	0.19	472	717
M_L, N	$\log N = -1.104\ M_L + 5.92$	0.12	534	718
M_L, N	$\log N = -0.650\ M_L + 2.71$	0.24	64	720
M_L, N	$\log N = -1.770\ M_L + 8.94$	0.31	474	353
M_L, N	$\log N = -1.166\ M_L + 5.84$	0.08	314	348
M_L, N	$\log N = -1.419\ M_L + 6.58$	0.23	105	347

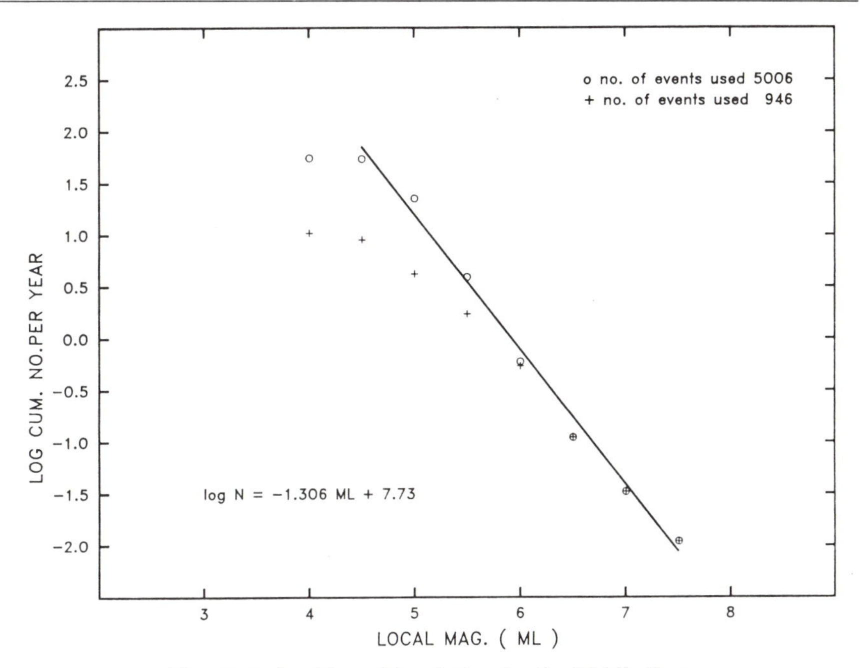

Figure 5. log N vs. M_L relation for the Middle East.

The value of b given by equation (7) is found in the high value range, which is related to young orogenic belts. Values of b ranging between 0.65 to 1.8 are obtained for different parts of the Middle East that have different geological ages and seismotectonic characteristics. For southern Iran, b is equal to 1.8.

The mean return period of earthquakes of a given magnitude or intensity could be calculated using the given equations. For the Middle East the following relation may be used for calculating the return period of an earthquake of magnitude equal to or larger than M_L:

$$t(M_L) = 10^{1.306\,M_L - 7.73} \text{ years.} \tag{9}$$

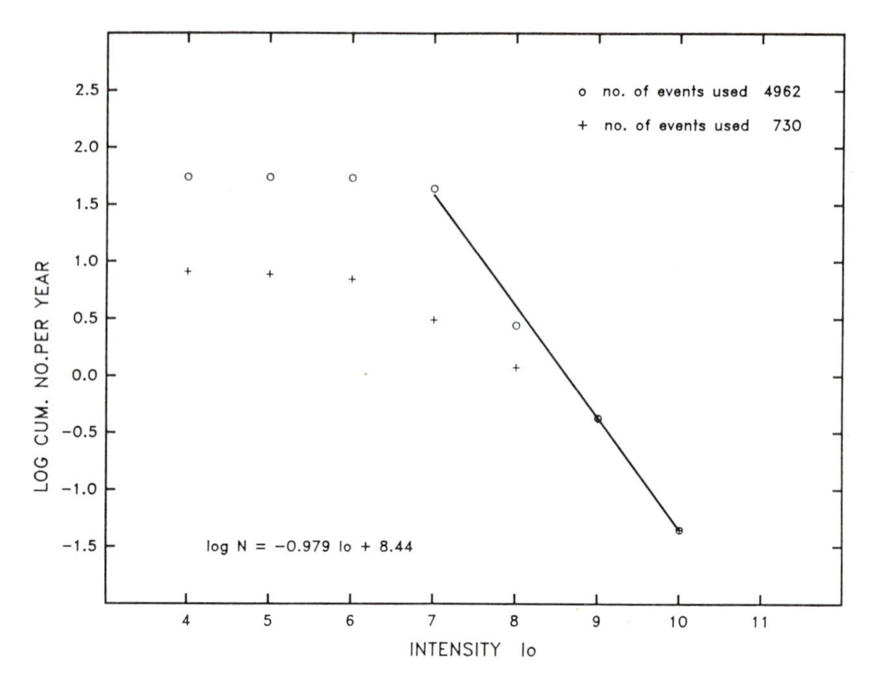

Figure 6. Log N vs. I_o relation for the Middle East.

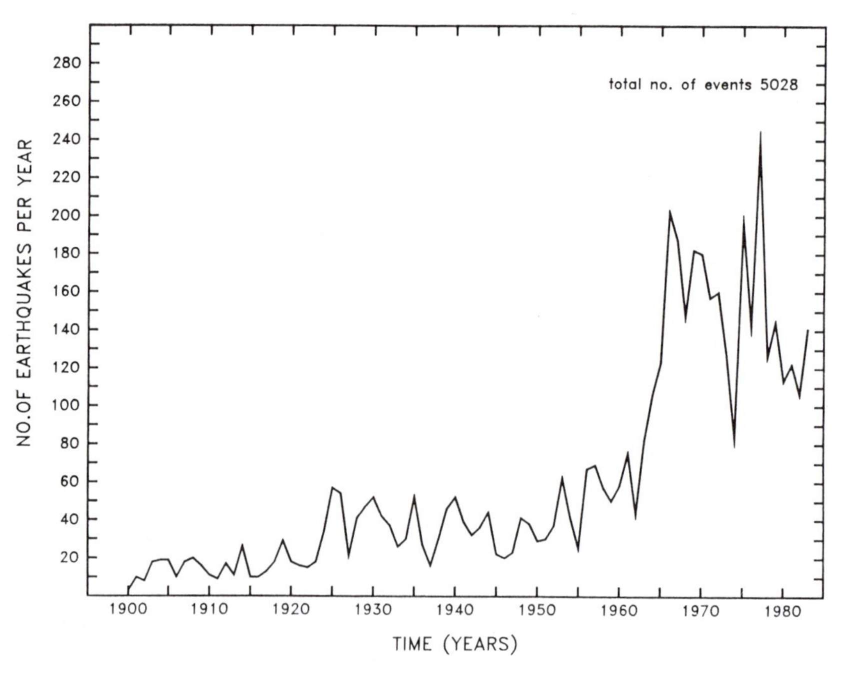

Figure 7. The number of earthquakes per year reported in the Middle East.

This relation gives a mean return period of 25 years for events of magnitude $M_L \geq 7.0$ in the considered time period.

The distribution of the number of earthquakes per year (Figure 7) in the Middle East shows no significant temporal behavior. The increased number of earthquakes after 1960 is probably because of the increased capability and number of recording stations all over the world.

ACKNOWLEDGEMENTS

The authors would like to acknowledge the assistance provided by Wilbur Rinehart, Mark Rockwell, Ron Smith, Jerry Coffman, and Susan Godeaux. This study was carried out while one of the authors (S.R.) was a visiting scientist at CIRES, University of Colorado, under Fulbright Grant #225.

REFERENCES

Aivazov, I. V. (1961). Zavisimost Mezhdu Ballnostyu, Intensivnostyu I Glubiny Ochaga Dlya Kaukaskikh Zemletryaseniv (The Relationship Between Magnitude, Intensity, and Focal Depth of Caucasian Earthquakes), *Soobshch. Acad. Sci. GSSR*, **26(2)**, 149-152.

Ambraseys, N. N. (1978). Middle East – a reappraisal of seismicity, *GSL Quart. J. Engin. Geol.*, **2**, 19-32.

Ambraseys, N. N. and C. P. Melville (1982). *A History of Persian Earthquakes*, Cambridge University Press, Cambridge, U.K., 219 pp.

Ambraseys, N. N. and C. P. Melville (1983). Seismicity of Yemen, *Nature*, **303**, 321-323.

Barazangi, M. and B. Rouhban (1983). Evaluating and reducing earthquake risk in Arab countries, *Nature and Resources*, **19(4)**, 2-6.

Bath, M. (1978). *Introduction to Seismology*, Halstead Press.

Ben-Menahem, A. (1979). Earthquake catalogue for the Middle East (92 B.C. - 1980 A.D.), *Boll. Geofis. Teorica Appl.*, **21(84)**, 313 pp.

Brazee, R. J. (1976). Final Report: An Analysis of Earthquake Intensities with Respect to Attenuation, Magnitude, and Rate of Recurrence, Revised Edition, NUREG/CR-1805, NOAA Tech. Memorandum EDS NGSDC-2, NOAA/National Geophysical Data Center, Boulder, CO, 105 pp.

Chiburis, E. F. (1981). Seismicity Recurrence Rates and Regionalizations of the Northeastern United States and Adjacent Southeastern Canada, New England Seismotectonic Study, (Patrick J. Barosh, Coordinator), Weston Observatory, Boston College, Weston, MA, 76 pp.

Flinn, E. A. and E. R. Engdahl (1965). A propcsed basis for geographical and seismic regionalization, *Rev. Geophys.*, **3**, 123-149.

Ganse, R. A. and J. B. Nelson (1981). Catalog of Significant Earthquakes 2000 B.C. - 1979, *Report SE-27, World Data Center-A for Solid Earth Geophysics*, U.S. Dept. of Commerce, Boulder, CO, 154 pp.

Gutenberg, B. and C. F. Richter (1942). Earthquake magnitude, intensity, energy, and acceleration, *Bull. Seism. Soc. Am.*, **32**, 163-191.

Gutenberg, B. and C. F. Richter (1954). *Seismicity of the Earth and Associated Phenomena*, Princeton University Press, second ed.; facsimile of 1954 ed. published in 1965 by Hafner Publishing Co., N.Y., 310 pp.

Gutenberg, B. and C. F. Richter (1956). Magnitude and energy of earthquakes, *Annali di Geofisica*, **9**, 1-15.

Kárník, V. (1969). *Seismicity of the European Area*, Part 1, D. Reidel Publishing Co., Dordrecht, The Netherlands.

Miyamura, S. (1962). Magnitude-frequency relation and its bearing to geotectonics, *Proc. Japan Acad.*, **38**, 27-30.

Mogi, K. (1962). Study of the elastic shocks caused by the fracture of heterogeneous materials and its relations to earthquake phenomena, *Bull. Earthq. Res. Inst.*, **40**, 125-173.

Shebalin, N. V. (1959). Opredelyniye glubinu ochaga zemletreseniya po evo magnitude m i makroseysmicheskim dannym (Determination of the depth of earthquakes using magnitude and macroseismic data), A.N. Grizomsloy SSR, *Trudy Inst. Geofiz.*, **18**, 159-169.

Solov'ev, S. L. (1961). Obshchiy Obzor Seysmichnosti SSSR, Zemletreyseniya v SSSR (General Review of Seismicity in the USSR, Earthquakes in the USSR), *Acad. Sci. USSR*, Moscow, 165-210.

Willis, D. E. (1974). Explosion Induced Ground Motion, Tidal and Tectonic Facies and their Relationship to Natural Seismicity, Dept. of Geological Science, University of Wisconsin.

Historical Seismicity and Earthquake Catalogues for the Indian Region

H. N. Srivastava and S. K. Das
India Meteorological Department
Lodi Road, New Delhi - 110003, India

ABSTRACT

Seismicity of the Indian region is discussed in the light of the historical data vis a vis expansion of seismological network in the country. The sources of epicentral data for this region are based on monthly seismological bulletins prepared by the India Meteorological Department, U.S. Geological Survey and International Seismological Centre, and catalogues of earthquakes by Tandon and Srivastava (1974) and many others. The limitations of these catalogues are discussed. The utility of a new catalogue of earthquakes in peninsular India for the period 1849 to 1900, based on felt reports through three newspapers (*Times of India*, *Statesman* and *Hindu*) is briefly discussed. This catalogue has brought out some interesting results. In particular, it has been noticed that in the region where significant earthquakes have occurred, tremors of felt intensity have been reported several years preceding the main events. Recently, earthquakes of similar intensity, as reported in the catalogue, have occurred near Hyderabad (30 June 1983; magnitude 4.8) and Bangalore (20 March 1985; magnitude 4.5) showing its utility in assessment of earthquake risk.

The India Meteorological Department is participating in the IASPEI Historical Seismogram Filming Project and more than 20,000 seismograms of Bombay have been microfilmed during the last two years. The recent techniques of source mechanism studies through digitization of old records (prior to 1962), as envisaged in the IASPEI Historical Seismogram Filming Project, will enable us to understand the nature of stresses in and around the Indian plate.

1. Introduction

The Himalayan mountain system has been developing since early Mesozoic time. It is attributed to the underthrusting of the Indian plate below the Eurasian plate. These movements are still continuing and are associated with active seismicity of the region where several great earthquakes (Assam 1897 and 1950, Kangra 1905, Bihar Nepal 1934) have caused huge loss of life and property. Although plate movements have been taking place for millions of years, a reliable history of great earthquakes in the Indian region extends back only 200 years. Of this time period, epicentral parameters of earthquakes based on instrumental data are available for less than a century.

The objective of this paper is to evaluate the importance of historical seismicity for the Indian region. This paper describes the limitations of earthquake catalogues for the region. The utility of the new catalogue of earthquakes for the period 1849 to 1900, based on felt reports for peninsular India, and the importance of the IASPEI Historical Seismogram Filming Project are discussed.

2. IMD Seismological Bulletins

Prior to 1938, seismological data from the Indian seismological stations were being published in the Annual Summary (Part D) of the Indian Weather Review. Beginning January 1938, a quarterly seismological bulletin was printed. This was later changed to a monthly seismological bulletin. It contains phase data from all of the permanent stations maintained by the India Meteorological Department (IMD). The epicentres as given by the U. S. Geological Survey (USGS) monthly listings generally are included. Occasionally, due to late receipt of USGS monthly listings, epicentres of earthquakes from Moscow Seismological Bulletins have also been used. For local events, epicentres are being determined by the HYPO 71 programme through IBM 360/44 (memory, 512K) using an appropriate velocity model (Lee and Lahr, 1975). For other earthquakes occurring in the Indian region, epicentres are reported using the programme given by Shaikh *et al.* (1982) based on the Jefferys-Bullen model. IMD Seismological Bulletins also contain felt earthquake reports and the microseismic data recorded by seismograph stations in India.

Although most of the data contained in IMD Seismological Bulletins are reported through the International Seismological Centre (ISC), U.K., the utility of the bulletin continues to remain for local seismicity studies. Even if the data is not sufficient to determine the epicentres for many events, it can still be used to study the trend of seismic activity from a single station. For example, Chaudhury and Srivastava (1976) reported 2,500 earthquakes within 4 degrees epicentral radius around Shillong Observatory for the period 1970 to 1973, highlighting the active seismicity of the region where the great Shillong earthquake of 1897 occurred. If we confine our attention to the epicentral map of the region (Figure 1), where earthquakes of magnitude 4.5 or more have been plotted, a high level of seismic activity is noticed near the Manipur-Burma border but not around Shillong as revealed from the local earthquakes recorded at one station. This is attributed to the sparsity of the seismological network. Of late, attempts have been made to lower the threshold of epicentral determination in northeast India with six permanent stations and a few others maintained by the India Meteorological Department, National Geophysical Research Institute, and Geological Survey of India (in collaboration with University of Roorkee). This has enabled us to report epicentres of many more events in Shillong plateau (Figure 2).

It may be mentioned that a number of "river valley project" seismological stations are being run in the country by several agencies. It has been noticed that their data, in combination with IMD Seimological Bulletins, have often enabled us to determine epicentres of many events which otherwise would have gone unrecorded.

Tandon and Srivastava (1974) have published a catalogue of earthquakes in the Indian region based on the following sources: (1) ISC, U.K.; (2) BCIS, Strasbourg (France); (3) Oldham (1883); (4) Gutenberg and Richter (1954); (5) IMD Seismological Bulletins and Catalogue; and (6) USGS Epicentral Cards.

Epicentral data has been divided into the following Indian regions: (a) Kashmir and Western Himalayas; (b) Nepal Himalayas; (c) Northeast India, extending up to Andaman-Nicobar Islands; (d) Indo-Gangetic plains and Rajasthan; (e) Cambay and Rann of Kutch; and (f) peninsular India.

This catalogue is fairly complete for earthquakes of magnitude ≥ 5 in the region. After 1964, data taken from the ISC bulletin includes earthquakes of magnitude $M < 5$ as well, up to the year 1970. In this catalogue however, H. M. Chaudhury

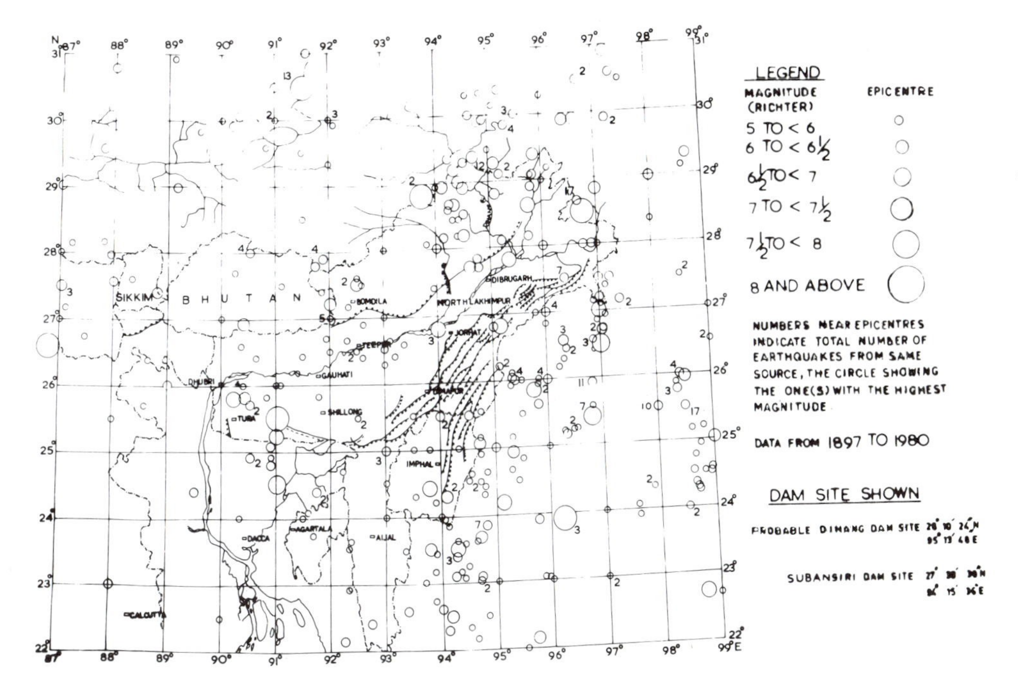

Figure 1. Earthquake Epicentres in Northeast India (1897-1980).

(personal communication) has pointed out one omission of a damaging earthquake on 18 October 1934 of estimated intensity VIII (M.M.), near Gorakhpur (Uttar Pradesh), which was reported in Indian Weather Review. For detailed studies pertaining to risk analysis of critical facilities to be made, there is a need to include earthquakes of $M < 5$ for the period prior to 1964.

The magnitudes of earthquakes in the above catalogue prior to the instrumental period were assigned on the basis of intensity-magnitude conversion formulae using Gutenberg and Richter's (1956) relation:

$$M = 1 + \frac{2}{3} I_o, \tag{1}$$

where M is magnitude, and I_o is epicentral intensity.

In the case of some past earthquakes, focal depth (h) of an event could be estimated from the relation given by Kárník (1969):

$$I_o - I_n = V \log \frac{D_n}{h}, \tag{2}$$

where I_o is the maximum intensity, I_n is a particular intensity, V is a constant which may take typical values of 3.0 or 4.5., and D_n is the distance for a particular intensity.

For earthquakes occurring during the instrumental era, magnitudes were assigned on the basis of distance up to which they have been recorded. At times, we have given a magnitude range of 5 to 6 instead of a single value for an earthquake, due

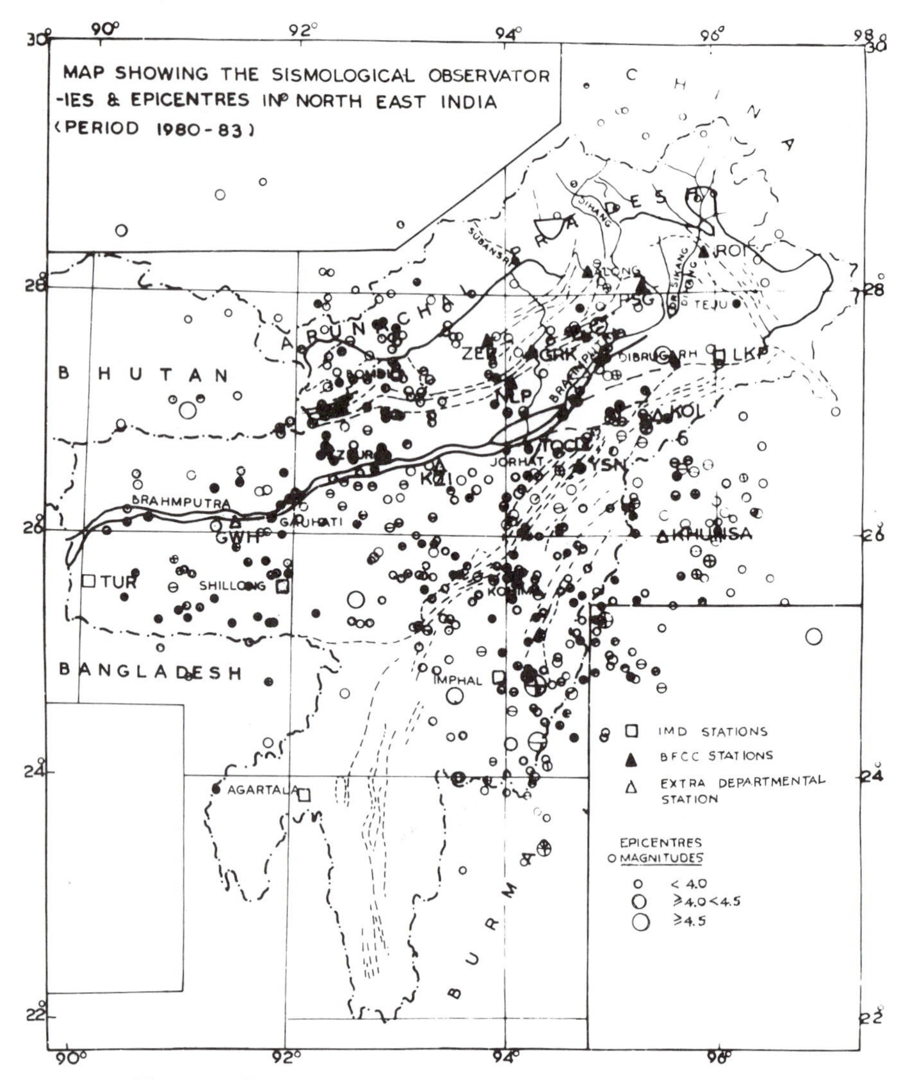

Figure 2. Earthquake Epicentres in Northeast India (1980-1983).

to ambiguities in mesoseismal or instrumental data. However, for the purpose of mathematical computations, mean value of magnitude has been assigned on the magnetic tape file of the IMD catalogue of earthquakes, which is being updated periodically with data from ISC, USGS, and IMD bulletins. This facility has made our consultancy services remarkably quicker.

Gubin (1968) compiled a catalogue of earthquakes for peninsular India after the Koyna earthquake of December 1967. It contained data on earthquakes in the region up to 1968. Later, Chandra (1977) prepared a catalogue of earthquakes for the region (5 – 28° N, 67.5 – 90° E) from the year 1594 to 1975 based on data from 33 sources. Besides updating the list of earthquakes to 1975, he also pointed out some inconsistencies in the dates and places of some earthquakes given by Gubin (1968). He has also expressed some doubt about the 26 May 1618 earthquake near Bombay which could possibly be due to a hurricane rather than an earthquake. However, this catalogue does not extend to earthquakes of $I <$ V which is important for delineating shear zones or other tectonic regions, particularly in lesser seismically-active areas such as peninsular India. Also, there are some omissions in the catalogue prepared by Chandra (1977), even for earthquakes of $I \geq$ V. For example, an earthquake of $M = 5$ which occurred on April 9, 1909 (Tandon and Srivastava, 1974) has not been reported in this catalogue.

Recently, the Indian Society of Earthquake Technology has published an earthquake catalogue compiled by Bapat, Kulkarni and Guha (1983) of Central Water and Power Research Institute, Pune, for the region bounded by Latitude 0 – 50° N and Longitude 50 – 100° E from the historical times to 1979. The catalogue gives an exhaustive bibliography for the years 1763 to 1979. The catalogue contains a few ancient earthquakes such as: Karachi earthquake of 326 B.C., based on sea level changes; Kalibagan earthquake of 1730 ± 100 B.C. in Rajasthan, from archaeological evidence of ruins; Brahminabad earthquake of 8th century B.C., based on archaeological ruins; and an earthquake in 56 B.C. in the Vindyan mountains. There are only 86 events reported from the 26th century B.C. to 1800 A.D. To quote the authors: "certainly there were other major earthquakes during long ancient and historical period in the area but no different evidences could be traced which would identify the events to be tectonic origin like earthquakes".

The catalogue by Bapat *et al.* (1983) also has the following limitations: (1) different magnitudes reported in the literature have not been standardized; and (2) from the year 1965, the epicentres have been taken from the U.S. Geological Survey instead of the ISC Bulletins which are based on more voluminous worldwide data.

After the occurrence of the 1967 Koyna earthquake, a project was undertaken by IMD to prepare a detailed catalogue of earthquakes for peninsular India based on three newspapers, *Hindu*, *Times of India*, and *Stateman*, published from Madras, Bombay, and Calcutta respectively. Of these, *Times of India* is the oldest. Earthquakes reported through these newspapers from the time of publication of the very first issues were noted until 1972 (Srivastava and Ramachandran, 1985). This catalogue has brought out some interesting features about the seismicity of peninsular India as noticed from the list of locally-felt earthquakes up to 1900 (Appendix). In this list, the intensity of earthquakes are assigned on the basis of the newspaper descriptions. This catalogue after scrutiny is being brought up-to-date so that shocks of smaller intensity which have significance for seismicity and risk analysis studies may be assigned due weightage. As many shocks in the new catalogue have been felt over a small area, their intensities were small. Significant errors in assigning their locations (epicentres) are not expected. Keeping this in view, assignment of geographical coordinates to the epicentres will enable us to utilize the data directly through the computer.

It may be of interest to point out the significance of the new catalogue for peninsular India, where small events have been reported several years prior to significant earthquakes. For example, near Jabalpur where an earthquake of $I =$ VI occurred on 17 May 1903 (Turner *et al.*, 1912), a felt earthquake has been reported on 25 November 1868. In the region of a damaging earthquake on 8 February 1900 near Coimbatore, an earthquake of I = IV was reported on 24 June 1865. Similarly, near Ongole, where an earthquake of $M = 5.4$ occurred on 27 March 1967, earthquakes of slightly lesser intensity were reported on 1 September 1869 and later. Two of the felt earthquakes reported in the new catalogue on 10 and 11 September were the aftershocks of the main event of 1 September but with decreasing intensity. Such a pattern of aftershocks with decreasing trend was characteristic of the tectonic activity of the region, and could be differentiated from swarm-type activity which is attributed to local crustal adjustments, as reported from neighbouring regions like Tambaran during 1966. The focal mechanism of the 1967 earthquake (Chandra, 1977) also supported its association with tectonic features, even though two possible fault plane solutions could be drawn due to paucity of data.

The extract of the new earthquake catalog of peninsular India was presented during the IASPEI/Unesco Workshop on Historical Seismograms, in Tokyo, December 1982. Since then, the following earthquakes have occurred of almost similar intensity as reported in the historical catalogues.

2.1. Hyderabad Earthquake

On 30 June 1983, an earthquake occurred near Hyderabad which caused minor damage such as cracks in roofs and walls in houses. Some portions of tiled roof and mud wall collapsed at villages of Kishtpur and Bandamailaram. Shaking was reported for about 20 seconds in the mesoseismal area which decreased to about 10 seconds at Hyderabad and Secunderabad. The earthquake was accompanied by loud sound. The maximum intensity was I = V (M.M.), and field data suggested that the epicentre was near Medchal village, about 30 km north of Hyderabad (Rastogi and Chadha, 1984). The shock was felt over an area of 450,000 mi^2. The epicentral parameters of this earthquake as determined by IMD are: latitude = 18.07° N; longitude = 78.43° E; origin time = 6h 59m 25.9s (GMT); focal depth = shallow; and $M = 5.0$.

About two months prior to the Hyderabad earthquake, IMD had communicated to the Geological Survey of India Hyderabad the description of an earthquake in 1876, based on the new catalogue of earthquakes prepared for peninsular India, and requested them to get more details from local Gazetteers. Although no additional information could be obtained, the occurrence of the 1983 event was not taken as a surprise because the 1876 event had already been brought to notice, which had its epicentre near Secunderabad and was of similar intensity. The historical event of 1876 was described in *Times of India* as follows:

> "Felt throughout the city, caused general alarm, glass panes broken in some of the houses, a number of sparrows found dead after the earthquake, barracks in contonment area were more or less in an oscillating condition for 55 to 60 seconds. Effects in different areas ranged from being thrown out of bed to as if rocked in a cradle. Punkah wires jingled, doors and windows shook, parrots screamed, dogs barked and men woke up by the rolling and loud noise accompanying the earthquakes, some private bungalows in and out of contonment were damaged, but no large scale destruction or loss of life."

2.2. Bangalore Earthquake

On 20 March 1984, an earthquake was reported felt at Bangalore, Mysore in Karnataka and Krishnagiri in Tamil Nadu at about 4 P.M. The earthquake was widely felt in the peninsular India.

Cracks were reported to have developed in old buildings at Bangalore (I = IV to V) but slightly more damage was reported at Kelamangalam (12.6° N, 77.88° E) and Kowthalam (12.58° N, 77.8° E) where intensity was assessed as VI (Iyengar and Meera, 1984). At Kelamangalam and Kowthalam, vertical cracks in the plaster were noticed in several structures of types A and B. At Kowthalam, light objects such as utensils overturned in most of the houses. A 2 meter tall papaya tree which had a circular diameter of about 30 cm was uprooted. In one of the houses, portions of a mud wall collapsed. The damage survey indicated that the epicentre lay in the village of Kowthalam. The intensity assessed by closer scrutiny of field data suggests the shock to be of the same order of magnitude as Hyderabad earthquake (1983). The epicentral parameters of this earthquake were as follows: origin time = 10h 45m 29.5s (GMT); epicentre = 12° 49.55′ N, 77° 26.27′ E; h = 21 km; M = 4.5.

A perusal of historical records has shown that the city of Bangalore and its neighbourhood had experienced the following earthquakes: (1) April 1882 – earthquake severly felt at Bangalore, also felt at Ootacaumund, Vercerd and Hosur; (2) 17 February 1891 – severe shock felt throughout Bangalore district; (3) 7 January 1916 – near Bangalore, M = 5; and (4) 3 May 1980 – south of Bangalore.

Both of the recent earthquakes of Hyderabad (1983) and Bangalore (1984) could be associated with northeast-trending lineaments identified from Land Sat images and corroborated by the pattern of isoseismals.

3. Seismological Network in India

The first seismological station in the country was established at Calcutta (Alipur) on 1 December 1898. During 1898-99, seismic observations were started at Bombay (Colaba) and Kodaikanal using Milne seismographs. After the great Kangra earthquake of 1905, a seismological observatory started functioning at Simla with an Omori-Ewing seismograph. In 1911-12 another instrument for recording vertical components of seismic waves, designed by Wiechert, was installed at Simla. This was moved to Agra in 1929. The observatory was also equipped with a Milne-Shaw seismograph. During the 1930s, seismological observatories also started functioning at Dehradun and the Nizamiah Observatory, Hyderabad. In 1941, the seismological observatory at Agra was shifted to Delhi. The number of seismological stations in the country increased to 8 in 1950, and later rose to 15 in 1960 when more sensitive instruments like Benioff, Sprengnether and Wood-Anderson seismographs were added. At present, the number of observatories under the National Network of stations in the country is 35 (Figure 3). The stations at Delhi, Poona, Kodaikanal, and Shillong were equipped with sensitive seismographs with known calibration curves during 1962-63 under the World-Wide Standardized Seismograph Network (WWSSN). An observatory with a similar set of instruments was started at the National Geophysical Research Institute, Hyderbad, in 1968. Prior to this, the Gauribidanur array station under the Bhabha Atomic Research Centre was also put into operation. 18 stations will be set up under the Department of Science and Technology project on "Seismicity and Seismotectonics of the Himalayan Region."

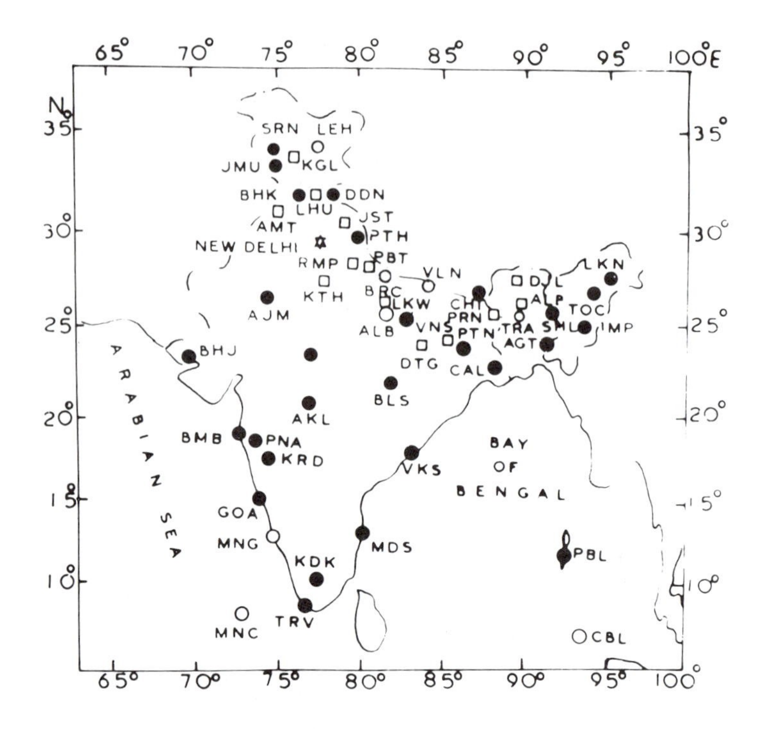

✱ SEISMOLOGICAL HEADQUATERS

● EXISTING SEISMOLOGICAL OBSERVATORIES UNDER NATIONAL NET WORK.

○ PROPOSED SEISMOLOGICAL OBSERVATORIES UNDER NATIONAL NET WORK DURING VI PLAN PERIOD.

□ PROPOSED OBSERVATORIES UNDER PROJECT HIMALYA.

Figure 3. Seismological observatories of the India Meteorological Department.

It may be worth mentioning that out of 55 stations maintained by IMD, as many as 11 stations have been opened during the two year period of 1982-83. It is expected that IMD will eventually have a network of 103 permanent stations during the next few years, so that the epicentral parameters of earthquakes of $M \geq 4$ may be reliably determined in the Indian region.

4. Seismicity Map of the Indian Region

Figure 4 shows the active seismicity of the Himalayan-Burmese-Andaman region, across which the more accepted boundary of the Indian plate extends. The following interesting features may be noted:

(1) Great earthquakes ($M > 8.0$) have occurred in 1819, 1897, 1905, 1934, 1941 and 1950, i.e. only during the last 175 years. In view of the proximity of the plate boundary, it is unlikely that great earthquakes have not occurred prior to 1800. It is therefore a limitation of the catalogues in the Indian region making it difficult to reliably assess earthquake risk.

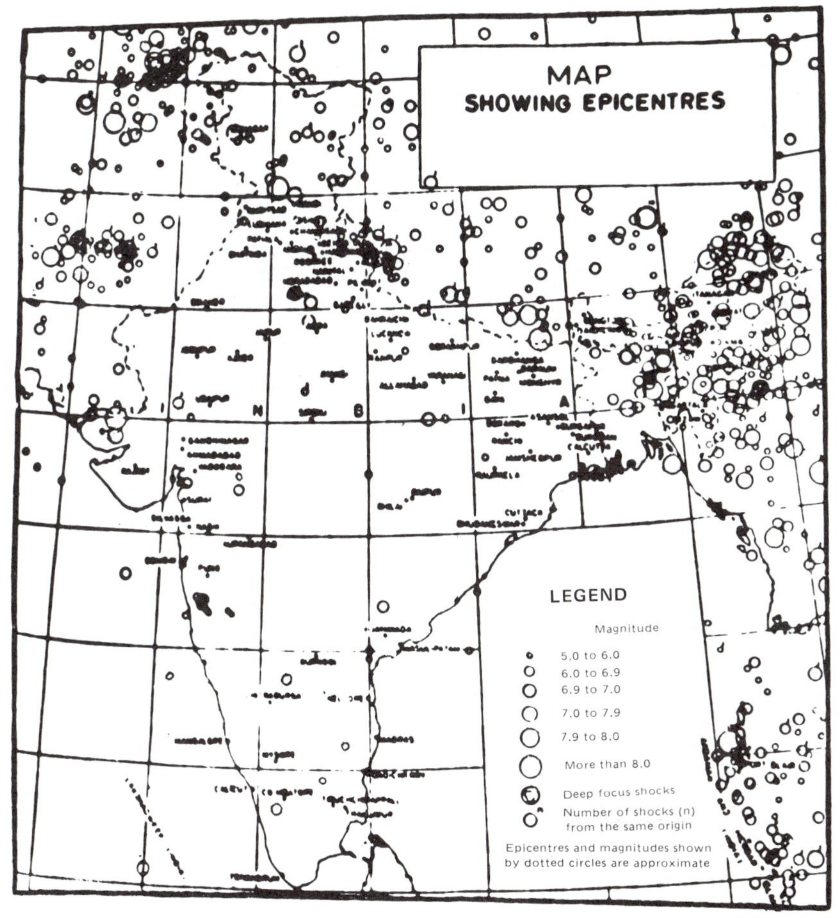

Figure 4. Earthquake epicentres in the India Region.

(2) A number of earthquakes have occurred within the Indian plate. Many of these are associated with transverse lineaments, which are broadly aligned parallel to the movement of the Indian plate. Collection of data for earthquakes of $M < 5$ will enable us to understand the extent and activity of these lineaments. Also, they may reveal the relationship between small tremors and bigger shocks, as brought out by the preliminary observations from the new catalogue for peninsular India presented here and supported by high-gain portable microseismograph observations.

(3) After the great Assam earthquake of 1950, there is a relative decrease of seismic activity in the Indian region. With the exception of Kinnaur earthquake of January 1975, the magnitude of earthquakes has rarely exceeded 6.0. This quiescence period, during which strain appears to be accumulating near the Indian-Eurasian plate boundary, cautions us to investigate the nature of stresses of past earthquakes in greater detail and to study the possibility of identifying premonitory changes (Srivastava, 1983).

5. IASPEI Historical Seismogram Filming Project

Seismograms of earthquakes recorded on photographic paper have a tendency to deteriorate with time under changing weather conditions. They require a large storage place, greater care in handling, and special precautions to prevent damage from white ants. Due to the rapid developments in modern seismological instrumentation of digital as well as analog type, a tendency has been noticed among some seismologists to undermine the utility of old records which were obtained through low-gain instruments like Omori-Ewing, Wiechert, Milne-Shaw, and Wood-Anderson. The recent development of new techniques of digitization and modern methods of analysis have increased the utility of old records to supplement results pertaining to source mechanism studies, crustal and upper mantle structure, etc. The IASPEI Historical Seismogram Filming Project will ensure the preservation and availability of old seismic records for a long time in the future. Among the Indian seismological stations, Bombay has been selected as one of the stations for participation in the Historical Seismogram Filming Project, and more than 20,000 seismograms have been microfilmed so far.

Table 1 lists the instruments and the seismological stations that were in operation during 1962. Because of storage of some seismograms at damp places like Calcutta and Bombay, some records have been lost due to white ants.

During the last few years, there has been a rapid advancement in understanding the nature of stress and source parameters of earthquakes in India and elsewhere (Ichikawa *et al.*, 1972; Tandon and Srivastava, 1975; Srivastava and Chaudhury, 1979; and many others). The Historical Seismogram Filming Project will enable seismologists to study these aspects in detail from near as well as far field angle based on similar records in Europe, India, Japan, and many other countries.

6. Significant Seismological Problems in the Indian Region

The following seismological problems are of immediate interest for the Indian region for which the Historical Seismogram Filming Project may provide additional data:

(1) The catalogue of earthquakes (Tandon and Srivastava, 1974) gives a range of magnitudes for many earthquakes in the Indian region. Refinement of methods of magnitude determination from historical seismograms will enable us to assign a more realistic value in different Indian regions from the point of view of risk analysis.

(2) A number of reliable fault-plane solutions of earthquakes in and around the Indian plate (Ichikawa *et al.*, 1972; Tandon and Srivastava, 1975) suggest refinement of regional plate tectonics models. A few studies have been made for some earthquakes in the Indian region prior to 1962 using body waves (Ichikawa *et al.*, 1972, and others). However, there is a need for detailed studies to evaluate stress drop, seismic moment and related parameters which may enable us to explain the difference of energy released between earthquakes near plate boundaries and within the plate, prior to 1962.

Table 1. List of Seismological Observatories (1962)

Station	Year	Lat. (N)	Long. (E)	Elev. (m)	Foundation	Seis	T_o sec	T_g sec	V
Kodaikanal	1932	10°14′	77°28′	2345	Rock	M-S	12.0	–	250
Bombay	1940	18°54′	72°49′	6	Deccan	M-S	12.0	–	250
	1949				trap.	Spr	8.0	8.0	5000
	1962					Ben	1.0	0.2	50000
Calcutta	1940	22°32′	88°20′	6	Alluvium	M-S	12.0	–	250
						O-E	19.0	–	300
	1949					SMS	6.8	6.8	1000
New Delhi	1942	28°41′	77°12′	230	Massive	M-S	12.0	–	250
	1951				Quartzite	W-A	0.8	–	1000
	1962					SMS	7.6	7.6	5000
Poona	1949	18°32′	73°51′	560	Deccan	M-S	12.0	–	250
					Trap	W-A	4.0	–	1100
						SMS	1.5	0.5	–
Chatra	1952	26°50′	87°10′	161	Sandstone	M-S	12.0	–	250
						W-A	0.8	–	1000
						Ben	0.6	0.45	1000
Madras	1952	13°00′	80°11′	15		SMS	1.53	1.56	–
						SMS	7.3	7.3	5000
Shillong	1952	25°34′	91°53′	1600	Quartzite	W-A	0.8	–	900
	1958				Sandstone	SMS	6.9	7.0	5000
Dehradun	1953	30°19′	78°03′	682	Gravel	M-S	12.0	–	250
						W-A	0.8	–	920
						W-L	1.3	1.55	–
Bhakra	1959	31°25′	76°25′	410	Sedim. rock	W-A	0.8	–	940
Visakhapatnam	1961	17°43′	83°18′	41	Khandellite	W-A	0.8	–	920
					rock.	IMD	1.6	0.5	25000
						SMS	6.8	6.8	5000
Bokaro	1960	23°47′	85°53′	298	Chotanagpur	SMS	7.2	7.2	5000
					Plateau.	W-A	0.8	–	4500
Port-Blair	–	11°40′	92°43′	79	Subsoil rock	M-S	12.0	–	250
						W-A	0.8	–	800
						SMS	7.4	7.3	4500
Sehore	1959	23°10′	77°05′	–		IMD	1.6	1.6	5000
						W-A	0.8	–	950
Tocklai	–	28°45′	94°45′	–	Alluvium	W-A	0.8	–	970

Notes:

T_o = seismometer free period; T_g = galvanometer period; V = peak magnification; Seis = Seismograph, as noted below.

M-S = Milne-Shaw; W-A = Wood Anderson; Spr = Sprengnether; Ben = Benioff; W-L = Wilson-Laminson; IMD = IMD Electromagnetic; SMS = Sprengnether Micro-seismograph.

7. Conclusions

(1) For reliable estimates of earthquake risk, there is an urgent need to scan the historical records in different Indian languages so that the catalogue of Indian earthquakes may be extended, similar to the Chinese and Japanese regions where data are available for 2000 to 3000 years.

(2) IMD is participating in the Historical Seismogram Filming Project of IASPEI which will enable us to apply techniques of analysis (presently developed or likely to be developed in the future) to the preserved original seismograms at an appropriate time. So far, more than 20,000 records have been microfilmed at Bombay which has been chosen for the programme.

(3) Exchange of historical seismograms for significant earthquakes will enable us to study seismological problems from near as well as far field angle for the Indian region and other seismic regions of the world.

REFERENCES

Bapat, A., R. C. Kulkarni, and Guha (1983). Catalogue of Earthquake in India and neighbourhood, from historical period up to 1971, *Ind. Soc. Earthq. Tech.*, Roorkee.

Chandra, U. (1977). Earthquakes of peninsular India – a seismotectonic study, *Bull. Seism. Soc. Am.*, **62**, 1387-1413.

Chaudhury, H. M. and H. N. Srivastava (1976). Seismicity and focal mechanism of earthquakes in northeast India, *Annal. Geofis.*, **29**, 41-57.

Gubin, I. E. (1968). Seismic zoning of Indian Peninsula, *Bull. Int. Inst. Seism. Earthq. Eng.*, **5**, 109-139.

Gutenberg, B. and C. F. Richter (1954). *Seismicity of the Earth and Associated Phenomena*, 2nd edition., Princeton University Press, Princeton N.J., 310 pp.

Gutenberg, B. and C. F. Richter (1956). Earthquake magnitude, intensity, energy, and acceleration (second paper), *Bull. Seism. Soc. Am.*, **46**, 105-145.

Ichikawa, M, H. N. Srivastava, and J. C. Drakopaulos (1972). Focal mechanism of earthquakes occurring in and around the Himalayan and Burmese mountain belt, *Papers Metor. Geophys.* (Tokyo), **23**, 149.

Iyenger, R. N. and K. Meera (1984). Earthquake in south India on March 20, 1984, *Bull. Ind. Soc. Earthq. Tech.*, (under publication).

Kárník, V. (1969). *Seismicity of the European Area, Part I*, D. Riedel Publishing Co., Dordredit, Holland, 364 pp.

Lee, W. H. K. and J. C. Lahr (1975). HYPO 71 (revised): A computer program for determining hypocentre, magnitude and first motion pattern of local earthquakes, *U.S. Geol. Surv. Open-file Rept. 75-311.*

Oldham, T. A. (1883). Catalogue of Indian earthquakes, *Mem. Geol. Surv. India*, **19**, 163-215.

Rastogi, B. K. and R. K. Chadha (1984). *Proc. Mid. Term Symposium*, BHEL Hyderabad, December 1984.

Shaikh, Z. E., H. N. Srivastava, and H. M. Chaudhury (1982). Epicentral parameters – a generalized software package, *Mausam*, **33**, 427-432.

Srivastava, H. N. (1983). *Forecasting Earthquakes*, National Book Trust, New Delhi – 16, 229 pp.

Srivastava, H. N. and H. M. Chaudhury (1979). Regional plate tectonics from Himalayan earthquakes and their prediction, *Mausam*, **30**, 181-186.

Srivastava, H. N. and K. Ramachandran (1985). New catalogue of earthquakes in Peninsular India, *Mausam*, **36**, 351.

Tandon A. N. and H. N. Srivastava (1974). *Earthquake Engineering (Jai Krishna Volume)*, Sarita Prakashan, Meerut, 1-48.

Tandon A. N. and H. N. Srivastava (1975). Focal mechanisms of some recent Himalayan earthquakes and regional plate tectonics, *Bull. Seism. Soc. Am.*, **65**, 963-970.

Turner H. H. *et al.* (1912). Seismological investigations, XII, Seismic activity 1899-1903 inclusive, *Brit. Assoc. Advan. Sci. 16th Report*, 57-65.

APPENDIX. New Catalogue of Earthquakes in Peninsular India

Date	Location	Intensity	Reference
1849 Jan 2	Erinporra 75 km from Abu followed by aftershocks	V	TOI
1849 Dec 26	Bombay	IV	TOI
1851 Mar 21	Kathiawar	III	TOI
1852 Jun 3	Deesa	II	TOI
1852 Jun 10	Deesa	II	TOI
1853 Feb 21	Vizag	IV	TOI
1856 Nov 2	Bhuj	III	TOI
1856 Dec 24	Buralpur	III	TOI
1861 Nov	Vijayanagaram	III	TOI
1862 Nov 18	Dhulia	III	TOI
1864 Aug	Gudur	III	TOI
1865 Jun 4	Mysore	III	TOI
1865 Jun 24	Coimbatore	IV	TOI
1865 Dec 31	Sholapur, Bombay	IV	TOI
	Lingasoogoor	III	
1867 Aug 3	Villupuram, South Arcot (Madras)	IV	TOI
1868 Nov 16	Juboulpore	III	TOI
1869 Sep 10	Ongole	V	TOI
1869 Sep 11	Ongole	III	TOI
1871 Jan 1	Deesa	V	TOI
1871 Jul 28	Surat	IV	TOI
1872 Jun	Amalsar (Near Surat)	III	TOI
1872 Oct 22	Ankleshwar	III	TOI
1873 Jun/Jul	Broach	III	TOI
1875 Jan 2	Ongole	III	TOI
1876 Jul 3	Udaipur	III	TOI
1876 Oct/Nov	Seconderabad	VI	TOI
1877 Dec	Bombay	II	TOI
1879 Jun 17	Hosur	IV	TOI
1879 Jul 30	Indore	III	TOI
1880 Nov 1	Gopalpur	II	TOI
1881 Mar	Tinnevely, Nangueri, Taluk	III	TOI
1881 Dec 27	Rajkot	III	TOI
1882 Feb 28	Ootacumund	III	TOI
	Calicut	IV	
	Felt in other places in Madras State		
1882 Apr	Bangalore, Ootacummund, Hosur	III	TOI
1882 Jun 10	Bhachoo (Kutch)	IV	TOI
	Wavania (Kathiawar)	III	TOI
1882 Jun 29	Eastern part of Kutch	III	TOI
1882 Dec 15	Mount Abu (1)	VI-VII	TOI
1883 Jul 27	Nagpur	IV	TOI
1884 Feb 5	Limda (Kathiawar)	III	STM
1886 Feb	Hirekarur, Dhaywar	III	TOI
1887 Nov 11	Rajkot	III	STM, TOI
1889 Mar 31	Mangalore	IV	TOI
1889 Aug 12	Madras	III	STM
1889 Nov 10	Askal, Ganjam Distt.	IV	STM
1890 Sep 27	Kathiawar	III	STM
1891 Feb 17	Bangalore	IV	STM
1891 May 6	Bangra & Koorla near Bombay	–	STM

continued

APPENDIX. *Continued*

Date	Location	Intensity	Reference
1892 May 6	Madras	III	STM
	Felt at Chingleput.		TOI
1896 Apr 30	Bombay	II	TOI
	Matheran	III	
	Lonavla	III	
1897 Jun 22	11 people died in Berhampore	VII	STM
1897 Jul 8	Berhampore	III	STM

Notes:

(1) Felt at Virumgam and Deesa. Felt over western Rajputana and western position of Narmada Valley. (2) TOI and STM refer to *Times of India* and *Statesman* respectively.

Pattern Analysis of Small Earthquakes and Explosions Recorded at Shasta, California

Robert A. Uhrhammer
Seismographic Station, University of California, Berkeley
Berkeley, California 94720

ABSTRACT

Seismograms recorded by the three-component short-period seismographs at Shasta Dam, California (SHS; 1951-1963) are analyzed for: 1) rate and distribution of seismicity; 2) characteristic patterns which identify small explosions; and 3) patterns of the distribution and orientation of earthquake foci. The high quality SHS station preceded the installation of a network of stations in the area by 40 years and local events (within 50 km) were recorded at a rate of approximately one per week. This study was conducted, in part, to determine how much information can be obtained from detailed analysis of records from a single three-component station.

The local seismicity is 4.2 earthquakes ($M_L \geq 2.5$) per year per 10,000 square kilometers. The seismicity is not uniformly distributed and the rate increases towards the Cascade Mountains to the northeast.

Small man-made explosions are identified, with approximately 99 per cent reliability, by the characteristic pattern of a relatively low-frequency maximum trace amplitude. The dominant frequency of the maximum trace amplitude is 2.9 Hz for explosions and 8.9 Hz for earthquakes.

Patterns in the distribution of earthquake foci indicate that: 1) the source depths are, for the most part, shallow (10-20 km); 2) the dominant mechanism is strike-slip with the tension axis oriented N 73° W; 3) few, if any, earthquakes occur within the Great Valley region; and 4) a few earthquakes in the Cascade Mountains may originate at a depth of 30 km or more.

1. Introduction

The University of California Seismographic Stations operated a high-magnification three-component short-period Benioff seismograph at Shasta Dam, California (SHS: 40°41.7′ N, 122°23.3′ W, 312 m elevation) from July 1952 until the station was closed in May 1963. SHS recorded many small man-made explosions and earthquakes which occurred in the vicinity of Lake Shasta.

The purpose of this paper is three-fold. First, to determine the characteristics of the seismicity in a geologically complex region of northern California in the vicinity of Lake Shasta (300 km north of San Francisco) in order to estimate, in part, the probability and rate of earthquake occurrence. Second, to determine characteristic patterns on the seismograms that will readily identify explosive sources in the region. Third, to determine the general characteristics of the local seismicity from detailed analysis of seismograms recorded at a single station. Extraction of a maximum amount of information is necessary in regions that are without dense seismographic networks (Uhrhammer, 1982).

Lake Shasta and vicinity are near the juxtaposition of the southeastern part of the Klamath Mountains with the Cascade Mountains and the Great Valley. This region exhibits a relatively low level of seismicity and no earthquakes with M_L larger than approximately 5 are known to have occurred in historical times (Townley and Allen, 1939; Bolt and Miller, 1975; Toppozada *et al.*, 1981; Toppozada and Park, 1982). Historical seismicity ($M_L \geq 3.0$) from Bolt and Miller (1975) is shown in Figure 1.

Seismographic records from the Shasta station (1952-1963) were examined, and 568 events were identified. The local magnitude (M_L), estimated from the SHS records, is calibrated using the standard Wood-Anderson seismograms recorded at the Mineral station (MIN; 85 km to the southeast in Figure 1). Approximately 80 per cent of the events recorded are small explosions (man-made blasts) associated with mining and quarry activity in the Shasta mining districts.

The following regional seismicity properties were inferred from the three-component short-period SHS seismograms: epicentral locations; the distribution of focal depths; a composite focal mechanism; and a depth cross-section for the local earthquakes. The inferred spatial characteristics of the earthquakes are subsequently compared the few other seismicity studies yet published for this region.

2. Instrumentation

The instrumentation at SHS consisted of three 50 kg moving coil Benioff seismometers coupled to galvanometers and recording on 35 mm film. The seismometer and galvanometer free periods were $T_S = 1.0$ sec and $T_G = 2.0$ sec, respectively, and the magnification was 50k (at T_S). Four 35 mm by 935 mm film strips were recorded each day, one for each of the three components and one for time. Examples

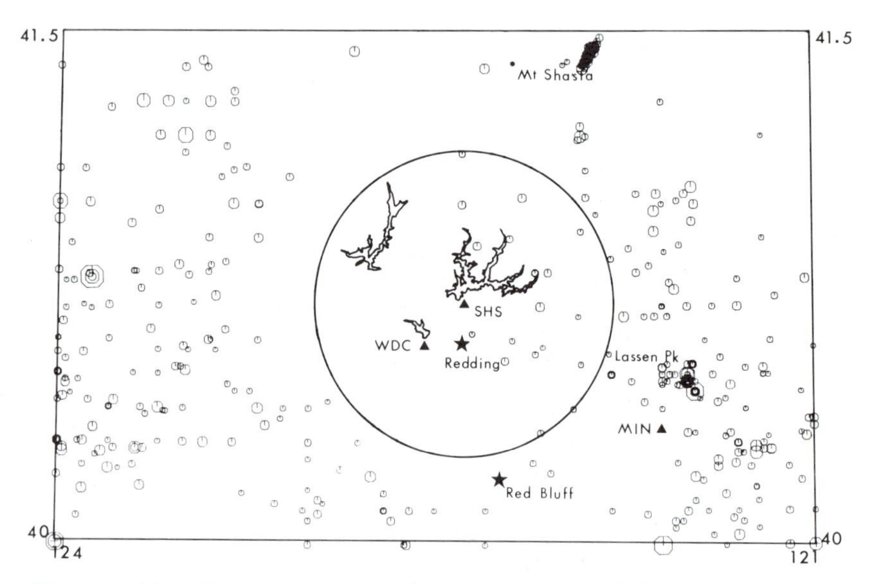

Figure 1. Map of historical seismicity ($M_L \geq 3.0$; 1928-1982). The 50 km radius circle centered on SHS defines the area studied in this paper.

of the three-component records from SHS are shown in Figures 2 and 3. Time marks were recorded on each film every 30 seconds and radio time was recorded on the time strip in order to determine the clock correction and absolute time. Bolt and Miller (1975) give a detailed account of the instrumental parameters.

3. Analysis of Seismograms

The film records from the Shasta (SHS) seismographic station are stored at Berkeley. These seismograms were examined and events with maximum trace amplitudes (A_m) of 0.5 mm or larger and with a differential S-P travel-time interval (t_{S-P}) of 7 seconds or less were recorded. The analysis included reading: 1) the onset times of the P- and S-waves; 2) the maximum trace amplitudes on each component; 3) the polarity of first-motion; and 4) the amplitudes of the first swing of the P-wave on

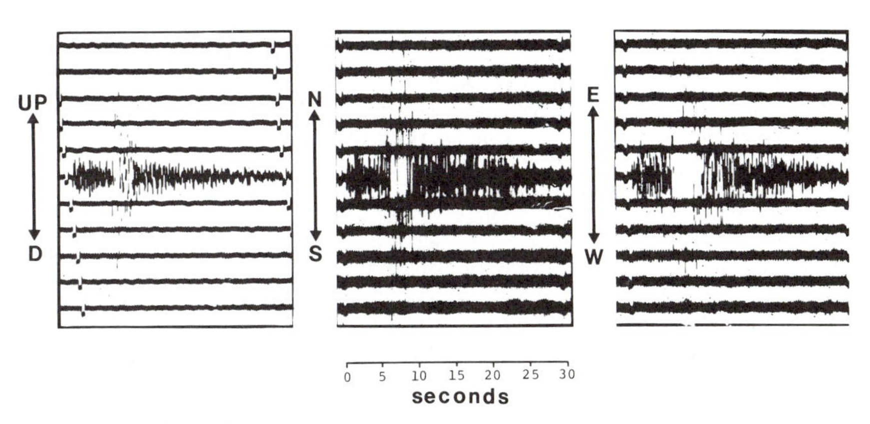

Figure 2. SHS recording of a natural earthquake ($M_L = 2.7$) which occurred at 12:33 UTC on June 29, 1953.

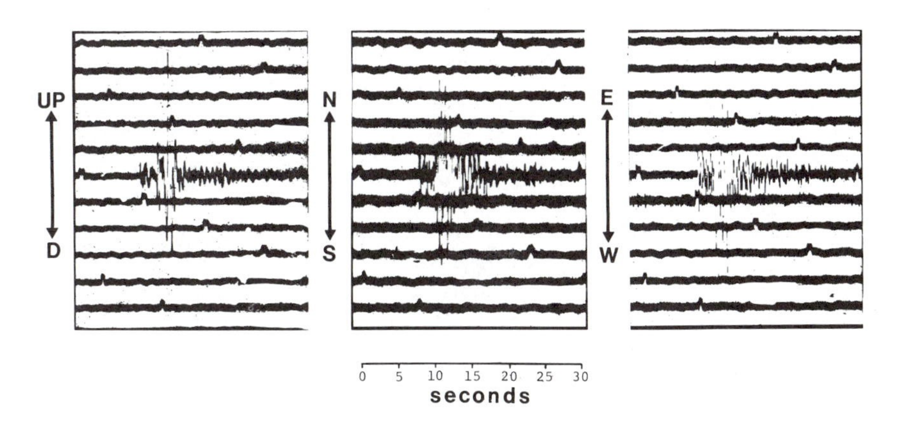

Figure 3. SHS recording of a small explosion (equivalent $M_L = 1.5$) which occurred at 19:31 UTC on August 24, 1962.

each component. Clock corrections, derived from radio time signals, allowed absolute time determination to an accuracy of 0.2 seconds. Any unusual characteristics of the recording were also noted.

Only those events with $t_{S-P} \leq 7$ seconds were selected in order to limit the number of events to those within a hypocentral distance of approximately 50 km. The amplitude threshold of 0.5 mm was chosen to provide sufficient signal amplitude (above the background noise level) to reliably pick the onset times and first motions. A record amplitude of 0.5 mm is equivalent to a ground motion of 10 nm at a frequency of 1 Hz.

The amplitude and time measurements were obtained using a comparator with a reticle resolution of 0.1 mm and lengths can be measured accurately to 0.05 mm. A length resolution of 0.05 mm corresponds to a time resolution of 0.2 seconds on the SHS seismograms (recorded at 15 mm/min). The distance resolution using t_{S-P} is thus approximately 1 km.

4. Magnitude Calibration

The local magnitude (M_L) of an event is determined from the maximum trace amplitudes (A) measured (in mm) on the horizontal component SHS seismograms using the standard formula (Richter, 1935):

$$M_L = \log A - \log A_o + \delta M_L, \tag{1}$$

where δM_L is the magnitude adjustment for SHS. M_L determined from the maximum trace amplitudes measured on the standard Wood-Anderson torsion seismograms recorded at Mineral (MIN; 85 km southeast of SHS) is adopted as standard to calibrate the magnitude adjustments for SHS. Thirty local events, recorded by the Benioff seismographs at SHS and also by the Wood-Anderson seismographs at MIN with usable trace amplitudes, were used to determine $\delta M_L = -0.60 \pm 0.12$ by weighted least-squares.

5. Separation of Small Man-made Explosions from Natural Earthquakes

A scan of the SHS seismograms immediately showed that a large majority of the recorded events (shown in Figure 4): 1) had compressional first-motion for the onset of the P-wave; 2) occurred during local daylight hours (PST); 3) were clustered in distance from SHS; and, 4) were approximately the same size. It was clear that most of these events were small man-made explosions (equivalent $M_L \leq 2$). The problem is to find a simple and effective method of culling the small explosions from the natural earthquakes.

A majority of the events are associated with commercial mining and quarry activity in the west Shasta mining district (Kinkel *et al.*, 1956) shown in Figure 4. The cluster midway between SHS and WDC in Figure 4 matches the location of the Iron Mountain Mine (the largest and most active mine in the district) and approximately 20 per cent of the events are within this cluster. Also, approximately 50 per cent of the events occurred within the mining district which occupies only 6 per cent of the total area. In addition, 97 per cent of the events occurring in the mining district occurred between 1800 UTC and 0200 UTC during local daylight hours (PST).

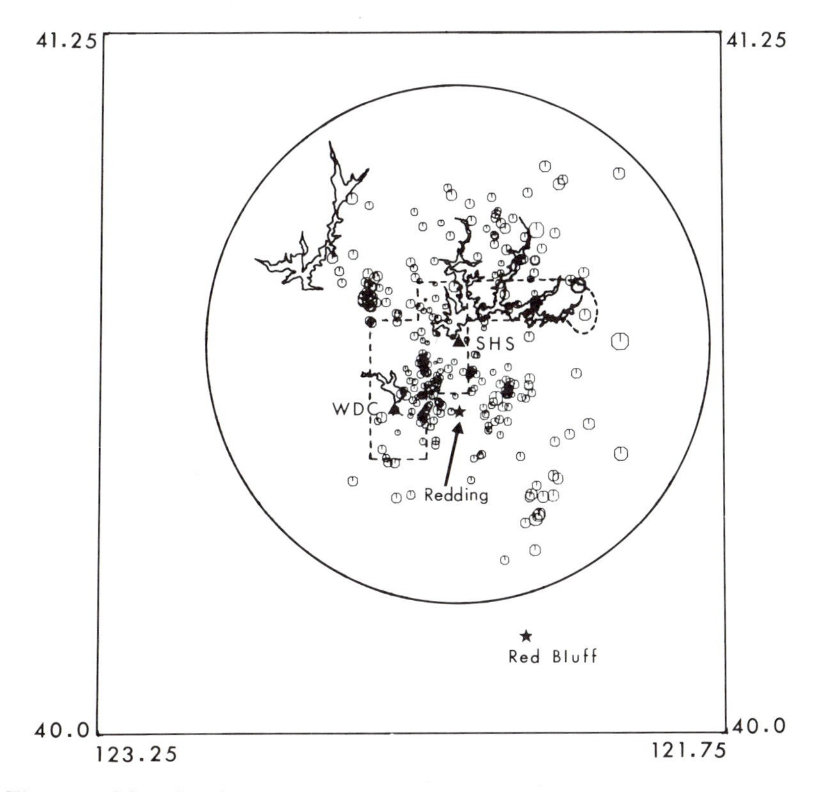

Figure 4. Map of study region, including east and west Shasta mining districts (dashed lines) and all events located by SHS. Notice the clusters of events within or near the mining districts. These clusters coincide with the locations of mines and quarries.

Based on the evidence above, we estimate the rate and distribution of natural seismicity in the region simply by using only those events with $M_L \geq 1.8$ which occurred between 0300 UTC and 1800 UTC as shown in Figure 5. The M_L threshold insures that the seismicity list is uniform for statistical purposes and the time window removes almost all of the small explosions. This method eliminates approximately 97 per cent of the blasts, but it has the disadvantage of also removing approximately 38 per cent of the earthquakes from the list. Removal of such a high percentage of earthquakes from the list may be unacceptable in those cases where the seismicity is very low and the sampling time is short, as there may be too few earthquakes left in the list to reliably estimate the rate and pattern of seismicity.

6. Characteristic Patterns for Small Man-made Explosions and Natural Earthquakes

An independent and more effective method of separation of explosions and earthquakes is to identify characteristics on the seismograms that are associated with small explosions. Example seismograms of a natural earthquake and a small explosion are shown in Figures 2 and 3, respectively.

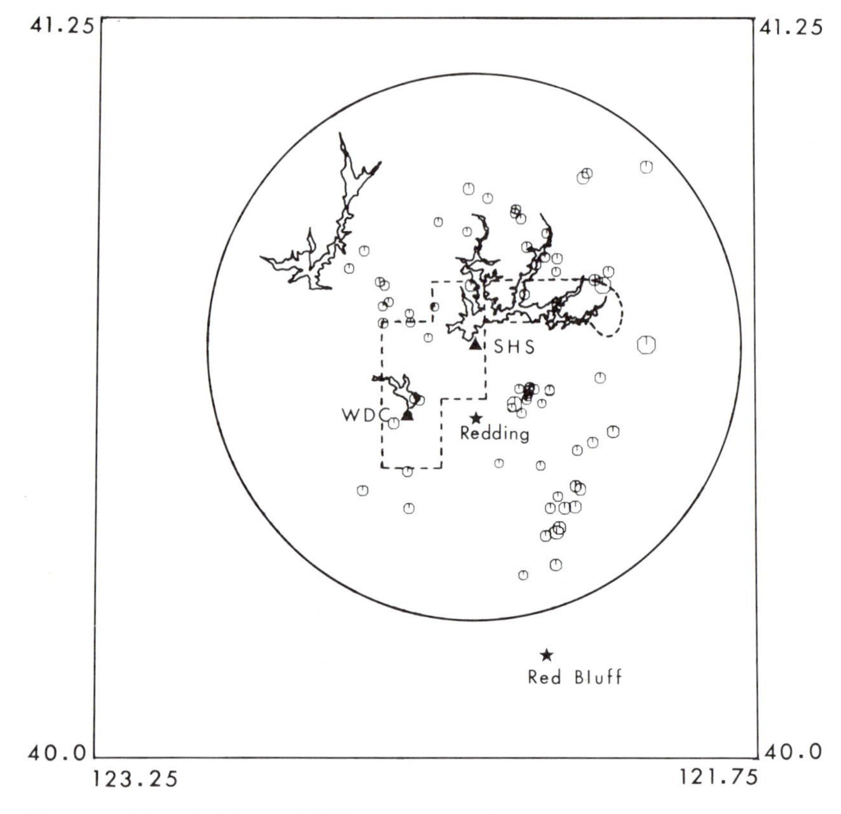

Figure 5. Plot of all located SHS events with $M_L \geq 1.8$ which occurred between 0300 UTC and 1800 UTC. These events are predominantly natural earthquakes.

We approach identification as a problem in pattern recognition (Gelfand *et al.*, 1976). A set of parameters and their corresponding thresholds, which describe characteristic patterns, needs to be identified in order to distinguish between two classes of events, namely, explosions and earthquakes. One difficulty is in selecting an easily estimated set of parameters that consistently distinguish small explosions from natural earthquakes with a high degree of reliability. We have examined the following parameters (measured on the vertical component SHS seismograms): 1) amplitude ratio of maximum trace to first cycle of P-wave (R); 2) rate of decay of P-wave coda (α); 3) dominant frequency of P-wave (f); 4) dominant frequency of maximum trace amplitude (F); 5) first-motion of P-wave (FM); 6) local magnitude of event (M_L); and 7) origin time of event (T). The primary goal is to ascertain characteristic patterns which reliably identify the explosions (also called blasts) so they can be culled from the list of events recorded at SHS in order to estimate the rate and pattern of seismicity in the region.

A summary of the probabilities for the seven parameters, discussed above, is given in Table 1. The dominant frequency of the maximum trace amplitude (F) has a much higher probability for identifying blasts (> 99 per cent) than does any

Table 1. Characteristic Pattern Parameters

Parameter	Threshold	Probability (percent)
R	> 3.7	63
α	$> .77$	76
f	< 3.8 Hz	57
F	< 5.5 Hz	> 99
FM	C	64
M_L	< 2.0	73
T	$1800 \leq T < 0300$	80

other individual parameter. Thus the simplest characteristic pattern, for identifying blasts, is described by the single parameter (F). This parameter probably applies more generally, whereas some of the others do not.

The dominant frequency (F) of the maximum trace amplitude (A_M) is estimated by the maximum likelihood method using the time differences between five consecutive peaks and troughs centered on A_M. The dominant frequency for a blast is $F_B = 2.9 \pm 0.7$ Hz and for an earthquake it is $F_E = 8.9 \pm 0.8$ Hz. For a threshold of 5.5 Hz, the probability that an event is a blast, if $F \leq 5.5$ Hz, is more than 99 per cent. The relatively low-frequency F_b is attributed to the excitation of a fundamental or higher mode Rayleigh wave (Mooney and Bolt, 1966; Cara and Minster, 1981). Theoretically, an explosive surface source will efficiently excite highermode Rayleigh waves at frequencies above 1 Hz in the relatively low velocity surface layers.

If F for an event is not available (owing to the trace being too faint to read, say) the next best characteristic pattern is described by a set of three parameters (FM, α, and f). If, simultaneously, FM = C, $\alpha > 0.77$ and $f < 3.8$ Hz for an event, the probability is 96 per cent that the event is a blast.

7. Rate of Seismicity

From July 1952 through May 1963, 109 earthquakes ($1.8 \leq M_L \leq 2.8$; $0300 \leq T < 1800$ UTC) were recorded within 50 km of SHS. The cumulative rate of seismicity is

$$\log N = 2.80 - 0.87\, M_L, \tag{2}$$

with a corresponding variance of

$$\sigma^2{}_{\log N} = 0.0428 - 0.0196\, M_L + 0.00798\, M_L{}^2, \tag{3}$$

where N is the cumulative number of earthquakes with magnitude $\geq M_L$. The coefficients in (2) and (3) are normalized to earthquakes per year per 10,000 km^2. The rate of earthquake occurrence ($r = 10^{\log N}$) and its standard error (σ_r), calculated from (2) and (3), are given in Table 2. Note that r cannot be reliably extrapolated above $M_L \geq 3.0$ where $r = 1.4 \pm 0.82$ earthquakes per year per 10,000 km^2.

Table 2. Rate of Seismicity

M_L	r (eq/yr)	σ_r (eq/yr)
1.5	31.	13.
2.0	11.	4.8
2.5	4.2	2.0
3.0	1.5	.82
3.5	.57	.35
4.0	.21	.15

8. Analysis of Three-Component Seismograms

The task is to determine the general characteristics of the local seismicity using only seismograms recorded at one station. A dense microearthquake network, consisting of many vertical-component short-period seismographs is commonly used to determine the hypocentral locations, magnitude and focal mechanisms of local earthquakes (Lee and Stewart, 1981). This is readily accomplished using easily measurable parameters on the records, such as polarity and onset time of P-wave first-motion and coda duration. When only one three-component station is available, it is still possible to determine the magnitude and hypocentral location of an event.

Magnitude is determined, in the usual way, from the maximum trace amplitude and the differential S-P travel-time interval (t_{S-P}). Hypocentral locations are determined by vectoring the first peak in the P-wave arrival into the apparent angle of incidence (i_a) and azimuth (Az) to define the raypath (for a given velocity model) and then using the differential travel time (t_{S-P}) to determine the location of the focus along the raypath. The precision of the hypocentral locations is approximately 3 km and, even if a few of the locations were grossly in error, trends in the relative locations of the hypocenters will be a robust indicator of the general characteristics of the local seismicity.

The focal depths of the events are estimated, from the records of a single three-component station (SHS), by plotting the apparent angle of incidence (i_a), measured from vertical, versus the differential S-P (t_{S-P}) travel-time difference for each event as shown in Figure 6. i_a is determined from the ratio of the horizontal (A_H) and vertical (A_V) component amplitudes of the first peak of the P-wave onset (Ewing *et al.*, 1957; p. 27)

$$i_a = \tan^{-1}(A_H/A_V). \tag{4}$$

The uncertainty in estimating i_a is about 5 degrees. The angle of incidence (i) of the P-wave is related to i_a by (Walker, 1919)

$$2 \sin 2i = \frac{\alpha^2}{\beta^2}(l - \cos i_a), \tag{5}$$

where α and β are the P- and S-wave velocities, respectively.

Contours of constant focal depth, assuming a half-space velocity model ($\alpha = 5.6$ and $\beta = 3.3$, $\alpha/\beta = 1.7$), are also plotted in Figure 6. The position of these contours

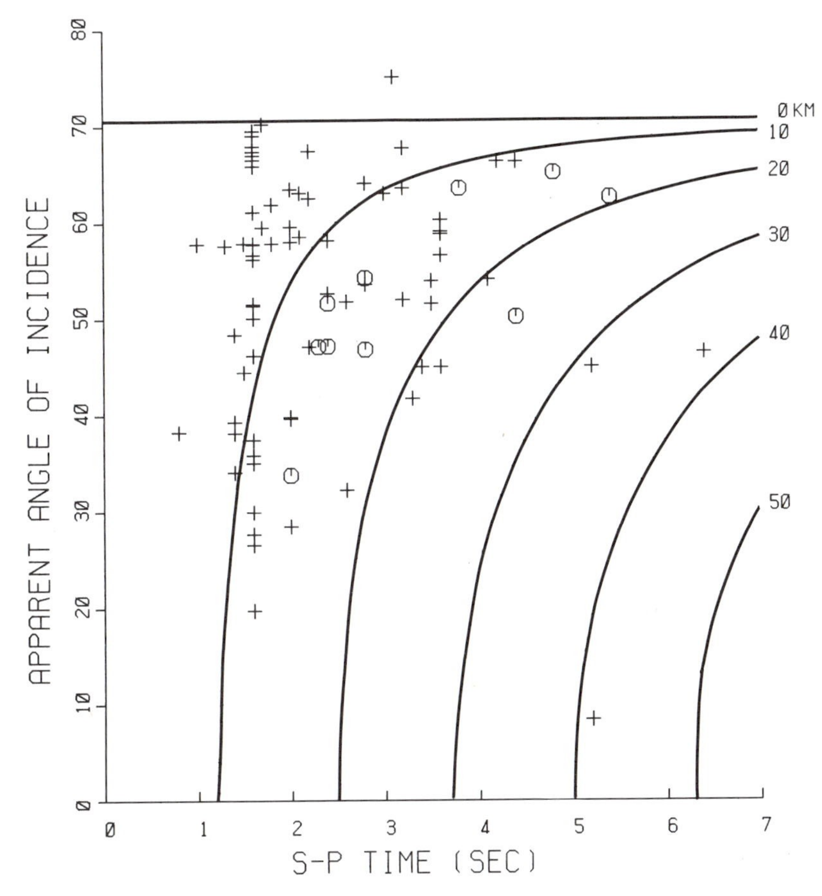

Figure 6. Plot of differential S-P travel-times versus apparent angle of incidence. The bold lines are depth contours. The plus symbols are for compressional and open circles are for dilatational first-motions, respectively. Notice the concentration of dilational events at shallow depths between 0 and 20 km.

represents an upper bound on the focal depth for a given i_a and t_{S-P}. Eight events lie below the $h = 20$ km contour and these events are candidates for deep focus earthquakes.

The apparent angle of incidence, azimuth, and first-motion were determined from the three-component SHS recording. The first-motion polarities are plotted on a lower hemisphere stereographic projection in Figure 7. When drawing nodal planes through the first-motion plot, it is assumed that the earthquakes in the region have a dominant mechanism. Although this composite plot has poorly constrained nodal planes, with a few compressional first-motions in the dilatational quadrants, the interpretation of a strike-slip mechanism is consistent with the local tectonics and with the few previous results of fault plane solutions for this region (Bolt *et al.*, 1968; Simila, 1981). The composite mechanism has a compressive stress orientation of about N 17° W and an associated east-west extension in a N 73° E direction.

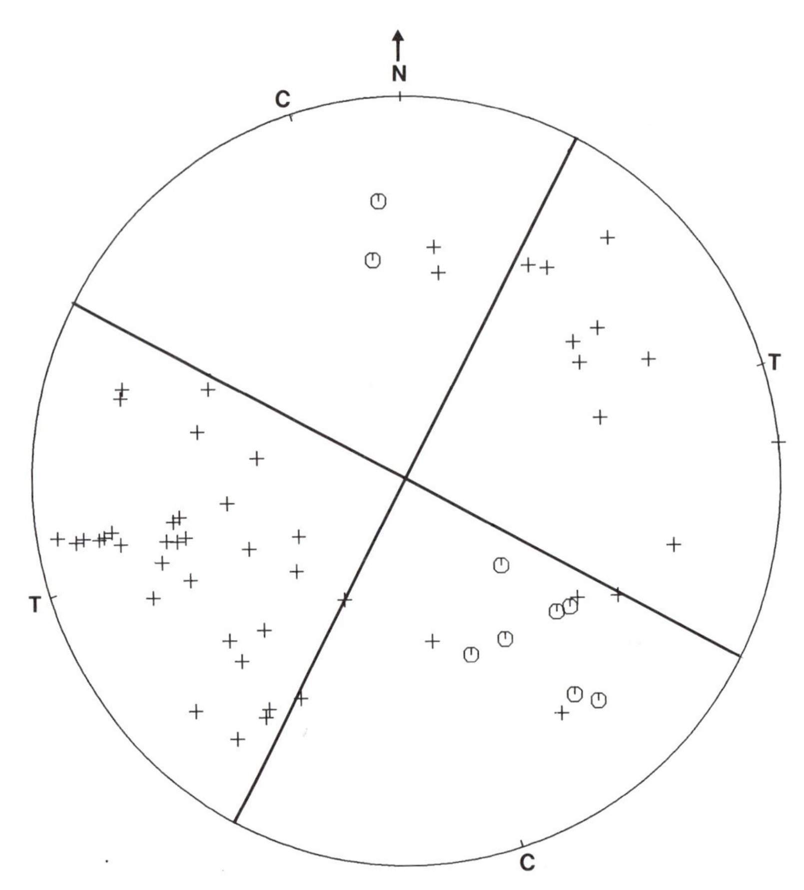

Figure 7. Lower hemisphere stereographic projection of compressional and dilatational first-motions of all events shown in Figure 6. The symbols are the same as in Figure 6. The compression axis is oriented N 17° W and the tension axis is oriented N 73° E.

A cross-section of depth versus distance from SHS along the tension axis (T-T in Figure 7) is shown in Figure 8. The tension axis is oriented N 73° E and it is approximately perpendicular to the trend of the major geological features in the region. The cross-section indicates that the focal depths of a majority of the earthquakes are relatively shallow (10–20 km). There is an apparent trend towards shallower focal depths in the N 73° E direction, although the trend is not significant considering the increased scatter in the focal depths in that direction. The clustering of the dilatational first-motions (in Figure 6) is an artifact of choosing the direction of the cross-section to coincide with the tension axis. Also note that the seismicity stops approximately 25 km to the west of SHS which is consistent with the observations of Bolt *et al.* (1968) and Simila (1981) that there are few if any earthquakes in the Great Valley region. The deeper earthquakes towards N 73° E in the Cascade Mountains-Modoc Plateau region are similar in depth distribution to the earthquakes observed in the Sierra Nevada foothills (Wong and Savage, 1983).

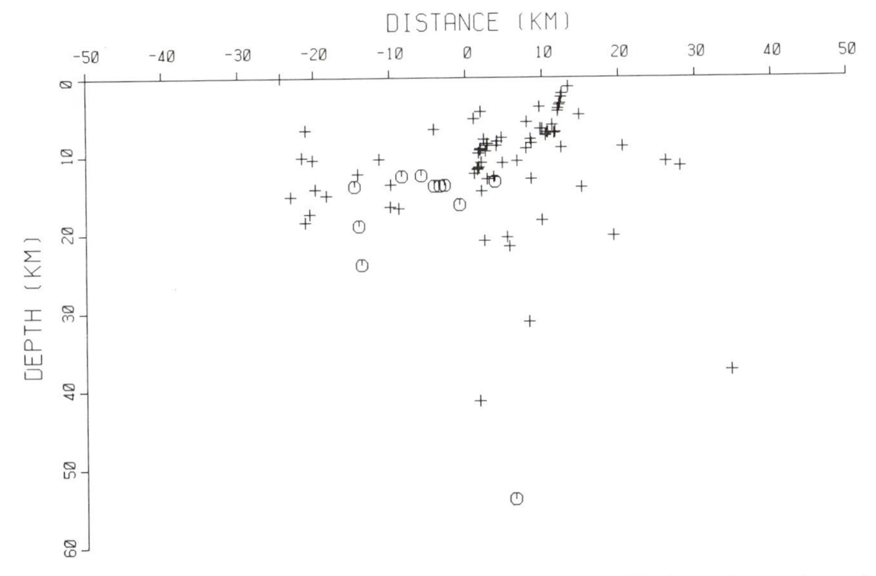

Figure 8. Cross-section of depth versus distance from SHS along the tension axis shown in Figure 7 (N 73° E). The symbols are the same as in Figure 6. The clustering of dilatational events is an artifact of choosing the direction of the cross-section to coincide with the tension axis. SHS is located at 0 km.

9. Conclusions

The seismographic records from SHS (1952-1963) are a crucial historical source of information for studying seismicity in extreme northern California. SHS preceded the installation of a dense local network in the Lake Shasta area by 40 years. Seismograms from a three-component short-period seismograph with high magnification can provide the general characteristics of seismicity in such a moderately seismic region. Observatories that operate or have operated such stations should ensure that the seismograms are preserved (especially if the rate of seismicity is very low; Uhrhammer, 1983). The rate of events observed within 50 km of SHS is approximately one per week.

The magnitude adjustment for SHS is $\delta M_L = -0.60 \pm 0.12$ as determined by adopting the M_L estimate from the Wood-Anderson seismographs at the MIN station, about 85 km away, as standard. Magnitude calibration is important for scaling the a-value (2.8) in the Gutenberg-Richter relation given in (3). The rate of seismicity ($M_L \geq 2.5$) in the vicinity of Lake Shasta averages 4.2 ± 2.0 earthquakes per year per 10,000 km^2. The actual rate decreases towards the Great Valley (southwest) and increases towards the Cascade Mountains-Modoc Plateau (northeast).

The search for parameters for a characteristic pattern which will reliably identify a small explosion from the subject population brought to light two particularly useful characteristic patterns for identifying small explosions. The first pattern requires only one parameter while the second requires three. Both characteristic patterns have a high probability (> 95 per cent) of identifying small explosions using only the information obtained from analysis of a single vertical-component seismogram. The best characteristic pattern is described by a single parameter, the dominant frequency of the maximum trace amplitude (F) recorded on the vertical-component seismogram. The characteristic pattern is easily recognizable by eye on the seismograms because the dominant frequency for a small explosion (2.9 ± 0.7 Hz) is approximately one-third of that for an earthquake (8.9 ± 0.8 Hz).

The general characteristics of the seismicity in the vicinity of Lake Shasta are determined using only the SHS seismograms. The major characteristics are: 1) the seismicity is predominantly shallow (10–km); 2) the rate of seismicity increases towards the Cascade Mountains-Modoc Plateau region and decreases to essentially zero in the Great Valley; 3) the dominant mechanism is strike-slip with the tension axis oriented N 73°E; and, 4) the variance in the depth estimate increases towards the Cascade Mountains-Modoc Plateau with a few earthquakes perhaps deeper than 30 km. These characteristics are confirmed by the data from the 10 station Shasta Dam Microearthquake Network which has been in operation since January 1983 (R. LaForge, personal communication, September 1985).

ACKNOWLEDGMENTS

The support provided by the Bureau of Reclamation (4PG8124250) to scan and analyze the Shasta seismograms is acknowledged. I also thank Drs. B. A. Bolt, L. A. Drake, and W. Stauder for enlightening discussions during this work and for critical comments on the manuscript.

REFERENCES

Bolt, B. A., C. Lomnitz, and T. V. McEvilly (1968). Seismological evidence on the tectonics of Central and Northern California and the Mendocino Escarpment, *Bull. Seism. Soc. Am.*, **58**, 1725-1768.

Bolt, B. A. and R. D. Miller (1975). *Catalogue of Earthquakes in Northern California and Adjoining Areas, 1 January 1910 - 31 December 1972*, Seismographic Stations, University of California, Berkeley, 567 pp.

Cara, M. and J. B. Minster (1981). Multi-mode analysis of Rayleigh-type L_g, Part 1. Theory and applicability of the method, *Bull. Seism. Soc. Am.*, **71**, 973-984.

Ewing, W. M., W. S. Jardetzky, and F. Press (1957). *Elastic Waves in Layered Media*, McGraw-Hill, New York, 380 pp.

Gelfand, I. M., S. A. Guberman, V. I. Keilis-Borok, L. Knopoff, F. Press, E. Y. Ranzman, I. M. Rotwain, and A. M. Sadovsky (1976). Pattern Recognition Applied to Earthquake Epicenters in California, *Phys. Earth Planet. Interiors*, **11**, 227-283.

Kinkel, A. R., W. E. Hall, and J. P. Albers (1956). Geology and Base-Metal Deposits of West Shasta Copper-Zinc District Shasta County, California. *U.S. Geol. Surv. Prof. Paper 285*, 156 pp.

Lee, W. H. K. and S. W. Stewart (1981). *Principles and Applications of Microearthquake Networks*, Adv. in Geophy, Supplement 2, Academic Press, 293 pp.

Mooney, H. M. and B. A. Bolt (1966). Dispersive characteristics of the first three Rayleigh modes for a single surface layer, *Bull. Seism. Soc. Am.*, **66**, p. 43-67.

Richter, C. F. (1935). An instrumental earthquake magnitude scale, *Bull. Seism. Soc. Am.*, **25**, 1-32.

Simila, G. W. (1981). Seismic Velocity Structure and Associated Tectonics of Northern California, *Ph.D. Thesis*, University of California, Berkeley, 180 pp.

Toppozada, R. T., C. R. Real, and D. L. Parke (1981). Preparation of Isoseismal Map and Summaries of Reported Effects for Pre-1900 California Earthquakes, *Calif. Div. Mines Geol., Open File Rept. 81-11*, 182 pp.

Toppozada, R. T. and D. L. Parke (1982). Areas Damaged by California Earthquakes, 1900-1949, *Calif. Div. Mines Geol., Open File Rept. 82-17*, 65 pp.

Townley, S. D. and M. W. Allen (1939). Descriptive Catalog of Earthquakes of the Pacific Coast of the United States, 1769 to 1928, *Bull. Seism. Soc. Am.*, **29**, No. 1.

Uhrhammer, R. A. (1982). The Optimal Estimation of Earthquake Parameters, *Phys. Earth Planet. Interiors*, **30**, 105-118.

Uhrhammer, R. A. (1983). Microfilming of historical seismograms from the Mount Hamilton (Lick) Seismographic Station, *Bull. Seism. Soc. Am.*, **73**, 1197-1202.

Walker, G. W. (1919). Surface Reflection of Earthquake Waves, *Phil. Trans. Roy. Soc. (London)*, **A218**, 373-393.

Wong, I. G. and W. U. Savage (1983). Deep intraplate seismicity in the Western Sierra Nevada, Central California, *Bull. Seism. Soc. Am.*, **73**, 797-812.

Historical Disastrous Earthquakes and Deep Fracture Zones in Ecuador

J. Vaněk and V. Hanuš
Geophysical Institute and Institute of Geology and Geotechnics
Czechoslovak Academy of Sciences, Prague, Czechoslovakia

ABSTRACT

A system of deep seismically active fracture zones was delineated in the region of Ecuador on the basis of the distribution of earthquake foci occurring in the continental South American plate. These zones were interpreted as fractures induced in the continental plate by the subduction process. Their existence and strike were confirmed by the distribution and shape of macroseismic fields of historical disastrous earthquakes observed in Ecuador since 1797. More than 80% of these earthquakes are bound to individual fracture zones, the rest being located in the subducted Nazca plate. This fact may be of decisive importance for an objective estimation of seismic hazard not only in Ecuador but also in other regions overlying active subduction zones.

1. Introduction

Regions of convergent plate margins are characterized by considerable seismic activity and, usually, by a high degree of seismic hazard. In areas with active subduction zones, the earthquake foci are distributed not only in the Wadati-Benioff zone but also in the overlying wedge of the continental plate. Such a distribution of earthquakes is observed in the whole region of Andean South America; in Ecuador, one third of earthquakes, including the strongest shallow events, occur in the continental wedge. It appears that these earthquakes are not distributed randomly, but are arranged into well-defined fracture zones induced or activated by the subduction process (Hanuš and Vaněk, 1987b).

This paper presents an attempt to coordinate individual historical disastrous earthquakes of Ecuador either to the Wadati-Benioff zone or to fracture zones in the continental plate. Such a genetic classification of historical disastrous earthquakes may be of decisive importance for an objective estimation of seismic hazard in Ecuador.

2. Materials

For the definition of the Wadati-Benioff zone and for the delineation of the seismically active fracture zones in Ecuador, the ISC data (Regional Catalogue of Earthquakes) for the period 1964-80 were used as basic material. All determinations with lower accuracy, characterized by errors greater than 0.2° in epicentral coordinates, were rejected. This basic data set was supplemented by data of the NEIS (Preliminary Determination of Epicenters) for 1981-83.

Macroseismic observations of strong historical earthquakes (epicentral macroseismic intensity $I_o \geq 8°$) in Ecuador since 1797 were taken from Egred *et al.* (1984). These data were supplemented by earthquakes observed in southern Colombia since 1827 (Ramírez, 1975).

3. Wadati-Benioff Zone and Seismically Active Fracture Zones In Ecuador

The depth distribution of earthquake foci in relation to the distance from the Peru-Chile trench axis reveals that a well-defined Wadati-Benioff zone exists in Ecuador. This fact is demonstrated by two vertical sections taken from the paper of Hanuš and Vaněk (1987a), which are shown in Figure 1. It can be seen that a considerable seismic activity can be observed in the continental plate above the Wadati-Benioff zone (compare also Stauder, 1975; Barazangi and Isacks, 1979). A detailed analysis of earthquakes occurring in the continental wedge shows that they are not distributed randomly and have a tendency to accumulate in well-separated fracture zones. These zones are interpreted as a set of deep seismically active fractures, induced or activated in the continental plate by the subduction process (Hanuš and Vaněk, 1987b).

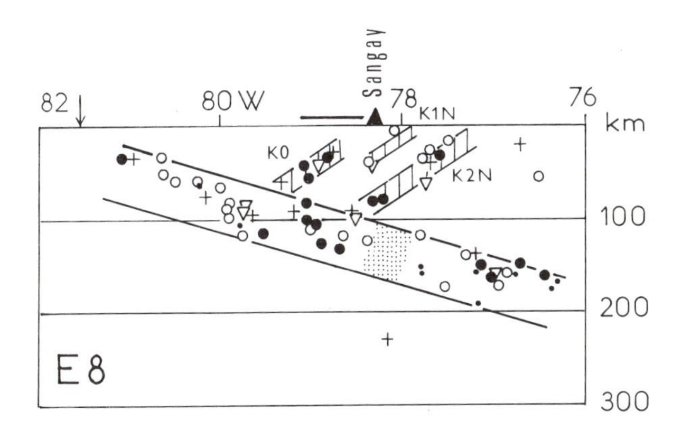

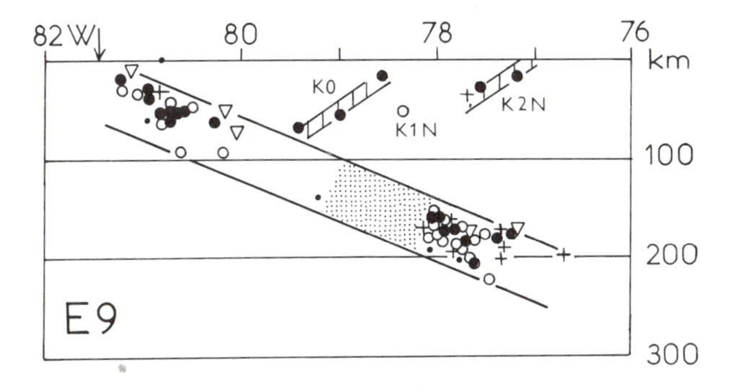

Figure 1. Two vertical sections giving the depth distribution of earthquake foci in relation to their distance from the trench; trench axis is denoted by an arrow, ISC foci by different symbols according to ISC magnitude ($\cdot m \leq 4.0$, ○ 4.1-4.5, • 4.6-5.0, ∇ 5.1-6.0), NEIS foci by crosses, Wadati-Benioff zone by heavy parallel lines, aseismic gap by a dotted area, and individual fracture zones by vertical hatching; section E8 covers the region between the parallels 2.0-2.5° S and section E9 between 1.5-2.0° S. For details see Hanuš and Vaněk (1987a).

In Ecuador, eight fracture zones of this type were delineated. The basic scheme of these fracture zones is given in Figure 2, which shows their position on the surface as determined by the shallowest earthquakes associated with the corresponding fracture zone. The most important parameters of the individual fracture zones are given below. Complete graphical documentation can be found in Hanuš and Vaněk (1987b); here, as an example, only sections for fracture zones KO, K1N, and K2N are shown (Figure 3).

(1) Otavalo-Umpalà fracture zone (F)

Outcrop position: (0.10° S, 79.15° W), (0.50° S, 78.65° W) - (7.15° N, 73.05° W), (6.70° N, 72.55° W); length of the active part: 1050 km; width on the surface: 70 km; dip 44° to southeast; maximum depth of associated earthquakes: 135 km.

(2) Pelileo-Zuñac fracture zone (KO)

Outcrop position: (1.00° S, 78.60° W), (1.00° S, 78.30° W) - (2.50° S, 78.60° W), (2.50° S, 78.30° W); length 165 km; width 30 km; dip 39° to west; maximum depth 70 km.

(3) Puyo-Yaupi fracture zone (K1N)

Outcrop position: (1.05° S, 78.05° W), (1.05° S, 77.70° W) - (2.90° S, 78.05° W), (2.90° S, 77.70° W); length 210 km; width 40 km; dip 38° to west; maximum depth 85 km.

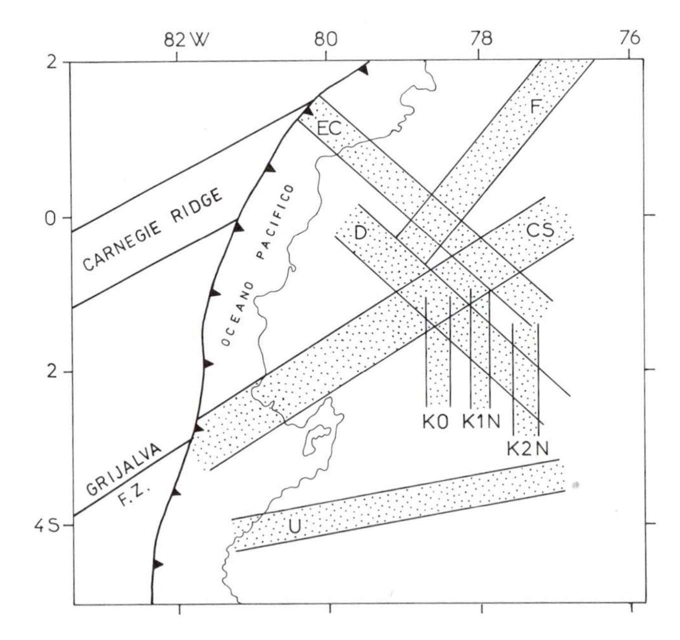

Figure 2. Scheme of seismically active fracture zones in Ecuador. Axis of the Peru-Chile trench is denoted by a serrated line.

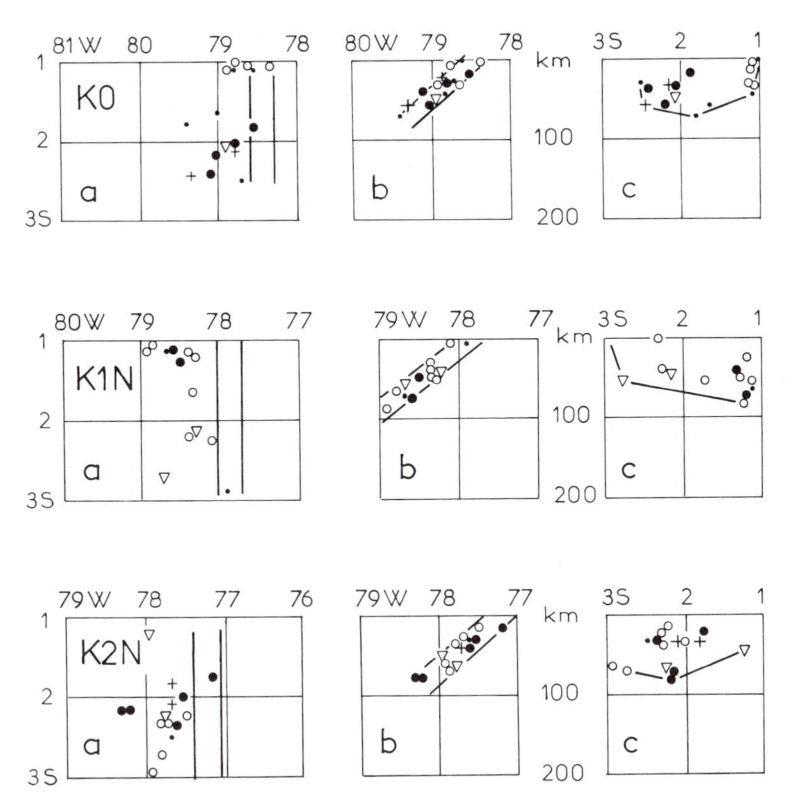

Figure 3. Location (a) of the Pelileo-Zuñac fracture zone (KO), the Puyo-Yaupi fracture zone (K1N) and the Tihuano-Huasaga fracture zone (K2N) with their associated earthquakes, cross-sections perpendicular to (b) and parallel to (c) the corresponding fracture zone. Symbols of earthquakes as in Figure 1.

(4) Tihuano-Huasaga fracture zone (K2N)

Outcrop position: (1.25° S, 77.40° W), (1.25° S, 77.05° W) - (2.95° S, 77.40° W), (2.95° S, 77.05° W); length 185 km; width 40 km; dip 41° to west; maximum depth 80 km.

(5) Mancora-Río Huitoyacu fracture zone (U)

Outcrop position: (3.90° S, 81.26° W), (4.30° S, 81.20° W) - (3.12° S, 76.93° W), (3.55° S, 76.84° W); length 490 km; width 45 km; vertical fault; maximum depth 130 km.

(6) Esmeraldas-Río Curaray fracture zone (EC)

Outcrop position: (1.65° N, 80.10° W), (1.30° N, 80.40° W) - (1.10° S, 76.80° W), (1.40° S, 77.10° W); length 475 km; width 50 km; dip 47° to northeast; maximum depth 130 km.

(7) Cojimies-Río Pastaza fracture zone (D)

Outcrop position: (0.20° N, 79.50° W), (0.25° S, 79.80° W) - (2.25° S, 76.60° W), (2.65° S, 76.95° W) length 420 km; width 60 km; dip 50° to northeast; maximum depth 85 km.

(8) Santa Helena-Santa Cecilia fracture zone (CS)

Outcrop position: (0.30° N, 77.15° W), (0.30° S, 76.80° W) - (2.70° S, 81.90° W), (3.25° S, 81.55° W); length 615 km; width 75 km; dip 45° to northwest; maximum depth 130 km.

4. Distribution of Historical Disastrous Earthquakes in Ecuador

The knowledge of the position and geometry of the Wadati-Benioff zone (Hanuš and Vaněk, 1987a) and the deep seismically active fracture zones in the continental plate (Hanuš and Vaněk, 1987b) enabled us to coordinate individual disastrous earthquakes, observed in Ecuador in the last two centuries, to geological structures along which they occurred. A list of disastrous earthquakes with epicentral intensity $I_o \geq 8°$ MCS observed in Ecuador since 1797 (Egred *et al.*, 1984) is given in Table 1;

Table 1. List of Disastrous Earthquakes in Ecuador and Southern Colombia

No.	Date	Time	Lat.	Long.	I_o	Macro-seismic field	Remark
1	1797 Feb 4	12:30	1.67S	78.64W	11	10-11	intersection of KO,CS,D
2	1859 Mar 22	13:30	0.22S	78.50W	8	7-8	intersection of F,EC
3	1868 Aug 15	19:30	0.81N	77.72W	8	7-8	F
4	1868 Aug 16	06:30	0.31N	78.18W	10	9-10	F
5	1923 May 12	00:22	0.50S	78.56W	8	7-8	intersection of CS,D N continuation of KO
6	1929 Jul 25	08:45	0.41S	78.53W	8	8	intersection of CS,D N continuation of KO
7	1938 Aug 10	02:02:06	0.31S	78.42W	9	8-9	intersection of F,EC
8	1942 May 14	02:13:18	0.75S	81.50W	9	8-9	subduction zone
9	1949 Aug 5	19:08:47	1.50S	78.25W	11	8-11	intersection of KO,CS,D
10	1955 May 11	11:04:00	0.00	78.00W	8	7-8	F
11	1956 Jan 16	23:37:37	0.50S	80.50W	9	7-9	subduction zone
12	1958 Jan 19	14:07:26	1.37N	79.34W	8	7-8	EC
13	1961 Apr 8	09:03:49	2.10S	79.10W	8	7-8	KO
14	1962 Nov 16	06:39:08	1.00S	78.60W	8	6-8	intersection of KO,CS,D
15	1964 May 19	23:03:38	0.84S	80.29W	8	6-8	subduction zone
16	1970 Dec 10	04:34:39	3.99S	80.72W	9	8-9	U
17	1976 Oct 6	09:12:39	0.76S	78.75W	9	7-9	intersection of CS,D N continuation of KO
18	1980 Aug 18	15:07:53	1.95S	80.02W	8	7-8	CS
					class		
19a	1827 Nov 16	23	1.8N	76.4W	III		F
19b	1827 Nov 17	16:10	1.8N	76.4W	III		F
20	1834 Jan 20	12	1.3N	76.9W	III		F
21	1885 May 25	20	2.5N	76.5W	III		F
22	1923 Dec 14	10:31:18	1.0N	77.5W	III		F
23	1947 Jul 14	07:01:00	1.2N	77.2W	II		F

the list is supplemented by disastrous earthquakes observed in the continental part of southern Colombia since 1827 (earthquakes of class II and III in Ramírez, 1975) to confirm the northern continuation of the fracture zone F.

The location and shape of fields of maximum macroseismic intensities for individual earthquakes are shown in Figures 4-6, together with the outcrops of the seismically active fracture zones. The range of intensities used for the construction of macroseismic fields is given in Table 1; epicenters of disastrous earthquakes in southern Colombia are also shown with crosses in Figures 4-5. It appears that all the disastrous earthquakes of Ecuador are located either in the fracture zones of the overriding South American plate or in the subducted Nazca plate. However, the majority of these earthquakes (83%) can be related to individual fracture zones in the continental plate, the rest being located in the subducted Nazca plate. The assignment of individual earthquakes to the respective fracture zone is also given in Table 1.

The most disastrous Ecuadorian earthquakes of 1797 and 1949 (Nos. 1 and 9 with $I_o = 11°$ MCS in Table 1 and Figures 4-5) occurred at the intersection of the fracture zones KO, CS and D, with their macroseismic fields elongated along the strike of the fracture zone KO (compare also Martelly, 1952). The macroseismic fields of other events appear also to be elongated along the strike of the corresponding fracture zones (compare fields of Nos. 1, 9, 14 with KO, Nos. 2, 3, 4, 10 with F, No. 16 with U, and No.18 with CS) or along the strike of the subduction zone (Nos. 8, 11, 15).

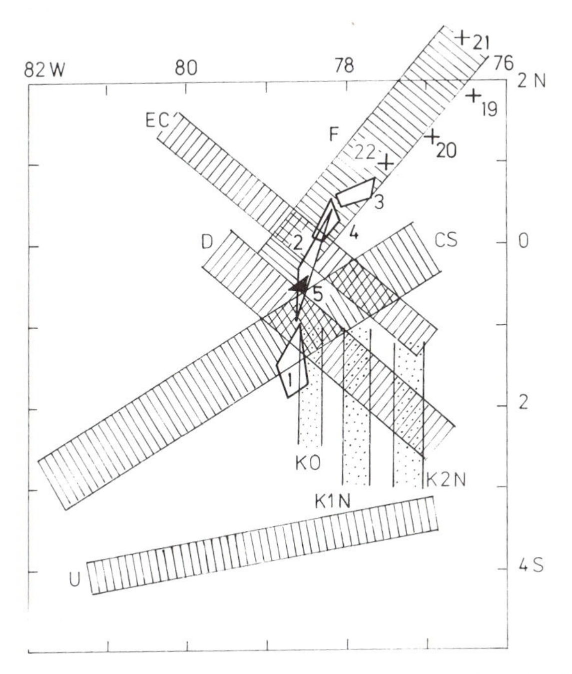

Figure 4. Position and shape of fields of maximum intensities for disastrous earthquakes observed in Ecuador from 1797 to 1923 (Nos. 1-5 in Table 1) in the pattern of the seismically active fracture zones (Figure 2); epicenters in southern Colombia (Nos. 19-22) are denoted by crosses.

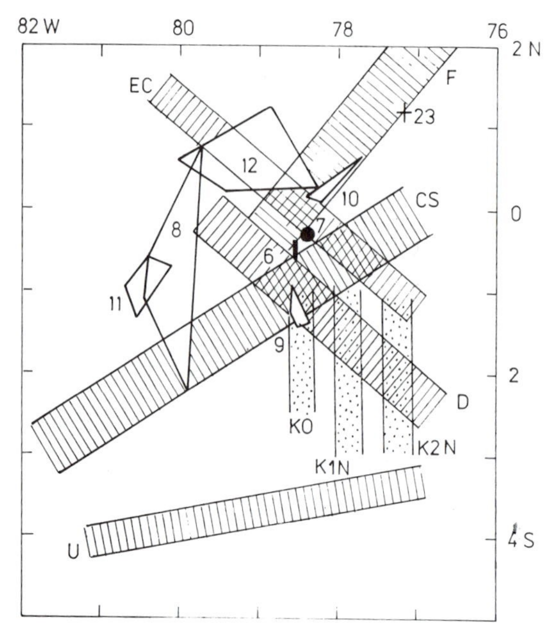

Figure 5. Position and shape of fields of maximum intensities for disastrous earthquakes observed in Ecuador from 1929 to 1958 (Nos. 6-12 in Table 1) in the pattern of the seismically active fracture zones (Figure 2); epicenter of No. 23 in southern Colombia is denoted by a cross.

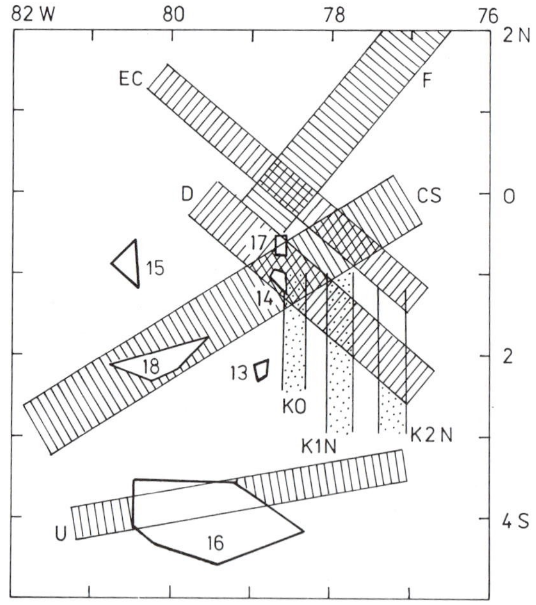

Figure 6. Position and shape of fields of maximum intensities for disastrous earthquakes observed in Ecuador from 1961 to 1980 (Nos. 13-18 in Table 1) in the pattern of the seismically active fracture zones (Figure 2).

It appears that in convergent plate margins with high seismicity a relatively short time sample of contemporary seismic activity is sufficient for defining the morphology of the Wadati-Benioff zone and for delineating seismically active fracture zones. However, historical earthquake observations are indispensable for an objective estimation of seismic hazard distribution because they enable us to determine the probability of occurrence of disastrous earthquakes and the relative degree of earthquake danger of individual fracture zones.

The historical data also reveal that in Ecuador the seismic hazard connected with active fracture zones in the continental plate is much higher than the hazard caused by earthquakes located in the subduction zone, in spite of the fact that two thirds of earthquakes, as recorded by the world network of seismic stations, occur in the subduction zone.

ACKNOWLEDGEMENTS

The present work was completed in the framework of the scientific cooperation programme between the Escuela Politécnica Nacional, Quito, and the Czechoslovak Academy of Sciences. We are indebted to P. Duque and M. Hall for support and stimulating discussions.

REFERENCES

Barazangi, M. and B.L. Isacks (1979). Subduction of the Nazca plate beneath Peru: evidence from spatial distribution of earthquakes, *Geophys. J. Roy. Astron. Soc.*, **57**, 537-555.

Egred, J., V. Caceres, and W. Costa (1984). Catálogo de intensidades de los sismos principales del Ecuador. Proyecto SISRA, Escuela Politécnica Nacional, Quito.

Hanuš, V., and J. Vaněk (1987a). Morphology of the Wadati-Benioff zone and volcanism in Ecuador and northern Peru. *Geofísica Internacional*, (in print).

Hanuš, V., and J. Vaněk (1987b). Deep seismically active fracture zones in Ecuador and northern Peru. *Studia Geoph. Geod.*, **3(1)**.

Martelly, J. (1952). Une méthode d'évaluation de la profondeur de l'hypocentre. Application au tremblement de terre de Pelileo (Ecuador, 1949). *Bureau Central Seism. Int.*, Trav. Sci. **A18**, 167-182.

Preliminary Determination of Epicenters (1981-83). *U. S. Geol. Surv., Nat. Earthq. Inform. Service*, Denver, CO.

Ramírez, J. E. (1975). Historia de los terremotos en Colombia. Instituto Geográfico Agustin Codazzi.

Regional Catalogue of Earthquakes (1964-80). *Int. Seism. Centre*, Edinburgh and Newbury.

Stauder, W. (1975). Subduction of the Nazca plate under Peru as evidenced by focal mechanisms and by seismicity, *J. Geophys. Res.*,**80**, 1053-1064.

VII. Filming and Processing of Historical Seismograms

Historical Seismogram Filming Project: Current Status

Dale P. Glover and Herbert Meyers
World Data Center-A for Solid Earth Geophysics
National Oceanic and Atmospheric Administration
Boulder, CO 80303, USA

ABSTRACT

The addition of important collections of seismograms from Hongo and Abuyama, Japan; Pasadena and Mt. Hamilton, California; St. Louis, Missouri; and Burlington, Vermont, has brought the number of records in the historical seismogram film archive to over a half-million microfilm copies. Seismograms from the Philippines, the USSR, and several locations in South America are now being added to the file. Station bulletins from 450 stations have also been filmed as part of this project. This report discusses the current status and future plans and presents an overview of the project since its beginning in 1979. By mutual agreement between World Data Center-A for Solid Earth Geophysics (National Geophysical Data Center) and U.S. Geological Survey (National Earthquake Information Service), the Survey will assume responsibility for the U.S. portion of the Historical Filming Project on October 1, 1985.

1. Introduction

Historical seismograms (pre-1963) are extremely important for determining seismicity, assessing earthquake risks and hazards, and conducting earthquake prediction programs and other studies where long recording periods are needed. Seismologists have long been concerned that the old seismograms were being lost or destroyed, deteriorating, or simply not generally accessible. Consequently, these concerns were brought to the attention of the General Assembly of the International Association of Seismology and Physics of the Earth's Interior (IASPEI) in 1977, and the following resolution was subsequently passed:

> "Noting that seismograms recorded at observatories around the world are basic for research on earthquakes and the structure of the Earth, and that many of the early seismograms have been lost in war, through natural hazards and deterioration and, therefore, it is essential that seismograms of significant earthquakes be systematically collected and preserved by making photographic copies at observatory sites, and be made available through the World Data Centres, IASPEI urges that seismological observatories around the world cooperate with a copying program by providing access to historical seismograms to be photographed on-site, and by preparing supporting observatory data to accompany the copies."

Following up this resolution, the Commission on Seismological Practice of IASPEI established a working group for copying and archiving historical seismograms during the 1981 joint IASPEI/UNESCO meeting of experts on historical seismograms in London, Ontario.

In 1982, a workshop on historical seismograms was convened in Tokyo, Japan, by the IASPEI/UNESCO Working Group on Historical Seismograms. The workshop emphasized collecting seismograms from Asia, Australia, and the Oceania area.

A second workshop was held by the IASPEI/UNESCO Working Group on Historical Seismograms at the International Union of Geodesy and Geophysics XVIII General Assembly at Hamburg, FRG, on August 18-19, 1983. The workshop was organized to discuss the status of historical seismological data from Latin America and Europe, to reaffirm prior resolutions, and to consider additional ones. The name of the working group was changed to IASPEI/UNESCO Working Group on Historical Seismograms and Earthquakes.

In response to these resolutions, World Data Center-A for Solid Earth Geophysics has been engaged since mid-1979 in collecting, microfilming, and archiving historical seismograms (dated prior to 1963) from selected stations and earthquakes worldwide, and has been encouraging and coordinating work done elsewhere. A group of stations was selected for which the entire chronological file of seismograms was filmed. Another group of stations was also chosen for filming the records of over 2,000 significant earthquakes (see Glover and Meyers, 1982).

The primary funding and support for the project has been provided by the U.S. Geological Survey (USGS) and NOAA's National Geophysical Data Center (NGDC). Supplemental funding has come from the U.S. Nuclear Regulatory Commission (NRC), the United Nations Education, Scientific, and Cultural Organization (UNESCO), and the Pan American Institute for Geography and History (PAIGH).

2. Current Status

The historical seismogram archive at WDC-A now contains over a half-million seismogram copies on 16- and 35-mm microfilm. An important collection of seismogram copies from the Hongo and Abuyama, Japan, observatories dating back to 1899 has recently been added to the archive. Seismograms from the stations at St. Louis, Missouri; Burlington, Vermont; and Mt. Hamilton (Lick) and Pasadena, California, have also been added during the past 2 years.

A camera was sent to the USSR about 2 years ago to film the records from nine stations. To date, seismograms from six of the stations have been received and more are expected in the near future. A camera has also been sent to the Helwan Observatory in Egypt and filming has started. Another camera was sent to the Regional Center for Seismology for South America (CERESIS) to film South American records. Copies of seismograms previously filmed by CERESIS have been received for some stations in Argentina, Bolivia, and Chile.

WDC-A is now filming the seismograms from the Baguio, Philippine Islands, station for the years 1952-62. Unfortunately, the records from earlier years were destroyed during World War II. WDC-A is also planning to film the historical seismograms from the Buffalo, New York (Canisius College) station.

Table 1 shows the present status of the historical seismogram film archive, and Figure 1 is a map showing locations of the seismograph stations that have furnished historical seismograms to WDC-A. A more detailed inventory is available from World Data Center-A for Solid Earth Geophysics (Glover *et al.*, 1985).

Table 1. Current Status of Historical Seismogram Filming Project

Code	Station name	Country	Years covered
AAA	Alma-Ata	USSR	1927-62*
ABU	Abuyama	Japan	1929-57
BAA	Buenos Aires	Argentina	1931-40*
BAG	Baguio City	Philippines	1952-62
BEC	Bermuda	Bermuda	1939-50*
BOZ	Bozeman, MT	United States	1931-60*
BUR	Burlington, VT	United States	1933-57
COL	College, AK	United States	1935-63
CPP	Copiapo	Chile	1940-48*
CSC	Columbia, SC	United States	1932-61*
DEN	Denver, CO	United States	1909-46*
HNG	Hongo	Japan	1899-1978
HON	Honolulu, HI	United States	1903-63
IRK	Irkutsk	USSR	1912-62*
LIN	Lincoln, NE	United States	1940-55*
LPA	LaPlata	Argentina	1935-59*
LPB	LaPaz	Bolivia	1913-59*
MHC	Mt. Hamilton, CA	United States	1911-62
PAS	Pasadena, CA	United States	1923-62
PUL	Pulkovo	USSR	1908-62*
RCD	Rapid City, SD	United States	1943-55*
SIM	Simferopol	USSR	1928-62*
SIT	Sitka, AK	United States	1904-61*
SJP	San Juan	Puerto Rico	1926-63
SLC	Salt Lake City, UT	United States	1939-55*
SLM	St. Louis, MO	United States	1933-43
STL	Santa Lucia	Chile	1950-59*
TAS	Tashkent	USSR	1912-62*
TIF	Tbilisi	USSR	1912-62*
TUC	Tucson, AZ	United States	1907-60
UKI	Ukiah, CA	United States	1931-55*
VQS	Vieques	Puerto Rico	1903-24

* Selected events only.

3. Historical Seismograph Station Bulletins

Saint Louis University, in conjunction with World Data Center-A for Solid Earth Geophysics, has compiled and microfilmed bulletins from about 450 seismograph stations as an adjunct to the seismogram filming project. Most of these bulletins were obtained from the collections held by St. Louis University and the National Geophysical Data Center (NGDC), collocated with WDC-A for Solid Earth Geophysics in Boulder, Colorado. This project is important because it provides additional data on seismograph stations and gives information on whether an earthquake was recorded at a particular station. The present collection of microfilmed bulletins at WDC-A, although not complete, represents a significant part of the seismological data available for the period 1900-65.

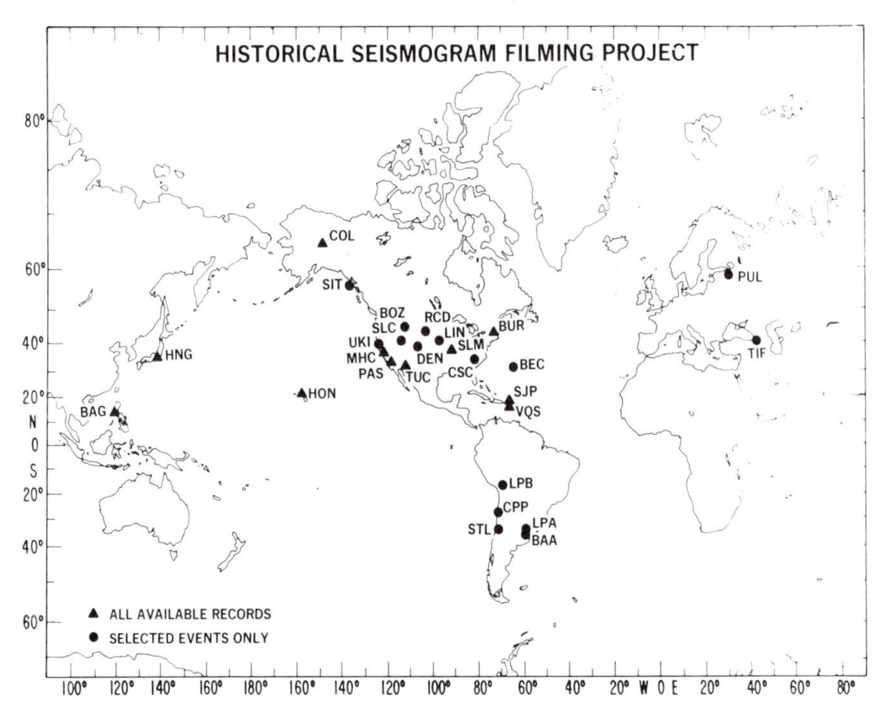

Figure 1. Seismograms available from World Data Center A for Solid Earth Geophysics.

4. Earthquake Catalogs

In 1976, a major effort was made to improve the collection of earthquake location data jointly held by WDC-A and NGDC. The file now contains over 500,000 event locations, consisting of earthquakes, explosions, rockbursts, and other Earth disturbances. The data were compiled from many catalogs contributed to WDC-A and also from the literature. Large historical earthquakes are included for earlier years (about 2100 B.C. to 1897 A.D.).

Another WDC-A/NGDC file contains about 140,000 earthquake intensity observations for the United States from 1638, which have been gathered from many sources. Each listing contains date, time, location, and Modified Mercalli intensity of the earthquake.

WDC-A and NGDC have also published a five-color map depicting the location and relative importance of 1,277 significant earthquakes from 1900 to 1979. This map was produced with support from the Office of U.S. Foreign Disaster Assistance, Agency for International Development.

An important catalog of significant earthquakes was published by WDC-A/NGDC in 1981. This volume identifies nearly 2,500 worldwide earthquakes since 2000 B.C., and includes quantitative estimates of damages and casualties.

World Data Center-A for Solid Earth Geophysics and the National Geophysical Data Center have also translated into English a very important study of earthquakes in the U.S.S.R. Methodologies are developed in the text to relate historical felt information to location, origin, date, time, intensity, and magnitude of about 6,000 earthquakes in the U.S.S.R. from 2100 B.C. through 1977 A.D. (Kondorskaya *et al.*, 1982).

5. Future Plans

It is expected that seismogram copies will be received in the near future from the Helwan Observatory, Egypt, and from CERESIS for a number of South American stations. World Data Center-B in Moscow will probably complete the filming of the U.S.S.R. stations within the next few months. WDC-A should also complete filming the Baguio records in the near future.

Plans are being formulated to prepare an inventory of historical seismograms that will not be filmed as part of this project. Funding restraints make it impossible to film seismograms from many important stations. A list of these seismograms and their locations can be an extremely important source of information. We, therefore, urge those who have knowledge of seismogram collections to send the information to WDC-A.

WDC-A will be working with the IASPEI/UNESCO Working Group on projects relating to global and regional earthquake catalogs. The work will involve encouraging additional nations to develop historical event catalogs for their region, recommending standards, and improving quality control.

6. Recommended Filming Techniques (from Glover and Meyers, 1982)

6.1. Seismogram Preparation

Seismograms should be inspected and repaired if necessary, before filming. If the records are damaged, they can be mended using transparent tape on the back side. Rolled or folded records can be laid out and pressed flat under heavy weights, and those on photographic paper can be rewetted and pressed between blotters to dry.

A monthly log that indicates which records are present and missing should be compiled for filming with the seismograms. Information that was written on the records at the time of recording (such as instrument type, component, on-off times, and polarity) should be recopied on the seismograms where necessary for readability. Additional technical information about the instruments (such as pendulum period, damping ratio, and magnification) and any other useful data (e.g., a brief station history) should be compiled if available and filmed or spliced on each roll of film. In addition, voluminous data such as station logs, time determinations, and earthquake reports if available can be filmed separately.

6.2. Cameras Used

Historical seismograms have been filmed at the WDC-A/NGDC facility using two cameras. A Micro Instrument Corporation 35-mm rotary camera was used to film station-years of seismograms. This camera was chosen for its ease of operation and its capability of filming many seismograms in a short time. The camera can film records as large as 24 in. (61 cm.) in width and of unlimited length at speeds up to 40 ft. (12.2 m) per minute. Reduction ratios are 16:1 and 8:1, depending on the size of the original records.

For filming selected events, a 16-mm Itek planetary camera that was built to specifications and specially modified for portability was used. The camera can film records as large as 14 by 40 in. (36 by 102 cm). Direct and transmitted light sources are provided, which permits a variety of records to be filmed. Most smoked-paper records in particular require both transmitted and direct lighting to give a good film image. The Itek camera simultaneously produces two 16-mm rolls of film at a reduction ratio of 32:1.

Portable "flow-through" or rotary 16-mm microfilm cameras are well suited for filming photographic, pen-and-ink, and heated-stylus type recordings if the width of the camera throat is sufficient to allow the seismogram to travel through the camera without being caught and torn. A camera of this type (a Bell & Howell Inter/Com Microfilm Recorder, Model 522A, modified for portability) was provided to California Institute of Technology (CIT) to film the Pasadena records.

Smoked-paper seismograms are the most difficult kind of records to film; many of these records are so fragile that they can be copied only with a planetary-type camera. WDC-A has successfully used a specially made light table with excellent results. Tests indicate, however, that by using white paper to back the smoked-paper seismograms, good results can sometimes be obtained with direct light. If the records are sturdy, even the small flow-through cameras can sometimes be used with the white backing.

6.3. Filming Procedures

Seismograms should be filmed in chronological order from the earliest date available. Information such as the monthly log, instrumental characteristics, and station history should be filmed at the beginning of each roll of film.

Seismograms from stations that have operated two or more sets of instruments should normally be filmed together; i.e., all records for 1 day regardless of type of record. The date, on-off times, instrument type and component, polarity, and magnification, if known, should be plainly indicated on each seismogram. Seismograms from remote sites may be filmed with those of the main operating observatory.

In filming seismograms from older stations, strict rules cannot be applied because the records from each station present their own unique problems. Ease of use, therefore, must be the primary guideline.

REFERENCES

Charlier, Ch. and J. M. Van Gils (1953). Liste des stations seismologiques mondiales, *Observatorie Royal de Belgique a Uccle*, Brussels, unpaginated.

Ganse, R. A. and J. B. Nelson (1981). Catalog of significant earthquakes, 2000 B.C. - 1979, *Report SE-27, World Data Center-A for Solid Earth Geophysics*, Boulder, CO, 154 pp.

Glover, D. P. (1977). Catalog of seismogram archives, Key to Geophysical Records Documentation No. 9, *National Geophysical Data Center*, Boulder, CO, 51 pp.

Glover, D. P. (1977). Directory of seismograph stations, *Report SE-7, World Data Center-A for Solid Earth Geophysics*, Boulder, CO, 143 pp.

Glover, D. P. (1980). Historical Seismogram Filming Project: second progress report, *Report SE-24, World Data Center-A for Solid Earth Geophysics*, Boulder, CO, 63 pp.

Glover, D. P. and H. Meyers (1981). Historical Seismogram Filming Project: third progress report, *Report SE-28, World Data Center-A for Solid Earth Geophysics*, Boulder, CO, 76 pp.

Glover, D. P. and H. Meyers (1982). Historical Seismogram Filming Project: fourth progress report, *Report SE-33, World Data Center-A for Solid Earth Geophysics*, Boulder, CO, 54 pp.

Glover, D. P., H. Meyers, R. B. Herrmann, and M. Whittington (1985). Inventory of filmed historical seismograms and station bulletins at World Data Center-A, *Report SE-37, World Data Center-A for Solid Earth Geophysics*, Boulder, CO, p. 74-220.

Gutenberg, B. (1956). Great earthquakes, 1896-1903, *Trans. Am. Geophy. Union*, **37**, 608-614.

Herrmann, R. B., M. Whittington, and H. Meyers (1983). Historical Station Bulletin Microfilming Project: preliminary inventory, *National Geophysical Data Center, NOAA*, Boulder, CO, 59 pp.

Kondorskaya, N. V., N.V. Shebalin, Ye. A. Khrometskaya, and A. D. Gvishiani (1982). New catalog of strong earthquakes in the U.S.S.R. from ancient times through 1977, *Report SE-31, World Data Center-A for Solid Earth Geophysics*, Boulder, CO, 608 pp.

McComb, H. E. and C. J. West (1931). List of seismological stations of the world, 2nd ed., *Bull. National Res. Council*, 119 pp.

Meyers, H. and W. H. K.Lee (1979). Historical Seismogram Filming Project: first progress report, *Report SE-22, World Data Center-A for Solid Earth Geophysics*, Boulder, CO, 68 pp.

Poppe, B. B. (1979). Historical survey of U.S. seismograph stations, *U.S. Geol. Surv. Prof. Paper No. 1096*, 389 pp.

Poppe, B. B., D. A. Naab, and J.S. Derr (1978). Seismograph station codes and characteristics, *U.S. Geological Survey Circular 791*, 171 pp.

Filming Seismograms and Related Materials at the California Institute of Technology

J. R. Goodstein
Institute Archives
California Institute of Technology
Pasadena, CA 91125, U.S.A.

P. Roberts
Kresge Seismological Laboratory
California Institute of Technology
Pasadena, CA 91125, U.S.A.

ABSTRACT

As part of a world-wide effort to create an international earthquake data bank, the seismology archive of the California Institute of Technology (Caltech) has been organized, labeled, described, and microfilmed. It includes a wide variety of original records, documents, and printed materials relating to local and distant earthquakes.

The single largest and most complex component of the task has been the preparation and microfilming of Caltech's vast collection of original seismograms. The original proposal envisioned a modest project in which a selected number of seismographic records at Caltech could be made more generally available to the scientific community. These single-copy records are stored at Kresge Laboratory and comprise thousands of individual photographic sheets – each 30 × 92 cm. In the end, we microfilmed both the Pasadena station records and those written at the six original stations in the Caltech network.

This task got underway in June 1981 and was completed in January 1985. In the course of the project, the staff sorted, arranged, inventoried, copied, and refiled more than 276,000 records written between January 10, 1923 and December 31, 1962. The microfilm edition of the earthquakes registered at the Seismological Laboratory at Pasadena and at auxiliary stations at Mt. Wilson, Riverside, Santa Barbara, La Jolla, and Tinemaha and Haiwee (in the Owens Valley) consists of 461 reels of film. The film archive is cataloged and available to researchers in Caltech's Millikan Library in Pasadena, at the U. S. Geological Survey in Menlo Park, CA and at the World Data Center A/NOAA in Boulder, CO.

1. Origins of the Seismology Archive

Seismology at Caltech arose from an "arranged marriage" between two different traditions – a European interest in global earthquakes and an American concern for local earthquakes (Goodstein, 1984).

German-born and Göttingen-trained, Beno Gutenberg brought to Caltech the European tradition of viewing seismology as a research tool. Rigorously trained in physics and mathematics, he used earthquake records to investigate the physical properties and structure of the earth's interior. Earthquake instruments installed in seismological stations around the world provided the data for his analysis. For Gutenberg, the entire globe was a scientific laboratory.

But Gutenberg's American colleagues, Harry Oscar Wood included, took a much more pragmatic view of the world. Few in number, and concentrated in California, the American seismological community saw their research in terms of finding a solution to the "California problem," as they called regional earthquakes. To a seismologist like Wood, Gutenberg's global problem shrank to the size of southern California (Figure 1).

In bringing these two men together in Pasadena in 1930, Robert A. Millikan, the head of Caltech, set in motion a chain of events that was ultimately to lead Charles Richter to develop the first earthquake magnitude scale.

Charles Richter took his Ph.D. in theoretical physics at Caltech. He became Wood's assistant in 1927, three years before Gutenberg's arrival. He spent the next several years measuring and filing seismograms. Much of the work was tedious and mechanical, and aside from Wood, Richter's contacts with seismologists remained limited. A theoretical physicist who fell into seismology more or less accidentally, Richter desperately needed a scientific mentor. Gutenberg fit the bill; indeed, if Gutenberg has not come to Richter, Richter would have gone to Gutenberg. As things turned out, they spent 30 years under the same academic roof.

American interest in seismology got its biggest boost from the 1906 San Francisco earthquake. At that time, no American university boasted a department of

Figure 1. Scientists taking part in an earthquake conference in Pasadena in 1929 pose on the steps of the Seismological Laboratory. Front row (left to right): Archie P. King; L. H. Adams; Hugo Benioff; Beno Gutenberg; Harold Jeffreys; Charles R. Richter; Arthur L. Day; Harry O. Wood; Ralph Arnold; John P. Buwalda. Top row: Alden C. Waite; Perry Byerly; Harry F. Reid; John A. Anderson; Father J. B. Macelwane.

seismology, let alone any professional seismologists. The quake triggered, among other things, the birth of the Seismological Society of America, an organization dedicated to stimulating interest in geophysical matters in general, and earthquake problems in particular. Meanwhile, at the state level, a commission under the direction of Andrew C. Lawson, a Berkeley geologist, investigated the temblor itself. The Carnegie Institution of Washington, a private philanthropic foundation, supplied the necessary funds when the Sacramento legislators, under pressure from the business community, refused to do so.

Lawson tapped Wood, who was then an instructor in the University of California, Berkeley, Geology Department, to study in detail the extent and nature of the earthquake damage within the city itself. Wood went into the exercise a field geologist and came out a seismologist.

It was Wood who brought seismology to southern California. His campaign began in 1916, with the publication of two papers. Stressing the importance of taking a regional approach to the study of local earthquakes, he suggested that the plan be tested on a modest scale in southern California.

Wood singled out southern California for two reasons: because the region had no recording instruments and because he expected the next large earthquake to occur there. The 1857 earthquake along the San Andreas fault had been the last great shock in southern California.

His research program also stressed the need for a new generation of instruments: there could be no hope of measuring short-period local earthquakes with instruments designed to measure long-period distant earthquakes. Finally, he emphasized the importance of field work, in particular, to locate weak shocks. Like most of his contemporaries, Wood was convinced that if geologists could identify the active faults associated with weak shocks, they could then "deduce ... the places where strong shocks are to originate, considerably in advance of their advent" (Wood, 1916). Big shocks, in other words, follow weak shocks. In time, he believed it would be possible to make, in his words, a "generalized prediction" of when and where to expect the next big quake.

The compelling reason for setting up branch seismological stations was therefore to detect and register the weak shocks systematically. But the routine registration of hundreds and hundreds of small California earthquakes did not confirm Wood's hypothesis. Nevertheless, the lure of that elusive idea drives seismological research to this day.

Like Lawson before him, Wood ultimately found a patron in the Carnegie Institution. Southern California's first seismological program began operation in Pasadena in June 1921, under Wood's direction. For the next six years, Wood ran the project from an office at the Mt. Wilson Observatory, located a short distance from the Caltech campus.

To succeed, the project needed the right instrument for recording nearby earthquakes. Fortune favored Wood in the person of John Anderson, one of Mt. Wilson's ablest astronomers. As part of the school's defense effort during World War I, Anderson had worked on submarine detection instruments sensitive enough to record short vibrations. Anderson's war-honed skills matched Wood's peace-time needs. The Wood-Anderson collaboration began immediately after Wood had settled into his office.

To do what Wood wanted it to do, the instrument had to be sensitive enough to record shocks having a period varying from 0.5 to 2.0 seconds. Seismometers designed for recording distant earthquakes typically have longer period response. In the case of Berkeley's station, the instruments in use had periods of 15 and 6 seconds respectively. In the early 1920's, instruments on the Atlantic seaboard could measure the time and place of California shocks with greater precision than could comparable instruments located in California.

In fall 1922, after several false starts, Wood and Anderson had designed a reliable, compact, portable instrument that, when placed vertically, consistently recorded the east-west and north-south components of the earth's motion during an earthquake. In practice, the Wood-Anderson torsion seismometer was an ideal instrument for recording the earth's horizontal movements over a short distance during an earthquake; it proved less successful for recording the earth's up-and-down motions. Shortly before Gutenberg arrived to take up his duties as professor of geophysics at Caltech in 1930, Hugo Benioff, Wood's assistant, designed and built a vertical seismometer to meet Wood's needs. Routine recording of local shocks using Benioff's instrument began in 1931, by which time Wood was predicting the new vertical-component seismometer would surpass any existing vertical then in use for the registration of distant earthquakes as well. Both the Wood-Anderson and Benioff instruments have since become standard equipment in seismic stations around the world.

The first Wood-Anderson instrumental records were written in December 1922; the first extant records date from mid-January, 1923. Ironically, the Wood-Anderson torsion seismometer did more than its creators intended. Wood had wanted a short-period instrument to register local earthquakes. But when the instrument was put to the test in 1923, he discovered that it also registered the first phases of distant earthquakes. Wood had unwittingly altered the course of his own program.

By the spring of 1924, the experimental torsion seismometers installed in the basement of the Observatory office and the physics building on the Caltech campus had recorded dozens of earthquakes, near and far, including the initial short-period phases of the devastating Japanese earthquake of September 1, 1923. The fact that Wood had recorded this event on an instrument designed to register local earthquakes was not lost on seismologists elsewhere. When Gutenberg heard the news in Germany, from a colleague who had attended an international gathering of geophysicists in Madrid, Spain, he held up the publication of his book on the fundamentals of seismology long enough to insert a diagram of the apparatus. By the end of the 1920's, 13 cities in the U.S., and one overseas, boasted Wood-Anderson instruments.

In 1925, Caltech started a geology program. The following year, Millikan formally invited the Carnegie Institution to conduct its earthquake research in the Institute's new Seismology Laboratory, located in the foothills of the San Rafael Mountains, a short drive from the campus. In January 1927, Wood left his temporary quarters at the observatory office and moved into the new building. The time to go earthquake hunting in earnest had begun. By 1929, six outlying stations, all within a 300 mile (480 km) radius of the central station in Pasadena were in place and working. Each station was equipped with a pair of horizontal component torsion instruments and recording drums, as well as with radio-timing equipment. Records were sent weekly to Pasadena for photographic processing, registration and interpretation.

2. Seismology Records in Microform

Caltech's seismology archive includes a wide variety of original records, documents, and printed materials relating to local and distant earthquakes. In 1979, the Institute Archives prepared, labeled, described, and filmed a group of published and unpublished items including the *Bulletin of the CIT Seismological Laboratory* (Pasadena and Auxiliary Stations), 1931-1968, various station clock corrections, Beno Gutenberg's annotated copy of the *International Seismological Summary*, 1918-1942, and the original Gutenberg-Richter worksheets for *Seismicity of the Earth* (1954). The notepads, more than 100 in all, include calculations and data (Figure 2) relating to the magnitude scales used by the two men in their catalog (Goodstein *et al.*, 1980).

Since then, we have concentrated on filming the original documentation on earthquakes registered at the Seismological Laboratory at Pasadena and at auxiliary stations at Mt. Wilson, Riverside, Santa Barbara, La Jolla, and Tinemaha and Haiwee (in the Owens Valley). In 1981, we completed the microfilm publication of the phase cards compiled at the laboratory and at the auxiliary stations belonging to the southern California network of seismological stations. There are 133 rolls of film, which cover the data from April 15, 1927–December 31, 1969. We also prepared a microfilm index of the collection.

In addition to the phase cards, the contents of five loose-leaf binders that were located in the laboratory's measuring room were also filmed. This material is contained on a separate roll of film marked, "Richter Notebooks: Local Shocks; Long Beach." Binder B is concerned exclusively with the Long Beach shock of March 10, 1933, and contains graphs and tabulations of readings from all stations recording the earthquake and its aftershocks, March 1933–June 1936. The other binders contain material relating to instruments and stations in the 1930's, tabulation of local shocks between October 1926 and December 1930, contemporary accounts of local and distant shocks between 1933 and 1935, and miscellaneous tables, news reports, and geological notes.

3. Some General Information about the Phase Cards

The microfilm edition of the phase cards mirrors the arrangement of the original cards in the shock file. Every card was filmed. Remarks, diagrams, calculations, and other information noted on the reverse of the card was also filmed.

The arrangement of the shock file, composed of guide and file cards, was established by C. F. Richter in 1929, and while changes in the earthquake measuring routine have occurred over the years, the phase card layout remains largely intact. The cards themselves are filed in chronological order within each drawer. Each shock is represented by a primary guide card, followed by a series of color-coded file cards. The primary guide card has a center, right, or left tab. Center tab cards contain information about local shocks; right-hand tab cards about teleseismic shocks. For a time, left-hand tabs indicated uncertainty as to whether the shock was teleseismic or local; this practice has been discontinued. By 1955, the left-hand tabs served as station markers, showing the point to which measurements for the auxiliary stations were completed. In addition, the primary guide card includes information pertaining to the character of the shock, epicenter, time (Pacific Standard Time and Greenwich Civil Time (GCT)), date, and recordings at other stations. Since January 1, 1951, only GCT is used.

1907 June 25 17h
81 1.4N 124.8E
M 1 N 126 E

Gött PP 104° 10 3 .7 7.3 8.0
Upsa Max 33° 40 4 1.0 (6.55) ≤7.5
Up pP' 97 30 8.6 7½ 7¾
Jena PP 104 12 6.7 7.3 8.0 7.7
Apia T 30 5 1.0 6.7 7.7 7.6
S 62° 20 10 .4 6.7 7.1 6.7

76

1.4N 124.8E
Szirtes

			P	S-P	PO (200)	O	54:40 (P-O)	Δ 200	δ	SKS
Manila	13.7	340	57:55	5:46	7:16	—	3:15	14.2	.5	
Bata	20.0	250	59:18	3:37	4:30	54:38	4:38	21.6	1.6	
Zikawei	30.0	340	00:19	2:30	3:14	54:05	5:39	28.2	−1.8	
Perth	37.9		00.0	6.5	8:13	—	5:20	26.0	—	
Calc	41.2		3.0	7.1	8:58	—	8.20	47.6	—	
Kodai	47.6		2.5				7.50	43.6	—	
Bomb	53.8		1.7				7:02	37.8	—	
Irk	54.8		1.6	1.6; 8.8	10:58	—	6:56	37.2	—	
Tash	63.8	300	4.4	9.6	11:47		9:44	59.3	−4.5	
Apia	64.6	100	4:40				10:00	61.6	−3.0	
Baku	77.6	320	6:14	10:02	12:12	54:02	11:34	76.8	−1.2	
Tiflis	81.6	320	6:32	9:45	11:56	54:36	11:52	80.2	−1.4	
Borzom	82.7		7:04	10:11	12:21	54:53	12:24	86.4	—	
Wien	101.7		8:44				14:04			
Graz	102.0		8:39	10:14	12:24		13:59	107.6		24:17 80
Zagreb	102.0		8:31	10:27	12:35		13:51	105.7		24:22 50
Jena	102.2	330	8:27				13:47	104.8	1.6	24:20 70
Triest	103.8		8:13				14:03			
Hamb	105.6		9.0				14:20			24:22 130
Kew	109.8		9.5				14:50			24:38 150
Gött	104.0									24:24 80
Jurjew	93.0									23:29 80
Pots	101									24:23 90
Strasb	106.									24:38 90
Ootaka			(14:00)							
Ups	97	340	8:15				13:35	102		

2, 0, −2, 90, 180, 270, pP, >29

#121a 17:54.6
1 N 127E M = 7½ ±
h = 200 ±
CCC

00643

Figure 2. One of the original Gutenberg and Richter worksheets used in figuring the magnitude of a large 1907 event for their textbook, *Seismicity of the Earth and Associated Phenomena*, 2nd Ed., Princeton University Press, Princeton, NJ, 1954.

Each station that registered the shock has a file card indicating the date and character of the shock. Each card contains the measurements of one instrumental record of one earthquake. The original file cards are colored buff, white, and blue, the color of the card indicating the direction of the particular instrument recording each shock. Although these colors do not show up on the black-and-white microfilm, the components of the direction are indicated on the card by a pair of letters: N-S, S-N, E-W, W-E, U-D, D-U (north-south...down-up).

To assist researchers in using the microfilmed phase cards, each roll includes a standard introductory section that contains the following information: site information, site instrumentation, a brief description of the records microfilmed, a station chronology, and a time-on/time-off index. The actual card index for the stations in the network was filmed alone. Richter's basic set of instructions governing the shock file and subsequent procedure manuals is also available.

After we had finished this project, we found, in the attic of the central station, records pertaining to the measurement of local and distant earthquakes for the period January 17, 1923–April 24, 1927. This group of records is now microfilmed and is available separately on four rolls of film.

4. Procedures for Preparing and Microfilming Seismograms

In size and complexity alone, the project to copy Caltech's archive of original seismograms surpasses any of the projects already described. These records, housed at the Kresge Seismological Laboratory, comprise close to 500,000 individual photographic sheets – each 30×92 cm. Seismograms are the principal source of information about earthquakes and earth's interior, but these records are also important because they span so much of the period for which instrumental data exist. We microfilmed both the Pasadena station records, as well as those written at the six outlying stations in the Caltech network. The project got underway in June 1981 and was completed in January 1985. During that time, we sorted, arranged, labeled, inventoried, copied, and refiled more than 276,000 records written between January 10, 1923 and December 31, 1962.

At the beginning of every roll of film, we inserted general information about the station site, station instrumentation, a description of the records, a historical chronology of the stations, and monthly seismogram inventory sheets. The inventory sheets indicate records filmed, missing, incomplete, and unreadable.

5. Guidelines and Experiences

Preparing the records for microfilming was by far the most time-consuming part of the project. The seismograms are filed chronologically in boxes and stored on shelves; seismograms for all stations are filed together by date. Because the microfilming was done by station, the seismograms had to be re-sorted. To prevent confusion, each station's records were kept in boxes labeled by date and station.

After the records had been organized by station, they were put in chronological order. Each day's seismograms were arranged by component direction. The sequence for "outside" stations generally is: NS, EW, Z (vertical component of the motion of the earth), and T (time) record, depending on the equipment at the stations. For example, in August 1932, HAI (Haiwee) and TIN (Tinemaha) have NS, EW, Z, T; but all the other stations have NS, EW, T. On any day that one or more of these components are missing, this information is noted on the inventory sheet, and those components remaining are arranged as closely as possible to the above sequence.

For the PAS (Pasadena) station, the sequence is different because there are more recording instruments. The response and direction of some of the instruments also changes occasionally. The daily sequence for the PAS seismograms includes records from the longest continuously recording seismograph first, and those running for shorter periods last. The different instruments in use are also listed at the top of every inventory sheet. As instruments are changed, the "older" records are deleted from the sequence, the remaining records are "moved up," and the "new" records are added to the inventory sheet.

Pasadena's seismograms are identified by letter, Roman numerals, or Arabic numbers on the back, along with date and component directions. The letter "G", for example, refers to an instrument. Originally, the instruments were identified by letters; as new instruments were installed, they were assigned Roman numerals; in more recent times, all the instruments were given Arabic numbers. Some instrument responses and types are also indicated on the back of the seismogram. What is written on the reverse side is, indeed, the principal source (the station information cards have been microfilmed separately) for the information written on the labels affixed to the front of the seismogram. If the instrument responses and types are not written on the back of the record, we have used the instrument number to determine what they are. On some records, this information is given on the first seismogram for the month, but not repeated for the rest of the month. The response and instrument type noted on the first record are repeated on subsequent records until changes are noted on the back of the seismogram. In effect, the monthly inventory sheets summarize the recording history of each instrument. Table 1 summarizes the instrumentation at each station.

Each seismogram has information written on the back. Since only the front of each record was filmed, the information on the back had to be transferred to the labels on the front. The labeling process involved circling the component directions; writing in the date, including the month, day-on/day-off, and year; and indicating the instrument responses. Occasionally, information on the reverse side conflicts with the standard printed notation on the label. Where the component directions were reversed, S-N rather than the standard N-S, for example, the printed letters were crossed out and the corrections written in. We have not filled in the space provided for time corrections, as this information is already available on microfiche. All labels reflect only the information originally written by Richter and others on the back of the record.

The labels are usually placed in the margin (where the paper overlaps when wrapped around the seismograph drum) at the left side, in the lower corner. If the paper was torn, curled, or handwritten information appeared in this area, the label was placed nearby.

The earliest records, in particular those for the years 1923-1927, were made with what were essentially experimental instruments. Not only were there many interruptions in the operation of the instruments, but there also were many changes and adjustments. Wood's own scientific correspondence in the 1920's is worth noting in this connection. To the seismologist who wrote and asked for records of the earthquake in China on May 22, 1927, Wood replied that he had no useful ones. He explained why in some detail:

Table 1. Historical Seismograph Stations in Southern California

Station	Latitude °N	Longitude °W	Seismometer *	Characteristics	Time of operation
Haiwee	36.1367	117.9467	Wood-Anderson NS, EW	Torsion/T_0=0.8	1929-1954
			Benioff Z	T_0=1, T_g=0.2	1929-1954
La Jolla	32.8633	117.2533	Wood-Anderson NS, EW	Torsion/T_0=0.8	1927-1952
			Benioff Z	T_0=1, T_g=0.2	1933-1952
Mt. Wilson	34.2238	118.0577	Wood-Anderson NS, EW	Torsion/T_0=0.8	1928-1951
			Benioff Z	T_0=1, T_g=0.2	1932-1956
Pasadena	34.1483	118.1717	Wood-Anderson NS, EW	Torsion/T_0=0.8	1927-
			Wood-Anderson NS	Torsion/T_0=6	1927-1932; 1941-1954
			Wood-Anderson EW	Torsion/T_0=6	1927-1954
			Benioff Z	T_0=1, T_g=0.2	1934-
			Benioff EW	T_0=1, T_g=0.2	1937-
			Benioff NS	T_0=1, T_g=0.2	1938-
			Benioff NS, EW, Z	T_0=1, T_g=90	1940-
			Linear strain NS	Strain/T_g=70	1938-1959
			Linear strain EW	Strain/T_g=70	1947-1958
Riverside	33.9933	117.3750	Wood-Anderson NS, EW	Torsion/T_0=0.8	1926-
			Benioff Z	T_0=1, T_g=0.2	1933-
			Benioff Z	T_0=1, T_g=90	1954-
			Benioff NS, EW	T_0=1, T_g=90	1955-
Santa Barbara	34.4417	119.7133	Wood-Anderson NS, EW	Torsion/T_0=0.8	1927-1952; 1964-
			Benioff Z	T_0=1, T_g=0.2	1933-1952
Tinemaha	37.0550	118.2283	Wood-Anderson NS, EW	Torsion/T_0=0.8	1929-
			Benioff Z	T_0=1, T_g=0.2	1929-
			Benioff Z	T_0=1, T_g=90	1951-
			Benioff NS	T_0=1, T_g=90	1951-
			Benioff EW	T_0=1, T_g=90	1954-

* Photographic paper. Does not include short term and experimental instruments.

> We have not yet any reasonably good time at the head station or at any of the outlying stations except the experimental pier at the Mt. Wilson Observatory office. Unfortunately the record ... [there] is defective for the day in question.
>
> There is not much use in determining constants until after the instruments have had a little opportunity to settle down ... to their environment. ... I have five graphs of the shock, two written with short-period local earthquake instruments operating at Riverside, two written with short-period local earthquake instruments operating at the head station, and the defective record written by instrument G. ... There are no time marks whatever on that, which is due to a change ... in the timing circuit. ... There is no value whatever to the time marks on any of the other four records. They are placed on the records only to indicate the character of the running, and the constants are only crudely approximate (Wood, 1927a).

Indeed, in marking the records in pencil on the back in the early years, Wood noted only the station, the date, and the letters E-W or N-S. He explained the meaning of the letters in this way:

> If you transpose them to the face of the record so that E on the face is at the same edge as E on the back, and so on, then a shift of the line of the seismogram towards the edge marked E means a shift of the earth towards the east (Wood, 1927b).

An inventory of the records is essential. The inventory form indicates records present, missing, unreadable, or incomplete. Unreadable seismograms are those that are either blank or black, or those that have faint lines or are fogged so badly that the lines are barely visible. Incomplete seismograms have lines that are readable, but have conspicuously fewer lines than "usual."

Occasionally, two sheets of paper were used per day per instrument component. An effort was made to determine the order of the sheets; they are then labeled sheet 1 and sheet 2, and two crosses are written on the inventory sheet for that day.

Seismographic records are microfilmed in chronological order with each station's records filmed on separate rolls of film. On average, 600 seismograms can be filmed on a 100 ft (30.5 m) roll. The deciding factor as to the actual number of seismograms filmed is the number of records for a full month that can fit on a roll. Each roll, in other words, starts at the beginning of a month and stops at the end of a month.

Because seismograms vary in darkness and quality of line and background, exposure corrections are made during the filming of each seismic record to compensate for these variations. A resolution chart with scales was also filmed, along with the roll number, station name, and starting and ending dates. Unreadable seismograms were filmed for completeness.

Afterward, an index and general appendix were prepared and filmed as a separate roll of film. This roll contains procedures and information on the project, seismogram recording and station technical data, notes made during the project, and the complete inventory of all seismograms filmed.

ACKNOWLEDGEMENTS

We thank John Lower, Erwin Morkisch, Graham McLaren, James Host, Richard Wood, and Yoram Meroz for providing valuable technical assistance. Funds for the filming of Caltech's seismograms was provided by the U. S. Geological Survey and the National Geophysical Data Center, NOAA. The microfiche publication project, the subsequent phase card project, and microfilming the seismograms of the Kresge Seismological Laboratory were initiated by Willie Lee and supported by Clarence Allen, Hiroo Kanamori, and Bob Yerkes.

REFERENCES

Goodstein, J. R. (1984). Waves in the earth: Seismology comes to southern California, *Hist. Studies Phys. Sci.*, **14**, 201-230.

Goodstein, J. R., H. Kanamori, and W. H. K. Lee, (eds.), (1980). Seismology microfiche publications from the Caltech archives, *Bull. Seism. Soc. Am.*, **70**, 657-658.

Wood, H. O. (1916). The earthquake problem in the western United States, *Bull. Seism. Soc. Am.*, **6**, 197-217, on p. 208.

Wood, H. O. (1927a). letter, Wood to William C. Repetti, 27 June 1927, Harry O. Wood Papers, Caltech Archives.

Wood, H. O. (1927b). letter, Wood to Repetti, 7 July 1972, Harry O. Wood Papers, Caltech Archives.

Digitization and Processing of the J.M.A. Strong Motion Records in the Period of 2 to 20 Sec from Nine Great Earthquakes

R. Inoue and T. Matsumoto
Faculty of Engineering, Ibaraki University
Ibaraki-Ken 316, Japan

ABSTRACT

17 seismograms (including 29 horizontal components) from 9 great earthquakes (M_S = 7.6 to 8.5) around the Japan arcs recorded by the J.M.A. (or C.M.O.) low-magnification seismograph network within an epicentral distance of several hundred kilometers were digitized and analysed.

Because the standard strong motion accelerograph (SMAC etc.) cannot accurately record the ground motions whose periods are longer than about 4 sec, this data base is very useful for seismic design of large-scale structures with natural periods around 10 sec.

The digitized data were corrected for instrument characteristics, and the high and low frequency noise in the data was removed by applying a bandpass filter (the upper cutoff period is 20 sec, and the lower is 2 sec).

Time history of corrected ground motions were obtained. Maximum ground displacement, velocity and acceleration are about 23 cm, 12 kine and 28 gal, respectively.

Velocity response spectra, Sv, for the period range of 2 to 20 sec with damping factors, h, of 0.1% and 2% were computed (h = 0.1% corresponds to oil storage tanks and 2% to steel structures such as suspension bridges). Here we denote $(Sv)_P$ as the peak value of Sv for each component. The results can be described as follows: (1) the maximum value of $(Sv)_P$ with h = 0.1% is about 190 kine, and there are 7 components whose $(Sv)_P$ with h = 0.1% are larger than 110 kine (the value 110 kine is the upper limit for Sv of the seismic design regulations for oil tanks in Japan); (2) the maximum value of $(Sv)_P$ with h = 2% is about 70 kine, and there are 16 components whose $(Sv)_P$ with h = 2% are larger than 40 kine (the value 40 kine is the upper limit for Sv of the seismic design specifications for suspension bridges in Japan). These results suggest that those seismic design specifications for large structures should be reexamined.

1. Introduction

In view of the recent increase of large-scale structures such as large oil tanks and long span suspension bridges whose natural periods are around 10 sec, the knowledge of the characteristics of ground motions in those period range from great earthquakes is becoming increasingly important for seismic design of such structures. For this purpose, three years ago we digitized the J.M.A. strong motion records (so-called for their low magnification) from the 1968 Tokachi-Oki event (17 seismograms, including 50 components) at the Public Work Research Institute (JSCE's Tank Committee, 1982). Following this work, we recently digitized and analysed the 17 seismograms (including 29 components which are all horizontal) from 9 great earthquakes around the Japan arcs recorded by the J.M.A. (or C.M.O.) low-magnification seismograph network. J.M.A. means the Japan Meteorological Agency and C.M.O. means the Central Meteorological Observatory. C.M.O. is the predecessor of J.M.A.

In the following sections we describe in some detail the digitization and correction procedures of the records and then compare the response spectra of the records with the Japanese seismic design specifications for large structures.

2. Records

Table 1 shows the list of the seismograms used in this analysis, and Figure 1 shows the locations of the stations and the epicenters of the earthquakes. As shown in Table 1, these earthquakes occurred from 1933 to 1983 and their magnitude, M_S, are between 7.6 and 8.5. The epicentral distances of the records are between 49 and 501 km.

The type of seismograph which has been operating since about 1950 at the J.M.A. stations is the J.M.A. strong motion seismometer, and those which operated during the period from about 1930 to about 1950 at the C.M.O. stations are Omori's, Imamura's and C.M.O. strong motion seismographs. Hereafter we will call the latter the old type seismographs in this paper. Table 2 shows the characteristics of each seismograph for each record. The range of the natural period of those seismographs is from 3.5 to 7.1 sec.

Table 1. List of the Seismograms

No.	Date	Earthquake	Lat. °N	Long. °E	D	M_S	Station	Δ	I
E1	1933. 3. 3	Sanriku-Oki	39.25	144.50	—	8.5	Sendai	332	5
E2	1938. 5.23	Iwaki-Oki	36.50	141.00	—	7.6	Mito	49	5
E3	1938.11. 5	Fukushima-Ken-Toho-Oki(1st.)	36.75	141.75	60	7.7	Mito	122	5
							Sendai	184	5
E4	1938.11. 5	Fukushima-Ken-Toho-Oki(2nd.)	37.25	141.75	60	7.7	Mito	150	4
E5	1938.11. 6	Fukushima-Ken-Toho-Oki(3rd.)	37.25	142.25	60	7.6	Sendai	164	4
							Mito	186	4
E6	1944.12. 7	Tonankai	33.75	136.00	—	8.0	Murotomisaki	178	4
							Mishima	313	4
E7	1946.12.21	Nankaido	32.50	134.50	—	8.2	Sumto	208	5
							Miyazaki	296	4
E8	1968. 5.16	Tokachi-Oki	40.90	143.40	9	8.1	Akita	246	4
							Skata	377	3
							Niigata	501	3
E9	1983. 5.26	Nihonkai-Chubu	40.36	139.08	14	7.7	Mori	231	4
							Niigata	272	3
							Tomakomai	328	2

No. = Earthquake number; Lat. = Epicenter latitude; Long. = Epicenter longitude; D = Depth in km; Δ = Epicentral Distance in km; I = Intensity (J.M.A. scale).

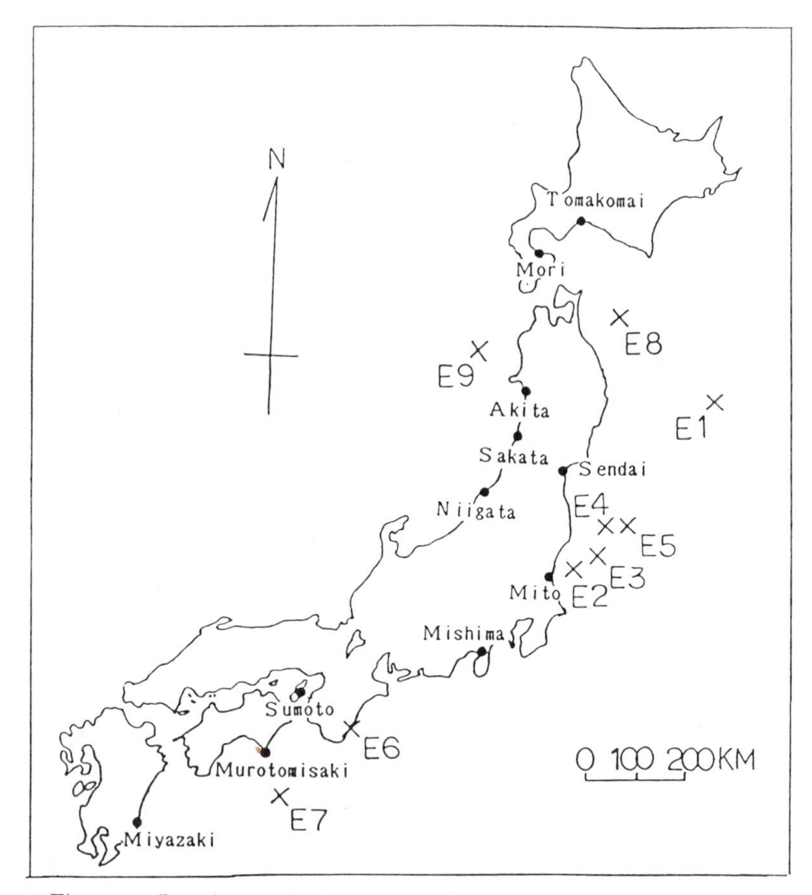

Figure 1. Locations of the stations and the epicenters of the earthquakes.

The seismogram at Akita from the 1968 Tokachi-Oki event was previously digitized 3 years ago with 16 other seismograms from that event (JSCE's Tank Committee, 1982). However, because the levels of response spectra and peak ground motions at Akita were the largest among the 17 seismograms from that event, we included this record in this analysis to compare the levels of ground motions with other seismograms in Table 1, which were newly digitized.

We collected more than 1000 J.M.A. (or C.M.O.) strong motion records during the large or great earthquakes from about 100 J.M.A. stations. The records only from great earthquakes were first selected. Among them, the records which are not extremely off-scale were finally used for digitization. Figure 2 shows an example of a seismogram recorded on smoked paper, however this example is a negative of the original.

Because the standard strong motion accelerograph (SMAC etc.) cannot accurately record the ground motions whose periods are longer than about 4 sec, this data base is very useful for seismic design of large-scale structures which have natural periods around 10 sec.

Table 2. List of the Seismograms

Earthquake	Station	#	V	Natural period		Damping factor (h)		Arm length		Paper speed
				NS	EW	NS	EW	NS	EW	
Sanriku-Oki (1933. 3. 3)	Sendai	O	2	4.0	—	.2155	—	207.1	—	.3971
Iwaki-Oki (1938. 5.23)	Mito	I	2	5.0	5.0	.4037	.3310	385.9	392.8	.5956
Fukussima-Ken-Toho-Oki (1938.11.5-6)	Mito									
	(1st)	C	2	4.5	3.5	.1665	.1480	402.6	390.5	.5834
	(2nd)	C	2	4.5	3.5	.1665	.1480	403.1	390.5	.5834
	(3rd)	C	2	4.5	3.5	.1665	.1480	392.1	408.7	.5834
	Sendai									
	(1st)	C	2	—	4.0	—	.2155	—	203.7	.5583
	(3rd)	C	2	—	4.0	—	.2155	—	203.7	.5736
Tonankai (1944.12. 7)	Murotomisaki	C	2	7.2	7.1	.3301	.3301	362.1	497.1	.5717
	Mishima	C	1	5.0	5.5	.1665	.1655	306.9	306.9	1.179
Nankaido (1946.12.21)	Sumoto	C	2	—	4.4	—	.3015	—	200.7	.3833
	Miyazki	C	2	—	6.2	—	.2155	—	284.7	.4372
Tokachi-Oki (1968. 5.16)	Akita	J	1	5.9	5.7	.5519	.5519	300.0	300.0	.5000
	Sakata	J	1	6.0	6.0	.5519	.5519	300.0	300.0	.5000
	Niigata	J	1	6.0	6.0	.5519	.5519	300.0	300.0	.5000
Nihonkai-Chubu (1968. 5.26)	Mori	J	1	6.0	6.0	.5519	.5519	300.0	300.0	.5000
	Niigata	J	1	5.9	6.0	.5519	.5731	300.0	300.0	.5000
	Tomakomai	J	1	6.0	6.0	.5519	.5266	300.0	300.0	.5000

# = Types of seismograph: O=Omori's, C = CMO, I = Imamura's, J = J.M.A.;
V = Magnification; Natural Period in sec; Arm Length in mm; Paper Speed in mm/sec.

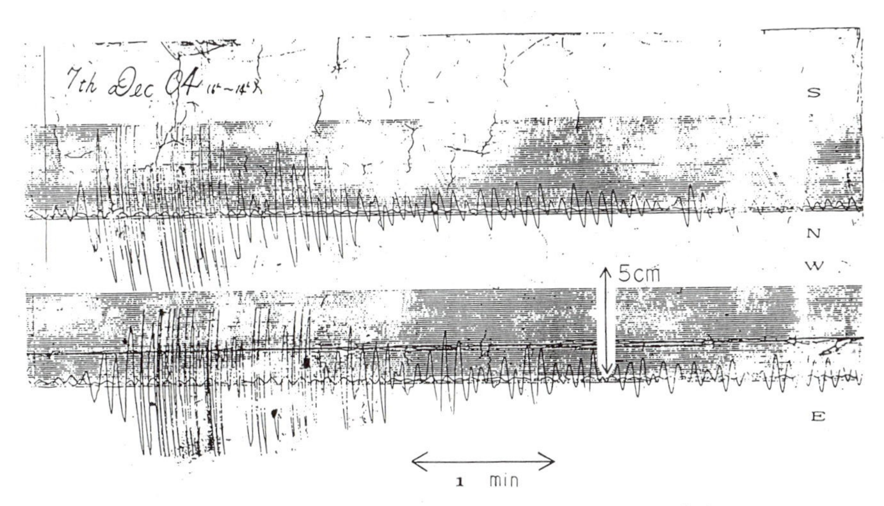

Figure 2. Seismogram at Mishima from the 1944 Tonankai event.

3. Digitization and Correction Procedures

To obtain the corrected data with equal time intervals, the following steps were performed on the original data: (1) The original seismogram was enlarged and printed on a polyester base. (2) The co-ordinates of peaks, troughs and points of inflection of the enlarged record were read by a digitizer. (3) Baseline corrections were performed on the data by the least squares method. (4) The effects of finite arm length and inclination of the arm were eliminated from the data. (5) The data were interpolated by a cosine curve and equally spaced data were produced. The time interval Δt was chosen as 0.25 sec in this analysis. (6) The corrections for instrumental characteristics on the data were performed. (7) A bandpass filter was applied to the data to remove the high and low frequency noise. As shown in Figure 3, the upper cut-off frequency is 0.5 Hz (2 sec), and the lower is 0.05 Hz (20 sec).

These correction procedures of the digitized data are similar to those outlined in Skoko and Sato (1973). The records of the J.M.A. strong motion seismometers were corrected by the above procedures.

However, in the case of the records by the old type seismographs, there are two additional problems to be solved. First, the mechanical arm length of the seismographs cannot be found in any of the documents now available. Second, there are a few records in which the apparent zero-line varies while the earthquake is being recorded (Figure 5). We solved these problems by the following procedures.

First, the procedures for estimating the arm length R are as follows (Figure 4): (1) A portion where the amplitude of the time history is very large was chosen. (2) An iso-time circle (Figure 4) which passes through the peak or trough of the large amplitude was constructed, and three points, P1, P2 and P3, on the iso-time circle were read by a digitizer. (3) The perpendicular bisectors of the line $\overline{P1 \cdot P2}$ and the line $\overline{P2 \cdot P3}$ were drawn, and the co-ordinates of the point C which is the intersection of the two bisectors was calculated. That point is the position of the pivot of arm. (4) The length of the line segment $\overline{P1 \cdot C}$ $(= \overline{P2 \cdot C} = \overline{P3 \cdot C})$ was measured. That is the arm length of the seismograph.

The values of the arm length for the old type seismographs in Table 2 were estimated by the above procedures.

Second, the procedures for the arm-correction of the above mentioned record are as follows (Figure 5): (1) For the given time history where the apparent zero-line varies in many stages (Figure 5a), the baseline correction in each section corre-

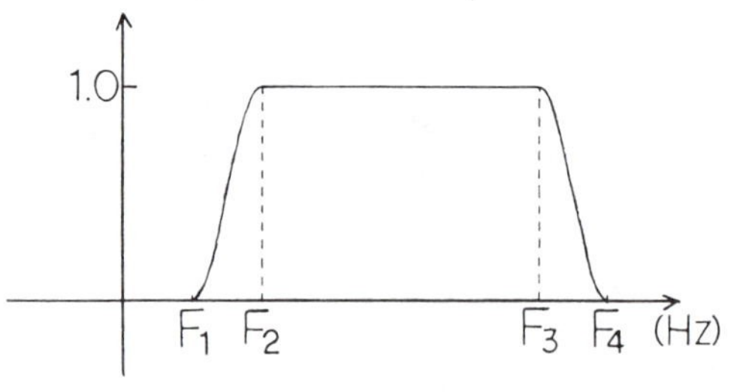

$F_1 = 0.035$ Hz (T = 28.6 sec); $F_2 = 0.05$ Hz (T = 20.0 sec);
$F_3 = 0.5$ Hz (T = 2.0 sec); $F_4 = 0.7$ Hz (T = 1.4 sec).

Figure 3. Characteristics of the bandpass filter.

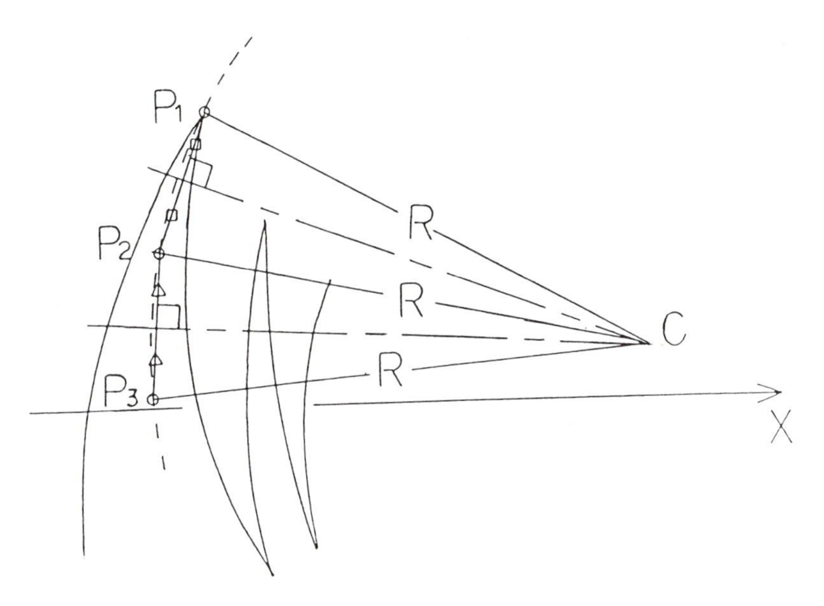

R : arm length.
C : position of the pivot of the arm.
P1, P2, P3 : points which co-ordinates are read by a digitizer.
—— — —— : perpendicular bisector.
– – – – – : iso-time circle (all points on the circle should have the same time-readings).

Figure 4. Method for estimating the arm length.

sponding to the each stage (Figure 5b) were performed. In Figure 5b, C1, C2 and C3 indicate the distance between the true and apparent zero-line in each section, and R indicates the arm length. (2) The effects of finite arm length and inclination of the arm were removed from the time history in each section (Figure 5c).

4. Peak Ground Motions

Following the procedures mentioned in the preceding section, the equally interpolated data of ground displacement for each component were obtained. Then, velocity and acceleration time histories were computed by once and twice differentiating the displacement time history. Peak values of ground motions for each component are shown in Table 3. In Table 3, the maximum value among the peak values is 23.3 cm for displacement, 12.8 kine for velocity and 27.7 gal for acceleration.

5. Velocity Response Spectra and Seismic Design Specifications

The velocity response spectra, Sv, of each component for the period range of 2 to 20 sec with damping factors, h, of 0.1% and 2% were computed. The value h of 0.1% corresponds to the sloshing of oil storage tanks and 2% to steel structures such as suspension bridges. Examples of the response spectra, Sv, whose levels are relatively large, are shown in Figures 6 to 8.

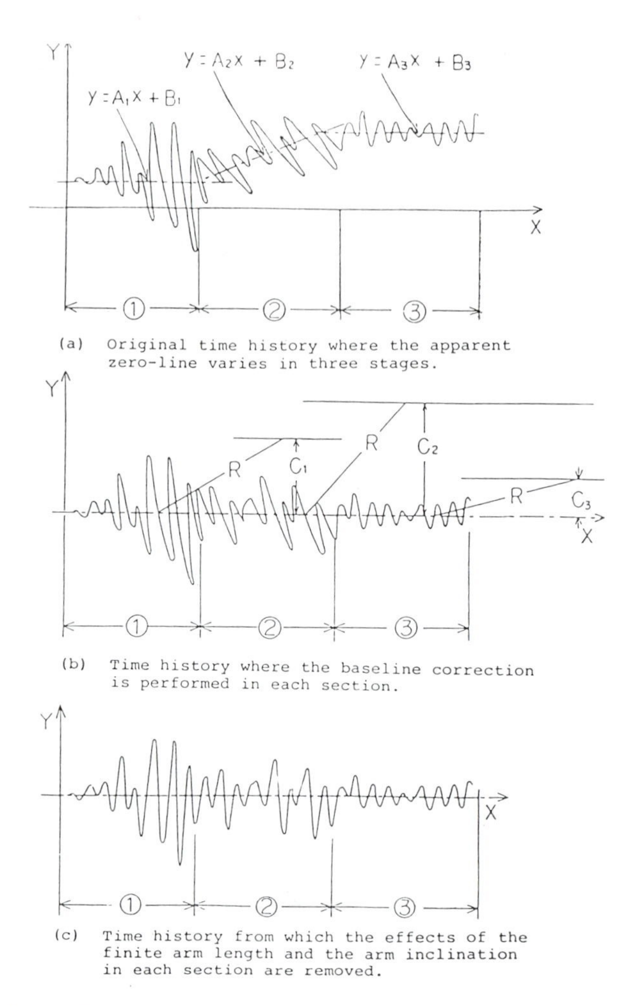

Figure 5. Arm-corrections done for an example seismogram.

As shown in Figure 9, we denote $(\mathrm{Sv})_P$ as the peak value of Sv in the period range of 2 to 20 sec for each component, and T_P as the corresponding period. The values of $(\mathrm{Sv})_P$ and T_P for each of the 29 components with damping of 0.1% and 2% are shown in Table 4, and the relation between $(\mathrm{Sv})_P$ and T_P for all components are shown in Figure 10. It is suggested from Table 4 that $(\mathrm{Sv})_P$ is not closely correlated with seismic intensity (J.M.A.) or epicentral distance Δ, but with the amplification characteristics at the station of the ground motions for the period around 10 sec.

The current Japanese seismic design specifications for oil storage tanks and long span suspension bridges are shown in Figure 11. It can be seen from Figure 11 that the upper limit for Sv of seismic design regulations for oil storage tanks, provided by the Japan Fire Defense Agency, is about 110 kine, and for the Honshu-Shikoku connecting bridges is 40 kine.

Table 3. List of Peak Values of Ground Motions for Each Component

Earthquake	Station	Δ (km)	*I*	N – S Component			E – W Component		
				Dis. (cm)	Vel. (kine)	Acc. (gal)	Dis. (cm)	Vel. (kine)	Acc. (gal)
Sanriki-Oki	Sendai	332	5	6.5	3.4	6.3	—	—	—
Iwaki-Oki	Mito	49	5	8.7	9.4	24.1	15.0	12.8	25.1
Fukushima-Ken-Toho-Oki	Mito								
	(1st)	122	5	14.2	7.0	8.1	17.7	9.4	10.8
	(2nd)	150	5	11.4	6.3	5.0	23.3	12.1	7.0
	(3rd)	164	4	7.3	5.5	7.8	20.8	11.3	8.8
	Sendai								
	(1st)	184	4	—	—	—	5.3	5.2	10.8
	(3rd)	164	4	—	—	—	2.7	2.1	3.7
Tonankai	Murotomisaki	178	4	6.9	9.7	27.7	4.8	3.6	6.6
	Mishima	313	4	5.4	4.0	8.1	7.5	9.6	26.0
Naknaido	Sumoto	208	5	—	—	—	18.9	8.9	19.7
	Miyazaki	296	4	—	—	—	5.3	5.9	16.9
Tokachi-Oki	Akita	246	4	8.0	10.2	23.2	9.8	11.2	25.2
	Skata	377	3	9.8	8.4	19.1	9.2	9.3	20.8
	Niigata	501	3	10.2	6.0	11.2	8.9	6.2	8.6
Nihonkai-Chubu	Mori	231	4	7.9	7.5	17.9	9.8	6.6	14.9
	Niigata	272	3	13.5	9.7	13.6	12.0	6.4	10.1
	Tomakomai	328	2	8.3	7.4	9.9	11.0	6.9	8.3

Δ = Epicentral distance in km; *I* = Intensity (J.M.A. scale).

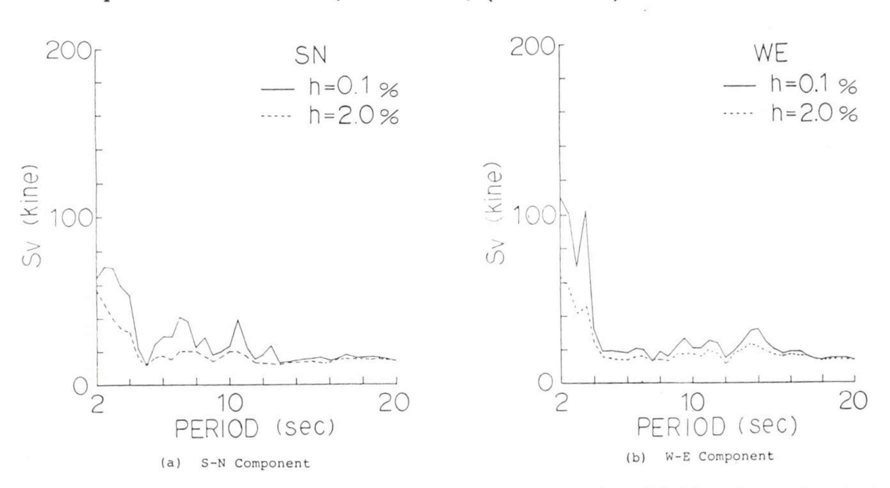

Figure 6. Velocity response spectra, Sv, for the record at Mishima from the 1944 Tonankai event.

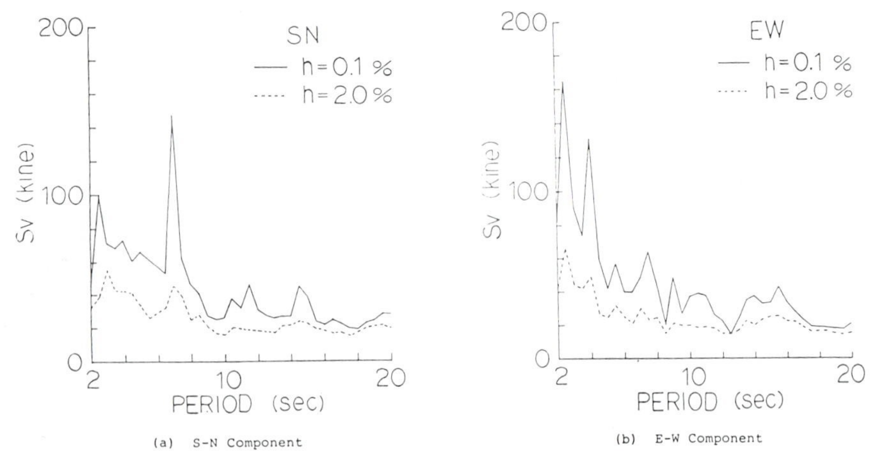

Figure 7. Sv for the record at Sakata from the 1968 Tokachi-Oki event.

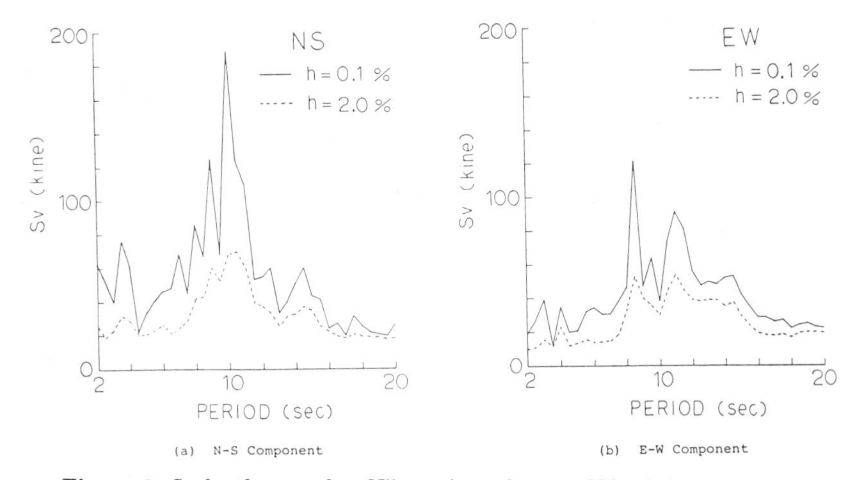

Figure 8. Sv for the record at Niigata from the 1983 Nihonkai-Chubu event.

From Table 4, Figure 10 and Figure 11, the following can be seen: (1) The maximum value of $(\mathrm{Sv})_P$ with h = 0.1% among the 29 components is 188 kine, and there are 7 components whose $(\mathrm{Sv})_P$ with h = 0.1% are larger than 110 kine. (2) The maximum value of $(\mathrm{Sv})_P$ with h = 2% among the 29 components is 70 kine, and there are 16 components whose $(\mathrm{Sv})_P$ with h = 2% are larger than 40 kine.

6. Conclusions

The results in the preceding section strongly suggest the need to examine those Japanese seismic design specifications.

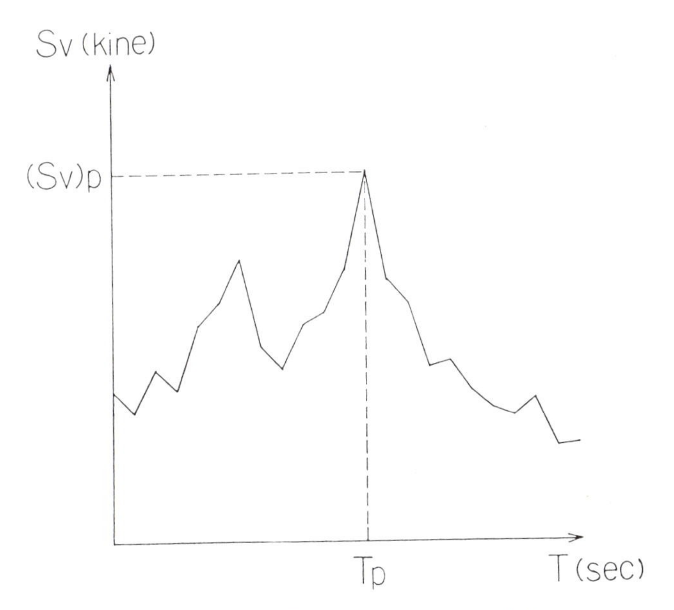

Figure 9. Definition of $(\text{Sv})_P$ and T_P.

Table 4. List of the Values of $(\text{Sv})_P$ and T_P for Each Component

Earthquake	Station	Δ (km)	I	N – S Component				E – W Component			
				h = 0.001		h = 0.02		h = 0.001		h = 0.02	
				$(\text{Sv})_P$	T_P	$(\text{Sv})_P$	T_P	$(\text{Sv})_P$	T_P	$(\text{Sv})_P$	T_P
Sanriki-Oki	Sendai	332	5	25	3.0	16	13.0	—	—	—	—
Iwaki-Oki	Mito	49	5	50	3.0	35	2.5	57	2.0	43	2.0
Fukushima-Ken-Toho-Oki	Mito										
	(1st)	122	5	43	6.0	31	6.5	62	4.0	41	8.5
	(2nd)	150	5	38	9.5	27	9.5	60	17.5	44	17.0
	(3rd)	164	4	62	6.0	34	6.5	81	11.0	45	10.0
	Sendai										
	(1st)	184	4	—	—	—	—	36	2.0	22	2.0
	(3rd)	164	4	—	—	—	—	21	2.0	10	2.0
Tonankai	Murotomisaki	178	4	25	2.0	17	3.5	25	3.0	16	2.5
	Mishima	313	4	71	2.5	57	2.0	112	2.0	64	2.0
Naknaido	Sumoto	208	5	—	—	—	—	87	2.5	46	2.5
	Miyazaki	296	4	—	—	—	—	42	2.0	32	2.0
Tokachi-Oki	Akita	246	4	79	2.5	41	2.5	102	3.0	47	4.5
	Skata	377	3	147	7.0	54	3.0	165	2.5	66	2.5
	Niigata	501	3	112	5.5	37	12.0	100	4.5	39	4.5
Nihonkai-Chubu	Mori	231	4	63	9.0	35	2.5	82	4.0	42	4.0
	Niigata	272	3	188	10.0	70	10.5	122	8.5	54	11.0
	Tomakomai	328	2	134	9.5	62	10.0	69	9.0	41	11.5

Δ = Epicentral distance in km; I = Intensity (JMA scale); $(\text{Sv})_P$ = velocity response spectra (peak value) in kine; T_P = period corresponding to $(\text{Sv})_P$ in sec.

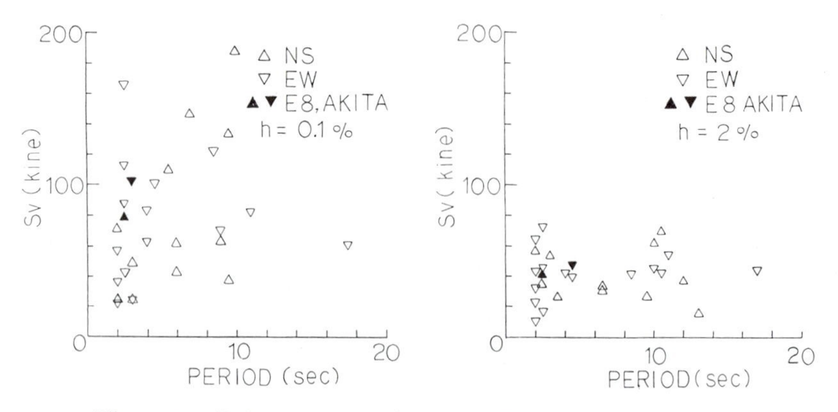

Figure 10. Relation between $(\mathrm{Sv})_P$ and T_P for all components.

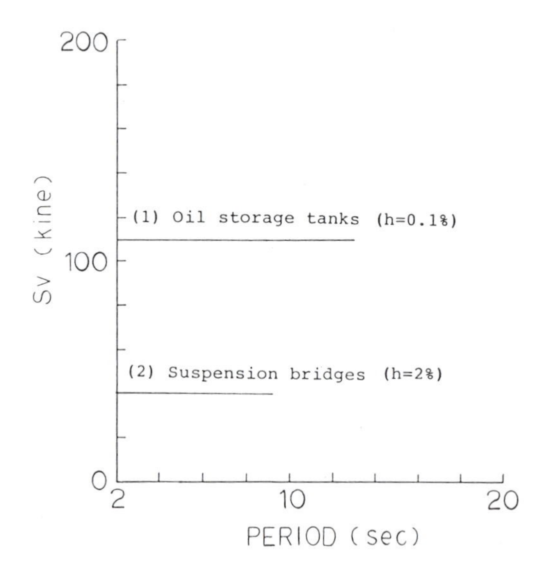

Figure 11. Current seismic design specifications for large structures: (1) Seismic design regulations for oil storage tanks, provided by the Japan Fire Defense Agency. (2) Seismic design specifications for the Honshu-Shikoku connecting bridge project.

ACKNOWLEDGEMENTS

We would like to thank Prof. Y. Fujino of the University of Tokyo for his valuable comments.

REFERENCES

JSCE's Tank Committee (1982). Report on the Research of the Seismic Safety of Above-Ground Storage Tanks (In Japanese).

Skoko, D. and Y. Sato (1973). Strong Motion Seismogramme of the Off-Boso Earthquakes on November 26, 1953, *Bull. Intern. Inst. Seism. Earthq. Eng.*, **11**, 43-66.

Some Remarks on Historical Seismograms and the Microfilming Project

S. Miyamura
Oki Electric Industry, Ltd.
Tokyo, Japan

ABSTRACT

For the selection of earthquakes in the seismogram filming project, different basic catalogs are compared. In particular, the catalogs prepared by World Data Center A (WDC-A) for the microfilming project are examined. The geographical inhomogeneity of magnitude threshold is discussed. The importance of selecting rare earthquakes far from the seismic zones is emphasized. Examples of rare earthquakes in the different parts of the Earth are shown in tables and maps.

Beside magnitude, damage is usually considered for the selection of earthquakes to be filmed. However, unlike magnitude catalogs, we have no accurate global damage catalog of destructive earthquakes. A provisional list of destructive earthquakes has been prepared, based on Rika Nenpyo (Science Almanac, Tokyo Astronomical Observatory), Båth (1979), and several regional catalogs.

In selecting stations for the filming project, it is necessary to adopt several stations in the same region, so as not to miss seismograms of any significant earthquakes. Finally, some problems related to the name, code, and coordinates of seismological stations are discussed.

1. Introduction

First, it must be emphasized that the seismogram filming project should not be used as justification for: (1) discarding the original seismograms that are selected for the filming; and (2) neglecting to keep seismograms at the stations that are not selected for the project. Even the best copies could not replace the original records. Moreover, it is possible that records not selected for the filming will be required in the future for some unforeseeable scientific or practical needs.

Seismograph stations also have to be selected for the filming project. However, seismograms at the stations not selected sometimes become important. Often, a station of high quality failed to record a certain earthquake, but a less qualified station in the same region might succeed in recording it. The information required to use the seismograms is expected to be furnished with the copies. However, providing complete information for each seismogram is not easy in many cases.

2. Selection of Earthquakes

2.1. Magnitudes in Different Catalogs

The common procedure for selecting earthquakes for the seismogram filming project is to define significant earthquakes by magnitude and damage. First, we study the discrepancies of magnitudes in the different standard catalogs. Gutenberg (1956), Gutenberg and Richter (1954), Rothé (1969), and Duda (1965) are considered as standard catalogs for the periods of 1896-1903, 1904-1952, 1953-1965, and 1897-1964, respectively. Recent contributions by Abe and Kanamori (1979), Abe (1981), Abe and Noguchi (1983a, b) and Abe (1984) provide revised catalogs.

Nevertheless, many ambiguous cases still remain. For example, Christoskov (1983) gives $M = 7.83$ for the April 4, 1904 Strumska, Bulgaria earthquake as the most important event for the seismic risk analysis of that country. Gutenberg and Richter (1954) [hereafter *GR*] gave $M_{GR} = 7.5$ and Abe and Noguchi (1983a, b) [hereafter *AN* and *AN'*] gave $M_S = 7.3$ and 7.1 respectively, while Duda (1956) [hereafter *DU*] did not list it as event of magnitude greater than 7. Abe (personal communication, 1984) informed me that M_{GR} was based only on $M_S = 7.3$ at Osaka (undamped seismograph record) and $m_B(P) = 7.4$ and $m_B(S) = 7.6$ at Göttingen ($\Delta = 13°$), which were not reliable. *AN* and *AN'* used undamped Milne seismograph records at 12 stations with revised calibration and seems to be more reliable. Because I do not know how Christoskov (1983) computed his magnitude, I cannot give a definite answer.

GR was considered "practically complete" for the a-class ($M \geq 7.75$) and b-class ($M \geq 7.0$) shallow earthquakes in 1904-1952 and 1918-1952, respectively, and "probably almost complete" for the large intermediate and deep earthquakes ($M \geq 7.0$, $h \geq 70$ km) in 1918-1952. However, Miyamura (1978) suggested the possibility of finding earthquakes with magnitudes around 7 that are not listed in *GR*.

DU added 105 earthquakes to *GR* for 1905-1917 and prepared a catalog of $M \geq 7.0$ earthquakes for 1897-1964. Many shocks $M_{GR} \geq 7$ in *GR* [e.g. April 4, 1904 (Region 30, No. 610), $M_{GR} = 7.5$; December 26, 1936 (Region 12, No. 190), $M_{GR} = 7\pm$], however, are omitted in *DU*.

Abe (1981) [hereafter *A*] prepared a revised magnitude catalog of earthquakes for 1904-1980. The revision was based on Gutenberg's worksheets, but no effort was made to add earthquakes not included in *GR*. *AN* and *AN'* revised the data in *A*, adding many earthquakes prior to 1917. M_S in *AN* and *AN'* are mostly smaller than M_{DU} (magnitude in *DU*), in particular, M_S for 37 out of 105 earthquakes added to *GR* by *DU* are less than 6.8, and are not shown in *AN* and *AN'*. However, 11 events are $M_{DU} \geq 7.5$ as listed in Table 1. Because I do not have the background data of *DU* and *AN/AN'*, I cannot say which are better.

Epicenters given in *GR* and *DU* for some earthquakes are very different. Epicenters given in *A* and *AN/AN'* follow those of *GR*. Relocation of some doubtful earthquakes appears to be necessary.

In selecting the earthquakes for the filming project, special attention should be given to include those earthquakes having doubtful location and magnitude (see Table 1, remarks 1 and 2; and Table 3, events with *).

2.2. WDC-A Catalog for the Filming Project

In the first report of the Historical Seismogram Filming Project, Meyers and Lee (1979) presented 3 lists of earthquakes to be used as a guide for the project. These lists include large worldwide events for the periods 1897-1899 and 1900-1962, and significant earthquakes in the United States and neighboring areas for 1887-1962. The lists are included in the second and third reports (Glover, 1980; and Glover

Table 1. Earthquakes in Duda (1965) (*DU*) with $M_{DU} \geq 7.5$, not listed in Gutenberg and Richter (1954) (*GR*) and Abe and Noguchi (1983a, b) (*AN*, *AN'*).

Year	Date	Time	Lat. °	Lon. °	Depth	M_{DU}	Rem.
1905	03.18	00:58	27.5 S	173.0 W	s	7.5	
	10.21	11:01	42.0 N	042.0 E	s	7.5	
1907	05.04	06:51	07.5 S	153.7 E	s	7.7	
1908	12.12	12:08	14.0 S	078.0 W	s	8.2	(1)
1909	12.09	23:23	10.0 S	165.0 E	s	7.7	(2)
1910	05.22	06:25	42.0 N	145.0 E	s	7.5	
1913	08.01	17:10	47.5 N	155.5 E	s	7.7	
1915	02.28	18:59	23.6 N	123.5 E	s	7.7	
1916	02.01	07:36	29.5 N	131.5 E	s	8.0	
	02.06	21:51	48.5 N	178.5 E	s	7.7	
1917	12.29	22.50	15.0 N	097.0 W	s	7.7	

Remarks: (1) *GR*: 12:54.9 26.5 N, 97 E, $M_{GR} = 7.5$ (26N360); $M_{AN} = 7.6$, $M_{AN'} = 7.0$. (2) *GR*: 23:28.8 12 N, 144.5 E, $h = 50$ km, $M_{GR} = 7.4$ (17N780); $M_{AN} = 7.1$, $M_{AN'} = 6.8$. cf. Remark (2) in Table 2.

and Meyers, 1981); but in the fourth report (Glover and Meyers, 1982), the global and U.S. lists have been combined into one catalog for 1900-1962.

The global catalog for 1900-1962 in Meyers and Lee (1979) [hereafter *SE22*] is a computer-readable version based on the earthquake catalog of Rika Nenpyo (Science Almanac) compiled by T. Usami (1975) [hereafter *RN*]. *SE22* stated that *RN* is based on Gutenberg (1956) [hereafter *G*], Gutenberg and Richter (1954) [*GR*], and Rothé (1969) [hereafter *RO*] for the periods 1900-1903, 1904-1952, and 1953-1962, respectively. However, the catalog is not so simple. It includes earthquakes with $M \geq 7.0$, and this M is the largest value among different sources. Unfortunately, the reference sources are not given explicitly. It is clear that the earthquakes in and near Japan prior to 1925 are taken from the catalog of the Central Meteorological Observatory (CMO) (1950), because it explains that: (1) M for those earthquakes are converted from the Kawasumi intensity magnitude M_k and generally are 0.5 units larger than the instrumental magnitude by the Japan Meteorological Agency (JMA), M_J; and (2) the number of earthquakes prior to 1925 is larger than the number of earthquakes in the later period. Japanese earthquakes after 1926 are taken from JMA data. In addition, damaging earthquakes having $M < 7.0$ are included, but neither the data source nor the selection criteria are given.

Computer output should be carefully examined visually. *SE22* includes the following errors:

(1) "1905 Feb 21 41.5N 142.5W 7.0 North Pacific Ocean" *should be* "1905 Feb 21 41.5N 142.5E 7.0 Hokkaido",

(2) "1906 Mar 16 23.5S 120.5E 8.5 Australia" *should be* "1906 Mar 16 23.5N 120.5E 8.5 Taiwan", and

(3) "1906 Apr 18 38.0S 123.0W 8.25 South Pacific Ocean" *should be* "1906 Apr 18 38.0N 123.0W 8.25 California".

Similar printing mistakes were found in *DU*, i.e. "W" *should be* "E" in the following earthquakes: February 1, 1938; November 10, 1942; May 3, 1943; and November 25, 1953.

Origin times of the following earthquakes in the Japanese area are not given in GMT and should be reduced by 9 hours:

(1) "1909 Oct 31 19:18" *should be* "10:18",
(2) "1913 Apr 13 15:40" *should be* "06:40",
(3) "1915 Aug 05 22:16" *should be* "13:16",
(4) "1915 Nov 18 13:04" *should be* "04:04",
(5) "1925 May 27 11:30" *should be* "02:30",
(6) "1931 Mar 09 12:49" *should be* "03:49".

Origin times of the following earthquakes should also be corrected as follows: "1935 Apr 20 20:01" *should be* "22:01", "1941 Nov 24 11:46" *should be* "21:46". All of these errors were found in *RN*.

GR is probably one of the main data sources for *RN* and *SE22*, but they do not include 29 earthquakes with $M_{GR} \geq 7.0$ for the period 1904-1917, 8 of which are $M_{GR} \geq 7.5$ (see Table 2). On the other hand, they include many earthquakes which are not found in *GR*, *DU*, *A*, *AN*, and *AN'*. They are mostly earthquakes in and near Japan taken from CMO (1950), but more than 20 are earthquakes outside the Japanese area (see Table 3). For example, the October 10, 1932 Solomon Islands earthquake ($M = 7.7$) and the March 19, 1951 Philippine Islands earthquake ($M = 7.8$) appear to be too large to be missed by *GR*. Their magnitudes may actually be less than 7.0, but we need amplitude data for verification.

Magnitudes in *RN* or *SE22* are the largest of the values from referred sources. However, M_{GR} of many earthquakes are larger than M in *SE22*, which indicates that *GR* was not the data source for those earthquakes (see Table 4).

The Glover and Meyers (1982) catalog [hereafter *SE33*] combined the global and United States catalogs in *SE22*. However, due to the lack of visual examination, *SE33* failed to include 15 events during October 16, 1906 to April 10, 1907, 6 events during December 26., 1949 to February 2, 1950, and 2 events during November 16, 1925 to January 25, 1926. Moreover, 3 earthquakes in *SE22* (November 5, 1904 in Taiwan; May 4, 1907 in Bonin Island; and May 30, 1929 in Argentina) are omitted in *SE33*, the reason of which is unclear. There are also minor errors in that several successive events are placed reverse in time, e.g., 1908 May 12 20:22 and 20:36; 1909 Jul 07 21:37 and 21:39, 1909 Aug 14 06:30 and 06:39. These errors again suggest the necessity of examining a computer prepared catalog carefully.

SE33 supplemented several earthquakes not included in *SE22* (Table 2) and revised many magnitudes in *SE22*, but the corrections are mostly less than 0.2, except for "1929 Mar 07 Aleutian Island" where $M = 8.1$ is revised to $M = 8.6$.

2.3. Geographical Inhomogeneity of Magnitude Threshold

The common procedure for selecting earthquakes having magnitude larger than a certain threshold introduces a regional bias of the number of selected events. *GR* lists 15 earthquakes with $M \geq 7.75$ in the Seismic Region 19 (Kamchatka – NE Honshu) and 6 events in Region 24 (Indonesia), but only 2 events in Regions 3 and 4 (California and Gulf of California), and 1 event in Region 29 (Iran – Ural). Further, no earthquakes with $M \geq 7.75$ are given in Region 32 (Atlantic Ocean), Region 44 (East Pacific Ocean), and Region 45 (Indian-Antarctic Swell). Therefore, it seems reasonable to change the magnitude threshold from region to region in selecting earthquakes. For *GR*, optimum levels of two alternatives are suggested as follows:

Table 2. Earthquakes in Gutenberg and Richter (1954) (*GR*), not listed in the catalog of WDC-A Report SE-22 (1979) (*SE22*).*

Year	Date	Time	Lat. °	Lon. °	h	GR No.	M_{GR}	M_{AN}	Rem.
1904	04.04	10:26	41.75 N	023.25 E		30N610	7.50	7.1	
	07.24	10:44	52.00 N	159.00 E		19N550	7.50	–	
	10.03	03:05	12.00 N	058.00 E		33N085	7.00	7.0	
1908	05.15	08:31	59.00 N	141.00 W		02N160	7.00	7.0	(1)
	12.12	12:54	26.50 N	097.00 E		26N360	7.50	7.0	
1909	12.09	23:28	12.00 N	144.50 E	50	17N780	7.40	6.8	(2)
1910	06.29	10:45	32.00 S	176.00 W		12N160	7.00	7.3	
	12.10	09:26	11.00 S	162.50 E	50	15N080	7.50	7.3	
1911	09.15	13:10	20.00 S	072.00 W		8N510	7.30	7.1	
	10.06	10:16	19.00 N	70.500 W		7N310	7.00	6.8	
	12.16	19:14	17.00 N	100.50 W	50	5N410	7.50	7.6	
	12.31	06:07	02.00 S	143.50 E		16N130	7.00	6.8	
1912	01.04	15:46	52.00 N	179.00 W		1N130	7.00	–	(1)
	06.10	16:06	59.00 N	153.00 W		1N471	7.00	6.9	(1)
	08.17	19:11	04.00 N	127.00 E		23N105	7.50	7.3	
	12.23	23:56	0	123.00 E		23N495	7.00	–	
1913	06.04	09:57	01.50 S	150.00 E		15N840	7.00	–	
	07.28	05:39	17.00 S	074.00 W		8N474	7.00	7.0	
1914	01.12	09:28	31.50 N	131.00 E		20N420	7.00	–	
	04.11	16:30	12.00 S	163.00 E	50	15N030	7.25	7.2	
	06.20	07:20	12.00 S	166.00 E	50	14N900	7.25	7.0	
1915	03.12	14:48	12.00 N	124.00 E		22N390	7.00	–	
1916	04.26	02:21	10.00 N	085.00 W		6N280	7.30	7.1	
1917	02.15	00:48	30.00 S	073.00 W		8N730	7.00	–	
	06.24	19:48	21.00 S	174.00 W	60	12N460	7.25	–	
	07.27	01:01	19.00 N	067.50 W	50	7N420	7.00	7.0	(1)
	07.27	02:51	31.00 S	070.00 W	60	8N800	7.00	–	
	07.29	21:29	03.50 S	141.00 E		16N200	7.60	7.7	

(*) N.B. 1906.04.18 California Eq. (03N490) is excluded (cf. text). Remarks: (1) Supplemented in *SE33*; (2) cf. Remark (2) in Table 1.

(1) No. 1 – No. 23, No. 46 (Circum-Pacific Seismic Regions):
for $M \geq 7.75, N = 64$; for $M \geq 7.5, N = 132$,
(2) No. 24 – No. 31, No. 47, No. 48 (Eurasian Seismic Regions):
for $M \geq 7.75, N = 28$; for $M \geq 7.5, N = 57$,
(3) No. 32 – No. 33 (Atlantic and Indian Ocean Seismic Regions):
for $M \geq 7.0, N = 26$; for $M \geq 6.75, N = 42$,
(4) Nos. 34, 37, 38, 41, 42, 51 (Continental Seismic Regions):
for $M \geq 7.0, N = 8$; for $M \geq 6.75, N = 22$,
(5) Nos. 35, 36, 39, 49, 50 (Continental and Oceanic Stable Regions):
for $M \geq 6.0, N = 11$; for $M \geq 5.5, N = 28$,

where N is the number of earthquakes selected above the threshold. The total numbers are 137 and 281, which are manageable for the limited time and budget. It seems more important to include smaller but rare earthquakes than to include many similar earthquakes that are expected to occur frequently.

Considering the importance of rare earthquakes, we must note that the Gutenberg-Richter Seismic Regions cover rather large areas, and we find earthquakes with magnitude below the suggested thresholds of the respective regions located far

Table 3. Earthquakes outside Japanese area, $M \geq 7.0$, in the catalog of WDC-A Report SE-22 (1979) (*SE22*), which are not listed in the standard catalogs: Gutenberg and Richter (1954) (*GR*), Rothé (1969) (*RO*), Duda (1965) (*DU*), Abe (1981) (*A*) and Abe and Noguchi (1983 a, b) (*AN*, *AN'*).

Year	Date	Time	Lat. °	Lon. °	h	M	Region	Rem.
1918	08.18		20.5 S	178.5 W	300	7.2	W. of Tonga Is.	
1927	12.04		34.5 N	121.5 W		7.3	Off Coast of Calif.	
1932	10.10		10.0 S	161.0 E		7.7	Solomon Is. *	
1932	11.02		32.0 S	131.5 W		7.5	N. Pacific Ocean	
1941	02.11	14:35	15.2 N	094.4 W		7.0	Near Coast of Oaxaca, Mexico	
1944	01.10	20:09	18.1 N	100.6 W		7.0	Guerrero, Mexico	
1945	04.15		35.0 S	178.0 W		7.0	E. of N. Is. NZ	
1945	06.03	13:05	08.3 N	082.6 W		7.0	Panama-Costa Rica border	
1946	05.16		01.0 S	098.0 E		7.0	S. Sumatra	
1948	06.18	00:53	06.5 S	155.0 E		7.0	Solomon Is.	
1948	08.11	10:36	17.7 N	095.2 W	30	7.0	Oaxaca, Mexico	
1949	05.30	01:32	20.8 S	069.0 W	100	7.0	N. Chile	
1950	01.02	15:14	11.5 S	165.4 E	150	7.0	Santa Cruz Is.	
1950	09.23		17.5 S	177.5 W	400	7.0	W. of Tonga Is.	
1950	11.06	22:22	07.2 S	155.3 E		7.0	Solomon Is.	
1951	03.19		09.5 N	127.3 E		7.8	Philippine Is. region*	
1951	04.14	00:45	24.0 S	066.5 W	250	7.0	Salta Province, Argentina	
1952	09.30	12:52	28.5 N	102.0 E		7.1	Szechwan, China	
1952	11.29	08:25	53.0 N	160.0 E		7.0	Near E. Coast of Kamchatka	
1954	04.29	11:34	29.5 N	029.5 W		7.5	N. Atlantic Ocean	
1954	11.21		41.0 S	081.0 E		7.0	Mid-Indian Rise	(1)
1957	07.01	19:30	25.0 N	094.0 E		7.3	Burma-India Border	

(*) doubtful location and magnitude. Remarks: (1) Oct. 21 41.3 S 80.2 E, $M_{RO} = 6.7$ (*RO*)?

Table 4. Earthquakes outside Japanese area with $M_{RN} < M_{GR}$ in the catalog of WDC-A Report SE-22 (1979) (*SE22*) based on Rika Nenpyo (1975) (*RN*).

Year	Date	Time	Lat. °	Lon. °	h	M_{RN}	M_{GR}
1907	04.18	23:52	13.6 N	122.9 E		7.20	7.60
1910	12.16	14:45	05.0 N	125.0 E		7.20	7.50
1912	07.07	07:57	63.3 N	157.4 W		7.10	7.40
1914	08.04	22:41	40.5 N	090.5 E		7.30	7.50
1915	11.01	07:25	38.9 N	143.1 E		7.50	7.70
1915	11.21	00:13	32.0 N	119.0 W		7.00	7.10
1917	02.20	19:29	19.0 N	080.0 W		7.10	7.40
1917	08.31	11:36	05.0 N	075.0 W		7.20	7.30
1919	01.01	02:59	19.5 S	176.5 W	180	7.75	7.88
1943	12.01	10:34	19.5 S	069.8 W	80	7.00	7.25

from the seismic belts. Table 5(a, b, c) lists several such rare earthquakes as an example. Figures 1 to 4 show the epicenter distributions of those rare earthquakes in SE Asia, Indian Ocean, Pacific Basin (1900-1965), and Pacific Basin (1964-1974), respectively.

Table 5a. Examples of Rare Earthquakes Far from the Seismic Zones in Gutenberg and Richter, 1954 (*GR*) and Rothé, 1969 (*RO*). (Earthquakes in Pacific Basin cf. Table 5b).

Region	Year	Date	Time	Lat. °		Lon. °		*h*	*M*	Ref. No.
South Coast of China	1918	0213	0607	24.0	N	117.0	E		7.3	*GR* 21N975
	1962	0318	2018	23.8	N	114.6	E	25	6.1	*RO* 25N018
	1962	0729	0857	23.6	N	114.3	E	65	d	*RO* 25N017
Northern Vietnam	1935	1101	1622	20.5	N	103.5	E		6.75	*GR* 25N400
	1961	0612	0958	21.6	N	106.0	E	n	d	*RO* 25N007
South China Sea	1929	1024	0634	22.0	N	118.0	E		6.5	*GR* 21N950
	1930	0721	1406	07.0	N	114.0	E		6.	*GR* 22N920
	1931	0921	1027	19.75	N	113.0	E		6.75	*GR* 25N320
	1932	0621	2259	16.5	N	112.0	E		d	*GR* 25N080
	1965	1007	0335	12.6	N	114.5	E	n	6.1	*RO* 24N101
Palawan Is.	1940	1219	1548	09.0	N	118.0	E		d	*GR* 22N840
Sulu Is.	1923	0811	0054	04.5	N	119.5	E		6.5	*GR* 23N015
	1932	0915	1113	06.05	N	120.75	E		6.25	*GR* 22N810
NE Borneo	1923	0419	0309	02.5	N	117.5	E		7.0	*GR* 23N165
	1958	1026	0217	05.3	N	117.1	E	58	d	*RO* 22N002
Sarawak	1965	0706	0448	03.7	N	113.5	E	40	d	*RO* 24N100
Philippine Sea	1935	1217	1917	22.5	N	125.3	E		7.2	*GR* 21N175
	1952	0113	0403	23.0	N	124.5	E		6.9	*GR* 21N335
	1954	0111	1709	22.7	N	125.8	E	n	6.	*RO* 20N001
	1955	1127	1930	23.1	N	124.8	E	n	d	*RO* 21N022
N. of Peninsular India	1927	0602	1637	23.5	N	081.0	E		6.5	*GR* 26N820
	1938	0314	0048	21.55	N	075.75	E		6.25	*GR* 26N940
	1957	0825	2104	22.0	N	080.0	E	n	d	*RO* 26N081
	1963	0409	0003	22.5	N	085.8	E	n	d	*RO* 26N060
	1963	0508	1415	21.7	N	085.0	E	n	d	*RO* 26N062
	1964	0415	1635	21.7	N	088.0	E	36	d	*RO* 26N052
E. Coast Peninsular India	1959	1012	1925	15.7	N	080.1	E	n	d	*RO* 26N079
	1963	1205	0407	17.3	N	080.1	E	n	d	*RO* 26N080
W. Coast Peninsular India	1965	0604	0337	17.0	N	073.4	E	n	d	*RO* 26N090
Off W. Coast of Ceylon	1938	0910	2223	07.5	N	079.0	E		d	*GR* 33N815
Maldive Is. Region	1935	0424	1552	00.55	N	074.25	E		6.	*GR* 33N200
	1938	0910	2223	07.5	N	079.0	E		d	*GR* 33N825
	1940	0316	2054	07.5	N	073.0	E		d	*GR* 33N150
	1944	0229	1628	00.5	N	076.0	E		7.2	*GR* 33N850
Bengal Bay	1927	0729	0003	15.0	N	087.0	E		6.5	*GR* 33N805
NE Indian Ocean	1913	0119	1705	02.0	N	086.0	E		7.0	*GR* 33N835
	1916	0509	1433	01.5	N	089.0	E		6.3	*GR* 33N840
	1936	0113	1810	04.0	S	085.0	E		d	*GR* 33N885
	1937	1130	0040	05.5	N	090.0	E		6.5	*GR* 24N990
Yemen	1941	0111	0831	17.0	N	043.0	E		6.25	*GR* 37N060
	1941	0204	0917	16.0	N	043.0	E		d	*GR* 37N120
	1960	1216	1649	14.7	N	042.6	E	n	d	*RO* 37N008
	1965	1017	2008	17.2	N	043.7	E	n	d	*RO* 37N004
Angola	1914	0514	1556	10.0	S	015.0	E		6.	*GR* 37N870
Gahna	1939	0622	1919	06.0	N	001.0	W		6.5	*GR* 37N810
Gabon	1945	0912	0051	02.0	N	015.0	E		6.	*GR* 37N840
Malvis Ridge	1933	0106	1910	22.0	S	001.0	E		6.	*GR* 32N115
Bermuda Is.	1931	0816	0806	29.0	N	065.0	W		d	*GR* 32N640
	1965	0329	1310	33.6	N	065.0	W		d	*RO* 32N140

(*continued*)

Table 5a. *Continued*

Region	Year	Date	Time	Lat. °		Lon. °		*h*	*M*	Ref. No.
Off N. Brazil	1933	0912	1253	08.0	N	049.0	W		d	*GRa* (1941)
Brazil	1939	0628	1132	27.5	S	048.5	W		d	*GR* 35N700
	1955	0131	0503	12.5	S	057.4	W	n	6.6	*RO* 35N002
	1955	0301	0146	19.9	S	036.7	W	n	6.	*RO* 35N004
	1964	0213	1121	18.1	S	056.8	W	n	d	*RO* 35N003
	1964	0619	0356	02.5	N	058.9	W	65	d	*RO* 35N001
SE Pacific Basin	1944	0805	0057	13.5	S	092.5	W		6.	*GR* 44N990
	1944	0805	0124	13.5	S	092.5	W		6.25	*GR* 44N991

Table 5b. Earthquakes in the Pacific Basin (*GR* Seism. Reg. No. 39), excluding Hawaiian Is. Region for 1900-1965.

No.	Year	Date	Time	Lat. °		Lon. °		*h*	*M*	Ref.	Seism. Reg./Remarks
01	1900	0702	2052	10.	S	165.	E	s	8.1	*DU*	14/39
02	1905	0217	1139	33.	N	152.	E	s	7.3	*DU*	39 near 18/19
03	1905	0614	1130	30.	S	159.	W	s	7.0	*DU*	39
04	1905	0630	1707	01.	S	168.	W	s	7.6	*DU*	39
05	1905	1210	1236	50.	N	180.		s	7.6	*DU*	1/39
06	1906	0410	2122	19.	N	138.	W	s	7.5	*DU*	39
07	1908	1106	0712	30.	N	160.	E	s	7.6	*DU*	39
08	1909	0410	0523	09.	S	180.		s	7.3	*DU*	39 near 13
09	1909	0410	1936	45.	N	168.	E	s	7.8	*DU*	39 Tenno Seamounts
10	1910	0909	0111	45.	N	170.	W	s	7.3	*DU*	39
11	1910	1126	0439	08.	S	167.	E	s	8.0	*DU*	39 near 14
12	1911	0411		07.	N	158.	E			*SIE*	39 *GRa*/strong in Ponape
13	1913	0331	0340	49.5	N	178.	W	s	7.3	*DU*	39/1
14	1913	0622	1349	48.	N	178.	W	s	7.2	*DU*	39 near 1
15	1916	0206	2151	48.5	N	178.5	E	s	7.7	*DU*	39 near 1
16	1916	1031	1530	46.5	N	160.	E	s	7.7	*DU*	19 near 39
17	1917	0617	08	34.	N	162.	W			ISS	39 *GRa*/epicenter ?
18	1918	0521	11	11.7	N	176.0	E			ISS	39 *GRa*/anywhere in 39
19	1918	0621	03	22.	S	141.	W			ISS	39 *GRa*/very doubtful
20	1918	1001	00	30.	N	174.	W			ISS	39 *GRa*/very unsatisfactory
21	1920	0121	06	05.	N	148.	E			ISS	18 *GRa*/poor
22	1921	0210	19	03.	S	177.5	W			ISS	39 *GRa*/? but maybe correct
23	1921	1006	15	43.	N	170.	E			ISS	39 *GRa*/deep in Kuril?
24	1923	0514	02	16.	N	153.5	E			ISS	39 *GRa*/data poor
25	1923	0919	08	12.5	N	168.	E			ISS	39 *GRa*/Tonga salient?
26	1924	0617	20	01.	N	129.	W			ISS	39 *GRa*/data hardly sufficient
27	1925	0516	10	09.	N	155.	E			ISS	39 *GRa*/can't be rejected
28	1926	0929	05	09.	N	155.	E			ISS	19 *GRa*/2 shocks, very rough
29	1927	0827	12	40.5	N	160.5	E			ISS	39 *GRa*/worthy to examine
30	1927	1125	19	01.	N	129.	W			ISS	39 *GRa*/data insufficient
31	1927	1128	10	03.	S	177.5	W			ISS	39 *GRa*/maybe in Tonga
32	1928	0303	17	09.	N	155.	E			ISS	39 *GRa*/data insufficient
33	1928	1011	23	26.8	N	172.0	E			ISS	39 *GRa*/twin, 52N, 175E ?
34	1928	1021	15	07.	S	178.	E			ISS	39 *GRa*/near Apia
35	1929	1224	04	03.	S	172.	W			ISS	39 *GRa*/very uncertain
36	1930	0421	21	40.5	N	160.5	E			ISS	39 *GRa*/very uncertain
37	1931	0806	1521	00.		151.	E		d	*GR*	15/39

(*continued*)

Table 5b. *Continued*

No.	Year	Date	Time	Lat. °	Lon. °	h	M	Ref.	Seism. Reg/Remarks
38	1933	0104	2110	28. N	126.5 W		5.5	*GR*	39 near 4
39	1933	1002	1359	10. S	166. E		6.0	*GR*	14/39
40	1938	0620	1402	06. N	119. W		d	*GR*	44 *GRb*/certainly in the Pacific
41	1941	0517	0224	10. S	166.25E		7.4	*GR*	14/39
42	1941	0723	0930	47.75N	152.5 E	120	6.0	*GR*	1/39 *GRb*/interm. in 19?
43	1945	0630	0531	17. N	115. W	s	6.75	*GR*	5
44	1949	1210	1915	04. N	129. W	s	5.75	*GR*	39
45	1953	0419	2247	50. N	179.5 W	n	d	*RO*	1/39
46	1955	1122	0324	24.4 S	122.6 W	n	6.6	*RO*	39 cf. Sykes (1963)
47	1959	0209	0442	50.0 N	177.6 W	n	6.2	*RO*	1/39
48	1961	0723	1437	06.8 N	123.5 W	n	5.9	*RO*	39
49	1963	0914	1616	33.6 S	126.7 W	n	5.5	*RO*	39
50	1963	1006	0848	21.9 N	127.4 W	n	e	*RO*	39 near 4
51	1965	0127	0144	18.9 N	176.6 E	n	e	*RO*	39
52	1965	0306	1110	18.4 S	132.9 W	35	5.9	*RO*	39

References: *GRa* = Gutenberg and Richter (1941); *GRb* = Gutenberg and Richter (1945); *SIE* = A. Sieberg, Hndb. d. Geophys. Bd. IV, p. 919.

Table 5c. Earthquakes in the Pacific Basin (*GR* Seism. Reg. No. 39), excluding Hawaiian Is. Region (FE Geogr. Reg. No. 613), Line Is. Region (FE Geogr. Reg. No. 620), Society Is. Region (FE Geogr. Reg. No. 628) and the areas outside FE Seism. Reg. No. 39, for 1964-1974. (data after ISC Bull., *FE*: Flinn-Engdahl).

No.	Year	Date	Time	Lat. °	Lon. °	h	m_b	Reg. No.	No. Obs.
01	1964	1120	0059	00.6 N	156.3 E	33	4.2	614	8
02	1965	0127	0144	18.9 N	176.6 E	33	4.4	611	8
03	1965	0306	1110	18.4 S	132.8 W	31	5.4	632	103
04	1966	0918	0640	18.4 S	132.9 W	67	5.0	632	46
05	1966	0924	0857	12.1 N	130.8 W	92	4.9	611	46
06	1967	0409	1752	39.0 N	163.0 E	33	–	611	14
07	1967	0523	1203	08.0 S	170.0 W	33	4.4	625	9
08	1967	1112	1528	25.6 S	169.9 W	33	4.3	632	7
09	1967	1228	0730	44.8 N	135.6 W	33	4.2	611	17
10	1968	0428	0418	44.8 N	174.6 E	36	5.5	611	157
11	1968	0428	0623	44.9 N	174.7 E	33	4.4	611	32
12	1969	0503	0822	08.3 N	175.6 W	82	5.0	611	62
13	1969	0708	0409	05.5 N	154.7 E	33	–	614	5
14	1969	0719	2051	08.3 S	174.6 W	33	4.3	625	8
15	1969	1029	0702	09.0 S	177.0 W	33	3.8	632	5
16	1969	1108	1705	12.0 S	140.3 W	33	4.3	630	6
17	1970	0101	0725	09.9 S	178.3 E	33	4.2	623	6
18	1970	1109	0204	02.8 N	150.4 E	33	–	614	5
19	1972	0302	0217	07.5 S	175.1 E	00	4.3	632	8
20	1972	0412	2142	06.2 N	130.3 W	00	4.4	611	14
21	1972	0603	1004	29.5 N	163.7 W	00	–	611	8
22	1973	0415	1059	21.0 N	154.5 W	00	4.2	612	9
23	1974	0312	1511	08.8 N	151.0 E	11	5.4	614	136
24	1974	0314	2322	08.7 N	151.1 E	00	4.8	614	10
25	1974	0502	2039	08.7 S	178.1 E	00	–	623	13
26	1974	0509	0952	05.2 S	129.2 W	33	–	632	5
27	1974	0518	1739	01.7 N	153.7 E	33	–	614	6

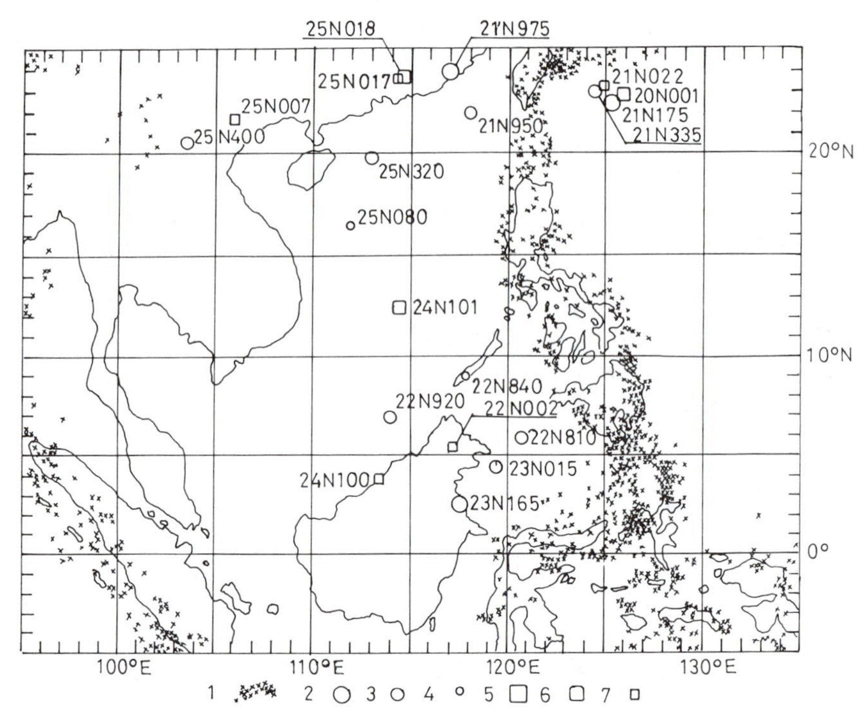

Figure 1. Rare earthquakes far from the seismic zones in South East Asia. (1) Seismic zones; (2) $M_{GR} \geq 7$; (3) $M_{GR} \geq 6$; (4) $M_{GR} < 6$; (5) $M_{RO} \geq 7$; (6) $M_{RO} \geq 6$; (7) $M_{RO} < 6$.

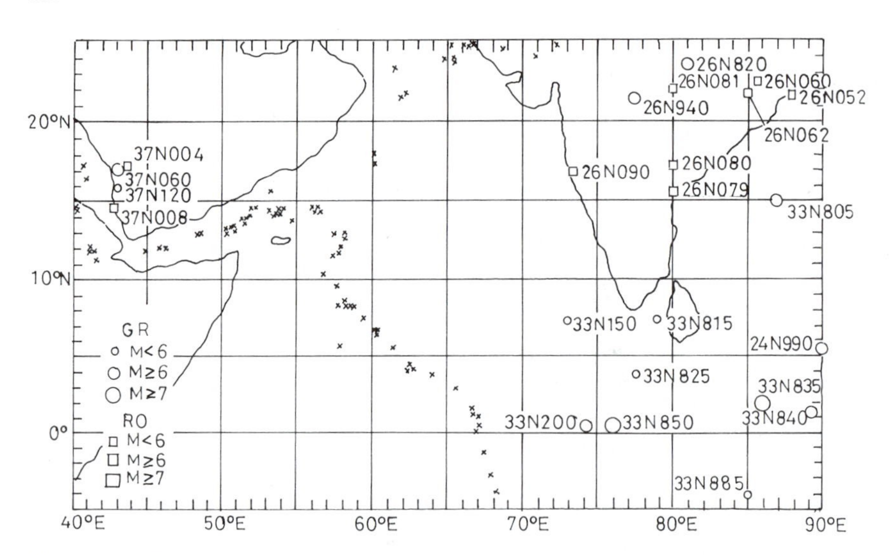

Figure 2. Rare earthquakes far from the seismic zones in Peninsular India, Indian Ocean and Southern Arabia. (Symbols are the same as in Figure 1).

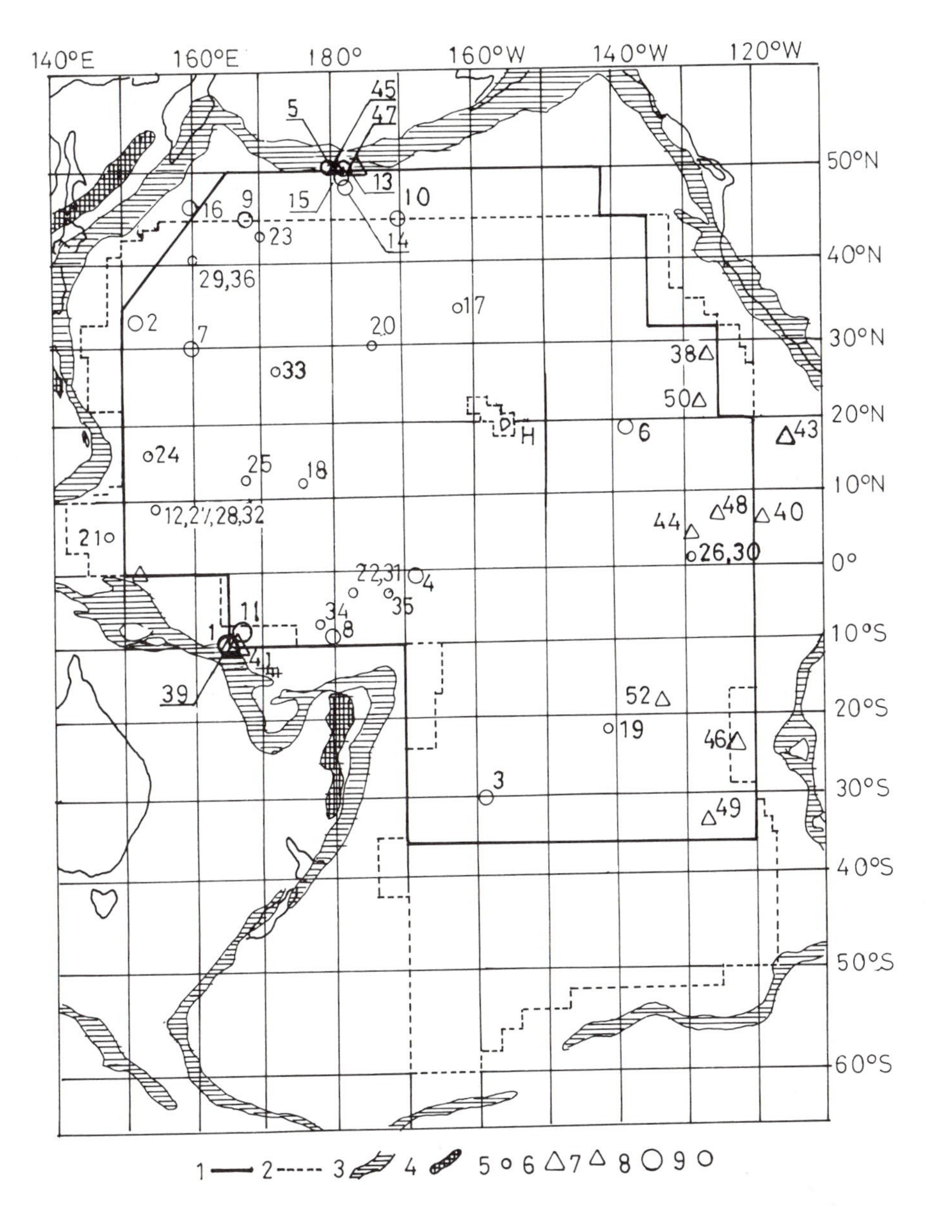

Figure 3. Rare earthquakes far from the seismic zones in the Pacific Basin (Seismic Region 39, excluding Hawaiian Is. Region) for 1900-1965. (1) Boundary of GR Seismic Region 39; (2) Boundary of FE Seismic Region 39; (3) Seismic zones (shallow earthquakes); (4) Seismic zones (deeper earthquakes); (5) Epicenter in ISS; (6) Epicenter in *GR* and *RO*, $M \geq 7$; (7) Epicenter in *GR* and *RO*, $M < 7$; (8) Epicenter in *DU*, $M \geq 8$; (9) Epicenter in *DU*, $M \geq 7$; Numbers with epicenters should be referred to Nos. in Table 5b.

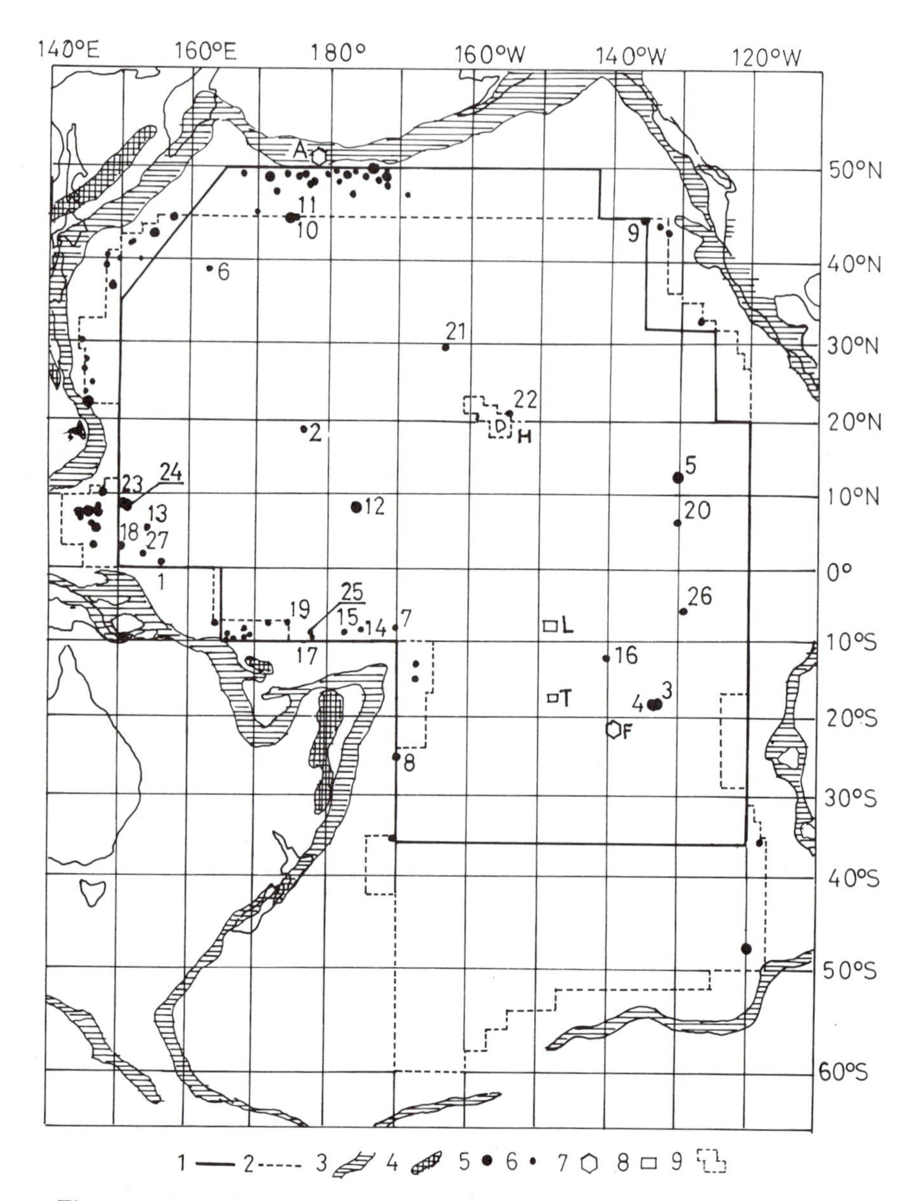

Figure 4. Rare earthquakes far from the seismic zones in the Pacific Basin (Seismic Region 39, excluding Hawaiian Is. Region, FE Geographic Region 613, Line Is. Region, FE Region 620 and Society Is. Region, FE Region 628) for 1964-1974 after ISC data. (1-4) same as in Figure 3; (5) Epicenter by ISC, $m_b \geq 4.5$; (6) Epicenter by ISC, $m_b < 4.5$; (7) Nuclear Weapon Test Site, A – Amchitka, F – Muruloa; (8) Occasional active area, L – Line Is., T – Society Is.; (9) H – Hawaiian Is., FE Region 613. Numbers with epicenters are referred to the Nos. in Table 5c.

Many earthquakes which were located far inside the Pacific Basin by the International Seismological Summary (ISS) were examined by Gutenberg and Richter (1941) [hereafter *GRa*], but their locations could not be definitely placed. However, *GRa* commented that some events were not fully rejected and worth examining (see Remarks in Table 5b). Moreover, considering that the International Seismological Center (ISC) located many events far inside the Pacific Basin as shown in Table 5c and Figure 4, it seems necessary to investigate the locations of all those given in Table 5, for which their seismograms would be important. It is hoped that the filming project will not exclude such rare earthquakes even though their magnitudes are smaller than the thresholds.

2.4. Selection of Destructive Earthquakes

It is well known that many small-magnitude earthquakes have caused catastrophic damage. The Agadir, Morocco earthquake (February 29, 1960; $M_{RO} = 5.9$; death toll about 15,000) was one of the most remarkable examples and it should not be forgotten in the filming project.

In selecting earthquakes for the filming, damage of earthquakes is also considered beside magnitude, and *SE22* includes many destructive earthquakes having magnitude less than 7. (Regrettably, the Agadir earthquake was not included.)

Although we have several standard magnitude catalogs, as mentioned in the previous sections, we do not have a standard earthquake damage catalog for the world. *RN* is probably one of the most comprehensive catalogs with many damaging earthquakes included. Båth (1979) prepared a catalog using the number of deaths as a common measure to compare the damage from different earthquakes. As shown in Table 6, deaths in *RN* and Båth (1979) are quite different for some earthquakes. *RN* gives no death toll for some events listed by Båth (1979), and vice versa.

Death tolls for some earthquakes not listed by *RN* and Båth (1979) are supplemented from a few local catalogs in Table 6. Geographical inhomogeneity is also observed in damaging earthquakes. Death tolls are not accurate, and differ from region to region. Because Table 6 uses a selection threshold of 100 deaths, it does not list the Accra, Ghana earthquake of June 22, 1939, $M = 6.5$, deaths 66 (*RN*). No global catalog except Ganse and Nelson (1981) [hereafter *SE27*] lists Cartago, Costa Rica earthquake of May 4, 1910, deaths 272. *SE27* is an important contribution, but it needs many corrections and additions. It is not advisable to use as is. I have sent some corrections and additions privately to WDC-A. Cooperation of many colleagues is needed to improve this catalog.

3. Selection of Stations

3.1. Necessity of Auxiliary Stations

In selecting a representative station of a region, it must be noted that one station may have been operated well in a certain period, and another station in another period. Of course, there are stations that have been operated almost without interruption during last several decades. However, they are rather exceptional. Many old stations had been interrupted in operations by the two World Wars and/or other unavoidable reasons. Studying the history of old stations is a prerequisite for

Table 6. Provisional list of destructive earthquakes for 1900-1962, with death toll over 100. M are $M_{AN'}$ for 1902-1903, M_{GR} for 1904-1952 and M_{RO} for 1953-1962. Otherwise reference is given in Rem. column.

Year	Date GMT	Region (Locality)	Number of deaths *RN*	Båth	Other Ref.	*M*	Rem.
1902	02.13	Kavkaz (Shemakha)	2,000	–		6.0	*K*
1902	04.18	Guatemala	2,000	–		7.5	
1902	08.22	Turkestan	2,500	–		7.7	
1902	12.16	Turkestan (Andijan)	700	4,500		6.4	*KS*
1903	04.28	Armenia	2,000	–		6.3	*K*
1903	09.25	Iran (Turshiz)	–	–	350 *AM*	5.9	*AM*,(d)
1904	11.05	Taiwan (Chiayi)	145	–		6.3	*HS*
1905	04.04	India (Kangra)	X	19,000		8.0	
1905	09.08	Italy (Calabria)	2,000	2,500		6.8	*AN'*
1905	11.08	Greece (Chalkidiki)	2,000	–		6.8	*AN'*
1906	01.31	Colombia (Tumaco)	X	1,000		8.6	
1906	03.16	Taiwan (Chiayi)	X	1,300	1,258 *HS*	6.8	*AN'*
1906	04.18	California (San Francisco)	600	700		8.25	(d)
1906	08.17	Chile (Valparaiso)	X	20,000		8.4	
1907	01.14	Jamaica (Kingston)	–	1,600		6.0	*B*,(d)
1907	10.21	Tajikistan (Karatag)	X	12,000		8.0	
1907	10.23	Italy (Calabria)	300	–		5.9	
1908	12.28	Italy (Messina)	110,000	83,000		7.5	
1909	06.03	Sumatra	200	–		7.6	
1910	05.04	Costa Rica (Cartago)	–	–	272 *GV*		(d)
1911	01.03	Turkestan (Verngi)	X	450		8.4	
1911	06.07	Mexico (Michoacan)	1,300	–		7.75	
1912	08.09	Turkey (Marmara Sea)	3,000	1,950	216 *ER*	7.75	
1913	06.14	Bulgaria (Tirnovo)	±500	–		6.75	
1913	11.04	Perú	120	–	150 *SI*		(a)
1914	05.08	Italy (Catania)	120	–		4.9	*K*
1915	01.13	Italy (Avezzano)	35,000	29,980		7.0	
1917	01.20	Indonesia (Bali Is.)	1,300	–			
1917	12.25	Guatemala	100	–		6.	*U*1
1918	02.13	S. Coast of China	200	–		7.3	
1918	08.15	Mindanao	100	–		8.25	
1920	01.04	Mexico (Vera Cruz)	4,000	–			
1920	09.07	Italy (Carrara)	556	–		5.75	
1920	11.26	Albania (Tepelene)	600	–		6.0	*K*,(b)
1920	12.16	China (Kansu, Shansi)	100,000	100,000		8.5	
1920	12.17	Argentina (Mendoza)	400	–			
1922	11.11	Chile-Perú (Atacama)	X	600		8.3	
1923	03.24	Szechwan, China	500	–		7.3	
1923	05.25	Iran (Turbat-Haklari)	±5,000	–	770 *AM*	5.8	*AM*,(c)
1923	09.01	Japan (Tokyo)	142,807	99,330		8.2	
1923	09.17	Iran (Bujnurd)	157	–	157 *AM*	6.5	
1923	09.22	Iran (Lalehzat)	–	–	200 *AM*	6.9	
1923	11.15	Shansi, China	1,500	–			
1923	12.14	Colombia-Ecuador	85	–	300 *RA*		
1924	09.13	Turkey (Erzurm)	200	–	X *ER*	6.75	
1924	11.12	Java (Wonosobo)	609	–			
1925	01.09	Kavkaz (Ardahan)	200	–		5.8	*K*
1925	03.16	Yunnan, China (Tali)	6,500	5,000		7.1	

(*continued*)

Table 6. *Continued*

Year	Date GMT	Region (Locality)	Number of deaths *RN*	Båth	Other	Ref.	*M*	Rem.
1925	05.23	Japan (Tajima)	428	–			6.75	
1926	06.28	Sumatra	222	–			6.75	
1926	10.22	Armenia (Erevan)	400	–	X	*ER*	d	
1927	03.07	Japan (Tango)	2,925	3,020			7.75	
1927	05.22	Tsinghai, China (Nanshan)	±50,000	200,000			8.0	
1927	07.11	Palestina	192	–			6.25	
1928	04.14	Bulgaria (Marushka)	103	–			6.75	
1928	12.01	Chile	218	–			8.0	
1929	01.17	Venezuela (Cumana)	200	–			6.9	
1929	05.01	Iran-USSR border	3,253	3,300	3,200	*AM*	7.1	
1930	05.05	Burma	6,000	–			7.3	
1930	05.06	Turkey-Iran border	3,000	–	2,500	*AM*	7.2	
1930	07.23	Italy (Ariano)	1,883	1,430			6.5	
1930	09.22	Tajikistan (Stalinabad)	175	–			6.25	
1930	11.25	Japan (North Izu)	272	–			7.1	
1931	02.02	New Zealand (Hawke's Bay)	285	225			7.75	
1931	03.31	Nicaragua (Managua)	2,500	–			d	
1931	04.27	Armenia (Nakhitchevan)	390	–			6.5	
1932	09.26	Macedonia (Chalkidihi)	491	–			6.9	
1932	12.25	Kansu, China	many	–			7.6	
1933	03.02	Japan (off Sanriku)	3,008	2,990			8.5	
1933	03.11	California (Long Beach)	–	–	115	*U2*	6.25	
1933	08.25	Szechwan, China	100	–			7.4	
1934	01.15	Nepal-India (Bihar)	7,253	10,700			8.3	
1935	04.20	Taiwan (Hsinchu-Taichung)	3,276	3,280	3,276	*HS*	7.1	
1935	05.30	Pakistan (Quetta)	many	30,000			7.5	
1936	12.19	El Salvador (San Vicente)	400	–				
1938	04.19	Turkey (Kirsehir)	–	–	155	*ER*	6.75	
1939	01.25	Chile (Concepcion)	30,000	28,000			7.75	
1939	12.26	Turkey (Erzincan)	32,741	30,000	X	*ER*	8.0	
1940	05.24	Perú	±500	–	179	*SI*	8.0	
1940	11.10	Romania	350	–			7.4	
1941	02.16	Iran (Muhammadabad)	–	–	680	*AM*	6.25	
1941	12.16	Taiwan (Chiayi)	X	–	358	*HS*	7.1	
1943	09.10	Japan (Tottori)	1,083	1,190			7.4	
1943	11.26	Turkey	X	–	4,000	*AM*	7.6	
1944	02.01	Turkey	X	–	4,000	*AM*	7.4	
1944	12.07	Japan (Tokaido)	998	1,000			8.0	
1945	01.12	Japan (Mikawa)	1,961	1,900			7.1	
1945	11.27	Pakistan	X	–	300	*AM*	8.25	
1946	04.01	Unimk Is.(Tsunami in Hawaii)	163	–			7.4	
1946	11.10	Perú (Ancash)	X	1,400	1,613	*SI*	7.25	
1946	12.20	Japan (Nankaido)	1,330	1,330			8.2	
1947	09.23	Iran (Muhammadabad)	–	–	570	*AM*	6.75	
1948	06.28	Japan (Fukui)	3,895	5,390			7.3	
1948	10.05	Iran-USSR border(Ashkhabad)	X	400			7.3	
1949	08.05	Ecuador (Ambato)	–	6,000			6.75	(d)
1950	05.21	Perú (Cuzco)	–	–	120	*SI*	6.0	*SI*
1950	08.15	India-Tibet (Assam)	574	1,530			8.6	
1950	07.09	Colombia (N. Santander)	–	–	106	*RA*	7.0	*RA*
1953	02.12	Iran (Mazandaran)	–	970	X	*AM*	6.5	(d)

(continued)

Table 6. *Continued*

Year	Date GMT	Region (Locality)	Number of deaths *RN*	Båth	Other Ref.	*M*	Rem.
1953	03.18	Turkey (Yenice)	244	240	224 *RO*	7.3	(f)
1953	08.12	Ionian Sea	455	460		7.1	
1954	09.09	Algeria (Orléanville)	–	1,250	1,243 *RO*	6.7	(d)
1955	03.31	Mindanao	432	430		7.6	
1956	06.09	Afganistan (Kabul)	±5	220	400 *RO*	7.6	
1957	07.02	Iran (Bandpay)	2,000	130	1,500 *AM*	7.1	
1957	12.13	Iran (Hamadan)	1,392	1,130	2,000 *RO*	7.2	(e)
1960	02.29	Morocco (Agadir)	–	15,000		5.9	(d)
1960	04.24	Iran (Lar)	–	450	400 *RO*	6.0	(d)(e)
1960	05.22	Chile	1,743	5,700		8.3	
1962	09.01	Iran (Qazvin)	10,000	12,230	12,200 *AM*	7.3	

In death toll column: (–) means the earthquake is not listed; (X) means the earthquake is listed but without death toll.

References: *AM* = Ambraseys and Melville (1982); *B* = Båth (1979); *ER* = Ergin *et al.* (1967); *GV* = Gonzalez-Viquez (1910); *HS* = Hsu (1980); *K* = Kárník (1968); *KS* = Kondorskaya and Shebalin (1977); *RA* = Ramirez (1975); *SI* = Silgado-F. (1978); *SS* = Sulstarova and Kociaj (1975); *U*1 = Espinosa (1976); *U*2 = Coffman and von Hake (1973).

Remarks: (a) Date 11.10 in *RN* should be 11.04 after *SI*; (b) Date 12.01 in RN should be 11.26 after *SS*; (c) Date 05.28 in *RN* should be 05.25 after *AM*; (d) not included in *SE*22; (e) deaths not given in *AM*; (f) deaths not given in *ER*.

preparing a global directory of seismic stations. The project of preparing a global directory of seismological stations by WDC-A is very important in this respect. See Poppe (1980), for U.S.A. and Canada, and Miyamura (1985) for Eastern Asia.

In addition, even during the well operated period, a good quality station sometimes failed to record an earthquake which was recorded by a less qualified station in the same region. For example, let us examine the ISS, October - December, 1953. Nanking and Zo-Se reported readings for about 40 earthquakes, but in addition, Nanking reported readings for about 10 more earthquakes, suggesting Nanking could be a representative station of the region. However, it seems advisable to select Zo-Se also as an auxiliary station. The readings of October 24, 1953, 23:19, 35.3 S, 180 E, Region 12, No. 001, $M_{RO} = 6.2$ are reported by Zo-Se but not by Nanking, although similar size earthquakes from the same region: (1) November 16, 1953, 17:17, 21.2 S, 168.5 E, Region 14, No. 026, $M_{RO} = 6.0$; (2) November 25, 1953, 17:48, 17.9 S, 176.5 E, Region 12, No. 210, $M_{RO} = 6.0$; and (3) November 27, 1953, 23:01, 18.3 S, 176.5 E, Region 12, No. 209, $M_{RO} = 6.2$ are reported by Nanking, but the last two are not reported by Zo-Se.

Praha and Cheb are also a similar pair of stations. For the last 3 months of 1953, Praha reported about 30 events in ISS, but Cheb reported only 20. However, for the earthquake October 13, 1953, 08:53, 30.0 N, 114.0 W, Region 4, No. 017, $M_{RO} = 6.2$, Cheb reported readings but Praha did not.

Another circumstance to necessitate the auxiliary stations will occur as follows. It often happens that the seismograms of an important earthquake cannot be found in the chronologically kept seismogram files, mostly because the seismograms were used by certain seismologists and have not been returned to the original file, while the seismograms at the auxiliary stations have been kept untouched. When we asked for the seismograms of the 1935 Hsinchu-Taichung, Taiwan earthquake, Stuttgart has sent us the seismograms of its auxiliary stations, Messtetten and Ravensburg, because the records of Stuttgart were lost.

Considering the above comments, it is recommended that an auxiliary station be selected in the same region of a representative station. It is probably easy to include a visit to the auxiliary station in the itinerary of the filming team.

3.2. Name, Code, and Coordinates of Stations

Name, code, and coordinates of the selected stations are, of course, basic for the filming project. International 3- or 4-letter codes in use are most convenient to identify the stations, especially when there are several stations having the same names, or if a station was moved its location several times.

The U. S. National Earthquake Information Service (NEIS) has a general policy to give a different code for a new station that is moved more than 1 km from the old station. This policy is reasonable for teleseismic studies, as far as arrival times are concerned. However, for the study of local earthquakes, 1 km may be too large. Moreover, if we study the seismogram, including amplitude and period characteristics, even a separation of less than 1 km may become important, especially where the foundations are different. It appears to be better to label seismograms of old and new stations with same name, but to give them different station codes.

In preparing the Directory of Seismographic Stations in Asia (Miyamura, 1985), I have tried to give tentative station codes to differentiate between old and new station sites.

Many networks use station codes rather carelessly, which eventually introduce confusion. To avoid duplicating codes, an up-to-date list of station codes should be published frequently, and institutions operating seismological stations should be asked not to use codes already registered by other stations.

In addition, ISC should examine duplicate station codes and cancel one of them, e.g., GZS and GOZ for Gozaisho, Japan; KRZ and KAZ for Karuizawa (not Kuruizawa), Japan; KMT and KNT for Kinomoto, Japan; AAI and AMO for Ambon, Indonesia, etc. Perhaps it is better to denote KAZ for the site (36.3333 N, 138.6000 E) used in 1941-49 near KRZ, and to denote KRZ for the site (36.3400 N, 138.5517 E) used since 1911 to the present. Similarly, we may use AMO for Ambon before World War II, and AAI for the new station in the same place since 1975.

Hamburg was one of the representative stations in Europe before World War II, but was destroyed by a catastrophic air raid in 1945. After the War, a new Hamburg station was established in Harburg, on the other side of the River Elbe. It is registered by ISC as Hamburg with station code HAM. A closed station of Hamburg, N.Y., U.S.A. is identified by the code HMB. But the historically important station at Hamburg in the city (reportedly moved once) is not registered in ISC, although its historical seismograms are kept at the Harburg station. Many famous old stations, including some of John Milne's worldwide network stations are not listed by ISC, although many closed stations of minor importance are registered with proper station codes.

In using the seismograms and/or station bulletins, careful examination of the station coordinates is necessary. For the April 20, 1935 Hsinchu-Taichung, Taiwan, earthquake, ISS lists the readings at the Leipzig station and I have requested a copy of the seismogram. The records reveal that from January 1, 1935, seismographs at the new outpost station of Leipzig, Collmberg (CLL), were used for the readings, but ISS did not change the station name (B. Tittel, 1985, personal communication). ISC lists Leipzig as LEI and mentions Leipzig Collm as an alternate name of Collmberg, and so there is no ambiguity.

In general, station coordinates are found in the station directories and station bulletins. We often find different coordinates for stations with the same name. However, the difference of coordinates does not necessarily correspond to the actual move of the seismometer sites. Sometimes the difference is due to printing mistakes, and sometimes it represents the move of the office building and not the seismometer site. The first Bergen station in the Bergen Museum prior to 1968 was 300 m north of the station in Vilari 9 old institute, and was never moved until 1968. However, station bulletins gave a couple of different coordinates, because of inaccurate determinations or printing mistakes (M. Sellevoll, 1977, personal communication).

4. Concluding Remarks

In this paper I have pointed out many errors in several valuable publications. I hope it is understood that the intention is not to degrade the importance of these publications, but rather to recommend the readers to use them properly.

In addition, I am afraid that this paper may include many mistakes due to my ignorance and/or carelessness. If the readers find any mistakes, please inform me directly and also publish the errors to avoid misleading the seismological community.

ACKNOWLEDGEMENTS

I would like to express my sincere gratitude to Drs. W. H. K. Lee, U.S. Geological Survey, and H. Meyers, National Geophysical Data Center, N.O.A.A., who gave me the opportunity to cooperate the microfilming project of WDC-A and to attend UNESCO/IASPEI workshop, 1983, Hamburg, and IASPEI Symposium No. 14 on Historical Earthquakes and Seismograms, 1985, Tokyo.

REFERENCES

Abe, K. (1981). Magnitudes of large shallow earthquakes from 1904 to 1980, *Phys. Earth Planet. Interiors*, **27**, 72-92.

Abe, K. (1984). Complements to "Magnitudes of large shallow earthquakes from 1904 to 1980", *Phys. Earth Planet. Interiors*, **34**, 17-23.

Abe, K. and H. Kanamori (1979). Magnitudes of great shallow earthquakes from 1953 to 1977, *Tectonophysics*, **62**, 191-203.

Abe, K. and S. Noguchi (1983a). Determination of magnitude for large shallow earthquakes, 1898-1917, *Phys. Earth Planet. Interiors*, **32**, 45-59.

Abe, K. and S. Noguchi (1983b). Revision of magnitudes of large shallow earthquakes, 1897-1912, *Phys. Earth Planet. Interiors*, **33**, 1-11.

Ambraseys, N. N. and C. P. Melville (1982). *A history of Persian earthquakes*, Cambridge Univ. Press, 219 pp.

Båth, M. (1979). *Introduction to Seismology*, 2nd ed., Birkhäuser Verlag, Basel-Boston-Stuttgart, 428 pp. (cf. p. 139-149)

Christoskov, L. V. (1983). *Razvitia na teoretichnite osnovi i prilodzhni aspekti v magnitudnata klasifikatsiya na zemetreseniyata. Avtoreferat na disertatsiya*, Bulgarska Akademiya na Naukite, Sofia, 54 pp. (cf. p. 36).

CMO (1950). Omona Jishin no Kibo-hyo (Magnitude list of main earthquakes), 1885-1950. Appendix to the *Seism. Bull. Central Meteorological Observatory* (the predecessor of Japan Meteorological Agency) for the year 1950.

Coffman, J. L. and C. A. von Hake (Editors) (1973). *Earthquake History of the United States*, NOAA, Boulder, CO, 208 pp.

Duda, S. J. (1965). Secular seismic energy release in the Circum-Pacific belt, *Tectonophysics*, **2**, 409-452.

Ergin, K., U. Guclu, and Z. Uz (1967). A catalog of earthquakes for Turkey and surrounding area (11-1964), T. U. Istanbul, 169 pp.

Espinosa, A. F. (Editor) (1976). The Guatemalan Earthquake of February 4, 1976. A Preliminary Report, *U.S. Geol. Surv. Prof. Paper 1002*, 90 pp.

Ganse, R.A. and J. B. Nelson (1981). Catalog of Significant Earthquakes, 2000 B.C.-1979, including quantitative casualties and damage, *World Data Center A Report SE-27*, 145 pp.

Glover, D. P. (1980). Historical Seismogram Filming Project: Second Progress Report, *World Data Center A Report SE-24*, 63 pp.

Glover, D. P. and H. Meyers (1981). Historical Seismogram Filming Project: Third Progress Report, *World Data Center A Report SE-28*, 76 pp.

Glover, D. P. and H. Meyers (1982). Historical Seismogram Filming Project: Fourth Progress Report, *World Data Center A Report SE-33*, 54 pp.

Gonzalez-Viquez, C. (1910). Temblores, terremotos, inundaciones y erupciones volcánicas en Costa Rica, 1608-1910, Tipografia de Avelino Alsina, San Jose, Costa Rica, 200 pp.

Gutenberg, B. (1956). Great earthquakes, 1896-1903, *Trans. Am. Geophys. Union*, **37**, 608-614.

Gutenberg, B. and C. F. Richter (1941). Seismicity of the Earth, *Geol. Soc. Am. Special Paper*, **34**, 131 pp.

Gutenberg, B. and C. F. Richter (1945). Seismicity of the Earth (Supplementary Report), *Bull. Geol. Soc. Am.*, **56**, 603-667.

Gutenberg, B. and C. F. Richter (1954). *Seismicity of the Earth and Associated Phenomena*, 2nd ed., Hafner Publ. Co., New York-London, 310pp.

Hsu, Ming-Tung (1980). Earthquake Catalogues in Taiwan (1964- 1979), Res. Center of Earthq. Eng., Taiwan Univ., 77 pp.

Kárník, V. (1968). *Seismicity of the European Area. Part 1*, Academy of Sciences, Praha, 364 pp.

Kondorskaya, N. B. and N. V. Shebalin (1977). *New Catalogue of Strong Earthquakes in USSR from old time to 1975*, Nauka, Moscow, 534 pp.

Meyers, H. and W. H. K. Lee (1979). Historical Seismogram Filming Project: First Progress Report, *World Data Center A Report SE-22*, 66 pp.

Miyamura, S. (1978). Magnitudes of large earthquakes not included in the Gutenberg-Richter's magnitude catalog, *Tectonophysics*, **49**, 171-184.

Miyamura, S. (1985). Directory of World Sesimograph Stations, Volume II: East Asia – China, Japan, Korea, and Mongolia, *World Data Center A Report SE-41*, 242 pp.

Poppe, B. B. (1980). Directory of World Seismograph Stations, Volume I: The Americas – Part 1, United States, Canada, and Bermuda *World Data Center A Report SE-25*, 465 pp.

Ramirez, J. E. (1975). *Historia de los terremotos en Colombia*, Instituto Geografico, Bogota, Colombia, 250 pp.

Rothé, J. P. (1969). *La séismicité du globe, 1953-1965*, UNESCO, Paris, 336 pp.

Silgado-F., E. (1978). Historia de los sismos mas notables ocurridos en el Perú (1513-1974), Instituto de Geologia y Mineria, Boletin No. 3, Lima, Perú, 130 pp.

Sulstarova, E. and S. Kociaj (1975). *Katlogu itermeteve te shqiperise*, Akademia e Shkencave, Tirane, Albania, 224 pp.

Sykes, L. R. (1963). Seismicity of the South Pacific Ocean, *J. Geophys. Res.*, **68**, 5999-6006.

Usami, T. (1975). Sekai Dai-jishin Nenpyo (Chronological list of world large earthquakes), in *Rika Nenpyo (Science Almanac of the Tokyo Astron. Obs.)*, Maruzen, Tokyo, 181-215.

Resources, Organization, and Microfilming of Historical Seismograms in China

Kexin Qu
Institute of Geophysics State Seismological Bureau
Beijing, China

1. Review

China, a country of high seismicity, has a long history of seismic observations. China also has abundant data on historical earthquakes. To make full use of these data, Chinese seismologists and historians, have compiled and edited the *Chronological Table of Chinese Earthquakes* (Academica Sinica, 1956). The editorial work of a revised and enlarged edition of the *Compilation of Historical Materials of Chinese Earthquakes* (Xie and Cai, 1983) was completed in 1982. This five-volume book is a general collection of historical earthquake data in China.

Instrumental data plays a main role in quantitative seismological analysis. Since the founding of the first seismic station in China in 1897, more than 400 seismic stations have been established. Although they have collected much data for seismic research, some of the data have been lost or damaged. Also, seismograms of strong earthquakes are frequently required simultaneously by several scientists, and it is difficult to make proper arrangements. An effective method of solving this problem is to microfilm the seismograms. Therefore, with the support of the Central Seismological Group, the Institute of Geophysics in 1975 began to cooperate with microfilm companies to manufacture the required equipment. Between 1975 and 1981, various pieces of equipment were produced, including camera, processor, duplicator, microfilm projector, digitizer and others.

The exchange of microfilmed seismograms which was included as a part of the Sino-U.S. Seismological Cooperation Project in 1981, did much to promote this work. At the Regional Workshop of the IASPEI/UNESCO Working Group on Historical Seismograms (Qu, 1982), it was decided that the Sheshan (formerly Zi-ka-wei), Lanzhou, and Taipei stations of China were three of the 30 global stations whose historical seismograms would be microfilmed. Because this project was supported by the Central Seismological Group, the work progressed smoothly.

2. A Brief History

More than 1800 years ago, the seismoscope was used for detecting earthquakes in China. However, normal seismograms were not recorded until the later part of the 19th century. In 1897, a seismograph station was first set up in Taipei, Taiwan Province, and in 1904, seismograph stations were established in Sheshan (formerly Zi-ka-wei, Shanghai) and Dalian. The Jiufeng seismic station (set up by Prof. S. P. Lee) began operation in 1930, and the Beijige (Nanjing) seismic station began recording in 1932. Seismological bulletins were compiled and published separately by both stations. Recording at these two stations was suspended in 1937 because of the war with Japan. After an interruption of 10 years, recording at the Nanjing seismic station was restored.

In 1954, many stations equipped with Type 51 mechanical seismographs were established in the Huanghe (Yellow River) Valley and other parts of China. The number of stations on the China mainland increased to 20 by 1957. In July of 1957, 12 of these stations were upgraded with SK Type (Kirnos Type) medium-period broadband seismographs, and SH Type (Halin Type) short-period seismographs.

Since 1959, networks consisting of individual stations having seismographs with electronic amplification have been set up around Beijing, Xinfengjiang and Xichang for microearthquake observations. By April 1966, a telemetered seismograph network had been installed around Beijing and had begun recording. Because of the successive occurrence of many strong earthquakes, the Chinese government and people have given tremendous support to seismology studies. The number of seismograph stations in China has increased rapidly and now exceeds 400. Smoked-paper and/or pen-and-ink recorders and quartz-crystal clocks are commonly used. To increase the ability for detecting microearthquakes in the plains regions, China has installed bore-hole seismographs at some stations.

3. Seismic Observation System

At present, most seismic stations in the country of China are engaged in comprehensive observations; i.e., besides seismic observations, they also make observations on geomagnetics, gravity, crustal deformation, and others. These stations can be roughly divided into three types, as shown in Figure 1:

(1) Standard seismic stations equipped with three-component short-period seismographs, three-component broadband seismographs, and three-component long-period seismographs; there are 22 stations of this type in operation.

(2) Regional seismic stations having three-component short-period seismographs; there are now about 70 such stations and over 300 local seismic stations.

(3) Regional telemetered seismograph networks; there are six such networks at present.

4. Instrument System

At present, the long-period seismograph used on the China mainland is Type 763; photographic recording is provided by a galvanometer (Figure 2). The broadband seismographs used are the SK (Kirnos) or GW (Galitzin-Wilip) Type, and also have photographic recording provided by galvanometers. The DK-1 seismograph, which has electronic amplification and pen-and-ink recording, is used only for monitoring regional seismicity and quick-reports of seismic events. Besides the DK-1, other types of short-period seismographs that also are used, can be summarized as follows:

(1) with photographic recording – Type 62, Type SH (Halin), Type VGK; and

(2) with pen-and-ink recording – such instruments consist of three units, the seismometer, the amplifier, and the recorder, as listed below (Figure 2):

Seismometer	Amplifier	Recorder
DD-1	DF-1	DD-1
Type 64	Type 63-A	Type 63-A
Type 65	Type 581	Type 768
VGK	Type 67	
BJ-1 (bore-hole)		

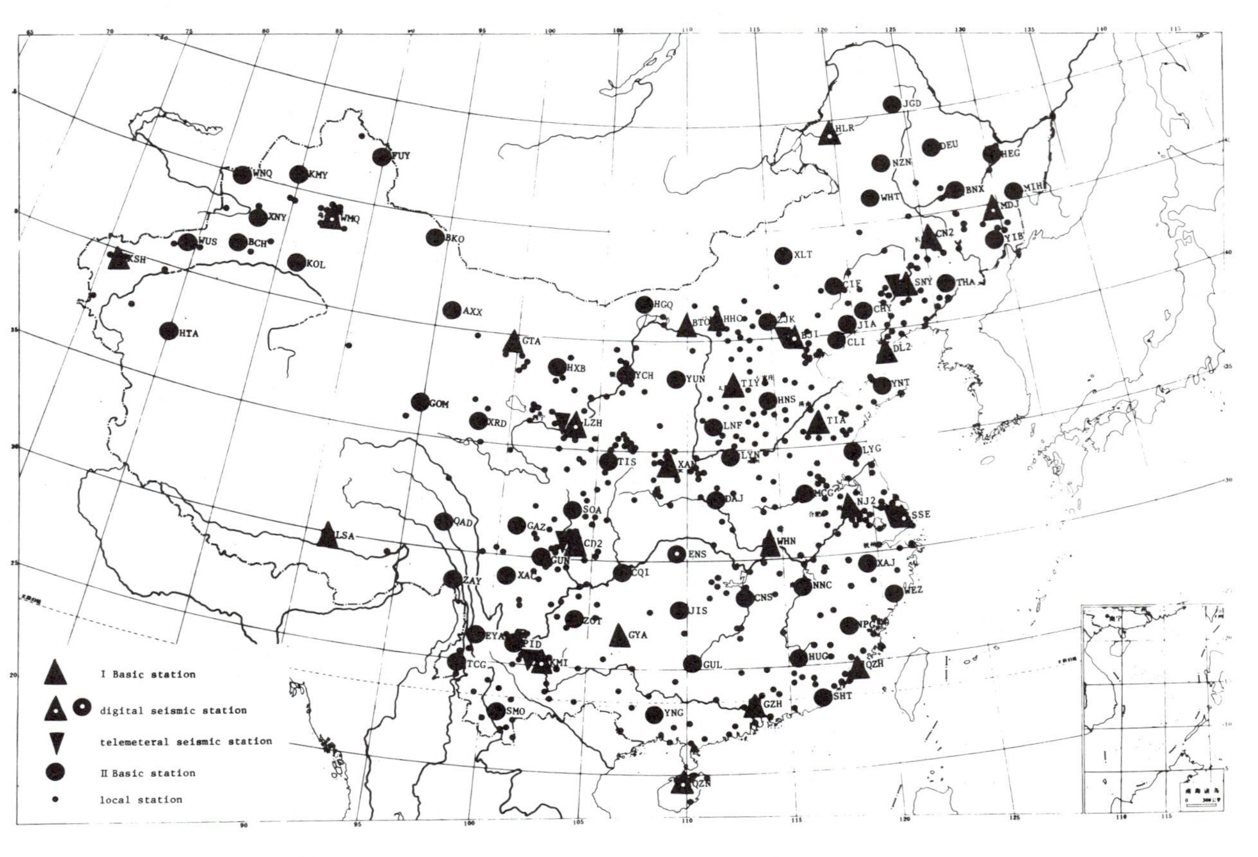

Figure 1. Seismic stations in the Chinese mainland.

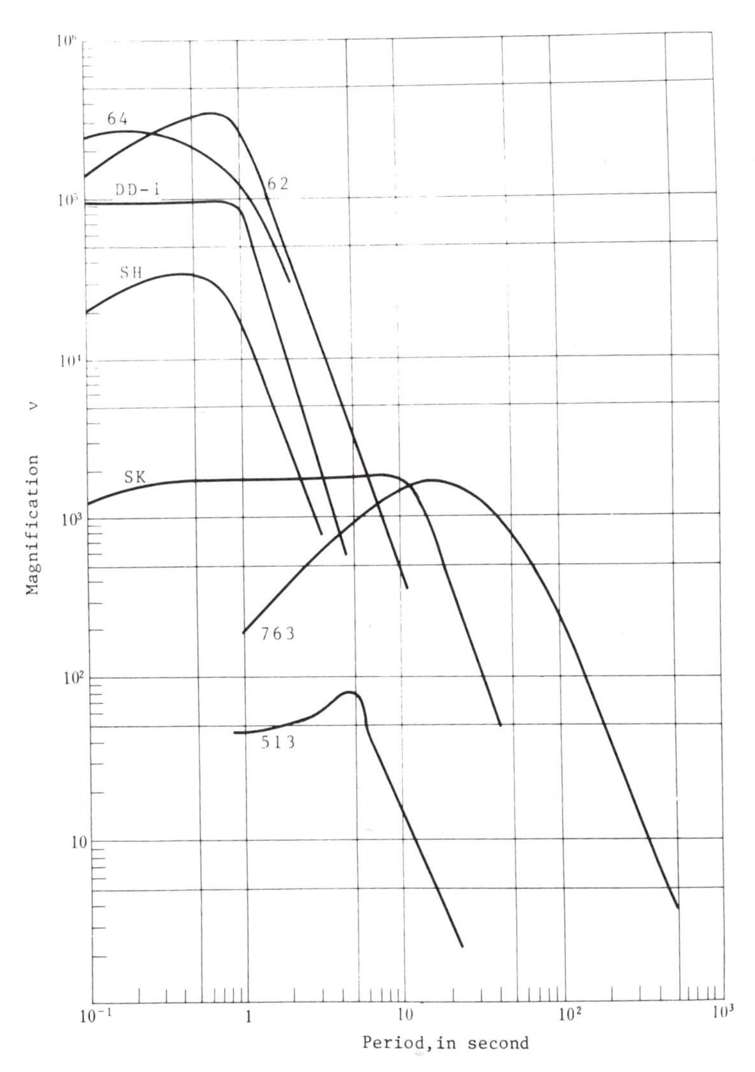

Figure 2. Magnification curves of instruments.

Combining different units with the proper matching impedance many configurations of observation systems can be formed. For instance, system 473 is a combination of a Type 64 seismometer plus a Type 67 amplifier and a Type 63-A pen-and-ink recorder. Such combinations are used chiefly at local stations.

Further, instruments used in the 1940s (such as the I-Type and Omori) and those used in the 1950s (such as the Type 51 and Type 54) are not longer in use. The SW (Wiechert) and Type 513 are still used at some stations.

To aid scientists in understanding the development of our seismic stations and in acquiring original observational data and the basic technical parameters of Chinese seismic stations, the State Seismological Bureau has decided to edit and publish a *History of Chinese Seismic Stations.* It is divided into 3 volumes and the first volume contains three parts: Part 1 concerns those basic stations participating in international exchange of seismic data; Part 2 concerns those not yet participating in international exchange; Part 3 concerns digital seismic stations. The second volume is about telemetered seismic networks; and the third volume is about networks of local stations. Part 1 of the first volume will be published by the end of 1985.

5. Timing System

In 1954, an astronomical clock with a second pendulum was commonly used in the seismic stations on the China mainland. To improve the time service, seismic stations built since 1958 were equipped with quartz clocks. From 1980 onward, the seismic stations that are included in the *Preliminary Seismic Report of Chinese Stations* have been using Type 768 clock stations or the AST system; and furthermore, it was stipulated that all seismic stations throughout the entire country use "BJUTC" instead of "$BJUT_1$" from 1980 on.

6. Results and Services of Seismic Observations

These seismic stations are all under the unified administration of the State Seismological Bureau of China. The Institute of Geophysics of the State Seismological Bureau is technically in charge of the quality examination of seismic records from the standard and basic seismic stations, analysis and compilation of seismological reports, microfilming of seismograms, management and exchange of seismic information and data, and improvement of observational techniques and equipment at the seismic stations. Local seismic stations under the administration of the provincial seismological bureaus (subordinate to the State Seismological Bureau) compile local seismic reports and seismic catalogs.

Seismograms and seismological reports are important observational results from the seismic stations. On the China mainland, seismograms recorded with some of the main types of instruments, such as SK, 763, Type 62, and others, are totally preserved whether or not they contain seismic events; but those seismograms without events (e.g., seismograms recorded with Type 513 or DK-1) are not preserved. Every station compiles its own seismic reports and every network compiles its own seismic catalogs monthly. The Institute of Geophysics of the State Seismological Bureau then compiles the *Preliminary Seismological Reports* and *Annual Seismological Reports of Chinese Seismic Stations.* These two reports are not only intended for use in China, but also for international exchange. We have already established exchange relations with 71 institutions in 49 countries or areas.

Recently, the number of persons who want to use our seismograms has increased greatly. To increase the rate of utilization of the seismograms and to ease the problems encountered in their management and storage, we are making microfilms of the seismograms and have started a microfilm exchange with World Data Center-A for Solid Earth Geophysics, Boulder, CO, USA.

Special emphasis was put on records from the 24 basic stations, including Sheshan, Lanzhou, Nanjing, Beijing, and Dalian. All seismograms from these basic seismic stations since 1969 were microfilmed, and microfilm copies of 39,324 three-component seismograms were sent to World Data Center-A.

The importance of early seismograms and the necessity of microfilming them are certainly obvious. However, besides the economic factors involved in the microfilming work, there are other problems. First, it is not easy to collect and bring together all the early seismograms, especially those from large earthquakes. Second, it is difficult to clarify the different annotations on the seismograms; for instance, the station where it was recorded, the type of instrument used, the direction of motion, the instrument constants, and the accuracy of the timing system.

Another difficulty encountered in the microfilming of early seismograms is how to enhance the distinctness of the traces to make the seismogram usable. Most early seismograms were recorded on smoked paper, and the traces have become indistinct owing to abrasion and inadequate storage. Also, the deterioration of the paper also makes the handling, storage, and microfilming difficult. Figure 3 shows two microfilms of the Tokyo earthquake of 1923 as recorded at Dalian station.

In 1982, we began the collation of some early seismograms on the China mainland, and found that many were missing, deteriorated, and damaged. Some preliminary statistics about the early seismograms from the Sheshan and Dalian stations are given in Table 1.

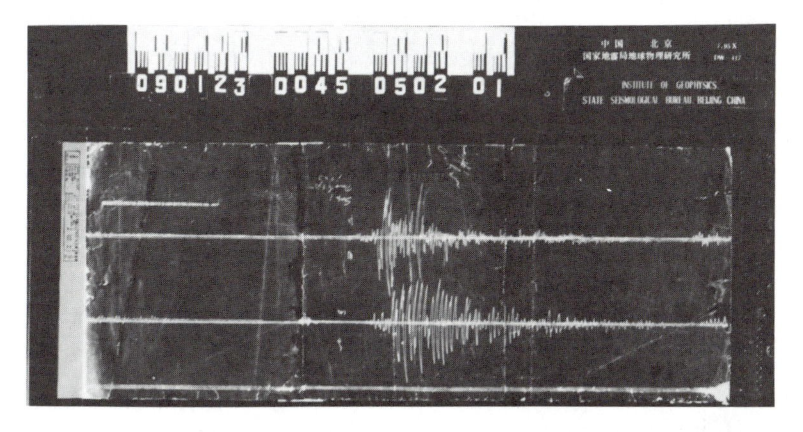

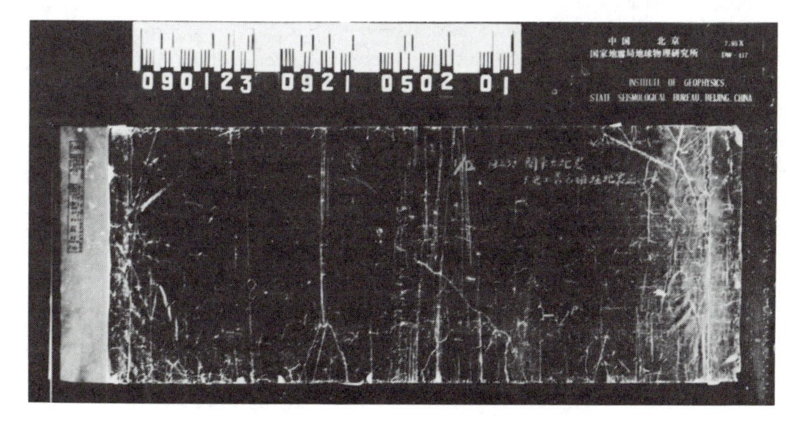

Figure 3. Microfilms of the 1923 Tokyo earthquake recorded at the Dalian station.

Table 1. Seismogram Statistics from the Sheshan and Dalian Stations

Dalian Station

Instrument type	Time of operation (year.month)	Number of seismograms	Monthly average
Omori*	1904 – 1943	301	1.0
Wiechert	1930 – 1945	304	1.68
Type 51	1955.08 – 1955.11	179	44.8
	1956.01 – 1963.12	2920	60
	1964.01 – 1965.09	243	11.6
Kirnos	1967.01 – 1970.12	1460	60
	1971.01 – 1971.12	35	2.9

Sheshan Station

Instrument type	Time of operation (year.month)	Number of seismograms	Monthly average
Omori	1904.01 – 1918.09	2380	13.4
Wiechert	1909.04 – 1950.03	4900	10.2
Galitzin	1915.04 – 1949.04	2789	6.8
Galitzin	1954.01 – 1956.09	701	21.2
Galitzin-Wilip	1932.04 – 1936.03	857	18.2
Wiechert	1950.04 – 1970	8518	36.1
Galitzin-Wilip	1950.08 – 1966.12	3406	17.3
Kirnos	1956.02 – 1981.12	10249	33.0
Type 54	1954.12 – 1957.06	770	17.9
Type 64	1969.08 – 1975.11	3206	42.2
DD-1	1973.07 – 1981.12	3643	35.7

(*) = The Omori seismograms are for the years 1918-1943.

7. Progress

Seismogram microfilming is undoubtedly an important factor in the establishment of a worldwide seismic database. Our work on the microfilming of seismograms was introduced briefly at the Tokyo Workshop in 1982 (Qu, 1982).

The work can be divided into two parts: (1) data sorting and labelling; and (2) microfilming. Seismic data sorting includes three parts: (a) Seismograms – the station code, recording date, and types of instruments are checked and then labelled on each seismogram. The Institute of Geophysics collects and sorts each record. As the seismogram storage place and the office in charge has changed many times, identification of historical seismograms is both difficult and time-consuming. (b) Seismograph parameters – the seismograph parameters and timing system of each station are checked and registered according to types of seismographs and different periods of recording. (c) The seismological bulletins compiled by individual stations are inventoried and sorted. After microfilming and careful checking, the seismograms are filed in chronological order.

Data sorting is completed at the Institute of Geophysics, State Seismological Bureau, and the Seismological Bureaus of Gansu, Yunnan, Jiangsu, and Shanghai as well as the 84 basic stations. We are happy to report that more than 800 Omori seismograms recorded at the Dalian station during 1928-1955 were recovered at Changchun. These seismograms were microfilmed.

To facilitate the use of our seismograms, the format of the microfilms are given in Figure 4.

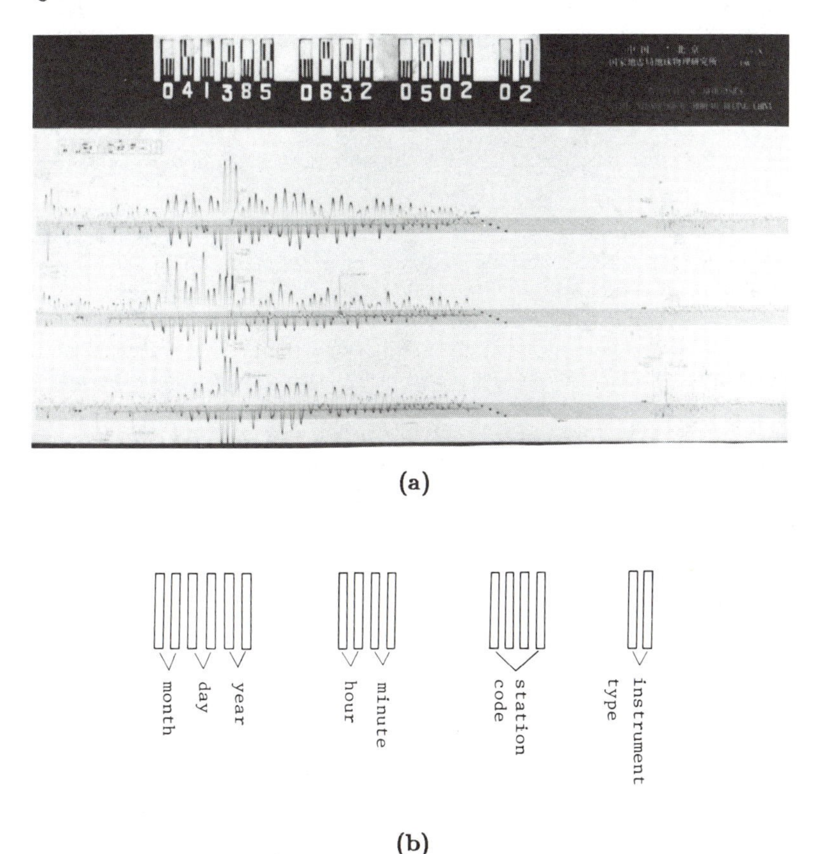

北京台 1980	Station	Code	Inst.	Comp.	Order N,E,Z,		N-E-U
	Mag-N	Mag-E	Mag-Z			Grnd Mot	⇧
	Date On	Beijing Time On		at	Corr	ms	
	Date Off	Beijing Time Off		at	Corr	ms	
	To Tg	Remarks:					Up

(c)

Figure 4. Format and notations for the use of microfilmed seismograms. (a) film size = 125 mm × 70 mm. The number 7.95× in the upper-right corner of the film indicates that the microfilm has been reduced by 7.95 times. (b) the binary code, located in the upper-left corner of the microfilm; the numbers below the code represent successively from left to right: date (month, day, year); time of mounting paper onto the recording drum (hour, minute); station code; and instrument type. (c) Annotations on the original seismogram, given in the usual format.

8. Plans for the Near Future

(1) After a thorough investigation of early seismograms, bulletins, and reports of early stations, and finding specifications of the instruments then in use, we plan to sort out the early seismograms of the Sheshan (Zi-ka-wai), Dalian, Beijing (Peking), Nanjing (Nanking), Lanzhou, and Kunming stations. We intend to renovate and annotate the early seismograms according to present requirements and standards. Then gradually microfilm those early seismograms that contain seismic events recorded at the above-mentioned stations.

(2) To improve the storage conditions and the efficiency in data management.

(3) To gain the consideration and sympathy and support from various sources for our work.

The 3,226 seismograms kept at Beijing Observatory still have to be identified and labelled with place of recording and type of instruments. The 8,798 historical seismograms from Lanzhou and Sheshan stations and the 1,147 seismograms of major earthquakes participating in the project are ready for microfilming.

Although we have made some progress in the sorting of seismograms, the work of microfilming historical seismograms of China proceeds slowly due to the limitation of equipment and lack of funds. We support the resolutions and suggestions of IASPEI/UNESCO, and are eager to actively participate in promoting this project.

REFERENCES

Academia Sinica (1956). *Chronological Table of Chinese Earthquakes*, edited by the Seismological Committee, Science Press, Beijing.

Xie, Y. S. and M. B. Cai (1983). *Compilation of Historical Materials of Chinese Earthquakes*, Science Press, Beijing.

Qu K. X. (1982). Seismic Observations and Service in China, presented at the Regional Workshop of the IASPEI/ UNESCO Working Group on Historical Seismograms held in Tokyo, Japan, Dec. 20-22, 1982.

MICROFILMING OF HISTORICAL SEISMOGRAMS IN THE USSR

O. Ye. Starovoit
Institute of the Physics of the Earth
USSR Academy of Sciences, Moscow, USSR

Following the recommendation of the International Association of Seismology and Physics of the Earth's Interior (IASPEI), supported by UNESCO, in 1981 the Soviet Union joined the international program for microfilming historical seismograms recorded at seismic stations from 1900-1963.

Nine seismic stations that operated for the longest period in 1900-63 were chosen for this program. Three stations are located in the European part of the USSR (Pulkovo, Sverdlovsk, Simferopol); two in the Caucasus (Tiflis and Baku); two in Central Asia (Alma-Ata and Tashkent); and one each in Siberia (Irkutsk) and the Far East (Vladivostok). These stations provide the coverage of nearly all seismic zones in this country. However, the network created by these stations is very sparse and thus is not representative. But the comparison of data provided by the Soviet stations to data provided by stations of other countries operated in the same period allows us to look at a global seismic network for recording and studying strong earthquakes ($M > 7.0$) of the past. The analysis of the efficiency of such a network can be made only after the Historical Seismograms Project is terminated, when we will know which stations recorded strong earthquakes. This project will be most important for the period prior to World War II.

Since 1981 the work on microfilming of historical seismograms has been coordinated by an IASPEI/UNESCO special Working Group, set up to develop scientific-methodological, technical and organizational recommendations as well as render assistance and support to countries participating in the program.

Microfilming of historical seismograms includes several stages:

(1) Selection of seismograms and preparation for microfilming. This is a tedious task which includes recovery of old and damaged seismograms, transfer of title blocks to the face of the seismograms, providing time marks, and converting local time to Universal Time. It was found that some seismograms lacked component names and information on polarity, but we managed to restore this information. Some seismograms were unreadable due to inadequate voltage and poor photographic treatment. Much time was spent on restoration of frequency characteristics and magnification curves of seismographs.

Seismograms recording earthquakes of $M > 7.0$ for the period 1900-1962 were selected. A catalog containing over 1,500 such events was published in Glover (1980, Table V).

(2) Microfilming process. A 16-mm Inter/Com Microfilm Recorder , 522A model, was used to microfilm the seismograms. This portable camera was kindly provided to us in 1983 by World Data Center A (USA) with the support of the IASPEI/UNESCO Working Group. This enabled us to microfilm the records on site.

Microfilms of seismograms were made on rolls of 16-mm film. At the beginning of each roll are several information frames containing the name and description of

Historical Seismograms
and Earthquakes of the World

a station, its three-letter code, the time period for which microfilms are available, a list of microfilmed seismograms, and the amplitude-frequency characteristics of the equipment.

(3) Arrangement and storage of microfilm sets and data exchange. Each roll is catalogued and stored in the Center of Seismic Information at the Institute of the Physics of the Earth in Obninsk, and is also available at World Data Center B (Moscow). The USSR and the USA exchange microfilms through respective WDC-B and WDC-A. In addition, copies of microfilm may be sent to researchers upon request.

(4) Preparation of catalog of microfilm and analysis of the quality of historical records of strong earthquakes. The main purpose here is to provide seismologists with information on microfilmed records of strong earthquakes recorded at Soviet stations, including their quality. This is the final stage, in which we develop quantitative criteria of record quality and further formalize the search and selection of these data using computers.

Presently in the Soviet Union, seismograms of six seismic stations are microfilmed. Table 1 gives data on the operation dates at these stations for which microfilms are available, as well as instrumentation history.

Table 1. Operation History of Seismic Stations in the USSR

Station	Operation dates	Instrument	Component	Magnification
Alma-Ata	1927-1949	Nikiforov	NS, EW	400
	1949-1962	Kirnos	Z, NS, EW	1000
Irkutsk	1912-1918	Galitzin	Z (from 1914)	700(Z)
			NS, EW	1000
	1923-1962	Galitzin	Z, NS, EW	1000
	1952-1962	Kirnos	Z, NS, EW	1200(Z);1800
Pulkovo	1908-1917	Galitzin	Z (from 1910) NS, EW	
	1923-1941	Galitzin	Z (from 1910) NS, EW	
	1950-1962	Galitzin	Z (from 1910) NS, EW	
	1951-1956	Kirnos	NS, EW 1000	
	1957-1962	Kirnos (long-period)	NS, EW	900
Simferopol	1928-1941	Nikiforov	NS, EW	400
	1946-1951	Nikiforov	NS, EW	400
	1951-1962	Kirnos	Z, NS, EW	1000
Tashkent	1912-1917	Galitzin	NS, EW	1000
	1925-1962	Galitzin	Z, NS, EW	700-1000; 1100-1800
	1961-1962	Kirnos	Z, NS, EW	1000
Tiflis (now Tbilisi)	1918-1939	Galitzin	NS, EW	50
	1944-1949	Galitzin	NS, EW	50
	1955-1959	Galitzin	NS, EW	50
	1912-1917	Galitzin	NS, EW	1000
	1926-1962	Galitzin	NS, EW	1000
	1955-1962	Kirnos	Z, NS, EW	1000

Data for the six seismic stations in Table 1 have been sent to WDC-A. Seismograms of the remaining three seismic stations at Baku, Vladivostok, and Sverdlovsk are prepared for microfilming; this task will be completed this year.

In 1986 we plan to start microfilming seismograms of two stations in the GDR (Moxa and Potsdam) and one station in Poland (Warsaw).

REFERENCES

Glover, D. P. (1980). Historical Seismogram Filming Project: Second Progress Report, *Report SE-24, World Data Center A for Solid Earth Geophysics*, Boulder, CO, 63 pp.

The Standardization of Seismological Data and the Filming Project of Seismograms in the International Latitude Observatory of Mizusawa

Yoshiaki Tamura, Hisashi Sasaki, Masatsugu Ooe,
and Kennosuke Hosoyama
International Latitude Observatory of Mizusawa
Mizusawa-shi, Iwate-ken, 023 Japan

ABSTRACT

Since 1902, seismological observations have been made at the International Latitude Observatory of Mizusawa (ILOM). Omori two-component horizontal seismographs were used during 1902-1970, and Nasu three-component seismographs were used during 1956-1970. Earthquakes were recorded on smoked paper. These seismographs were replaced by Sax-type instruments (long-period, electromagnetic) in 1970. Several important historical facts about these observations include: (1) Mizusawa was one of the few stations in Japan before a seismological observation network was installed; (2) all original seismograms are available; and (3) accurate time corrections have been kept since the beginning stage of observations.

To permit these important data to be used more fully, we conducted a project to standardize about 25,700 seismograms recorded before 1967. In this project, two kinds of tasks were performed: (1) film copies were made of about 3,000 seismograms for earthquakes occurring near Japan having magnitudes greater than or equal to 6.0, and for large earthquakes occurring throughout the world. A 98 × 70 mm film format was used to obtain a suitable resolution; (2) all seismological data were compiled in a standard format, and were made machine readable. Completion of these tasks make it easy to use the data for analyzing traveltime residuals and reestimating magnitudes of historical earthquakes.

Though our original project has been completed, the number of filmed seismograms represents only about 10% of all the earthquakes observed at Mizusawa and the film format may not match the standard format recommended by the International Association of Seismology and Physics of the Earth's Interior (IASPEI).

Now we are planning a new filming project for all seismograms according to IASPEI recommendations. The number of seismograms to be filmed in this project is about 60,000. We tested many kinds of film, and Kodak Technical Pan 2415 film with Technidol LC developer has proved to be the best. This combination of film and developer provides high-resolution copies and moderate contrast for the smoked-paper records. Though the reduced scale is about 1/20 (original seismogram size is 45 × 75 cm), filmed records have at least 0.2-mm resolution when converted to their original size. This resolution is sufficient to measure the arrival times or to digitize the waveforms.

1. Seismological Observations

The ILOM was established to observe latitude variations for investigating the mechanism of polar motion and the rotation of the Earth. The amplitude and phase of the polar motion appear to change with time. Though some remarkable part of the excitation of the Chandler Wobble is considered to be due to atmospheric and synchronized phenomena, the remaining part may be caused by seismic activity and accompanying crustal movements. At the end of the 19th century, an important problem was to determine the internal structure of the Earth by observing earth tides, deflection of the vertical, and velocities of seismic waves.

The ILOM began seismological observations in 1902 using Omori horizontal pendulums, which recorded on smoked paper. Nasu seismographs, which consisted of three components, were also installed in the same room in 1933. Though all components of the Nasu instruments recorded on the same paper, readings from only the vertical component were reported along with readings from the horizontal components of the Omori seismographs. All seismographs had no damping devices until 1956. And after that time, magnetic dampers were installed on each seismograph except the N–S component of the Omori seismograph. As the instrumental constants were sometimes slightly changed, the values for 1970 are shown in Table 1 as an example. Time marks were provided by electric contacts each minute using a pendulum clock. Time corrections were made by referring to signals given by wireless (~1902), telephone (~1906), and radio (~1924) time service. The adopted time-mark and the recording systems are shown in Figure 1.

In 1966, these instruments were moved into a new building located about 100 m from the previous station. Then, in 1970, they were replaced by Sax-type long-period electromagnetic seismographs and new recording systems. The location of the new station and the characteristics of the new seismographs are shown in Tables 2 and 3, respectively.

2. Standardization of Seismological Data

A 3-year project to standardize seismological data recorded by the Omori and Nasu seismographs was begun in 1981. As shown in Figure 2, the main tasks were repairing and filming the seismograms and standardizing the observed data. The standardized data were published in March 1984.

Table 1. Constants of Old Seismographs for the Year 1970

	Omori Seismograph		Nasu Seismograph
	E–W	N–S	U–D
Proper period (sec)	10.8	33.0	4.8
Statical magnification	100	20	25
Coefficient of friction	3.8	2.5	1.7
Damping ratio	2	(no damper)	2
Mass of weight (kg)	40.5	17.6	4.4
Paper speed (mm/min)	22	25	50

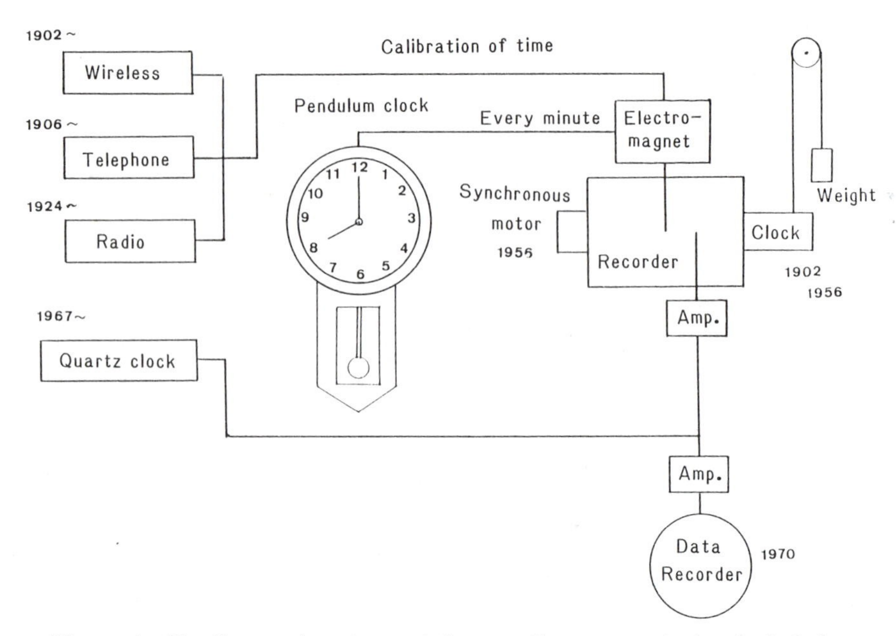

Figure 1. The time mark system and the recording system of seismological observations.

Table 2. Location of the Present Observation Station

Latitude	39.1355° N
Longitude	141.1318° E
Height	59.81 m above mean sea level
Depth	1.8 m below ground surface

Table 3. The Constants and Capacities of the Present Seismographs

	Horizontal	Vertical
Maximum period, T_o (sec)	> 30	> 30
Mass of pendulum, M (kg)	13 ± 1	7 ± 0.5
Center of gravity of pendulum, H (m)	0.26 ± 0.01	0.31 ± 0.01
Moment of inertia, I (kg m^2)	1.1 ± 0.2	0.82 ± 0.2
Equivalent length of pendulum, L (m)	0.31 ± 0.01	0.38 ± 0.01
Electromagnetic constant: main coil, G_M (volt-sec)	> 65	> 80
Electromagnetic constant: test coil, G_T (volt-sec)	> 6.5	> 8.0
Sensitivity of main coil, S_{VM} (volt/kine)	> 2.0	> 2.0
Sensitivity of test coil, S_{VT} (volt/kine)	> 0.20	> 0.20
Main coil resistance, R_M ($K\Omega$)	2.4 ± 0.3	2.4 ± 0.3
Test coil resistance, R_T ($K\Omega$)	0.24 ± 0.03	0.24 ± 0.03
Allowance of pendulum displacements (mm)	± 8	± 8
External damping (he)	< 0.2	< 0.2

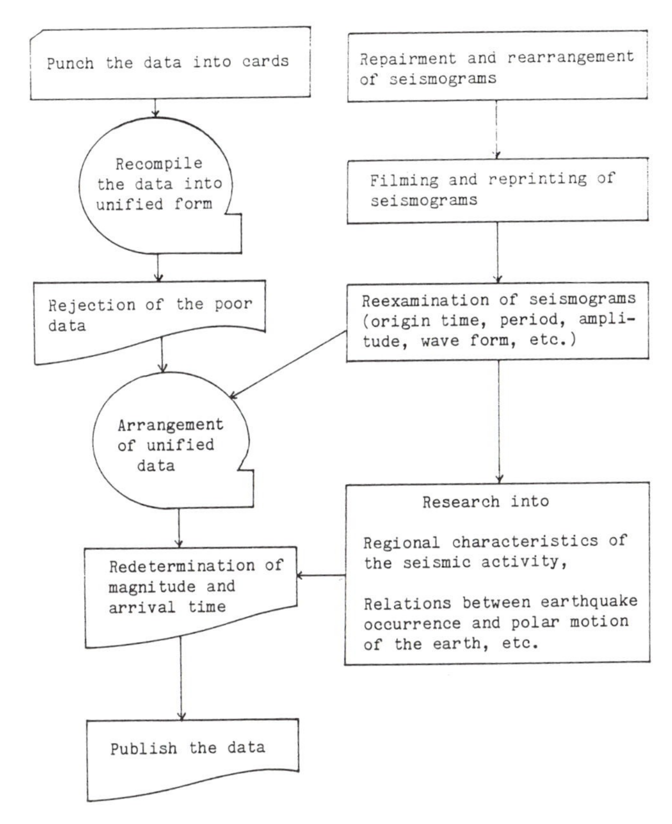

Figure 2. Block diagram for the standardization of seismological data.

2.1. Filming the Seismograms

Seismograms to be filmed were restricted to records from significant earthquakes greater than or equal to magnitude 6.0, which had occurred in and near Japan between 1902 and 1970; and to records from large earthquakes, which had occurred elsewhere in the world and were listed in "Rika Nenpyo" (1975, 1981). The total number of seismograms to be filmed was about 3,000.

The size of the recording paper of the E–W and N–S components of the Omori seismographs is 350 × 750 mm. All three components of the Nasu seismograph are recorded on the same seismogram, which is 450 × 750 mm in size. The seismic trace on the recording paper is 0.05 mm in width. The reading accuracy limits are 0.1 sec in time and 0.1 mm in amplitude. Therefore, it is necessary to maintain the sharpness of the trace in the photographing procedure. We compared 35-mm and 70-mm film copies, and found that 70-mm film reproduces more accurately and covers the wide range of contrast on the smoked paper better. The width of the traces after enlarging the film to its original size was 0.23 mm and 0.08 mm for the 35-mm and 70-mm film, respectively. Also, the 70-mm film is kept in separate film holders, which allows seismograms to be retrieved easily. Because of these factors, 70-mm film is considered to be more suitable for filming our seismograms.

2.2. Standardizing Observed Data

The timing system has been very accurate because the time was checked using astronomical observations. Therefore, the observed seismological data were often used for determining earthquake epicenters and magnitudes. The seismological data were sent to the International Seismological Center in the United Kingdom every year, and were also published in the annual report of the ILOM. The catalog formats of the annual report are given in Figure 3, which shows that the formats changed at various times. For example, both a UTC (Universal Time Coordinated) timing system and Japanese standard time were used; the measuring units varied; and "amplitude" sometimes means "double amplitude". These nonstandard formats made it difficult to use the seismological data. Therefore, the data were recompiled into one standardized format by reexamining observational notes and related materials, and were also made machine readable.

No.	Date	Time of Occurrence † (NS)	Time of Occurrence † (EW)	Duration of Total Earthquake	Maximum Range of Motion (NS)	Maximum Range of Motion (EW)	Character of Motion	Intensity	Remarks
		h m s	m s	m	mm	mm			
1	1902 Jan. 1	2 27 4 pm	26 51	50.8	0.22	0.01	Slow	Feeble	
2	2	0 15 22 am	15 20	2.6	0.17	0.08	Quick	,,	
3	4	6 9 51 am	10 0	7.2	0.39	0.25	,,	,,	
4	12	6 30 18 am	30 40	3.0	0.13	0.07	,,	,,	
5	13	7 29 1 am	28 56	2.2	0.06	0.03	Slow	,,	

No.	Date 1916	Time of Occurrence † E W	Time of Occurrence † N S	Beginning of the Principal Potion E W	Beginning of the Principal Potion N S	Duration of Total Earthquake (mean)	Maximum Range of Motion E W	Maximum Range of Motion N S	Character of Motion	Intensity	Remarks
		h m s	m s	h m s	m s	m	mm	mm			
1	Jan. 1	22 28 56	28 57	22 35 23*	35 30*	161.5	0.63	?	Slow	Feeble	
2	7	22 09.5	— —	22 09 50	— —	1.8	0.01	—	,,	,,	
3	12	5 54 24	54 27	5 54 53	54 56	2.9	0.03	0.03	,,	,,	
4	13	15 26 27	26 27	15 32 43*	32 46*	75.5	0.08	0.40	,,	,,	
5	13	17 28 37	28 37	17 35 02*	34 56*	143.6	?	1.20	,,	,,	

番號	月 日 1942	P 東西動	P 南北動	S 東西動	S 南北動	L 東西動	L 南北動	最大動振幅 (ミクロン) 東西動	最大動振幅 (ミクロン) 南北動	總振動時間	震度	震央
		時 分 秒	分 秒	分 秒	分 秒	分 秒	分 秒	μ	μ	分 秒		
1	I 1	11 15 25	- -	15 41	- -	- -	- -	-	-	2 01	0	
2	1	e13 37 40	- -	37 58	- -	- -	- -	-	-	2 29	0	
3	4	e18 01 14	e01 15	01 36	01 37	- -	- -	— 9	-	3 02	0	
4	6	15 36 27	36 26	36 49	36 50	- -	- -	— 30	+ 50	6 07	0	福島縣東方沖
5	7	e12 58 10	- -	58 45	- -	- -	- -	— 4	-	4 26	0	

No.	Date 1956	P E W	P N S	P Z	S E W	S N S	S Z	Maximum Amplitude E W	Maximum Amplitude N S	Maximum Amplitude Z	P ~ S	P ~ F	Intensity	Epicenter and Remarks
		h m s	m s	m s	m s	m s	m s	μ	μ	μ	s	m s		
1	Jan. 1	6 15 01	15 01	— —	15 23	15 23	— —	77	100	—	16	9 26	0	41.4°N, 142.0°E [70]
2	2	e 19 26 24	— —	— —	26 46	e 26 45	26 46	5	—	4	22	2 21	0	36.8N, 141.8E [50]
3	3	6 34 22	— —	e 34 24	34 36	— —	34 37	2	—	—	14	1 28	0	
4	3	19 43 00	43 01	— —	44 06	44 06	44 04	15	13	8	66	4 26	0	43.5N, 147.5E [S]
5	4	0 44 35	e 44 52	— —	47 00	e 47 06	47 04	3	5	—	145	7 58	0	48.5N, 155.0E [S]

SEISMOMETRICAL DATA FOR THE YEAR 1970

NO	DATE	I	ARRIVAL TIME		MAXIMUM MOTION N-S A/T		MAXIMUM MOTION E-W A/T		MAXIMUM MOTION U-D A/T		INITIAL MOTION N-S	INITIAL MOTION E-W	INITIAL MOTION U-D	SUPPLEMENTS DELTA	SUPPLEMENTS AZM	SUPPLEMENTS O-P	SUPPLEMENTS DEPTH
			h m s	m s	*	s	*	s	*	s	*	*	*	km	deg	s	km
1	Jan 1	0	ep 4 5 24.6	is 8 13.3	387	9.6	47	6.2				W 2					
2	3	0	ep 16 39 19.0	es 39 37.2	13	1.2	19	1.2	10	1.0		E 1	D	148.4	36	29.1	0
3	5	0	ep 2 7 44.1	es 13 24.8	43	12.3	27	14.4			N	E 5	U				
4	5	0	ip 5 42 31.3	is 42 44.9	73	1.0	111	1.5	146	0.9	S 7	W 2	D 12	126.8	98	20.6	40
5	5	0	ep 20 31 12.6	es 31 29.6	5	0.5	6	0.5	6	0.7	S	E	U	140.0	40	26.2	0

Figure 3. Catalogue forms of seismological data published as the Annual Report of the ILOM.

Preliminary analysis of traveltime residuals was made using the above standardized data (Figure 4). Information on hypocenters, origin times, and a standard traveltime curve were taken from data compiled by the Japan Meteorological Agency (JMA). In this analysis only the post-1951 records were used, because JMA began estimating the origin time to the nearest 0.1 sec after 1951. An analysis such as this would be impossible unless the data were standardized and machine readable.

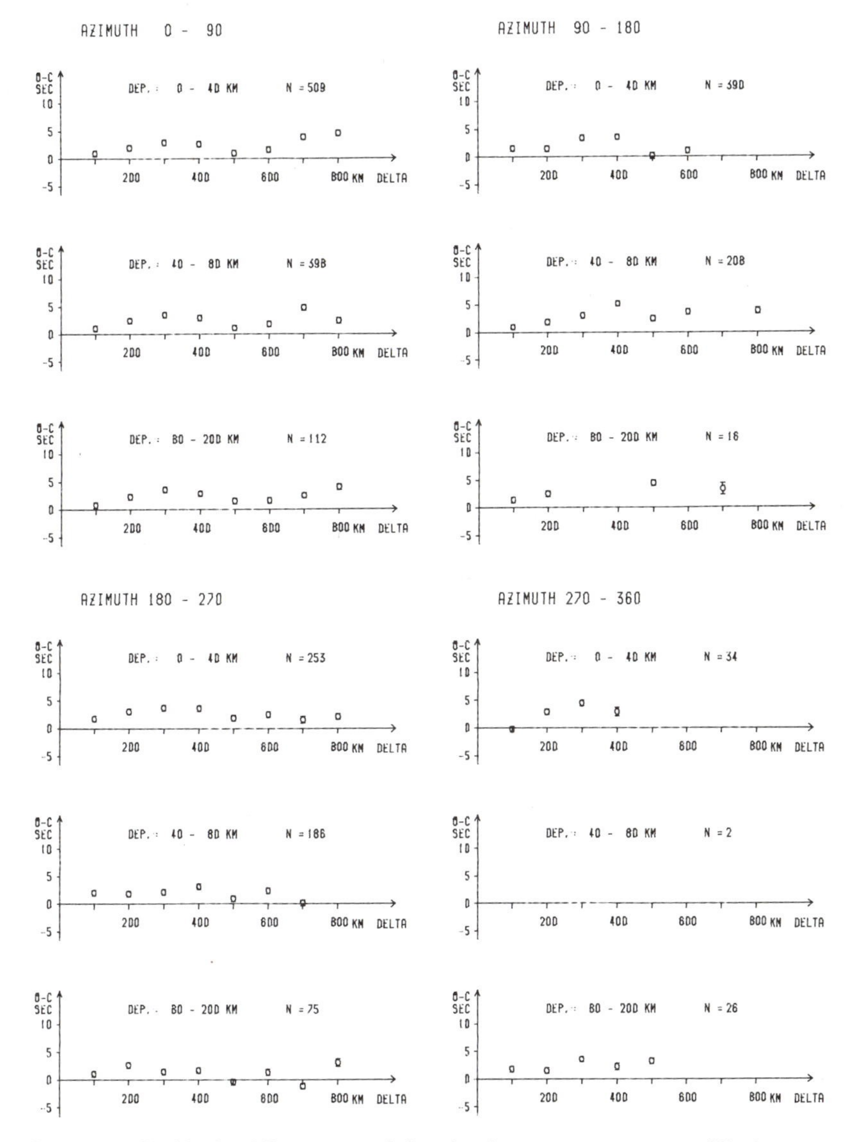

Figure 4. Residuals of P-wave travel time for the years 1951 to 1966. The hypocenters are divided into three layers by focal depth and into four areas by azimuth in every 90 degrees from the north.

3. New Filming Project

Our original filming project was carried out using only the seismographs which registered large earthquakes. We selected a 70-mm film format to ensure high-resolution copies. The actual operations were carried out by an expert commercial laboratory, even though the cost was rather expensive. This project was possible because the number of filmed records was only about 10% of all the observed seismograms.

Now we are carrying out a new filming project for all seismograms on 35-mm film according to the recommendations of IASPEI. About 60,000 seismograms, including those on which no earthquake is recorded, will be filmed. This project requires a budget larger than 5 million yen.

As the 35-mm film size is rather small compared to the original record, we have selected the copying film carefully, finally choosing Kodak's Technical Pan 2415 film and Technidol LC developer. This film has fine granules and high resolution in spite of its 35-mm size, and the developer provides moderate contrast for smoked-paper records. The filmed records have at least 0.2-mm resolution and can be used to measure amplitude to $\pm$ 0.1 mm and time to $\pm$ 0.1 sec when enlarged to their original size. The reduction scale is about 1/20. The above resolution may be necessary to measure arrival times and to digitize the waveforms.

We hope that these filming procedures will safely preserve these valuable seismograms. All data in the ILOM will be available to any researcher for studies of historical earthquakes.

REFERENCES

Rika Nenpyo (Chronological Scientific Tables) (1975). geophysical part, pp. 181-212, ed. Tokyo Astronomical Observatory, Maruzen Co., Ltd.

Rika Nenpyo (Chronological Scientific Tables) (1981). geophysical part, pp. 204-221, ed. Tokyo Astronomical Observatory, Maruzen Co., Ltd.

Seismological Observations at Mizusawa for the Period between 1902-1967 (1984). Int. Latit. Obs. Mizusawa.

Microfilming of Historical Seismograms at Abuyama Seismological Observatory, Kyoto University

Yasuhiro Umeda and Katsuyoshi Ito
Abuyama Seismological Observatory, Kyoto University
944 Nasahara Takatsuki, Osaka 569, Japan

1. History of Seismological Observations

Abuyama Seismological Observatory of the Faculty of Science, Kyoto University, was established in 1930. In the previous year, the experimental seismological observations began with a two-component horizontal Wiechert seismograph. Continuous observation with a three-component Wiechert seismograph began in 1932. The Wiechert seismograph is an excellent instrument which has a response of relatively long-period and high-gain. In those days, many good records were obtained. However, for the moderate and large earthquakes occurring in and around Japan, the instrument is unstable because of its complex structure.

In 1934, two types of strong-motion seismographs were designed and operated by the staff of our observatory. One of them is a set of long-period (T_o = 24 - 27 sec), low-magnification (V = 1.1) two-component horizontal seismographs. The length of the pendulum is 1.7 m, and the recording stroke is about 30 cm. This instrument has recorded destructive earthquakes occurring near the Abuyama Observatory and great earthquakes such as the 1960 Chile and the 1964 Alaska earthquakes. The other instrument is a set of short-period (T_o = 1.4 - 6.0 sec), strong-motion (V = 5 - 15) three-component seismographs. Those two sets of seismographs constitute the low-gain broad-band seismograph system.

Historical seismograms have been obtained mainly using the above three kinds of seismographs. The Wiechert seismograms are recorded daily and saved, but the seismograms of the latter two sets of instruments are saved only when events are recorded.

2. Microfilming in the First Stage

In response to the resolutions of the 1982 Tokyo Workshop on Historical Seismograms, we began preliminary work in preparation for microfilming. On one hand we applied for financial support, and on the other hand we arranged our archived seismograms in chronological order and listed all missing records. In 1984, this microfilming project was supported by the Science Research Fund of the Ministry of Education of Japan. Because this support will continue until March 1987, we have made a three-year plan for the microfilming project as shown in Figure 1.

In the first fiscal year (from April 1984 to March 1985), 20,346 sheets of Wiechert seismograms were arranged and filmed on 30 rolls of microfilm. An average of 500 frames are filmed on one roll 30.5 m long. The total number of frames is 14,196.

3. Filming Format

Every roll of film is kept in a small box. The roll number, the kind of instrument, and the dates of the records are given on the face of this box.

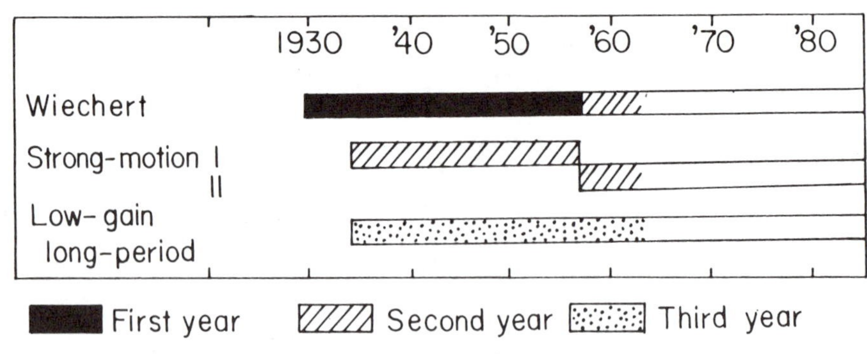

Figure 1. Microfilming schedule.

The first frame of the microfilm is a resolution chart for the focus and the exposure corrections. The second to the eighth frames contain information on how to use the microfilm. The second frame shows the roll number and the dates of the seismograms being filmed. The third frame shows the station name, address, latitude, longitude, etc. The fourth frame (Figure 2) shows the name of the instrument and its constants along with the response curves. The fifth frame (Figure 3) shows photographs of the Wiechert seismographs; the direction of ground displacement on the original seismogram is also drawn on the right side of this frame. The original seismograms are smoked paper records.

The sixth frame (Figure 4) indicates the filming format. The seismogram of the Wiechert instrument is a ring-shaped paper with a total length of 180 cm. Each seismogram is folded in two, as shown in the upper left of Figure 4. Two sides of the seismogram, i.e. A- and B-plane, are filmed in two frames of the microfilm. One side of the three-component seismogram is filmed on one frame, and the other side is filmed on the successive frame, as shown in the bottom of Figure 4. Generally, the side on which some information is described on the seismogram is chosen as the A-plane. The three-component seismogram is arranged from top to bottom in the order of vertical, east-west, and north-south components. If the east-west component is missing, the middle place in the frame is blank, as shown in Figure 4. A small label (10×3.5 cm) is affixed in the margin of each seismogram. The station, instrument, A- or B-plane, year, component, data on/off time, and a scale indicating the original record length per minute are given on this label. Since the drive speed of the recording sheet is 3 cm per minute, the length of one record line (180 cm) is one hour. Time is Japanese Standard Time (JST), and Universal Coordinate Time (UTC) is JST minus 9 hours.

The seventh frame, sometimes 7th to 9th frames, is an inventory sheet (Figure 5) indicating whether the seismograms are present or missing. A blank column indicates that the records are saved, whereas an oblique line in a column indicates a missing record. When large amplitude seismic waves were recorded and/or the instrument was adjusted, a new recording sheet was set on twice or three times in a day. In this case, the number of seismograms per day is described in the column. "C" in the column indicates that the calibration signals, i.e. free period, damping ratio and the static magnification of the instrument are recorded on the seismogram.

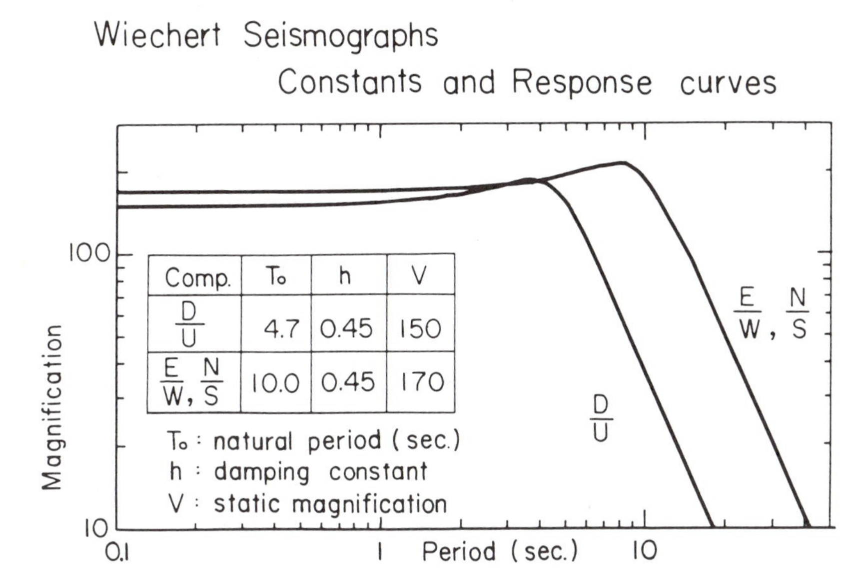

Comp.	T_0	h	V
D/U	4.7	0.45	150
E/W, N/S	10.0	0.45	170

Figure 2. Response curves of Wiechert seismographs.

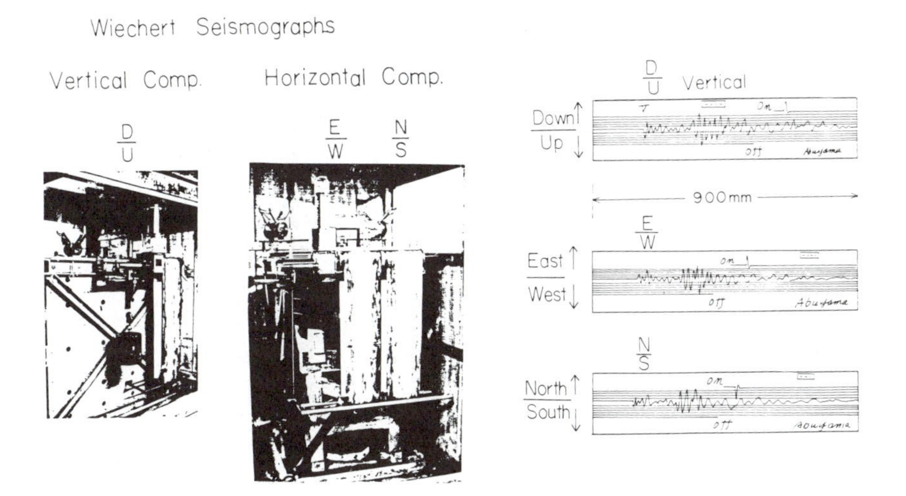

Figure 3. Photographs of instruments and seismograms.

Format

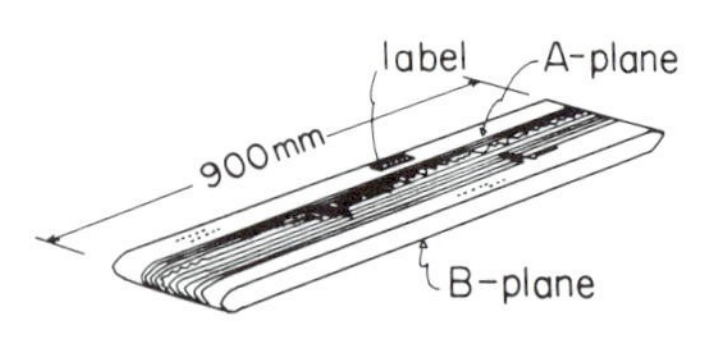

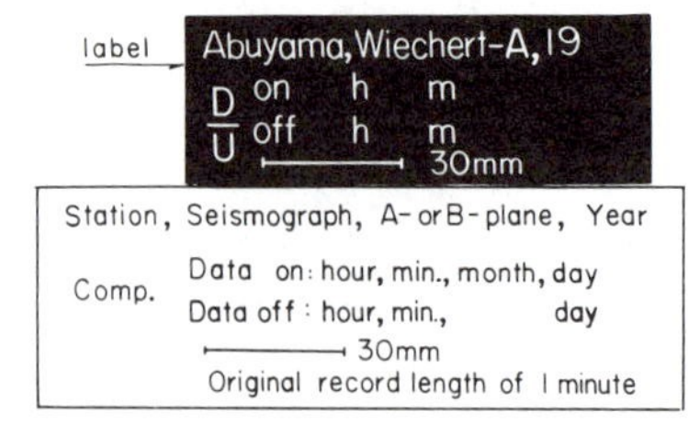

Original Seismogram : Smoked paper

Paper speed : 30mm/minute

Time : JST (Japanese Standard Time)

UTC = JST − 9hour

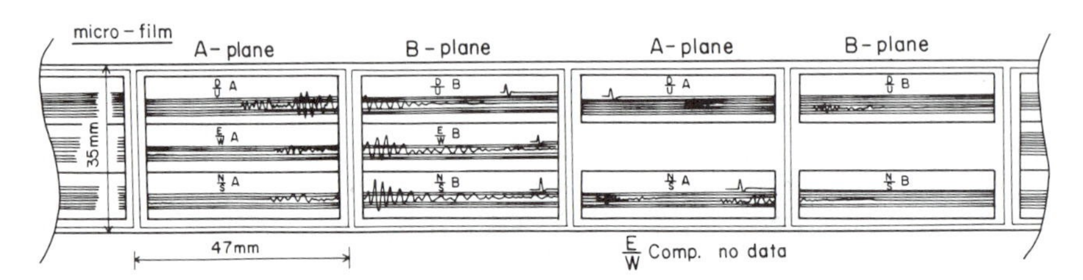

Figure 4. Microfilming format and label.

1933 Wiechert Seismograms Abuyama Seis. Obs.

Date	Jan. D/U	Jan. E/W	Jan. N/S	Feb. D/U	Feb. E/W	Feb. N/S	Mar. D/U	Mar. E/W	Mar. N/S	April D/U	April E/W	April N/S	May D/U	May E/W	May N/S	June D/U	June E/W	June N/S
1 — 2																		
2 — 3															2			
3 — 4																		
4 — 5																		
5 — 6								2	2									
6 — 7	/	/	/															
7 — 8																		
8 — 9																		
9 — 10																		
10 — 11																		
11 — 12																	2	
12 — 13																		
13 — 14																		
14 — 15																		
15 — 16																		
16 — 17																		
17 — 18																		
18 — 19																/	/	/
19 — 20																		
20 — 21																		
21 — 22																		
22 — 23																		
23 — 24											2	2						
24 — 25																		
25 — 26																		
26 — 27																		
27 — 28																		
28 — 29																		
29 — 30																		
30 — 31																		
31 — 1																		

Date	July D/U	July E/W	July N/S	Aug. D/U	Aug. E/W	Aug. N/S	Sep. D/U	Sep. E/W	Sep. N/S	Oct. D/U	Oct. E/W	Oct. N/S	Nov. D/U	Nov. E/W	Nov. N/S	Dec. D/U	Dec. E/W	Dec. N/S
1 — 2							/	/	/									
2 — 3																		
3 — 4							/	/	/									
4 — 5							/	/										
5 — 6							/	/	/									
6 — 7																		
7 — 8																		
8 — 9							/	/	/					2				
9 — 10							/	/	/									
10 — 11							/	/	/									
11 — 12							/	/	/									
12 — 13								/	/		2							
13 — 14																		
14 — 15									/		2	2						
15 — 16																2		
16 — 17					2	2												
17 — 18																		
18 — 19																		
19 — 20																2		
20 — 21	2							/	/									
21 — 22					2	2												
22 — 23																		
23 — 24																		
24 — 25																		
25 — 26																		
26 — 27																		
27 — 28																		
28 — 29												2						
29 — 30																		
30 — 31														2	2			
31 — 1				/	/	/										/	/	/

/ : no data, 2 : two seismograms per day, C : calibration seismogram

Figure 5. An inventory sheet.

The reduction ratio of microfilm to original seismogram is 40.5/900 = 0.045, which is invariable for all microfilm of the Wiechert seismograms. The exact reduction ratio can be obtained by the scale filmed in the eighth frame. At the beginning of each month, one blank frame is inserted to permit convenient access to the required seismogram. Examples of microfilms are shown in Figures 6 and 7.

4. Other Information and Future Plans

The time mark is indicated on seismograms by an upward deflection of the trace. The seismic trace breaks off by a short length every minute and every hour. The minute marks are about 1 mm in length and hour marks are about 2.5 mm in length. For the time correction, Δt, data are not available until 1951. Since 1952, the Seismological Bulletin of Abuyama Seismological Observatory has been published, and time corrections are included there.

The microfilm (30 rolls) has been sent to the World Data Center A for Solid Earth Geophysics, Boulder, CO, U.S.A., and the Earthquake Research Institute of the University of Tokyo, Tokyo, Japan. In the second year of this project, we will film the Wiechert seismograms from 1958 to 1962 and the strong-motion seismograms. The low-gain, long-period seismograms will be filmed in the final year. We also plan to publish in the final year all inventory sheets together with the instrument constants and other information on the microfilmed seismograms.

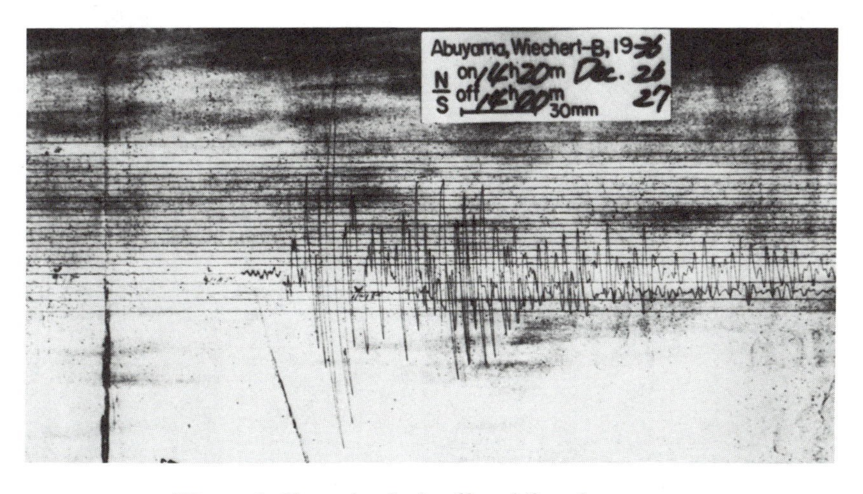

Figure 6. Example of microfilm of the seismograms.

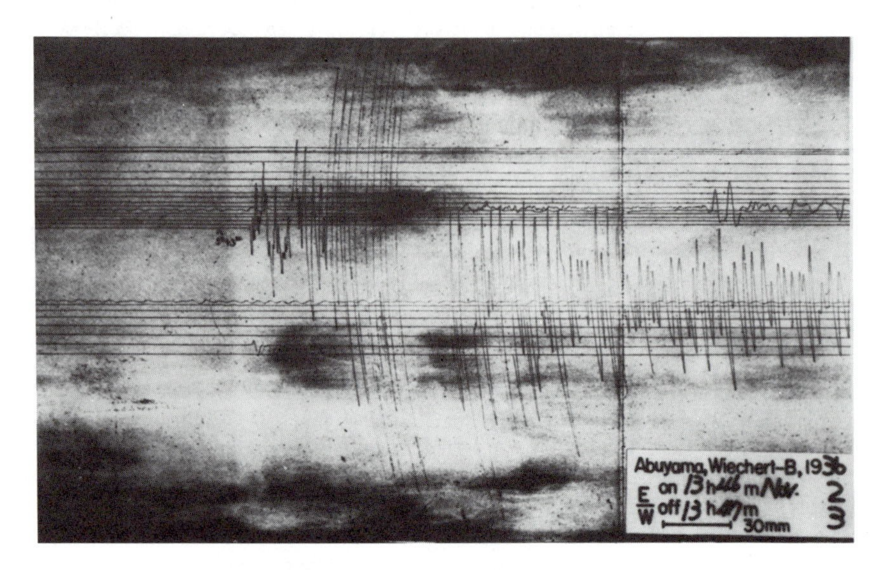

Figure 7. Example of microfilm of the seismograms.

VIII. Historical Seismograms from Various Observatories

Historical Seismograms in Italy

R. Console and P. Favali
Istituto Nazionale di Geofisica
Rome, Italy

ABSTRACT

Instrumental seismology began in Italy before 1890, and historical seismograms recorded at the end of last century are still available. In particular, the Geophysical Observatory of Rocca di Papa (Rome) operated from 1888 to 1936.

Seismograms recorded at Rocca di Papa are kept at the Istituto Nazionale di Geofisica (Rome), where an inventory is being carried out to prepare a catalog of the records.

The seismograms are supposed to be reliable for reevaluation of magnitudes (using both amplitude and duration) and, for seismograms that were recorded at a high paper speed with special devices, digitation for computer-aided waveform analysis.

Several other seismic stations were also operating in Italy around the beginning of this century. Some of them, such as Florence, Naples, and Foggia, maintain an archive of old seismograms.

We intend to cooperate with the International Historical Seismograms Filming Project, acting as the national agency for coordination in this field, and to study techniques for the analysis of historical seismograms.

1. Introduction

Italy is well known for its contributions to seismology. The names of De Rossi, Mercalli, and Cancani are recognized all over the world. Since 1874, the *Bollettino del Vulcanismo Italiano* (edited by De Rossi) reported on seismic events recorded in Italy.

In 1885 the task of carrying out seismological observations was given by law to the Royal Central Office of Meteorology (renamed the Royal Central Office of Meteorology and Geodynamics in 1887). This organization soon founded three new observatories, in addition to those already operated in Italy by private institutions. The new observatories were at Catania, Casamicciola (Ischia Island), and Rocca di Papa (near Rome); the latter was directed until 1936 by Professor G. Agamennone.

The *Bulletin of the Italian Seismological Society* (BISS), founded in 1895 by Tacchini, systematically reported on seismic events in Italy and their instrumental recordings.

As an example, the BISS reported that, in 1909, more than 30 seismological observatories were operating in Italy, though their geographical distribution and instrumental quality were not homogeneous. Most of the material recorded by this impressive network has been lost, but in some observatories, e.g., Rocca di Papa, Florence, Naples, and Foggia, many important seismograms still exist.

2. The Archives of the Rocca di Papa Observatory

The Royal Geodynamics Observatory of Rocca di Papa was installed in 1888 by a governmental decree. It operated with a variety of seismographs, most of them designed by its director, Professor G. Agamennone. A new law transferred the task of seismological observations from the Central Office of Meteorology and Geodynamics to the new Istituto Nazionale di Geofisica (ING) in 1936. All the historical records and the library, as well as the observatory building, became part of ING, and the old seismographs were replaced by relatively new equipment (Wiechert mechanical seismographs).

Fortunately, Prof. Agamennone kept voluminous books (each of them covering about 1 year) that contained information about earthquakes. Essentially, they contained systematic "readings" of seismograms and periodically revised instrumental parameters and the most significant records themselves. Because of this, the ING can access a large set of historical seismograms and have a good idea of the kind of instruments that recorded them.

Not all the volumes from 1888 to 1936 are available at present. Table 1 shows the sequence of dates covered by the existing volumes.

Table 1. Seismological Observations at Rocca di Papa

Volume	Dates (from)	(to)	Volume	Dates (from)	(to)
I			XIV		
II			XV	23 Oct 1907	16 May 1908
III			XVI	17 May 1908	22 Sep 1909
IV			XVII	Oct 1909	Oct 1910
V	17 Aug 1899	31 Dec 1900	XVIII		
VI			XIX	4 Jan 1912	30 May 1913
VII	1 Sep 1901	31 Dec 1901	XX	Jun 1913	Dec 1914
VIII	1 Jan 1902	11 Jun 1902	XXI	Jan 1915	Aug 1915
IX	14 Jun 1902	31 Dec 1902	XXII	Sep 1915	Jul 1916
X	1 Jan 1903	31 Dec 1903	XXIII		
XI	1 Jan 1904	31 Dec 1904	XXIV		
XII	1 Jan 1905	6 Nov 1905	XXV		
XIII			XXVI	Jan 1919	Dec 1920

3. Examples of Existing Historical Seismic Records

Figures 1, 2 and 3 illustrate 3 strong Italian earthquakes recorded at the beginning of the century at Rocca di Papa. Table 2 gives the origin time, coordinates, depth, magnitude, maximum intensity, and epicentral distance of the 3 earthquakes.

Table 3 gives the instrumental constants for the seismographs that provided the records in Figures 1-3 and Table 2. Damping factors were not considered to be essential at that time. These seismographs differed from the usual mechanical instruments recording on smoked paper in that they remained on standby until triggered by ground motion.

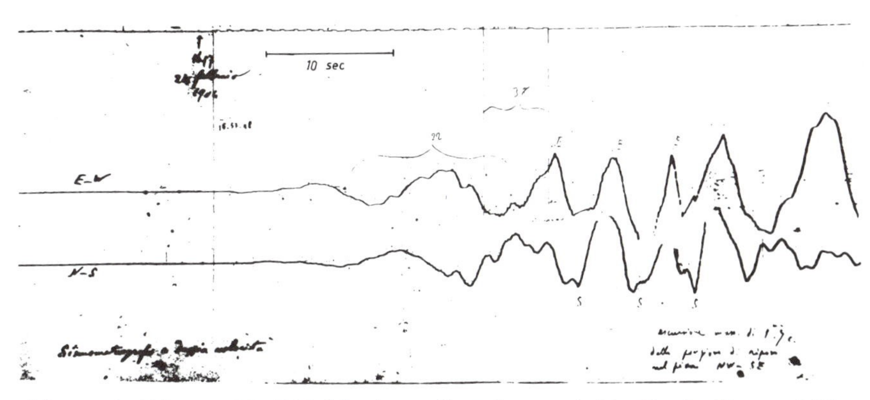

Figure 1. February 24, 1904, Marsica earthquake recorded by the double-speed "Agamennone" seismograph at Rocca di Papa ($M = 5.6, \Delta = 60$ kms).

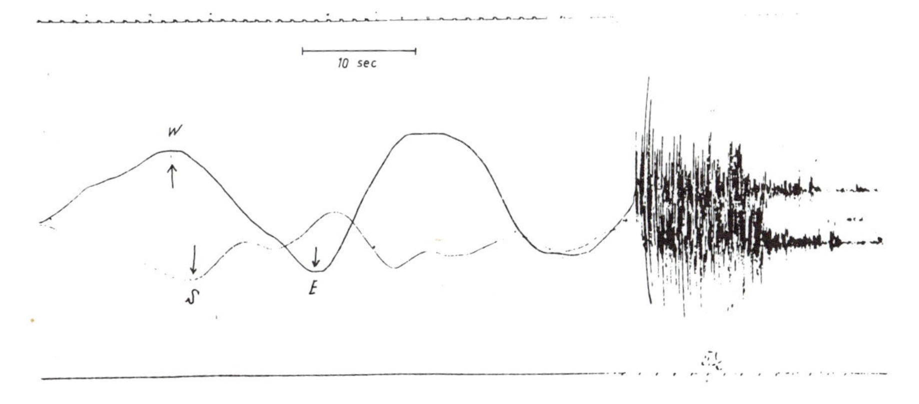

Figure 2. December 28, 1908, Messina earthquake recorded by the double-speed "Agamennone" seismograph at Rocca di Papa ($M = 7.2, \Delta = 470$ kms).

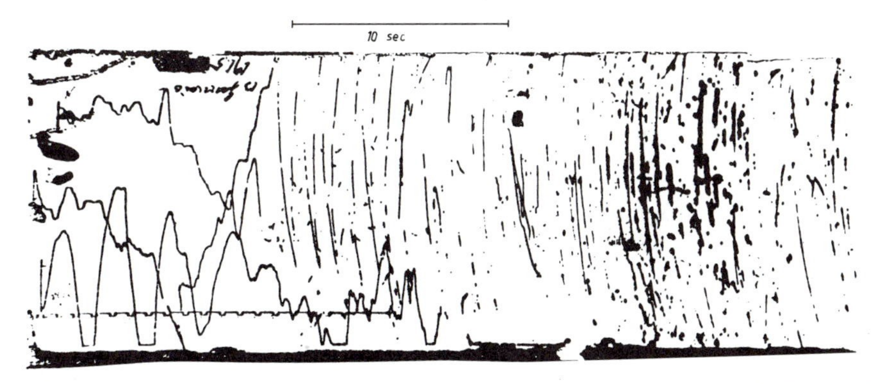

Figure 3. January 13, 1915, Avezzano earthquake recorded by the triggered "Brassart" seismograph at Rocca di Papa ($M = 6.6, \Delta = 60$ kms), S-waves out of scale.

Table 2. Some Examples of Earthquakes Recorded at Rocca di Papa (RDP)

Date year mo da	Time hr:mn:sc	Inst.	Epicenter		h	M_S	I_o	Δ	Location
			Lat.	Long.					
1904 02 24	15:53:26	A	42°07′ N	13°20′ E	6	5.6	9	60	Marsica
1908 12 28	04:20:27	A	38°10′ N	15°35′ E	10	6.9	11	470	Messina
1915 01 13	06:52:43	B	41°59′ N	13°36′ E	8	6.6	11	60	Avezzano

Instruments: A = Agamennone, B = Brassart.

Table 3. Instrumental Constants

Sismometrografo "Agamennone" a doppia velocitá
$M = 200$ kg, $T = 4.4$ sec, $A = 14$

Sismometrografo "Brassart" a tre componenti con lastra di vetro
$M = 26.4$ kg, $T = 3.5$ sec, $A = 10$

The first instrument (designed by Agamennone) recorded two horizontal components on a strip of white paper (sometimes blue and red ink were used for N-S and E-W components, respectively). Usually the paper speed was about 5 mm/min, but when an external pendulum produced an electric contact, the speed increased to about 5 mm/sec and the timemarks accelerated from 1 min to 1 sec. After 1 minute, the paper driver returned to normal.

The second seismograph (designed by Brassart) is actually a parent of our modern strong-motion recorders. It recorded all three components on a plate of smoked-glass, which moved at a speed of 6 mm/sec when an external seismoscope detected a light movement and triggered the motor. The seismograms available in our file are contact-print photocopies of the original smoked plates.

4. Conclusions and Suggestions for Future Work

Our program for reevaluating the old records consists of the following steps: (1) Compilation of a catalog of existing records, including all available information on instrumental parameters; (2) Study of the problem of obtaining magnitude values from maximum amplitudes and/or durations; (3) Digitization of seismograms recorded at a high rate of paper speed; and (4) Study of the problem of analyzing waveforms to obtain ground displacement and spectra from seismograms.

If adequate financial support and manpower are available, we intend to carry out the above program not only for the Rocca di Papa Observatory, but possibly also for other old observatories on a national scale. Cooperation with international initiatives like the International Historical Seismograms Filming Project is also planned.

Historical Seismograms From Australia

David Denham
Bureau of Mineral Resources, Geology and Geophysics
P.O. Box 378, Canberra City 2601, Australia

ABSTRACT

Permanent seismographs were established on the Australian continent shortly after the turn of the century. Milne instruments were set up in Perth (1901), Melbourne (1902), Sydney (1906), and Adelaide (1909) and a Weichert was installed at Riverview (Sydney) in 1909. The magnification of the early Milne instruments was very low (≈ 6), however, and it was not until Milne-Shaw instruments were installed in the 1920's that a useful network of four stations was established.

Most of the seismograms for these four stations are available from the 1920's onward. The records are still stored in their original photographic format, however, and a significant number of the early seismograms have deteriorated. Expert conservation will be required if information on the early seismograms is to be preserved.

1. Introduction

The historical record in Australia goes back only about 200 years. Since Australia was first settled by Europeans, however, reports of earthquakes and their effects have been common. The first reported earthquake was felt at Port Jackson (Sydney) in June 1788, when "The 22nd of this month (June) we had a slight shock of an earthquake; it did not last more than 2 or 3 seconds. I felt the ground shake under me, and heard a noise that came from the southward, which I at first took for the report of guns fired at a great distance" (Phillip, 1788).

In 1837 the first settlers in South Australia were made aware of the existence of earthquakes when, "there was a loud rumbling noise that lasted 20 seconds. The earth shook and trembled. It was an earthquake" (Blackett, 1907). Similarly, in the early histories of Melbourne (1841; Underwood, 1972) and Perth (1849; McCue, 1973) earthquakes were felt. Fortunately, there is no record of anybody being killed by an earthquake in Australia and the total damage to buildings and installations since 1950 has amounted to a modest 50 million in 1985 dollar values.

Although the level of Australian seismicity is comparatively low, the Australian continent forms an excellent platform for studies of global seismicity and earth structure. Active seismic zones range in distance from 5° to 180° from the continent, and the locations of earthquakes in the southwest Pacific and in southeast Asia are constrained by seismic observations from Australia.

2. Early Seismological Stations

In spite of Australia's global seismological importance, it is somewhat ironic that the first seismic instruments were set up there to record local earthquakes. These instruments were constructed in Launceston, Tasmania, by Biggs (1885) to record the larger earthquakes from the swarm of over 2,000 events that occurred off the northeast coast of Tasmania between 1883 and 1885. Biggs (1885) constructed his instruments "to arrive at an estimate of the actual magnitude of the surface-motion of the earth; and to gain some idea of the position of the source, or focus, of the disturbance". Biggs built a three-component instrument that recorded on smoked glass; unfortunately, none of these records are known to be extant.

In 1888, H. C. Russell, government astronomer at Sydney, obtained a Ewing seismograph, and in the same year, a Gray-Milne was installed at the Melbourne Observatory (Doyle and Underwood, 1965). Records from these instruments also are not known to be extant.

The seismological committees of the Australian and British Associations for the Advancement of Science sponsored a program to establish the first world network of seismographs, and in 1901 Milne instruments were obtained for Melbourne, Perth and Sydney. These were installed in 1901, 1902, and 1906, respectively, and in 1909 an additional Milne was installed in Adelaide (Doyle and Underwood, 1965). Because the magnification of the Milne instruments was low (≈ 6) and they were underdamped, few events would have been recorded by them.

In the 1920's the situation improved considerably when Milne-Shaw seismographs were installed at Perth (1923), Adelaide (1924), and Melbourne (1928). These instruments had a magnification in the range 150-250 and had magnetic damping. The valuable records from these instruments form the core of historical seismograms from Australia. Meanwhile at the Riverview station near Sydney, which began operation in 1909, the instrumentation was gradually improved to make it the best known, and the best equipped, station in Australia for many years.

Wiechert N-S and E-W seismographs operated from March 1909 until June 1955; a Wiechert vertical seismograph operated from May 1909 until January 1944; Mainka N-S and E-W seismographs operated from August 1910 until December 1962; and Galitzin vertical, N-S and E-W seismographs operated from January 1941 until December 1962. All these instruments were calibrated regularly, and Drake (1985) has made a thorough study of their response characteristics. In his 1985 paper Drake describes (1) where to find the instrumental constants of the seismographs, (2) how to find the magnification and phase lead from these constants, and (3) how the constants were determined.

3. Historical Seismograms

Seismograms from the Melbourne, Perth and Riverview observatories form the best historical sets. The Melbourne and Perth records have been properly archived and readily available for several years. However, the Riverview records have only recently been cataloged and archived. None of these seismograms have been microfilmed and many are fragile.

Adelaide seismograms, which had been presumed lost for several years, were recently found but they are not properly cataloged (S. Greenhalgh, Flinders University, South Australia, pers. com. 1985).

Table 1 lists the periods of operation of the early seismographs and the availability of the records. The data from this table are based mainly on Doyle and Underwood (1965). Figure 1 shows locations of the early seismograph stations. The historical seismograms from Australia are reasonably well cataloged and available. However,

Table 1. Australian Seismograph Stations, pre-1960.

Station	Station codes	Operation dates	Instrument	Component	Curator of seismograms
Adelaide I		1909-1941	M	E	SI
		1924-1950	MS	N	SI
Adelaide II	ADE	1958-1962	B	Z, N, E	SI
Avon	AVO	1958-	B	Z	ANU
Brisbane I		1937-1951	MS	N, E	UQ
		1943-1951	B	Z, N	UQ
Brisbane II	BRS	1953-1963	B, MS	Z, N, E	UQ
Cabramurra	CAB	1959-1972	B	Z	ANU
Canberra	CAN	1958-	B	Z	ANU
Charters Towers	CTA	1957-	B	Z, N, E	UQ
Fort Nelson	FNT	1957-1962	W	Z	UT
Geehi	GEE	1958-1968	B	Z	ANU
Hall's Lagoon	HLA	1959-1970	B	Z	ANU
Heard Island	HII	1951-1954	WA	N	BMR
Inveralochy	INV	1959-1972	W	Z	ANU
Jenolan	JNL	1959-	B	Z	ANU
Jindabyne	JIN	1958-1972	B	Z	ANU
Kuranda	KDA	1959-1964	WL	Z	UQ
Macquarie Island	MCQ	1951-1960	WA	N, E	BMR
		1956-1960	Gr	Z	BMR
Mawson	MAW	1956-1959	LB	Z, N, E	BMR
Melbourne	MEL	1902-1927	M	E	BMR
		1928-1962	MS	E	BMR
		1955-1956	WA	N, E	BMR
		1955-1956	Gr	Z	BMR
		1956-1962	B	Z, N, E	BMR
Mundaring	MUN	1959-1962	B	Z, N, E	BMR
Perth	PER	1901-1937	M	E	BMR (1)
		1923-1964	MS	N	BMR (1)
		1956-1964	C	Z	BMR (1)
Riverview	RIV	1909-1955	Wt	Z, N, E	BMR
		1910-	Ma	N, E	BMR
		1941-	G	Z, N, E	BMR
Sydney	SYD	1906-1948	M	E	BMR
Townsville	TVL	1956-1965	S	E	UQ
Wambrook	WAM	1957-1958	W	Z	BMR
		1958-	B	Z, N, E	ANU
Watheroo	WAT	1958-1959	B	Z	BMR (1)
Werombi	WER	1959-	B	Z, N, E	ANU
Wilkes	WIL	1957-1966	PE	Z, N, E	BMR

B = Benioff
C = Columbia LP
G = Galitzin
Gr = Grenet
LB = Leet-Blumberg
M = Milne
Ma = Mainka
MS = Milne-Shaw
PE = Press-Ewing
S = Sprengnether
W = Willmore
WL = Wilson-Lamison
WA = Wood-Anderson
Wt = Wiechert

ANU = Res. Sch. Earth Sci., Australian Nat. University, P.O. Box 4, Canberra, 2601.
BMR = Bureau of Mineral Resources, P. O. Box 378, Canberra, 2601.
BMR (1) = Bureau of Mineral Resources, Mundaring, WA, 6073.
SI = Sutton Institute of Earthquake Physics, P.O. Box 151, Eastwood, SA 5063.
UQ = Dept. of Geology and Mineralogy, University of Queensland, QLD, 4067.
UT = Dept. of Geology, University of Tasmania, Hobart, 7001.

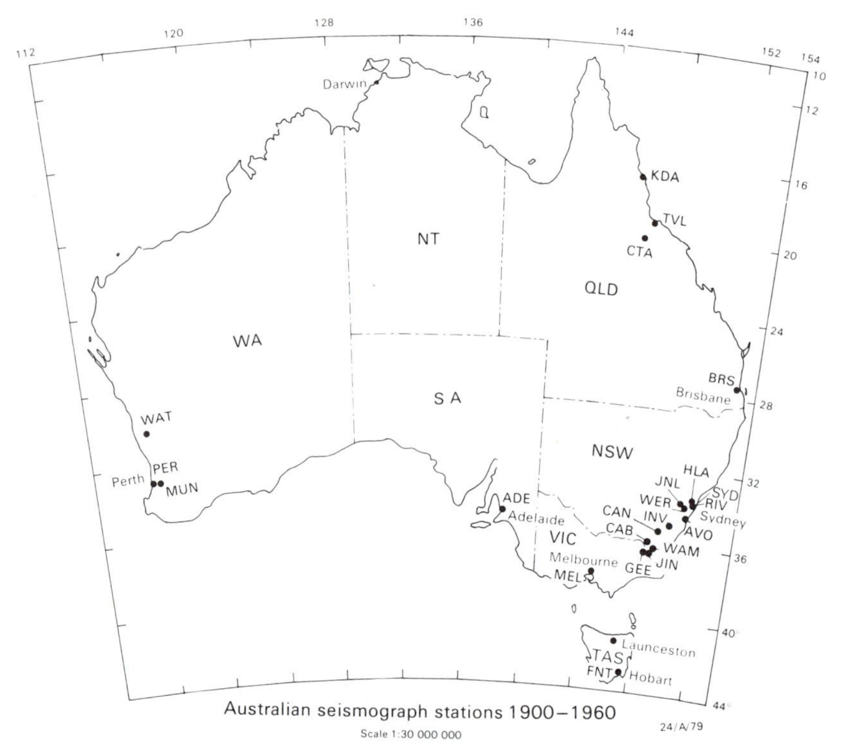

Figure 1. Australian seismograph stations, 1900-1960.

many of the early records are in poor condition, and these should be microfilmed as soon as possible. Furthermore, the whereabouts of the Sydney seismograms are unknown, and so valuable information has already been lost.

REFERENCES

Biggs, A. B. (1885). The Tasmanian Earth Tremors, 1883-4-5, *Papers and Proceedings of the Royal Society of Tasmania*, 325-334.

Blackett, Rev. J. (1907). *Early History of South Australia*, Vardon & Sons Ltd. 393 pp.

Doyle, H. A. and R. Underwood (1965). Seismological Stations in Australia, *Australian Journal of Science*, **28(2)**, 40-43.

Drake, L. A. (1985). The Response and Calibration of Seismographs at Riverview College Observatory, New South Wales, 1909-1962. *BMR Journal of Australian Geology amd Geophysics*, **9(4)**, in press.

McCue, K. J. (1973). Relative seismicity of Western Australia, Parts I and II. Report to the Western Australian Public Works Department and the Commonwealth Department of Works, 25 pp + 11 figs. + 5 appendices, 37 pp. + appendices.

Phillip, Governor A. (1788). Letters to Nepean, in *Historical Records of Australia*, Series 1, Vol. 1, p. 50, Library Committee of the Commonwealth Parliament, 1914, Melbourne, 822 pp.

Underwood, R. (1972). Studies of Victorian Seismicity, *Proceedings of the Royal Society of Victoria*, **85(1)**, 27-48.

Seismograms Made Before 1963 at Stations in the South-West Pacific

G. A. Eiby
Seismological Observatory
P.O. Box 1320, Wellington, New Zealand

ABSTRACT

This paper outlines the history of the seismograph stations in the South-West Pacific and neighbouring parts of Antarctica and gives particulars of the present custody and condition of the records. Instrumental records were made in New Zealand in 1884, with a horizontal pendulum installed at the Colonial Museum in Wellington, and there are earlier references to the design and construction of seismoscopes in the country. The oldest surviving records come from Milne seismographs established in Wellington and Christchurch in 1900 and 1901, and from a Wiechert instrument installed at the Apia Observatory, Western Samoa, a year later. About 1923 a third Milne instrument began recording at Suva, Fiji. The number and quality of New Zealand stations steadily improved, and the record archive in Wellington is now by far the largest in the region. In 1957 the first stations of a comparable network covering New Caledonia and the New Hebrides (now Vanuatu) were installed. The International Geophysical Year led to the improvement of many stations, and the installation of new ones in Antarctica and on islands in the Pacific. Many of these stations have continued to operate. Whether records made in the Ross Sea Dependency by the Scott Polar Expedition in 1902 and Byrd Expedition in 1940 still exist cannot be confirmed. Most station files are incomplete. The missing records are usually early ones or those of large teleseisms. In almost every case they have been interpreted and some data published, though not always in the detail that a modern seismologist could wish.

1. Introduction

This paper outlines the history of instrumental seismology in the South-West Pacific up until the year 1963, and documents the present location and condition of the surviving records. The area considered extends from Fiji and Samoa in the north to Adélie Land and the Ross Sea Dependency in the Antarctic (Figure 1). There are many earthquakes in the region, but it is mainly ocean, with no great land-masses or large centres of population. Nevertheless, by the end of the period covered in this paper extensive networks of seismographs had been established in New Zealand and in French Pacific territories, and single stations set up in the Antarctic possessions of those countries, and on a number of the smaller and widely scattered islands, the most important being Samoa, Fiji, Raoul Island, and Chatham Island. The largest archives of seismograms are in Wellington, at the Geophysics Division of the New Zealand government's Department of Scientific and Industrial Research, and in

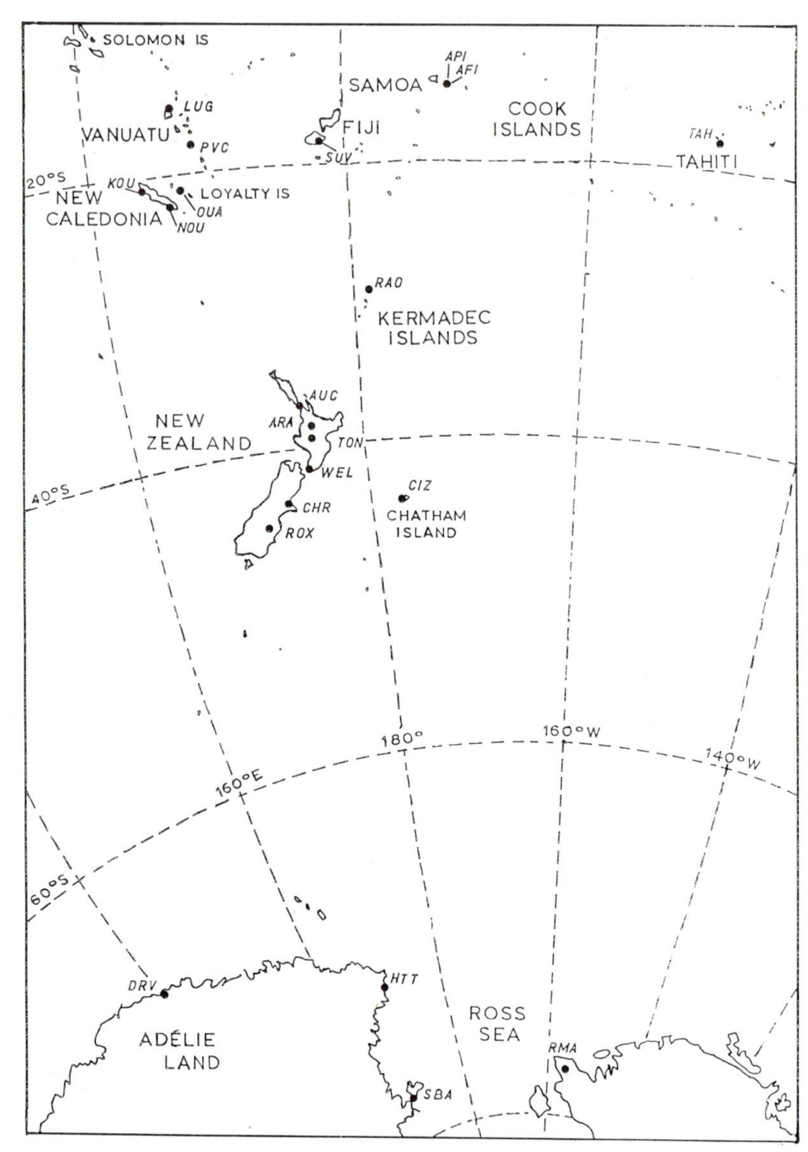

Figure 1. Seismograph stations of the South-West Pacific operating before 1963. For reasons of clarity, fifteen New Zealand stations intended principally for local earthquake study have not been shown. Details of these, and periods of operation of all stations are given in the text.

New Caledonia, at the Nouméa headquarters of the Office de la Recherche Scientifique et Technique Outre-mer (ORSTOM). Other records are widely scattered, but are preserved in Germany, France, and the United States of America.

2. New Zealand

New Zealand possesses the largest and most complete collection of records made in the region. Most of them are to be found in the archives of the Seismological Observatory of the Geophysics Division of the government Department of Scientific and Industrial Research in Wellington. (P.O. Box 1320, Wellington, New Zealand). Most of the records are in good condition, but some of the early ones have become brittle, and others have suffered from poor photographic processing. Gaps in the series are mainly the result of borrowers failing to return records, and seismograms of some important teleseisms are missing. Most stations of the New Zealand network were primarily intended for the study of local earthquakes, and have only a single short-period instrument. Before 1963 it was with few exceptions a horizontal component. However, many large teleseisms have been recorded, and several stations, including Wellington and Christchurch (the first to be established) were equipped to record teleseisms. The history of the individual stations under New Zealand control and their instrumentation has been summarized by Smith (1981).

2.1. Wellington (WEL)

In 1884 an instrument built locally to designs supplied by John Milne began recording at the Colonial Museum (Young *et al.*, 1984). No surviving records have been found. Records from a Milne pendulum erected in 1900 are to be found in the DSIR archive, and in the files of the National Museum, but those dating from before 1915, when the instrument came under the control of the Dominion Observatory, are incomplete and not readily identifiable. It is believed that some early records were sent to Milne at Shide, and may still be found somewhere in England. In 1923 a Milne-Shaw instrument was installed. A second component was added in 1925, and a Galitzin-Wilip vertical in 1930. Wood-Anderson and other short-period instruments came in 1931, including a Jones vertical geophone which ran from 1937-46. A short-period vertical Benioff ran from 1955-57, and World-Wide Standard Seismographs were installed in 1962 May. Timing in Wellington was derived directly from the national time-service, and is of very high reliability from the earliest years. Time was distributed throughout the country by telegraph, and radio time-signals were broadcast from about 1914, and were used to control timing at other New Zealand stations. By the 1930s they were impressed on the records by hand-key, and automatic recording was introduced in 1955.

2.2. Christchurch (CHR)

Regular operation of a Milne pendulum began in 1902, but records of some aftershocks of the large earthquake at Cheviot on 1901 November 16 exist. In 1930 a three-component set of classical Galitzin instruments replaced the Milne, and a Wood-Anderson was added the following year. In 1947 March the instruments were moved to Wairiri (WAI), and returned about a year later, with some interruption to recording. At the end of 1956 the Galitzins were moved to Roxburgh (ROX) and the Wood-Anderson to Gebbies Pass (GPZ) and the station closed.

2.3. Arapuni (ARA)

A Milne pendulum operated from 1930-49, and a Wood-Anderson seismometer from 1949 November until 1950 October, when the station was closed because vibration from the near-by power station greatly reduced the usefulness of its records.

2.4. Auckland (AUC)

A Milne-Shaw instrument was installed in 1941. The station was seriously affected by tilt, and overlapping traces sometimes make the time-signals difficult to distinguish and the clock correction uncertain, but many teleseisms are well recorded.

2.5. Roxburgh (ROX)

Three-component classical Galitzin instruments began recording in 1957, but calibrations are uncertain. A Willmore short-period vertical was added in 1962.

2.6. Tongariro (TON)

A short-period Jones vertical geophone recorded with some interruption from 1952 until 1962, when it was replaced by a Wood-Anderson. During its operation the Jones was probably the most effective instrument in New Zealand for recording the P-phases of teleseisms. Some records from this station were destroyed by fire, and others are badly charred at the edges.

2.7. Chatham Islands (CIZ)

A Milne instrument operated with frequent and often long interruptions between 1932 and 1941.

2.8. Raoul Island (RAO)

This station is strategically placed with respect to Tonga-Kermadec seismicity, but is troubled by a high background of microseisms and intermittent volcanic tremor. The short-period Willmore vertical began operation in 1957. Instrumental breakdown caused an interruption from 1958 April until December.

2.9. Other New Zealand Stations

The following stations operating before 1963 were equipped with Wood-Anderson instruments. The dates are those when recording began: Chateau (CNZ), 1961; Cobb River (COB), 1949; Gebbies Pass (GPZ), 1956; Kaimata (KAI), 1942; Karapiro (KRP), 1951 (replaced by Willmore vertical, 1958); New Plymouth (NPZ), 1931; Onerahi (ONE), 1954; Tuai (TUA), 1939 (A Jaggar shock-recorder operated from 1932-39, The Wood-Anderson was replaced by a Willmore vertical in 1961).

At various times Imamura, Jaggar, and other low magnification instruments were in use at Bunnythorpe (BUN, 1932); East Cape (ECZ, 1934-37); Greymouth (1932-42); Monowai (MNW, 1935); Napier (NPR, 1953-55); Stratford (STZ, 1934-39); Takaka (TAK, 1931-51).

In the period 1930-50 several privately constructed stations operated. No continuous files of records from these stations are known to have survived, but the Wellington archive contains a few seismograms of large events.

3. Samoa

The University of Göttingen established the Samoan Observatory at Apia (API) at the end of last century. A Wiechert horizontal instrument began recording in 1902, and a vertical was added in 1914. Seismograms covering the period from 1902 December to 1913 April are in Göttingen, in the care of the Institut für Geophysik of the University (P.O. Box 876, Göttingen 3400). No analyses are known to have been made or published. A list of 93 shocks recorded between 1913 and 1920 was published by Angenheister (1922). The records for this period have not yet been located, but are most probably in the Wellington archive, where all later records are filed. New Zealand assumed responsibility for the station in 1920. A Wood-Anderson instrument was installed in 1948, and a second component added in 1957, when the station at Afiamalu (AFI) was established on a better site farther inland, and the Wiechert instruments dismantled. Afiamalu was provided with a short- and long-period vertical and a long-period horizontal Benioff seismometer. In 1962 Afiamalu became a World-Wide Standard Seismograph Station.

4. French Territories in the Pacific

4.1. The ORSTOM Network

Earthquake recording in the French territories of the Pacific and in the Condominium of the New Hebrides (now Vanuatu) during the period of this report was the responsibility of the Office de la Recherche Scientifique et Technique Outre-mer (ORSTOM), and the seismograms are stored at the regional headquarters in Nouméa. They are reported to be in excellent condition. All stations have Coulomb-Grenet instruments. A single vertical component began recording at a site near Nouméa in 1954, and in 1957 horizontal components were added and the installation moved to a new site. A second three-component station was established at Port Vila (PVC) in the New Hebrides in the same year, and a vertical component at Koumac (KOU), New Caledonia, in 1959. The stations at Luganville (LUG) in the New Hebrides and Ouanaham (OUA) in the Loyalty Islands were established only at the very end of 1963. Each has a short-period vertical instrument, and Luganville has a long-period instrument as well.

4.2. Tahiti (TAH)

In 1957 Dr. Claude Blot installed two Labrouste horizontal pendulums at a site on the isthmus of Taravao, as part of the International Geophysical Year programme. The station was destroyed by hurricane, but some records are believed to be held at the Institut de Physique du Globe in Strasbourg.

5. Fiji

A Milne seismograph was set up in Suva township (SUV) in 1923, and operated until 1934, the records being sent to Wellington for interpretation and storage. Data were published with those from New Zealand stations. No further recordings were made until New Zealand initiative resulted in a Milne-Shaw instrument being placed in the care of the meteorological service at Laucala (pronounced and sometimes written as Lauthala) Bay. The station continued to be known as Suva. Records were interpreted in Wellington as before, and are in storage there. In 1954 May the clockwork driving the recording drum broke down, and resisted the efforts of local mechanics to repair it. Two years later an enquiry from Lamont Observatory about the availability of Fijian records produced a reply from the Fiji Department of Lands, Mines, and Survey stating that the instrument was "of uncertain ownership" and "has been out of action for at least 18 months". However, a new clock had been ordered, and Suva readings re-appear in the New Zealand Seismological Report for 1956 July.

At the end of 1957 Dr. Claude Blot of the Institut Français d'Océanie, acting on behalf of Lamont Observatory, installed a set of long-period instruments on the same pier as the Milne-Shaw. At this period responsibility seems to have passed to the Department of Lands, Mines, and Survey. In 1958 March, Lamont Observatory was notified that control had passed to the Geological Survey, but this seems merely to reflect an administrative change within the Fijian public service. A file of long-period seismograms from 1957 December 12 to 1962 April 1 is held at the Lamont-Doherty Geological Observatory (Palisades, New York 10964, U.S.A.). Milne-Shaw records are in Wellington.

On 1962 November 3 Lamont Observatory and the Fijian Geological Observatory both withdrew, and the Meteorological Service again took over the Milne-Shaw on behalf of Wellington. In 1964 February it was replaced by a Willmore short-period vertical instrument. Lamont Observatory began further studies in Fiji and Tonga in late 1964.

6. Antarctica

6.1. Ross Sea Dependency

The first seismograms made in Antarctica were recorded by a Milne pendulum operated by members of the Scott expedition of 1902. Many South Pacific shocks were recorded at a site on Ross Island close to the present Scott Base. Records were taken to England, but it has not proved possible to find out whether they were preserved.

The Byrd expedition obtained two months of recordings at Rockefeller Mountain (RMA) in 1940 November and December. Their subsequent history has not been traced.

The New Zealand station at Scott Base (SBA) began operating in 1957 March with a six-component Benioff instrument recording on 35-mm film. The records are preserved in the Wellington archive.

The joint New Zealand - U.S. base at Hallett (HTT) also began recording in 1957, with long-period Press-Ewing instruments and a short-period Willmore vertical. World-Wide Standard Seismographs were installed in 1963, and in 1964 February the station was destroyed by fire. The previous year's records are believed to have been recovered after the fire, but lost in shipment. It is therefore just possible that they still exist, perhaps in the U.S.A., but have not been identified. Preliminary readings had been sent to Wellington by radio, and have been published. After interpretation in Wellington, the rest of the records were sent to Lamont Observatory for storage. The Wellington archive contains a file of full-sized copies of the long-period records of teleseisms in 1961 January and 1961 August.

6.2. Adélie Land

Some confusion exists with respect to recording in French Antarctica before 1955. In 1950 the privately organized Missions Paul-Emile Victor set up a base on a small island off the coast near Pointe Géologie. It operated for three seasons, and was then evacuated until 1955, when it was re-occupied. Earthquake recording began in 1951 March, apparently with a short-period vertical seismometer. Details of the equipment and of the fate of the records are now uncertain. If they still exist they are presumably in France.

The base at Dumont d'Urville (DRV) was established for the International Geophysical Year, and operated three short-period APX instruments between 1957 April 14 and 1962 December 12. The Institut de Physique du Globe in Strasbourg has records of 236 shocks made during this period. All are in good condition.

ACKNOWLEDGEMENTS

I wish to thank the many observatories who have answered enquiries, especially those who fruitfully re-directed the course of my search. In particular I must instance Dr. Robin Adams of the International Seismological Centre, M. R. Louat of ORSTOM, Nouméa, and those members of the staff of the Lamont-Doherty Observatory who unearthed correspondence containing details of operations in Fiji, incidentally throwing light on obscure parts of the Pacific.

REFERENCES

Angenheister, G. (1922). List of the most important earthquakes registered at the Observatory, Apia, Samoa, from 1913 to 1920, Nachr. Ges. Wissensch. Göttingen, Math-phys. Klasse pp 1-3.

Eiby, G. A. (1985). Documenting New Zealand earthquakes, this volume.

Smith, W. D. (1981). A short history of New Zealand seismograph stations, *N. Z. Dept. of Scientific and Industrial Research, Geophysics Division Report 171*, 27 pp.

Young, R. M., J. H. Ansell, and M. A. Hurst (1984). New Zealand's first seismograph: the Hector seismograph 1884-1902, *J . Roy. Soc. N. Z.*, **14**, 159-173.

Historical Seismograms Recorded in Venezuela

G. E. Fiedler B.
Chief, Seismological Institute CAR (ret.)
Apartado postal 5407 Carmelitas, Caracas 1010-A, Venezuela

1. Introduction

The Cagigal Observatory in Caracas, Venezuela, has been recording earthquakes since 1933. This observatory was created by governmental decree on September 8, 1888, basically for astronomy and meteorology research. Figure 1 gives an idea of this historical building. After the large and destructive Caracas earthquake of October 29, 1900, seismology was included by name in the mission of the observatory. The director, Dr. Luis Ugueto, asked for financial help to buy two Agamennone seismographs for about Bolivares 260 each at that time, but he could not raise the funds. Venezuela, therefore, had to wait until 1932-33, when a vertical and a horizontal mechanical seismograph designed by Emil Wiechert were first installed at the observatory.

The vertical Wiechert seismograph had a mass of 80 kg, an equivalent pendulum length of 9.91 m, and a static magnification of 62. Unfortunately, because of problems in the spring-suspension system from the beginning, this instrument never produced acceptable records.

Like the vertical, the Wiechert horizontal astatic mechanical seismograph came from the Spindler and Hoyer factory of Göttingen, Germany (Figure 2). According to Centeno-Grau (1940, p. 162), "This inverted pendulum had a mass of 80 kg, the amplification of the indicator was 101, and the equivalent pendulum length was 9.85 m for one of the two components and 9.01 m for the other." From these data we derive that the free period of the pendulum was about 6 to 7 sec (the factory gives 4 to 12 sec). Other important data like damping and dynamic response of the system have to be estimated from the records. The speed of rotation of the recording drum (paper speed) was 5.44 sec/mm record or 0.184 mm/sec.

The Wiechert seismographs are no longer in operation. They were removed in 1953-54 when the observatory building was razed. Seismographic recordings in Venezuela continued in 1955-56 in the newly constructed Seismological Institute of the "Observatorio Astronómico Sismológico Geomagnético Juan Manuel Cagigal." The new instruments were 3 Galitzin-Wilip, 5 Sprengnether, 1 Benioff vertical, and a 20-ton horizontal Wiechert pendulum recording the ground motion in the two components EW and NS, on smoked paper. The intrinsic constants of this instrument, also made by Spindler and Hoyer, installed and calibrated by the author of this paper, are the following: mass of the pendulum 19,600 kg, free period 1.62 sec, equivalent pendulum length 0.64 meters, indicator-length 1577 meters (lever-arm system and torsions), static magnification about 1800, damping ratio for half period (overshoot ratio) 4:1, resonant gain 2400. This instrument is still recording. It was not until April 1962, when the World-Wide Standardized Seismograph Network (WWSSN) crystal clock timing system replaced the mechanical Riefler clock. This invar-steel pendulum clock was held within the accuracy of ± 2.5 sec per day. The problem is, before 1955, no time marks like minute or hour marks were connected to

Figure 1. The historical building of the former Observatory "Juan Manuel Cagigal," Caracas, seen from south-west towards the mountain Avila, which is covered by clouds. The building, which has resisted well the Caracas earthquake of 29 October 1900, unfortunately was razed in the years 1953-54.

the recording device of the Wiechert seismometer, and just handwritten time marks were noted on the smoked paper. It is surprising, that besides the high development of exploration geophysics in Venezuela during the years 1930 to 1950 or earlier, – and besides the occurrence of about 5 Venezuelan earthquakes with $M \geq 6$ (PAS) between 1900 and 1950, – nobody from inside or outside the country has shown any effective interest for the improvements of the Venezuelan earthquake seismology, during the first half of this century.

2. Geographical and Geophysical Constants of the Seismological Institute of the Cagigal Observatory, Caracas (CAR)

For investigation or comparison, a few important constants of CAR may be useful. The data were computed by Cambridge University, Wisconsin University, the National Cartographic Office, the Ministry of Energy and Mines, and CAR itself.

The 4-meter-deep seismometer piers rest on strongly weathered schist of Upper (Late) Cretaceous age.

The legal time in Venezuela (HLV) before January 1, 1965 was Universal Time Coordinated (UTC) minus 4 hours 30 minutes. But from that date to the present it is UTC minus 4 hours, which refers to the 60° W meridian. Other data:

Geographic longitude	66.9230° W
Geographic latitude	10.5056° N
Geocentric latitude	10.4375° N

Figure 2. The Wiechert horizontal astatic mechanical seismograph installed in the historical Cagigal Observatory, Caracas. One observes on the right the inverted pendulum and one of the two air damping devices, adjustable from zero to aperiodic viscous damping. On the left the recording drum with the two recording pens. The price of this instrument in 1931 was 1640 German Marks or US$ 500-600.

Elevation (above sea level)	1, 035 meters
Gravity (pendulum)	978.03965 gals
Maximum variation, earth tides	0.00026 gals (approx. observed)
Geomagnetic declination, 1960.0	4° 44.2′ W
Variation per year of declination	+6.3′
Geomagnetic inclination, 1960.0	41° 04.5′ N
Geomagnetic horizontal intensity	29, 300 γ
Geomagnetic vertical intensity	25, 937 γ
Geomagnetic total intensity	39, 130 γ

3. Description of the Historical Seismograms from Venezuela

The absence of time marks and test records for instrumental calibration poses a large problem for accurate interpretation of these historical records. Another problem is the interpretation of the seismograms themselves, as shown by the following example: In Centeno-Grau (1940, p. 308), we find that the horizontal, two-component astatic Wiechert seismograph at Cagigal (OCV) recorded the first earthquake on November 4, 1933, at 04:15 HLV. In the NNE-SSW direction, the duration of the record was about 1 minute. The earthquake was felt in the Barquisimeto-Cúcuta-Merida-San Cristobal-Cucuta-Pamplona region. Comparing this data to that from the Earthquake Data File of the U.S. National Geophysical Data Center (1972), we have:

OCV	1933 11/04 04:15	Felt Cúcuta-Pamplona, and Venezuelan Andes						
CGS	1933 11/04 08:41:15.0	9.000 N	72.000 W	000	0.0 M	101		
GUT	1933 11/04 08:41:17.0	8.500 N	72.000 W	000	6.0 M_S	101.		

Adding 4h 30m to the arrival time of an unknown phase recorded at OCV, we must subtract 3m 45s to obtain the origin time given by CGS, or 3m 43s for the GUT epicenter, which is located 55 km south of the CGS epicenter.

The total duration of the seismogram recorded by OCV was 1 minute and the epicentral distances are 580 km (CGS), 600 km (GUT) and 700 to 750 km (region Cúcuta-Pamplona, where the real epicenter was first assumed). The duration of 1 minute excludes the possibility that P-phases were recorded, because just the time intervals of the arrival (Sg − Pg) or (Sn − Pn) for epicentral distances of 580 km or more are longer than the mentioned duration of 1 minute. Probably the small P-phases were not recognized by the station operator of that time and the given onset therefore corresponds to the Sg-wave with a travel time of 2m 55s corresponding to the velocity of propagation of 3.3 km/sec. The error in origin time calculated from that Sg-wave is 50 sec and this corresponds to only 9.19 mm on the record, which is not bad if there were no minute marks. The main failing was in the interpretation of the seismogram, and because of this, it appears that practically all seismograms from the former Cagigal Observatory (OCV) must be reanalyzed, if these important records are to be meaningfully used with other network data to improve seismic maps and risk calculations.

Between November 1933 and December 1953, the two-component Wiechert seismograph at OCV recorded about 149 earthquakes, of which about 68 events (or 46 percent) can be related to earthquakes published in the data files of GUT, CGS, and others.

Table 1 gives the arrival time at Cagigal (OCV in this paper) in UTC, and the epicentral region of earthquakes for which the Pasadena (PAS) magnitude is 7 or larger. Origin times are taken from the Earthquake Data File of the U.S. National Geophysical Data Center (1972). The earthquakes listed in Table 1 occurred in the region 5°S to 21°N and 55°W to 100°W. Because the OCV arrival times are not in all cases P-wave arrivals, reevaluation of the records is essential.

Table 1. Earthquakes with $M \geq 7$ Recorded at Cagigal (OCV)

Date year/mo/da	Origin time hr:mn:sec	Arrival time hr:mn:sec	Epicenter Lat.	Epicenter Long.	M (PAS)	Approximate epicentral region
1934/07/18	01:36:24.0	01:49:06.0	08.00 N	082.50 W	7.7	Panama
1935/12/14	22:05:17.0	22:32:04.0	14.75 N	092.50 W	7.3	Mexico
1937/12/23	13:17:56.0	13:29:26.0	16.75 N	098.50 W	7.5	Mexico
1938/02/05	02:23:24.0	02:26:06.0	04.50 N	076.25 W	7.0	Colombia
1939/12/21	20:54:48.0	21:18:07.0	10.00 N	085.00 W	7.3	Costa Rica
1941/12/05	20:46:58.0	20:48:48.0	08.50 N	083.00 W	7.5	Panama-Costa Rica
1942/05/14	02:13:18.0	02:18:10.0	00.75 S	081.50 W	8.3	Ecuador
1942/08/06	23:36:59.0	23:37:00.0	14.00 N	091.00 W	8.3	Guatemala
1942/08/24	22:50:27.0	22:57:19.0	15.00 S	076.00 W	8.6	Peru
1943/05/02	17:18:09.0	17:19:24.0	06.50 N	080.00 W	7.1	Malpelo-Isl.
1943/07/29	03:02:16.0	03:06:00.0	19.25 N	067.50 W	7.9	N. Puerto Rico
1944/06/28	07:58:54.0	08:12:48.0	15.00 N	092.50 W	7.0	Guatemala-Mexico
1946/05/21	09:16:42.0	09:18:48.0	14.50 N	060.50 W	7.0	E Martinique Isl.
1946/08/04	17:51:05.0	17:53:19.0	19.25 N	069.00 W	8.1	Dominican Rep.
1946/08/08	13:28:28.0	13:30:41.0	19.50 N	069.50 W	7.9	Dominican Rep.
1946/10/04	14:45:26.0	14:48:18.0	18.75 N	068.50 W	7.0	Mona-Passage
1948/04/21	20:22:02.0	20:24:12.0	19.25 N	069.25 W	7.3	Dominican Rep.
1950/10/05	16:09:31.0	16:14:07.0	11.00 N	085.00 W	7.7	Costa Rica-Nicaragua
1950/10/23	16:13:20.0	16:23:25.0	14.50 N	091.50 W	7.1	Guatemala
1953/03/19	08:27:53.0	08:26:45.0	14.10 N	061.21 W	7.3	Sta. Lucia-Martinique
1953/12/12	17:31:25.0	17:36:27.0	03.40 S	080.60 W	7.7	Ecuador

4. Conclusion

The smoked-paper seismograms from the first, or historical, Cagigal Observatory in Caracas were recorded on a two-component horizontal Wiechert seismograph. These records may contribute significantly to the relocation of several important and large earthquakes that occurred from 1934 to 1953. A careful reevaluation of the dynamic response curve of the seismograph and the records must be made to permit this information to be used with later data from CAR and other international seismological observatories, to improve knowledge on earthquake activity and tectonics.

AKNOWLEDGEMENTS

The author of this paper likes to thank Dr. W. H. K. Lee and his staff for their technical help.

REFERENCES

Centeno Grau, M. (1940). Estudios Sismologicos, Litografia Comercio, Caracas.

U.S. National Geophysical Data Center (1972). Earthquake Data File Summary, Key to Geophysical Records Documentation No. 5, Boulder, CO., USA.

Venezuela, Cagigal Observatory (1953). Notes of recorded earthquakes at the Cagigal Observatory, 1933-1953, handwritten manuscript, not published, Caracas.

The Historical Seismograms of Colombia

J. Rafaél Goberna, S.J.
Instituto Geofísico, Universidad Javeriana
Carrera 7a. No. 40-76 - Apartado Aereo 56710, Bogotá, Colombia

1. Introduction

The history of the seismological station of Bogotá has to be divided in two different periods, not only because of the change of its location and the kind of its instruments, but also because of its time and duration.

The first period comprises the years from 1923 to 1940, during which the station was situated in the San Bartolomé College at the central part of Bogotá; its instruments were of mechanical type recording on smoked papers.

The second period comprises the years from 1942 to 1960, and later years. The station site was the same as the present one in the San Bartolomé College at the northeast border of the city in the section called La Merced. In this period, besides the mechanical Wiechert seismograph, the station was equipped with electric and photographic instruments.

2. First Period (1923-1940)

2.1 Instruments

In the year of 1916, after finishing their scientific studies at the Colégio Máximo, which the Society of Jésus had in the town of Oña, Burgos province (Spain), Fr. Enríque Pérez Arbeláez and Fr. Carlos Ortíz, Colombian Jesuits, took a short course in Seismology with the Rev. Fr. Manuel Ma. Sánchez-Navarro Newman, Director of the Seismological Observatory at Granada (Spain). As a consequence of this course they ordered to the workshop "Automática" of Torres Quevedo in Madrid, the construction of a seismograph of bifilar type entirely similar to those functioning at the Granada seismic station. For different reasons this seismograph did not reach Bogotá until some years later. However, it still remained stored for some time, because no adequate site could be found, which would be sufficiently protected and removed from traffic and other machinery, that was functioning on nearby places and even in the same building of San Bartolomé College.

During the yearly vacations and thanks to the activity and enthusiasm of F. Arbeláez, the instrument was provisionally installed in the ground floor of San Bartolomé College at a site near the principal court and to its north side (today little plaza Camilo Torres, Carrera 7, calle 10). Contrary to what one would expect, the seismograph showed more or less insensitive to the short period vibration, coming from traffic and motors functioning in the College building, but it was highly sensitive to long period oscillations produced by seismic events.

On the 17th of January, 1922, the first seismogram of a distant earthquake was recorded. It was exhibited on the display windows of the Merss. José Manuel Rodríguez shop for public admiration. Unfortunately, this first recorded seismogram at the Bogotá seismic station lacked time marks, which we may deplore today. However, after a cursory examination about its date, we found out that this seismogram was the record nearest to the epicenter of a very deep and great earthquake occurred at the Perú-Colombian border on the night of the 16th to 17th of January of 1922, at $H = 03h\ 50m\ 33.0s$ GMT with the epicenter about 800 km to the southeast of Bogotá at 2.5° N and 71.0° W, according to Gutenberg's determination; its depth was about 650 km and its magnitude about 7.6. It was selected and studied by Turner (1922) to prove the occurrence of very deep earthquakes and it was the object of a very detailed analysis made by Vicente Inglada (1934).

The trace amplitude was large with a peak of 20 cm. This led Fr. Carlos Ortíz to calculate the magnification of the seismograph, making use of Wiechert formulas and of the mass, period, friction, damping, etc. of the instrument. The result was a magnification of 4.7 K. In the following year, Rev. Fr. Sarasola, Director of the San Bartolomé Meteorological Observatory, installed this instrument in a better protected and permanent underground place at the southeast corner of the same building (Carrera 6 and Calle 9), although this site was not yet entirely satisfactory. The seismometer mass weighted 200 kg, but other constants and specifications were unknown, except for the provisionally determined magnification of about 4,700. The components were orientation NNE-SSW and ESE-WNW.

The clock used for time marks as well as the method for its correction were also unknown. There are two old pendulum clocks with their weight chains and minutes contacts, which might be used for recording time marks. Although Rev. Sarasola usually kept the time in the Observatory adjusted through astronomical observations by an electric clock of Riefler type, the seismic clock did not keep time with same precision, as can be seen from notes found sometimes on the seismograms indicating corrections of two or more minutes.

The drum was of metal and its size was 28 cm in length by 18 cm of radius with recording paper of 60 cm in length. Its motion was through a clockwork with rotation varied from 1.5 cm to over 2.0 cm. Its displacement was obtained through a wormscrew, keeping a distance between lines of about 2 mm; this is the only piece that today remains at the station and maybe also its clock. This equipment was kept functioning from July, 1923, until the end of 1925. From 1926 until April of 1928, the station operation was suspended for unknown reasons. However it was operated again in April of 1928 until the end of the same year.

In 1928 a new seismograph was acquired in Europe, constructed by Spindler and Hoyer of Göttingen (Germany) of the Wiechert type; this instrument was kept functioning until to the present time with some occasional interruptions. Its original constants were determined at the factory before being shipped to Bogotá. It is a horizontal astatic and mechanical pendulum with a mass of 200 kg. One component has a period of 5.78 sec and the other one of 5.63 sec; its magnification is about 92 and 96, and its damping is about 4.5 and 4.6 respectively. The components' orientations were N-S and E-W, but their polarization was unknown. However, it was determined later since its position was never changed.

Again in this case, it is not known with certainty the type of clock used for time marking. Most probably, it was a clock made by the same Spindler and Hoyer factory for the Wiechert seismograph, but it can not be found at the station now. The Wiechert drum was also somewhat different from the previous one. It was a bit shorter in length, but a bit longer in radius, so that its records are 21.5 cm in width and 62 cm in length. Its rotation is also slower and is obtained through a wormscrew moved by a weight. The rotation is also of opposite direction: it is from right to left in the Granada type, opposite to the clock hands movement, and it is clockwise in the Wiechert type. As a result, the seismogram reading is from left to right for the Granada-type records, and it is from right to left for the Wiechert-type records.

The first seismogram of the Wiechert instrument is of June 16, 1928, and both instruments were operated simultaneously for some months, so that the records of both instruments were obtained for the same days and for the same seismic events. Therefore, by comparison, the Granada-type instrument's magnification might be determined. It looks somewhat greater than Wiechert's magnification, but not so much as that assigned by Rev. C. Ortíz. At the end of 1928, both seismographs ceased functioning for unknown reasons. The first seismograph made in Madrid at the Quevedo factory was dismantled, and only some of its parts are kept at the Bogotá station. However, the Wiechert instrument after some occasional operation on January of 1930, was back in operation on October of the same year.

Besides this, on the month of June, a new mechanical horizontal seismograph, made at the Granada seismic station under the direction of Rev. Fr. Sánchez-Navarro and with the same specifications of the Cartuja-type, was received and installed in Bogotá station. The mass of this instrument was 1,000 kg but other constants are unknown. Its magnification seems also somewhat greater than that of Wiechert. It might be determined by comparison with that of Wiechert because both instruments were kept functioning simultaneously for a long time in the same place. Its clock seems to have been the same one used with Wiechert seismograph, and therefore, its precision should be the same. However, the time correction was not written down on the records.

The drum was like that of the Wiechert's. Its rotation was in same direction, and the records were of same size: 21.5 cm in width and 61.8 cm in length. But the rotation was more rapid in the Cartuja-type, varying between 2 and 3 cm; while in the Wiechert-type, it varied between 1 and 1.5 cm. The spacing between lines was much more irregular; in one of its components the spacing was much greater than in the other. For the Wiechert seismograph, spacing was more regular in both components. The components orientation was normal: N-S and E-W, but its polarity is also unknown. Both instruments were kept functioning in the same place until 1940, when both were dismantled because the Government took the building of San Bartolomé. The Wiechert was removed and installed later in a new site.

The geographical position of this first site of the Bogotá seismic station was: 4° 35′ 50″ N and 74° 04′ 52.65″ W, and its altitude was 2,600 m above the mean sea level.

2.2 Seismograms

The records of this period are all on smoked paper; most of them are on bad paper, although well fixed and quite well preserved. Their size depends upon the recording drum of the respective instrument. Time marks are frequently missed in some of the components. It is especially difficult to identify the hour marks in some of the seismograms, because usually the time was not written down at the beginning and at the end of the recording. In some cases, the identification of the date might be difficult, because most of the time it was written on the back of the records, but usually only the last date of recording, and even that was not regularly observed. In addition, the time correction was usually not written down, or only the hour, minutes and seconds at the beginning of a seismic event were written.

The seismograms of the first period have not been analyzed, and therefore, the recorded events are not yet identified. This is a job that should be made before or at the same time of being photocopied, especially in the case of having to select the seismograms which contain valuable seismic information. It should be also known that not all records of the period have been preserved. It seems that only those records, which the occurrence of some seismic event have been noted or identified, have been preserved. The polarity of the components is not known, but it may be possible to determine by the same construction of the instruments.

The seismograms corresponding to the years 1923 to 1925 were recorded on ordinary, but thin paper, and therefore they are brittle and easily perishable. Their size is 28 cm in width and 59 cm in length. The smoke coating was very thin and so was the recording. Consequently, it is often very difficult to identify the beginning of a seismic event and some of its phases. This happens very often in one of the components, and sometimes the recording of one component is also missing. This surely will make copying of these seismograms very difficult.

The quantity of seismograms for these two and half years is quite small, and the amplitudes of the events recorded are also small or moderate. This makes dubious the magnification of 4,700 supposedly assigned to this instrument by Fr. C. Ortíz. Only 35 seismograms are preserved for 6 months of 1923, 40 seismograms for 1924, and 22 seismograms for 1925 (a total of 97 seismograms).

No seismograms are preserved for 1926 and 1927, probably because the seismograph was damaged. It started functioning again in April of 1928, but only with one component. On June 16, 1928, the Wiechert seismograph made its first record; and for the rest of the year, both instruments were operating simultaneously. Thin ordinary paper was used in both instruments, but with better smoke coating, so that the time marks and the seismic traces are clearer. Therefore, copying these seismograms can be made without any difficulty. The size of paper for the Wiechert recording is narrower and somewhat longer than that for the Madrid instrument: 21.5 cm in width and 62 cm in length. The drums rotation is opposite; the Wiechert drum rotates from right to left, and the other from left to right. Hence, the seismogram reads from left to right for the Wiechert instrument, and from right to left for the Madrid instrument.

No records are preserved for 1929, and the reasons are unknown. The Wiechert seismograph had not been functioning, except for occasional days in January, until the month of October in 1930. However, on June of this year, a new mechanical horizontal seismograph of the Cartuja-type was installed; its mass was of 1000 kg. Since the month of October, 1930, both the new Cartuja and the Wiechert seismographs were operating simultaneously at many times. The paper used from 1930 onwards was somewhat thicker and more solid, and therefore better preserved. The smoke coating was often very dense so that reading of these seismograms is sometimes very difficult, and this may cause problems in copying. The paper size is the same for both instruments; the direction of drums' rotation is also the same. However, the rotation speed of the Cartuja instrument is a bit greater so that with the spacing between lines and time marks are larger. Also the line spacing in one of its components is greater than the other, as if each component were running in different drum. Consequently, part of one record is sometimes superposed on the other, making it difficult to read and identify phases in the seismic events.

The quantity of records for most of these years was greater at the beginning of the period, but decreasing appreciably and gradually for the last years. The total number of seismograms is about 554 for the last 12 years, and with the 97 seismograms recorded in the first 5 years, we have a total of about 651 for the first period of the Bogotá seismic station from 1923 to 1940.

3. Second Period (1942-1961)

The second period of the seismic station at Bogotá began with the establishment of the Instituto Geofísico de Los Andes Colombianos on September, 1941, in the San Bartolomé College "La Merced", situated to the north of the former and to the northeast border of the city of Bogotá.

The geographic situation of the new site is given by the following coordinates: 4° 37′ 23″ N and 74° 03′, 54″ W at an elevation of 2,653 m above the mean sea level.

For the new station, a place was selected as isolated as possible from the city traffic, and the site was prepared for the best conditions according to the specifications of modern seismology. By taking advantage of a small cavity in a rocky escarpment within the College campus, a cave was excavated in the rock, and the place was prepared so that the seismic instruments were installed on cemented bases over the rocky ground. Of the three previous mechanical seismographs, only the Wiechert-type was kept and installed in the new place. The other two instruments were dismantled, and most of their parts were dispersed or destroyed; only one recording system, the Cartuja mass of 1000 kg, and the clocks were preserved. In addition, an electromagnetic seismograph was acquired: a short-period, photographic, Benioff model, constructed under direction of the same author at Pasadena, California.

The seismic station was completed with an electric clock of the Standard Electric Time Co., one radio transmitter by Johnson Viking, and a Hallicrafter receiver to receive the international time. A three-component electromagnetic and photographic seismograph of the Galitzin-type was ordered from the Masing-Wilip workshop in Estonia, but it was never received because of the Second World War. Later on, a seismograph of two horizontal components was ordered from Sprengnether at St. Louis, Mo. (U.S.A.).

The new station was ready to start operation at the end of 1941, but its regular function did not begin until the month of March, 1942. The period of the 100 kg Benioff vertical seismograph was 1 sec and that of its galvanometer was 0.3 sec; its damping was critical and its static magnification was 1,200, but its polarity is unknown. Its drum is 32 cm in length and its circumference 91 cm; its rotation is 1 mm/sec and the spacing between lines is 2.5 mm. The size of the recording photographic paper is the same as that used in the Vela project.

The Wiechert seismograph started operation again at the beginning of 1943. The horizontal seismograph (ordered from Sprengnether to replace the Masing-Wilip) reached Bogotá in 1946; one component had to be repaired, and therefore the whole instrument installation was made in July of 1947.

This equipment is composed of two horizontal components, Sprengnether model H, galvanometers by Leeds & Northrup of Philadelphia (U.S.A.), and a triple drum for photographic paper of 30 cm by 90 cm. The drum rotation is 30 cm per minute. The natural period of this seismograph was 16 sec, and the components orientation was N-S and E-W; other constants, such as damping, polarity, and magnification, were unknown.

3.1 Seismograms

On account of the station being dismantled at the end of 1940, the instruments being transferred to a new site, and the new place being prepared, the installation of the seismographs was delayed until April 1942, when the new photographic vertical component began operation. Therefore, no seismic records exist for these two years. The operation of the Benioff seismograph was somewhat irregular during the year because it was difficult to get photographic paper during the war. For some months the paper available was only 23.5 cm in width, which did not last the 24 hours, and therefore, some seismic events were missed. This condition lasted for several months, and even at the end of 1942, it was impossible to get any kind of photographic paper. The problem was solved by March, 1943.

The Wiechert seismograph was reinstalled on July of 1943 and its normal operation continued until the end of this period. Its recording paper was quite thin, but the seismograms are well preserved. The smoke coating was at times too thick so that the seismograms are difficult to read or copy. Often the time marks were missing, but the seismogram readings can be made from the photographic records; the clock used was the same for both instruments. The time correction was automatically marked on the seismogram by radio signals received from the Annapolis station (U.S.A.) or from Panama.

Since July 1947, the records of the horizontal Sprengnether seismograph are also preserved, and therefore, the amount of the photographic records was greatly increased. During the early years the drum was kept running at the rate of 1 mm/sec and at 5 mm of spacing between the lines. In this way the photographic paper did not last for 24 hours, and sometimes seismic events were missed. This problem was not solved until 1950, when the rotation speed of the recorder was changed to 0.5 mm/sec, but the vertical component was kept running at 1 mm/sec.

The amount of seismograms for this period is not exactly known as yet, but from the time period and the daily papers used, a minimum of records can be calculated. The minimum amount of photographic records may be about 15,000, and the amount of smoked records may be about 6,000.

3.2. Analysis

During this second period, the analysis of the seismograms were made every day, and the data of all seismic events were published in the Institute Seismic Bulletins and sent to several other institutions, some of which usually published them in their seismic bulletins or used them for determining the parameters of the seismic events. Thus, they were published in the Mensuel Bulletin of the B.C.I.S. and in I.S.S. since 1943. The first seismic data published by the Instituto Geofísico were those of June and July 1942, and after a few months of interruption they resumed publication on April 1943, until the end of this period.

4. Other Colombian Stations

When the seismic station of Bogotá became well established and its permanent operation secured, its Director tried to find funds for establishing some other seismic stations so that with their data one could determine the parameters of Colombian earthquakes. Two trading and financial institutions, the National Federation of

Colombian Coffee Planters and the Division of Salinas of the Bank of the Republic, became interested in helping the aims of the Institute Director. With their financial support, installation of two seismic stations, one in Chinchiná (4° 58′ N, 75° 37′ W, elevation 1,360 m), and other in Galerazamba (10° 47′ 08″ N, 75° 15′ 44″ W, elevation 21 m), was approved in 1945.

The new stations began operation at Galerazamba on April and at Chinchiná on August of 1949, and since that time, they were kept in operation until the end of this period. The instruments of both stations were similar: photographic seismographs from Sprengnether and galvanometers from Leeds & Northrup of Philadelphia; one short-period vertical-component of the DM type and two horizontal components of the H type; the short period was 1.5 seconds and the long period was 16 seconds. The equipment was completed with a triple drum, one electric clock from the Standard Time Clock Co., and one transmitting and receiving radio. The orientation of the components was N-S and E-W; other constants could be determined; however, they were not taken into account for parameters determination. Time correction was taken from the international WWV time station and automatically marked on the records. The drums rotation was kept at 0.5 mm/sec, and the spacing between the lines at 5 mm. The recording photographic paper was of standard size: 30 cm in width and 90 cm in length.

Both stations were kept functioning until 1960 and onwards with three components, and therefore, the quantity of records are about 4,000 each, for a total of 8,000 during this period. However, the useful records from Galerazamba are appreciably less, because the interference of sea waves in this station is too great.

The seismograms of both stations have been analyzed with those from Bogotá and their data published in the same Bulletins.

As a final note, it should be mentioned here that in 1957, a new seismic station was established in Fuquene Island, about 100 km to the north of Bogotá, to cooperate in the investigations of the International Geophysical Year. The seismograph used was a photographic vertical component with a period of 6 sec, Galitzin-Wilip type, made by the Askania workshop.

REFERENCES

Turner, H. H. (1922). On the arrival of earthquake waves at the antipodes, and the measurement of the focal depth of an earthquake, *Mon. Not. Roy. Astron. Soc. Geophys. Suppl.*, **1**, 1-13. (also see *International Seismological Summary*, 1921 and 1922).

Inglada Y Ors, Vicente (1943). Contribución al Estudio del Batisismo Suramericano de 17 de ener de 1922, *Inst. Geogr. y Catast. - Memorias del I.G.y C., Tomo 16*, **8**, Madrid.

HELWAN HISTORICAL SEISMOGRAMS

R. M. Kebeasy
Helwan Instutute of Astronomy and Geophysics
Helwan, Cairo, Egypt

Seismographic recording began at the Helwan station in September 1899 using an E-W Milne seismograph. A N-S Milne-Shaw seismograph began operation in 1922, a vertical component Galitzin-Wilip was added in 1938, and a three-component short-period Sprengnether was installed in 1951. All these instruments continued to operate until 1962.

In 1962 the Helwan station became a World-Wide Standard Seismograph Station, and three-component short- and long-period seismographs were installed. In 1972, a three-component short-period electromagnetic Japanese-type seismograph was added to the system. A 12-channel short-period frequency analyzer, which was connected in parallel with the vertical component, operated continuously from 1972 to 1980.

In addition to the Helwan station, three other stations have operated intermittently since 1975 in Aswan, Abu-Simbel, and Mersa Matrouh. A radio-telemetry network of nine stations was established around the northern part of Aswan Lake in 1982. This network was expanded in 1985 to include four additional stations and a digital recording system.

The history of the Egyptian National Seismograph Network, its instrumentation and operational specifications, is listed in Table 1.

Table 1. Egyptian Seismograph Network

Station location	Instrument type	Comp.	T_o sec	T_g sec	Speed mm/min	Magnification	Operation period
Helwan	Milne-Shaw	E-W	12.0	11.1	10 & 15	250	1899-1962
Helwan	Milne-Shaw	N-S	12.0	—	10 & 15	250	1922-1962
Helwan	Galitzin-Willip	V	11.2	11.1	15	1,000	1938-1962
Helwan	Sprengnether	3-comp	1.5	1.05	60	3,000	1951-1962
Helwan	Benioff	3-comp	1.0	0.74	60	50,000	1962-pres.@
Helwan	Sprengnether	3-comp	15.0	100.	30	3,000	1962-pres.@
Helwan	Japanese	3-comp	SP	—	120	30,000	1972-pres.
Matrouh	MSK(CKM-3)	3-comp	SP	—	60	10,000	1975-pres.*
Aswan	MSK(CKM-3)	3-comp	SP	—	60	10,000	1975-pres.*
Aswan	MSK(CK-D)	3-comp	LP	—	30	2,000	1975-pres.*
Abu-Simbel	MSK(CKM-3)	3-comp	SP	—	60	20,000	1975-pres.*
Aswan Telemetry Network		11 V 2 E-W 2 N-S 2 L-G	1.0		Analog and Digital	30,000 to 100,000	1982-pres.$

Station Codes: Helwan = HLW, Matrouh = MMT, Aswan = ASW, Abu-Simbel = ASL.
@ – World-Wide Standard Seismograph Station.
* – Russian Standardized Seismograph System; not continuously operated.
$ – The network is under expansion and upgrading (Simpson *et al.*, 1984) and is located in Northern Nasser Reservoir Area.

Although seismic recording in Helwan began in 1899, seismograms are available only from 1907 to the present. Phase readings from seismograms are available in monthly bulletins, and a yearly seismological bulletin officially began in 1936. A detailed report on near and local earthquakes recorded by the Helwan station from 1903 to 1950 was published by Ismail in 1960.

In July 1984, a microfilm camera and 20 rolls of film were provided to the Helwan station through the IASPEI/UNESCO workshop on historical seismograms. When received, the Inter/COM Microfilm Recorder was installed and inspected by a Bell and Howell representative in Egypt, who provided the necessary training to a technician hired specifically for the microfilming operation. However, to proceed with the seismogram copying, the representative suggested that a microfilm reader-processor be provided. The Helwan Institute has no budget to do this at present.

Seismograms covering 5 years of data have now been copied, processed by the Bell and Howell representative, and found to be acceptable. In addition, it has been decided to copy all smoked paper seismograms and also to make arrangements for copying the photographic paper records. Because the Bell and Howell office in Egypt is a sales office only and does not process microfilm, we cannot be certain that all seismograms are copied properly. In this regard, a reader-processor is necessary.

Planned Studies

(1) Because the seismographs that operated at Helwan from 1899 to the present have different natural periods, a study of the spectral magnitude of local and regional earthquakes, is planned.

(2) International cooperation is invited to reevaluate the magnitudes of local, near and distant earthquakes for hazards studies.

REFERENCES

Ismail, A. (1960). Near and local earthquakes of Helwan (1903-1950), *Bulletin of Helwan Observatory*, No. 49.

Simpson, D. W., R. M. Kebeasy, C. Nicholson, M. Maamoun, R. N. H. Albert, E. M. Ibrahim, A. Meghed, A. Gharib, and A. Hussein (1984). Aswan Telemetered seismograph network, *Proceedings of the third annual meeting of the Egyptian Geophysical Society*.

SEISMOLOGICAL MEASUREMENT IN HONG KONG

H. K. Lam
Royal Observatory
Nathan Road, Kowloon, Hong Kong

1. Introduction

Seismological measurements in Hong Kong began in 1921 by the Royal Observatory, Hong Kong. Apart from the period 1941-1950, continuous recordings have been made.

2. Seismographs

Table 1 shows the seismological instrumentation used since 1921. Before World War II, the instrumentation consisted of two horizontal-component Milne-Shaw seismometers each having a natural period of 12 seconds. The recording was made photographically using a galvanometer of similar period, the exact values of which are no longer on record. The seismometers were located on a pillar in a double-walled cellar about 6 meters below the Royal Observatory grounds. These instruments were removed during the war, and so no records are available for the period 1941-1950.

Seismological recordings resumed in 1951 with the acquisition of two horizontal-component and one vertical-component Sprengnether seismometers. The natural period of the horizontal seismometers was 14.5 seconds, and that of the vertical seismometer was 1.5 seconds. Recordings were made photographically using mirror galvanometers having periods of 18.0 seconds and 1.5 seconds, respectively, for the horizontal and vertical components. The operation of this set of seismographs was terminated in 1976 due to frequent seismometer and galvanometer faults.

As part of the programs for the International Geophysical Year, a set of three-component long-period seismographs with 15-second Press-Ewing seismometers and 105-second galvanometers was set up in 1958 in a cooperative experiment with the Lamont-Doherty Geological Observatory of Columbia University. It operated until 1976 when the supply of spares and consumables from Lamont ceased. The Royal Observatory later acquired electronic recording equipment for the Lamont seismometers. Pen-and-ink recordings resumed in 1979.

In 1963, a set of three-component long-period seismographs with 15-second Sprengnether seismometers and 100-second galvanometers and another set of three-component short-period seismographs with 1-second Benioff seismometers and 0.75-second galvanometers were installed as part of the U.S. Worldwide Standardized Seismograph Network (WWSSN). These sets are currently operational.

In recent years, there has been increased interest in local seismology in connection with the design of many extensive engineering projects that might be subject to the earthquake hazards. This interest was heightened by the occurrences of disastrous earthquakes in China, such as the Tangshan earthquake in 1976, and by earthquake predictions made at various times. Despite such interests, the Royal Observatory site was increasingly affected by ground noise from urban development.

In order to monitor in more detail the location and magnitude of earthquakes in the vicinity of Hong Kong, a three-station network of short-period seismographs was installed in 1979. The stations, each equipped with a Teledyne Geotech S-13

Table 1. Seismological Instrumentation in Hong Kong

Station code	Seismo-meter	T_o (sec)	T_g (sec)	Speed ($\frac{mm}{h}$)	Rate ($\frac{mm}{h}$)	Type of record	Mag at T_o	Operation period
HKC	M-S, E	12.0	*	475	7	Photo	150	Oct 1921–Dec 1941
	M-S, N	12.0	*	475	7	Photo	150	Dec 1922–Dec 1941
	SPR-H,N	14.5	18.0	900	5	Photo	1,200	Aug 1951–Nov 1976
	SPR-H,E	14.5	18.0	900	5	Photo	1,200	
	SPR-H,Z	1.5	1.5	900	5	Photo	2,500	
	P-E LP,N	15.0	105	900	10	Photo	*	Jan 1958–Jun 1977
	P-E LP,E	15.0	105	900	10	Photo	*	
	P-E LP,Z	15.0	95	900	10	Photo	*	
	Ben 1051,Z	1.0	0.75	3,600	10	Photo	12,500	Since May 1963
	Ben 1101,N	1.0	0.75	3,600	10	Photo	12,500	
	Ben 1101,E	1.0	0.75	3,600	10	Photo	12,500	
	SPR LP,Z	15.0	101	900	10	Photo	750	
	SPR LP,N	15.0	99	900	10	Photo	750	
	SPR LP,E	15.0	100	900	10	Photo	750	
	P-E LP,N	15.0	/	900	3.5	Pen/ink	1,500	Since Jun 1978
	P-E LP,E	15.0	/	900	3.5	Pen/ink	1,500	
	P-E LP,Z	15.0	/	900	3.5	Pen/ink	1,500	
CCHK	Geo S-13,Z	1.0	/	3,600	10	Pen/ink	4,000	Since Nov 1979
THK	Geo S-13,Z	1.0	/	3,600	10	Pen/ink	17,200	
YHK	Geo S-13,Z	1.0	/	3,600	10	Pen/ink	20,000	

(1) Speed and Rate refer to the recording drum.
(2) * = data not available; / = not applicable.
(3) The photographic Press-Ewing LP seismographs were operated for the Lamont-Doherty Geological Observatory.
(4) The Benioff and Sprengnether LP seismographs were operated as part of WWSSN.
(5) M-S = Milne-Shaw; SPR = Sprengnether; P-E = Press-Ewing; Ben = Benioff; Geo = Geotech.

vertical-component short-period (1.0 second) seismometer, are located in remote areas as shown in Figure 1. The signals are amplified electronically and transmitted to a central recording station at the Royal Observatory headquarters. The signals are registered on three pen-and-ink recorders, and also are digitized at a rate of 20 samples per second and sent to the memory of a microcomputer. The memory holds the latest 10 seconds of data. Should the amplitudes of the signal exceed a certain threshold, the contents of the buffer, together with data for the next 5 minutes, are recorded on a floppy disk. Thus, only significant tremors are recorded. The digital records are retrieved for analysis on a graphic terminal using a combination of visual and correlation methods.

3. Seismological Records

Figure 2 summarizes the seismological records available at present. Although seismological recordings began in 1921, a large number of the seismograms were lost during the war. Only 90 seismograms as listed in Table 2 are available and most of them have recorded events. In an archival program initiated in 1975 by the Royal Observatory, these records were microfilmed and placed on aperture cards.

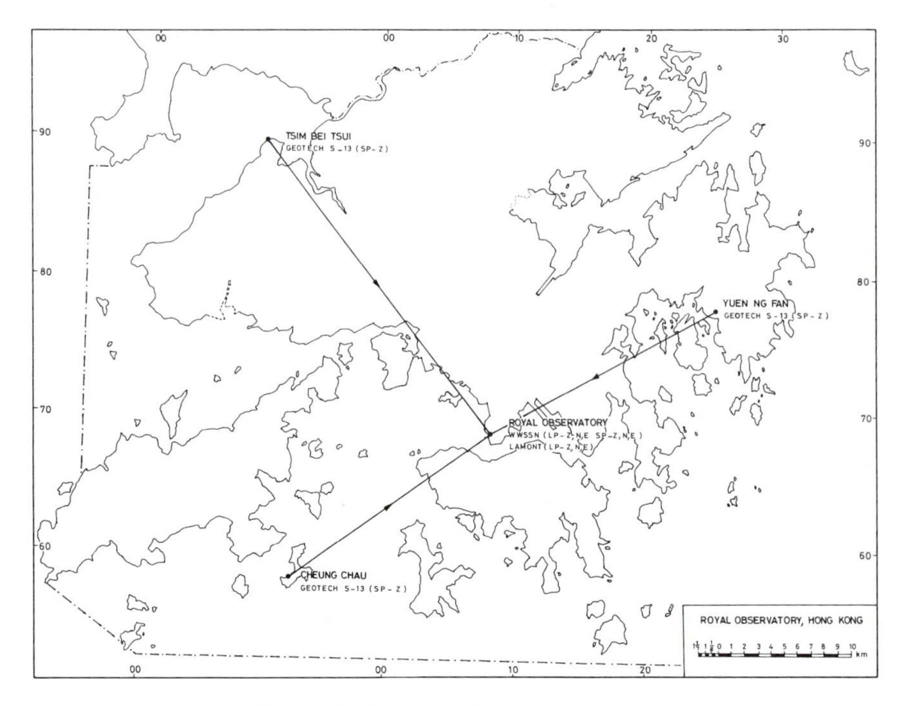

Figure 1. Seismographs in Hong Kong.

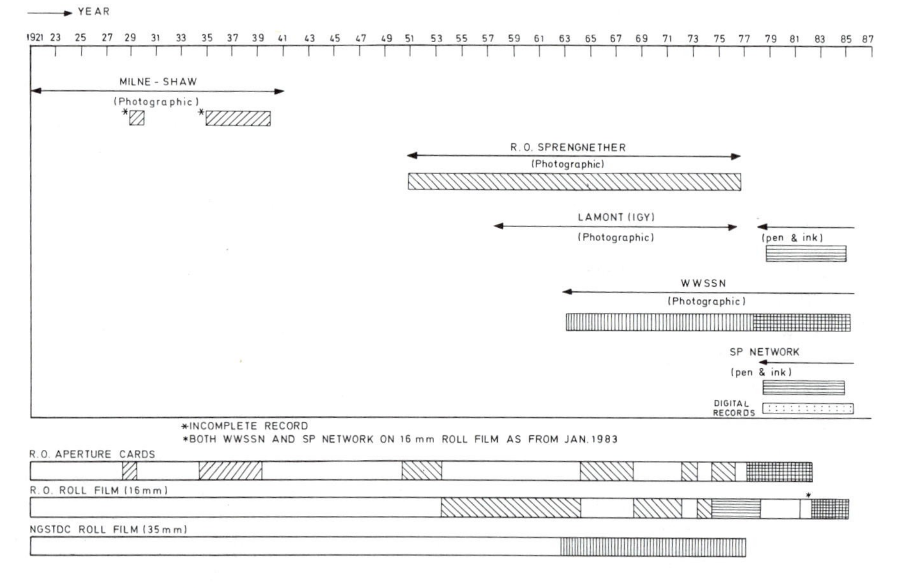

Figure 2. Seismological Measurement in Hong Kong.

Table 2. Available Pre-war HKC Seismograms on Aperture Cards

Year	Mon.	Day	Comp.	Year	Mon.	Day	Comp.	Year	Mon.	Day	Comp.
1929	Jan	12	N	1929	Jun	13	N	1929	Nov	12	N
		13	N			15	N			14	N
		16	N			17	N			15	N
		17	N			19	N			16	N
		20	N			23	N			17	N
		21	N			27	N			18	N
		24	N			29	N			22	N
		30	N		Jul	05	N		Dec	06	N
	Feb	01	N			07	N			09	N
		14	N			14	N			17	N
		22	N			26	N			18	N
		26	N			31	N	1935	Jul	11	N
	Mar	06	N		Aug	08	N	1936	Feb	12	N
		09	N			17	N		May	08	N
		21	N			18	N	1937	May	12	E
		23	N			19	N		Aug	11	E
		25	N			20	N	1938	Jan	18	N
		31	N			21	N		Feb	01	E,N
	Apr	08	N			28	N		Apr	14	N
	May	01	N		Sep	02	N		May	08	N
		02	N			11	N		Aug	18	N
		05	N			17	N		Oct	20	N
		19	N			21	N	1939	Jun	13	E
		21	N		Oct	05	N		Jul	10	E,N
		26	N			06	N		Dec	20	E,N
	Jun	02	N			16	N	1940	Jun	18	E,N
		03	N			19	N				
		04	N			22	N				
		09	N			24	N				
		12	N		Nov	05	N				

All the post-war seismograms made by the Royal Observatory Sprengnether seismographs (1951-1976) were microfilmed in the form of aperture cards or 16-mm roll film. The originals are no longer available.

The Lamont seismograms from 1957 to 1976 were sent to the Lamont-Doherty Geological Observatory monthly without being microfilmed. After changing to electronic recording in 1979, all the pen-and-ink records were microfilmed in 16-mm roll film. The original seismograms were disposed of after microfilming.

The WWSSN seismograms were also sent away monthly since 1963. A copy of these seismograms from 1963 to 1977 was later obtained from the U.S. National Geophysical Data Center in 35-mm roll film. Beginning from 1978, the WWSSN seismograms were also microfilmed at the Royal Observatory and put on aperture cards before they were dispatched.

All the seismograms of the three-station short-period network since its inception in November 1979 were microfilmed on 16-mm roll film before they were disposed of. Digital recordings of significant events within 320 km of Hong Kong were kept on floppy disks. The first 25 seconds of these records were also plotted by a plotter using an expanded time scale of 1 second per 5 mm and kept as permanent records.

All the seismograms were in satisfactory condition when they were microfilmed. There are two copies of microfilm kept by the Royal Observatory: a master copy and a duplicate copy. The master copy is kept separately from the duplicate copy under temperature and humidity controlled environment. The duplicate copy is normally the working copy. Information such as station name, times of on and off, polarity, and timing corrections were normally recorded on the seismograms. These and other additional information on instrumental constants were also available on station log books and/or seismological bulletins for the periods from 1921 to 1940 and from 1951 to present. These records, however, were not microfilmed.

A Brief History of the Seismological Observatories in the British Isles, 1896-1960

G. Neilson and P. W. Burton
Natural Environment Research Council, British Geological Survey
Murchison House, West Mains Road, Edinburgh EH9 3LA, UK

1. Introduction

The earliest known application of instruments to the study of British earthquakes was the result of the work of a committee of the British Association for the Advancement of Science, which was set up in 1840 to study the earthquake sequence then in progress at Comrie, Scotland (British Association, 1841-1844). The secretary of this committee, David Milne (later Milne-Home) coined the word "seismometer", to describe the invention of another member of the committee, Professor J. D. Forbes (Forbes, 1844). This instrument, the basis of which was an inverted pendulum, would now be described as a seismoscope. A number of these instruments, together with other, simpler types such as common pendulums, were placed around Comrie, but the results produced were disappointing. With the decrease in the frequency and severity of the earthquakes, the committee lapsed after 1844.

In 1869, following signs of renewal of seismic activity in Comrie, a new committee of the British Association was set up to investigate the earthquakes of Scotland. This committee reported to the Association in 1870 to 1876 (British Association, 1870-1876). This committee discovered that at least one of the Forbes instruments was still working but decided that it was not sensitive enough. A simpler device, designed by Robert Mallet was adopted (Davison, 1924). A series of upright cylinders of varying heights and diameters was set up in two lines at right angles to each other on a bed of sand in the "earthquake house" in Comrie. The theory was that a shock would upset the cylinders, the orientation of the overturned cylinders would give the direction of the shock, and the mass of the largest cylinder overturned would indicate the strength of the disturbance. Bryce, the Chairman of the committee, died in 1876, and after his death the committee lapsed. Mallet's apparatus is reported to have only been disturbed once, but the source of the disturbance was the antics of local boys, not an earthquake.

About the same time, E. J. Lowe erected a simple pendulum instrument to detect shocks at his observatory in Nottinghamshire (Burton and others, 1984).

Credit for first demonstrating the value of seismographs to seismology belongs to workers such as Gray, Ewing and Milne who developed various types of instruments in Japan from about 1880 to 1895 (Dewey and Byerly, 1969). The most influential member of this group was John Milne. On his return to Britain in 1895 he became secretary of the British Association Seismological Investigations Committee. This committee tested various types of seismographs and eventually decided that the Milne horizontal pendulum was the best instrument. It also set up a worldwide network of stations equipped with these instruments to record large earthquakes and report their observations to Milne's Observatory at Shide, Isle of Wight. Here, the epicentres of the shocks were computed and the data circulated to contributing stations.

After Milne's death in 1913 the headquarters of this organization, now known as the International Seismological Summary (ISS), moved to Oxford under the direction of Professor H. H. Turner.

British stations formed a vital part of the new worldwide network set up by Milne. Research was performed on site noise, microseisms, etc., by comparing the seismograms written at the various stations in Britain, and applying the results to all British Association stations.

The scientific results from the worldwide organizations were of enormous importance to seismology because they covered topics as diverse as deep-focus earthquakes, submarine disturbances caused by earthquakes, seismicity studies and epicenter-determination methods, and others.

Seismology in Britain was considerably disrupted by World War I, although most of the stations managed to continue operating throughout the war. After 1918 the situation became normal again fairly quickly. The period between World Wars I and II saw a great advance in seismology in Britain, the most significant contribution possibly being the production of the Jeffreys-Bullen traveltime tables (Jeffreys, 1939), but only two new observatories were established, Jersey and Durham.

By the end of World War II few British observatories were operating purely in the interests of seismology. Jersey, under the German occupation, had great difficulty finding materials to continue recording. Edinburgh and Durham had ceased analysing the records for the duration of the war, although the seismographs at Edinburgh were kept in operation throughout.

After World War II, the stations in the British Isles were equipped with obsolete equipment, efficiency at the stations had declined, and a few stations had closed altogether. In 1960, only Kew, Durham, Jersey, and Aberdeen were operating and Kew and Aberdeen closed within a few years.

The Geneva Test Ban negotiations, from 1958 onwards, led to a re-awakening of interest in seismology and provided the catalyst necessary for the renovation of instrumental seismology. In Britain this led to the establishment of the World Wide Standard Station at Eskdalemuir and of a large array station nearby (Truscott, 1964). In Ireland a World Wide Standard Station was set up at Valentia. Edinburgh was re-equipped with modern instruments recording on magnetic tape (Willmore and Connell, 1963). An array station was set up at Rookhope, near Durham, and many modern stations have been deployed over all the British Isles in the years since 1960 (Crampin and others, 1970; Browitt and others, 1985).

This report deals with the period beginning with the return of Milne from Japan and ends with the renaissance of instrumental seismology in Britain, which began in the 1960's.

2. The History of the Establishment of Seismological Observatories in the British Isles

The principal observatories established in Britain and Ireland during 1896-1960 are shown in Figure 1 and listed in Table 1. The first observatory in Britain equipped with a seismograph was Milne's at Shide, Isle of Wight, set up in 1895. Milne also installed a temporary station at Carisbrooke Castle to compare records. Within two or three years other stations such as Edinburgh, Kew, Paisley and Bidston were established. All these places had astronomical and/or meteorological observatories to provide an accurate source of timing for seismographic recording.

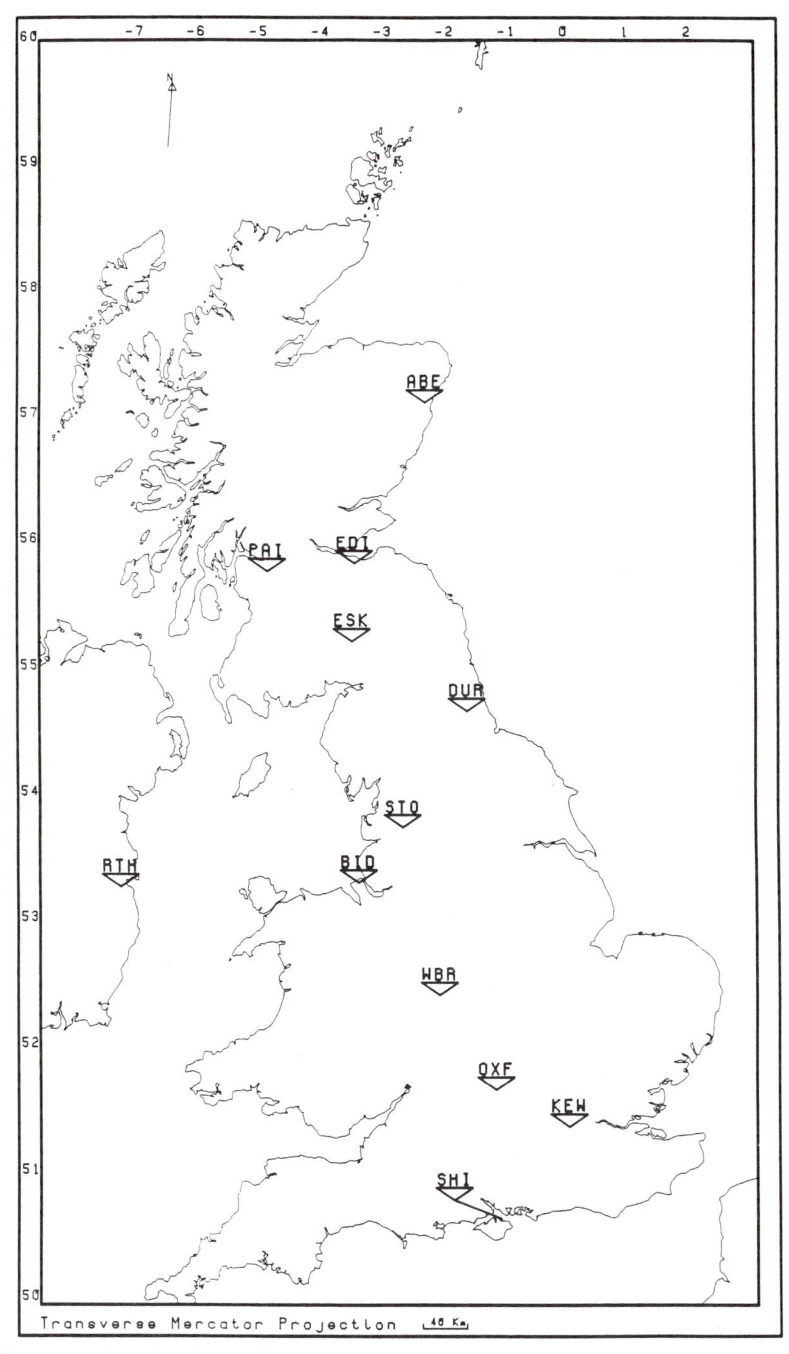

Figure 1. The locations of some historical Seismological Observatories in the British Isles. Station names are listed in Table 1.

Table 1. The Locations of Some Historical Seismological Observatories in the British Isles

Station	Name	Latitude (°N)	Longitude (°W)	Elevation (m)
ABE	Aberdeen	57.217	2.167	53
BID	Bidston	53.401	3.072	54
DUR	Durham	54.767	1.583	103
EDI	Edinburgh	55.925	3.184	131
ESK	Eskdalemuir	55.312	3.206	263
JRS	Jersey	48.192	2.099	53
KEW	Kew	51.468	0.313	6
OXF	Oxford	51.767	1.250	61
PAI	Paisley	55.846	4.431	31
RTH	Rathfarnham (Dublin)	53.300	6.283	52
SHI	Shide	50.688	1.286	15
STO	Stonyhurst	53.844	2.470	111
WBR	West Bromwich	52.517	1.983	156

The first two decades of the 20th century saw the addition of more stations in both Britain and Ireland, the most important of which was probably West Bromwich. This observatory was established by J. J. Shaw, primarily to test instruments that he had built or modified. The original instrument used to equip the British Association observatories was the Milne horizontal pendulum (Milne, 1906). The performance of this instrument was limited owing to its lack of damping. Shaw introduced damping to the Milne design and increased its gain, thereby greatly enhancing the usefulness of these instruments (Shaw, n.d.).

Following a series of rather large earthquakes that were felt in Jersey during the 1920's, a station was established at Maison St. Louis, St. Helier, Jersey, in 1935. A station had been established at Durham by E. F. Baxter a few years previously, also with the intention of studying local earthquakes. Unfortunately, it was not possible to equip either station with an instrument of short enough period to make them suitable for such studies.

During World War II, Kew Observatory was used to monitor bombing, especially the V1 and V2 attacks in 1944.

After the war new stations were not installed in the British Isles until the 1960's, when organizations such as the United Kingdom Atomic Energy Authority and the Institute of Geological Sciences (now the British Geological Survey) installed several high-gain modern instruments, recording on magnetic tape, at various sites in the British Isles.

3. History of Individual British Seismological Observatories

3.1. Bidston

The first instrument in operation at Bidston was a Darwin bifilar pendulum, the readings from which were first published in the British Association Report for 1898. By the year 1901 a Milne instrument also had been installed. A vault was specially constructed in the observatory to house the instruments. The vault contained two piers on which the seismographs rested, one made of brick and the other a large, cement-filled drainpipe.

A Milne-Shaw seismograph was installed in 1914 (no. 1) and an improved version of the same instrument replaced it in 1924 (no. 32). The static magnification of these instruments was 150. A second component was added in 1934. The station ceased recording in 1957.

3.2. Durham

Durham University Seismological Observatory was set up in 1930, with the installation of a Milne-Shaw (N-S component) seismograph which started operation in January, 1931. A second Milne-Shaw (E-W component) was added in January 1938. The station ceased operation in July, 1939 and restarted in January, 1946.

A vault was constructed in 1955 and a Wilson-Lamison vertical instrument was installed, together with the two Milne-Shaw seismographs, in October 1956. The station was still in operation in 1985.

3.3. Dyce/Aberdeen

The first seismological observatory to be established in the Aberdeen area was at Parkhill House, Dyce. This observatory was founded by J. E. Crombie. Recording was begun in 1914 using two Mainka horizontal seismographs and an Agamennone instrument. The Mainkas had a free period of 10 seconds and a static magnification of 150. In 1918 a Milne-Shaw seismograph was added. All these instruments ran until Crombie's death in August 1932, when the Agamennone and one of the Mainkas was donated to the Science Museum. The Milne-Shaw, which was donated to Aberdeen University, was set up for testing in the basement of King's College, Aberdeen in 1932, and eventually became the E-W component of the Aberdeen station which operated continuously from December 1936 to June 1967. A Milne-Shaw N-S component was added in 1938. The Milne-Shaw instruments ceased recording in 1966 (Neilson, 1981).

3.4. The Royal Observatory, Edinburgh

The first instrument to operate in Edinburgh was a Darwin bifilar pendulum, which was originally installed in the Calton Hill Observatory in March 1894 and moved to the Royal Observatory, Blackford Hill, in October 1895. Photographic registration apparatus was added in August 1896 and a second pendulum was purchased in May 1898. In 1900 a Milne pendulum was installed. New recording drums were added in 1909, which gave a faster paper speed of 240 mm per hour.

A Milne-Shaw seismograph (no. 3), formerly at Eskdalemuir, was moved to Edinburgh and began recording on July 4, 1919. The E-W component had a static magnification of 250.

After World War II, E. Tillotson took over the reading of the Edinburgh seismograms. During the period 1939-1940 various interruptions to recordings occurred owing to defects of the drum drive and to quarry blasting.

The Milne-Shaw instrument ceased operation in April 1962, when it was transferred to the Royal Scottish Museum.

3.5. Eskdalemuir

The first instruments at Eskdalemuir were a Milne twin boom instrument which began operating in 1908 shortly after the observatory opened, and a Wiechert was installed in 1909. The installation of an Omori followed in 1910. In the same year Professor Schuster of Manchester University presented the observatory with two Galitzin horizontal seismographs, which were set up as a N-S and an E-W component. Two years later he donated a vertical instrument, and these three instruments remained for many years the only three component set of seismographs in operation in the United Kingdom. The instruments were made in St. Petersburg by H. Masing and their installation was supervised by Prince Galitzin. Recording with the Galitzins began in July 1910 (Jacobs, 1964). In July 1915 a Milne-Shaw instrument was installed for comparison with the Galitzin. This instrument was transferred to the Royal Observatory, Edinburgh, in June 1919. The Omori was donated to the Science Museum in 1921 and the Galitzins were transferred to Kew Observatory in 1925 (Scrase, 1969).

After 1925 seismology studies ceased at Eskdalemuir until the World Wide Standard Station was installed in 1963 and began continuous operation in 1964.

3.6. Jersey, Channel Islands

After the plans fell through to move one of J. E. Crombie's Mainka instruments from Dyce Observatory to St. Louis Observatory, St. Helier, Jersey, M. E. Rothé of the Institut de Physique du Globe, Strasbourg, lent an instrument of the same type to St. Louis Observatory. This seismograph was set up in Jersey in 1935. It has a mass of 450 kg, a static magnification of 140, a free period of 13.7 seconds, and a damping ratio of 2.9.

The station began operating in June 1936 and recording was interrupted in 1940 by the German occupation of the Channel Islands. Recording recommenced in June 1946 and the station is still in operation.

3.7. Kew

Initially the body responsible for seismology at Kew Observatory was the National Physical Laboratory. Later the responsible body became the Meterological Office, which was also responsible for seismology at Eskdalemuir Observatory. The first seismograph installed at Kew was a Milne seismograph, which began operation in 1898. This instrument, the ninth to be manufactured, was set up as an E-W component.

Operation of the Milne instrument was discontinued in 1925. In the same year, the Galitzin three components were transferred from Eskdalemuir to Kew and were installed in the basement of the main building. This location was not satisfactory, because the instruments suffered severely from strong, wind-induced noise (Lee, 1939).

Two Wood-Anderson torsion seismographs were constructed at the Observatory in 1933 and 1935 and put into operation. The year 1936 saw the construction of a new building designed to house all the seismographs and in February 1937 the Wood-Anderson instruments were moved into it, followed by the Galitzin horizontals in April and the Galitzin vertical in September of that year. A short period vertical instrument, having a free period of 1.5 seconds was installed in 1938. This instrument was manufactured in the observatory workshops.

The observatory ceased seismological recording in 1969 and the instruments were transferred to the Science Museum, London.

3.8. Oxford

This station was set up in 1918 to take over the work done by Milne at Shide until his death in 1913. The International Seismological Summary was also transferred to Oxford at this time. Both organizations were under the direction of Professor H. H. Turner. The first instrument, a Milne-Shaw seismograph, was set up in the basement of the Clarendon Laboratory on October 8, 1918 as an E-W component. Milne's library was also moved to Oxford in 1919, and most of it was housed in the Student's Observatory.

A new vault was constructed, at the University Observatory in 1927; it was named the "Crombie Basement", after J. E. Crombie who donated the two Milne-Shaw seismographs (nos. 1 and 4) with which it was equipped. These instruments had free periods of 12 seconds and a damping ratio of 20:1. After some initial problems of settlement of the 8 $\times$ 4 feet pier, the two instruments were installed in October 1928.

The seismographs and the Milne Library were transferred from Oxford to Downe House, Kent, on January 22, 1947.

3.9. The Coats Observatory, Paisley

A description of the observatory is given in Anderson (1901). The first seismological instruments installed in the observatory were a Ewing seismoscope and a Milne seismograph. The latter instrument (no. 18) was installed in July 1898. The original building is now demolished and an extension of the Paisley Art Gallery and Museum covers the site including the piers on which the instruments stood.

In 1912 an "improved Milne", probably a Milne-Shaw seismograph, was installed. The observatory carried out continuous seismographic recording until at least June 1918.

3.10. Rathfarnham

A Milne-Shaw seismograph was installed at Rathfarnham Castle, Dublin, in 1919. Prior to this, Fr. W. O'Leary constructed an inverted pendulum-type of seismograph in 1916 to 1917. This instrument had a mass of 1,700 kg and recorded both N-S and E-W components of ground motion. The static magnification was 300 and the free period was 18 seconds. The theoretical description of the instrument is given in Ingram and Timoney (1954).

In 1950 a short period vertical instrument having a free period of 1.2 seconds was installed. Recording ceased at Rathfarnham in 1964.

3.11. Shide, Isle of Wight

This observatory was founded in 1895 by John Milne on his return from Japan. A brick pier was erected on a concrete foundation in an old stable. The first instrument was in operation by August 16, 1895. In 1896 another instrument was installed, for comparison, at nearby Carisbrooke Castle. A laboratory for testing equipment was added in 1900. The number and type of instruments running at Shide varied from time to time; for example, in 1902 seven horizontal components of various types were in operation together with a vertical instrument. The principal type of seismograph that operated at Shide was the Milne horizontal pendulum. The observatory closed in 1913.

3.12. Stonyhurst College

The original instrument with which this observatory was equipped was the Milne pendulum, which was used in the Antarctic by the "Discovery" expedition. It was installed in 1908 and set up as an E-W component. Recording with the Milne ceased in 1924, and it was replaced by a Milne-Shaw instrument in 1928. This was also set up as an E-W component. Recording was suspended in 1947.

3.13. West Bromwich

This observatory was established by J. J. Shaw in 1909. The first instrument in use was a twin boom Omori horizontal pendulum and the two components were N-S and E-W. The static magnification was 60 and the free period was 12 seconds. By 1916 two Milne-Shaw instruments were in operation. Shaw used this observatory mainly for experimental work on instruments. (He was responsible for introducing damping to the Milne design thus producing the Milne-Shaw.) The observatory ceased operation shortly after Shaw's death in 1946.

ACKNOWLEDGEMENTS

This work was supported by the Natural Environment Research Council and is published with the approval of the Director of the British Geological Survey (NERC).

REFERENCES

Anderson, A. (1901). *The Coats Observatory, Its History and Equipment*, J&R Parlane, Paisley.

British Association (1841-1844). Report of the British Association for the Advancement of Science, 1841, 46-50; 1842, 92-98; 1843, 120-127; 1844, 85-90.

British Association (1870-1876). Reports of the Committee on Earthquakes in Scotland, Report of the British Association for the Advancement of Science, 1870, 48-49; 1871, 197-198; 1872, 240-241; 1873, 194-197; 1874, 241; 1875, 64-65; 1876, 74.

Browitt, C. W. A., T. Turbitt, and S. N. Morgan (1985). Investigation of British earthquakes using the national monitoring network of the British Geological Survey, in *Earthquake Engineering in Britian*, Thomas Telford, London, 33-47.

Burton, P. W., R. M. W. Musson, and G. Neilson (1984). Studies of Historical British Earthquakes, *Brit. Geol. Survey Glob. Seism. Unit, Report No. 237.*

Crampin, S., A. W. B. Jacob, A. Miller, and G. Neilson (1970). The LOWNET Radio-linked Seismometer Network in Scotland, *Geophys. J. R. Astr. Soc.*, **21**, 207-216.

Davison, C. (1924). *A History of British Earthquakes*, Cambridge University Press.
Dewey, J. and P. Byerly (1969). The Early History of Seismometry (To 1900), *Bull. Seism. Soc. Am.*, **59**, 183-227.
Forbes, J. D. (1844). On the Theory and Construction of a Seismometer, or Instrument for Measuring Earthquake Shocks and Other Concussions, *Trans. Roy. Soc. Edin.*, XV, pt. 1, 219-228.
Ingram, R. E. and J. R. Timoney (1954). Theory of an Inverted Pendulum with Trifilar Suspension, Dublin Institute of Advanced Sciences, School of Cosmic Physics, *Geophysical Bulletin No. 9.*
Jacobs, L. (1964). Seismology at Eskdalemuir Observatory, *Met. Mag.*, **93**, 289-294.
Jeffreys, H. (1939). Seismological Tables, *Mon. Not. R. Astr. Soc.*, **99**, 397-408.
Lee, A. W. (1939). Seismology at Kew Observatory, Meteorological Office, *Geophysical Memoirs No. 78*, H.M.S.O., London.
Milne, J. (1906). *On the Installation and Working of Milne's Horizontal Pendulum Seismograph*, R. W. Munro, London.
Neilson, G. (1981). Historical Seismological Archives 2: Report on a visit to the Aberdeen University Department of Natural Philosophy, *Brit. Geol. Survey Glob. Seism. Unit, Report No. 144.*
Scrase, F. J. (1969). Some Reminiscences of Kew Observatory in the Twenties, *Met. Mag.*, **98**, 180-186.
Shaw, J. J. (n.d.). "Milne-Shaw" Seismograph Handbook, J. J. Shaw, West Bromwich.
Truscott, J. R. (1964). The Eskdalemuir Seismological Station, *Geophys. J. R. Astr. Soc.*, **9**, 59-68.
Willmore, P. L. and D. V. Connell (1963). A New Short-Period Seismometer for Field and Observatory Use, *Bull. Seism. Soc. Am.*, **53**, 835-844.

Historical Seismograms of the Manila Observatory

Sergio S. Su, S.J.
Manila Observatory
P.O. Box 1231, Manila, Philippines 2800

The Manila Observatory was established in 1865, during the Spanish regime, originally for weather observations. It began as a private institution run by the Spanish Jesuits and supported by donations from merchants and businessmen. Twenty years later the King of Spain raised it to the status of Royal Meteorological Observatory, thus making it a government institution. When the Americans took over the rule from the Spaniards in 1899, they saw the need and advantage of keeping the unification of the Manila Observatory and the Philippine Weather Bureau. This was maintained until the Japanese invasion in 1941. For the next ten years the Manila Observatory was non-existent, while the Philippine Weather Bureau was revived around 1945, independently of the Jesuits. In 1952, the observatory was revived by the Jesuits, once more as a private institution, not in Manila (even though the name was retained) but in Baguio, 260 kilometers north of Manila. In 1962, the Manila Observatory moved from Baguio to Quezon City which is now part of Metropolitan Manila.

Oddly enough, the only extant historical seismograms of the observatory (i.e. originals and not just copies) are those of the period from 1952 to 1962 when the observatory was not in Manila but in Baguio. These seismograms of Baguio pre-date its establishment as a WWSSN station in 1962.

Other historical seismograms that were lost or destroyed during World War II included those that dated back to 1877 when regular uninterrupted seismographic recordings began (intermittent seismographic records started as early as 1866). These included records from the Vicentini, the Gray-Milne, and the Cecchi seismographs. Maso (1895) showed some pictures of these instruments. Also lost during the Second World War were the records from provincial stations such as those of the Cotabato station in Mindanao and the Ambuklao station in Luzon.

Besides the above-mentioned original historical seismograms of BAG, there are copies of seismograms of historical earthquakes contained in articles and earthquake reports in the Manila Observatory archives. Repetti (1946) gives a fairly complete catalogue (with accounts and descriptions) of earthquakes from 1589-1899. Another very probable source of information on historical earthquakes in the Philippines is the archives of Seville, Spain.

Table 1 summarizes all the records (copies of seismograms) of historical earthquakes that have been gathered so far from monographs, articles and reports found in the archives of the Manila Observatory. Table 2 contains descriptions of seismographs whose records are found in Table 1.

Table 1. List of Historial Seismograms

01. Luzon earthquake recorded in Manila: Time: 1880 July 18, 12h 40m Description in Repetti (1946, p. 249). See Figure 1. *	02. Luzon earthq. rec. in Manila: Time: 1880 July 20, 15h 40m Description in Repetti (1946, p. 251). See Figure 1. *
03. Luzon earthq. rec. in Manila: Time: 1880 July 20, 22h 40m Description in Repetti (1946, p. 252) See Figure 1. *	04. Luzon earthq. rec. in Manila: Time: 1880 Jul 25, 04h 02m Description in Repetti (1946, p. 254) See Figure 1. *
05. Luzon earthq. rec. in Manila: Time: 1880 July 15. 00h 53m Description in Repetti (1946, p. 248) See Figure 1. *	06. Zamboanga earthq. rec. in Cotabato: Time: 1897 September 21, 03h 25m Description in Repetti (1946, p. 314) See Figure 2. * Amplitude $A = 21^\circ$ W 10° N Amplitude $B = 11^\circ$ N 10° W Duration = 52 seconds
07. Zamboanga earthq. rec. in Cotabato: Time: 1897 September 21, 13h 32m Description in Repetti (1946, p. 314) See Figure 3. * Amplitude $A = 09^\circ$ 20' N 45° W Amplitude $B = 13^\circ$ N 45° E Duration = 12 seconds	08. Polloc earthq. rec. in Cotabato: Time: 1893 June 3, 06h 58m Description in Repetti (1946, p. 299) See Figure 4. * Amplitude $A = 07^\circ$ 15' S 50° W Amplitude $B = 15^\circ$ 07' S 10° E Duration = 12 seconds
09. Mongolian earthq. rec. in Manila: Time: 1905 July 9, 17h 48m 13s. See Figure 5. †	10. Mongolian earthq. rec. in Manila: Time: 1905 July 23, 10h 53m 58s See Figure 5. †
11. Indian earthq. rec. in Manila: Time: 1905 April 4, 08h 58m 25s See Figure 6. †	12. Luzon earthq. rec. in Manila: Time: 1882 March 24, 08h 04m 19s Description in Repetti (1946, p. 262) See Figure 7, C-1 ∓
13. Luzon earthq. rec. in Manila: Time: 1882 September 11, 23h 08m 09s Description in Repetti (1946, p. 264) See Figure 7, C-2 ∓	14. Luzon earthq. rec. in Manila: Time: 1882 November 29, 05h 03m 37s Description in Repetti (1946, p. 266) See Figure 7, C-3 ∓
15. Luzon earthq. rec. in Manila: Time: 1883 February 6, 12h 19m 29s Description in Repetti (1946, p. 267) See Figure 7, C-4 ∓	16. Luzon earthq. rec. in Manila: Time: 1883 July 14, 04h 25m 35s Description in Repetti (1946, p. 267) See Figure 7, C-5 ∓
17. Luzon earthq. rec. in Manila: Time: 1883 July 27, 17h 51m 36s Description in Repetti (1946, p. 268) See Figure 7, C-6 ∓	18. Luzon earthq. rec. in Manila: Time: 1884 October 29, 04h 09m 45s Description in Repetti (1946, p. 270) See Figure 7, C-7 ∓
19. Luzon earthq. rec. in Manila: Time: 1884 November 11, 16h 35m 26s Description in Repetti (1946, p. 270) See Figure 7, C-8 ∓	20. Luzon earthq. rec. in Manila: Time: 1884 December 20, 20h 39m 10s Description in Repetti (1946, p. 271) See Figure 7, C-9 ∓
21. Luzon earthq. rec. in Manila: Time: 1885 November 16, 23h 21m 35s Description in Repetti (1946, p. 275) See Figure 7, C-10 ∓	22. Luzon earthq. rec. in Manila: Time: 1885 November 18, 02h 37m 57s Description in Repetti (1946, p. 275) See Figure 7, C-11 ∓

continued

Table 1. *Continued*

23. Luzon earthq. rec. in Manila: Time: 1885 November 19, 21h 31m 05s Description in Repetti (1946, p. 275) See Figure 7, C-12 ∓	24. Luzon earthq. rec. in Manila: Time: 1886 May 21, 12h 18m 00s Description in Repetti (1946, p. 276) See Figure 7, C-13 ∓
25. Luzon earthq. rec. in Manila: Time: 1886 August 2, 03h 58m 57s Description in Repetti (1946, p. 277) See Figure 7, C-14 ∓	26. Luzon earthq. rec. in Manila: Time: 1889 May 26, 02h 23m 13s Description in Repetti (1946, p. 283) See Figure 7, C-15 ∓

* = Seismoscope record; † = Vicentini Seismograph; ∓ = Cecchi Seismograph

Table 2. Instruments for the Records of Table 1

1. Vicentini Seismograph	2. Seismoscope	3. Cecchi Seismograph
Pendulum length = 1.5 m Weight of bob = 100 kg Period = 1.2 sec Magnification = 50.	Pendulum length = 1.0 m? Magnification = 15?	(horizontal component only) Magnification = 15?

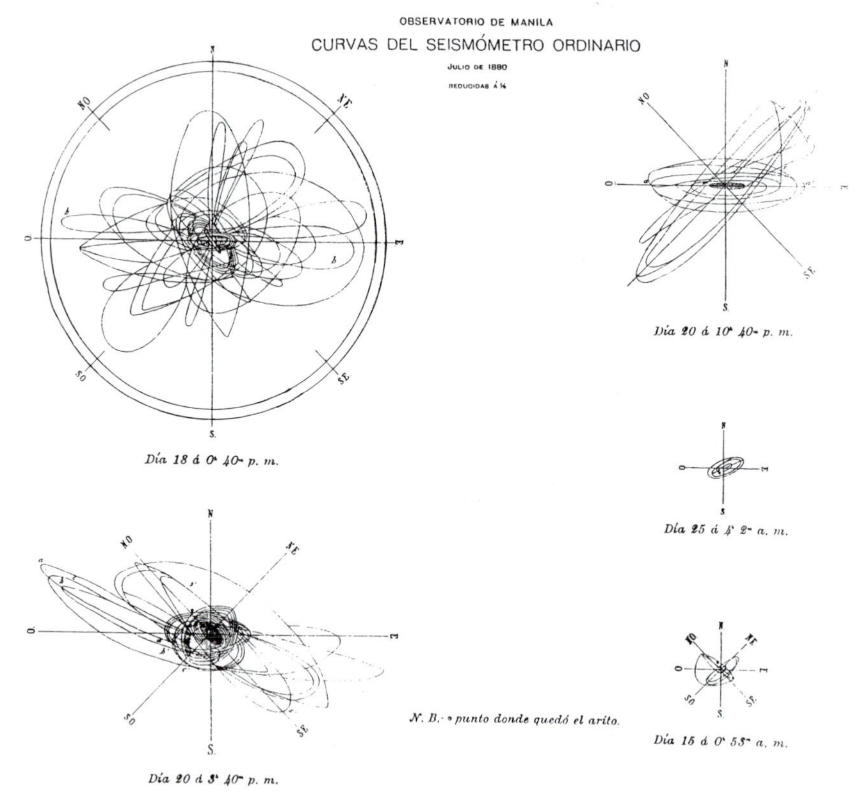

Figure 1. Seismoscope record of Luzon earthquakes of 1880 July. See Table 1, 1-5.

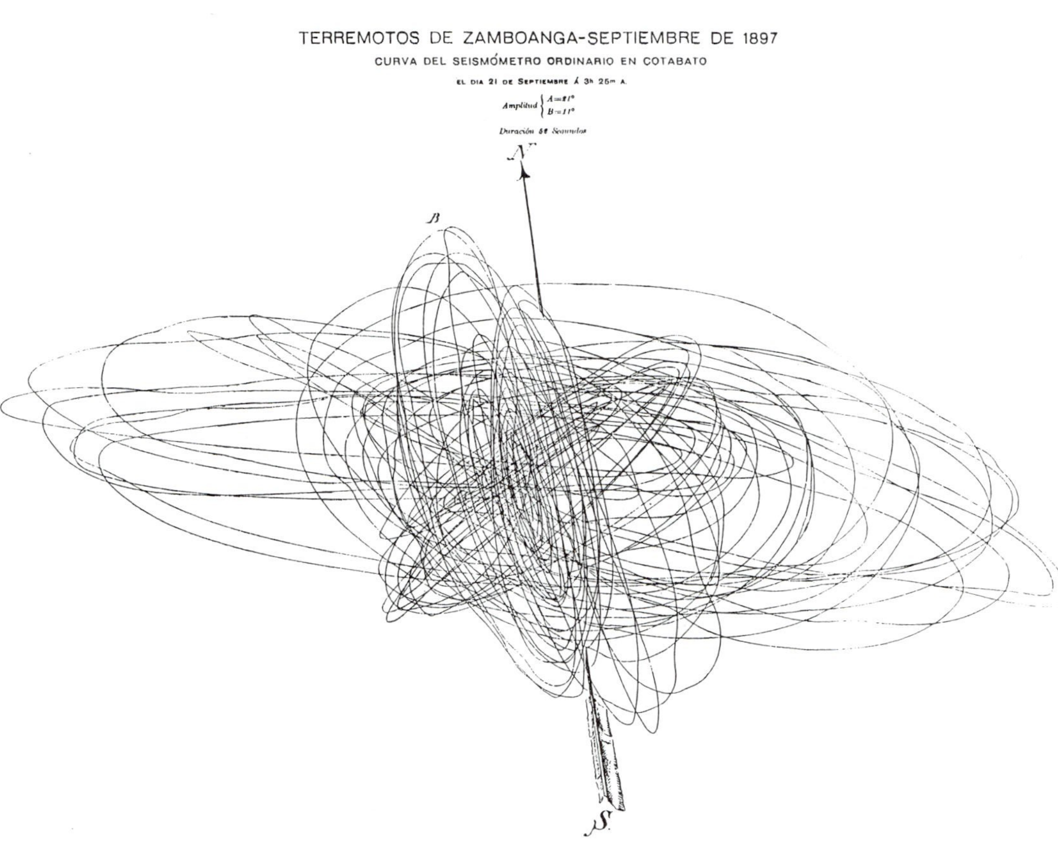

Figure 2. Seismoscope record of Zamboanga earthquake of 1897 September 21, 03h 25m. See Table 1, 6.

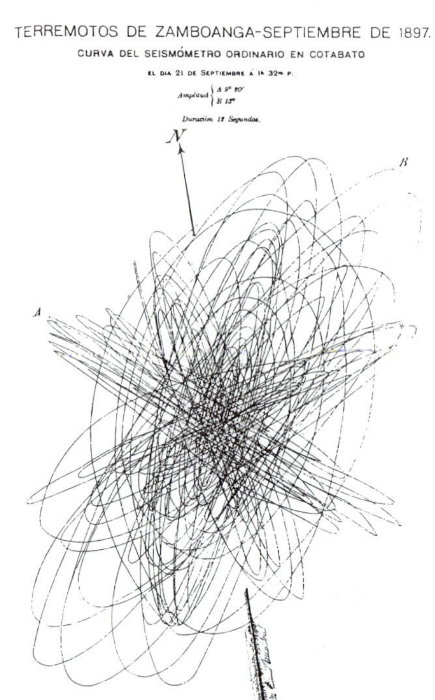

Figure 3. Seismoscope record of Zamboanga earthquake of 1897 September 21, 13h 32m. See Table 1, 7.

Figure 4. Seismoscope record of Polloc earthquake of 1893 June 3, 06h 58m. See Table 1, 8.

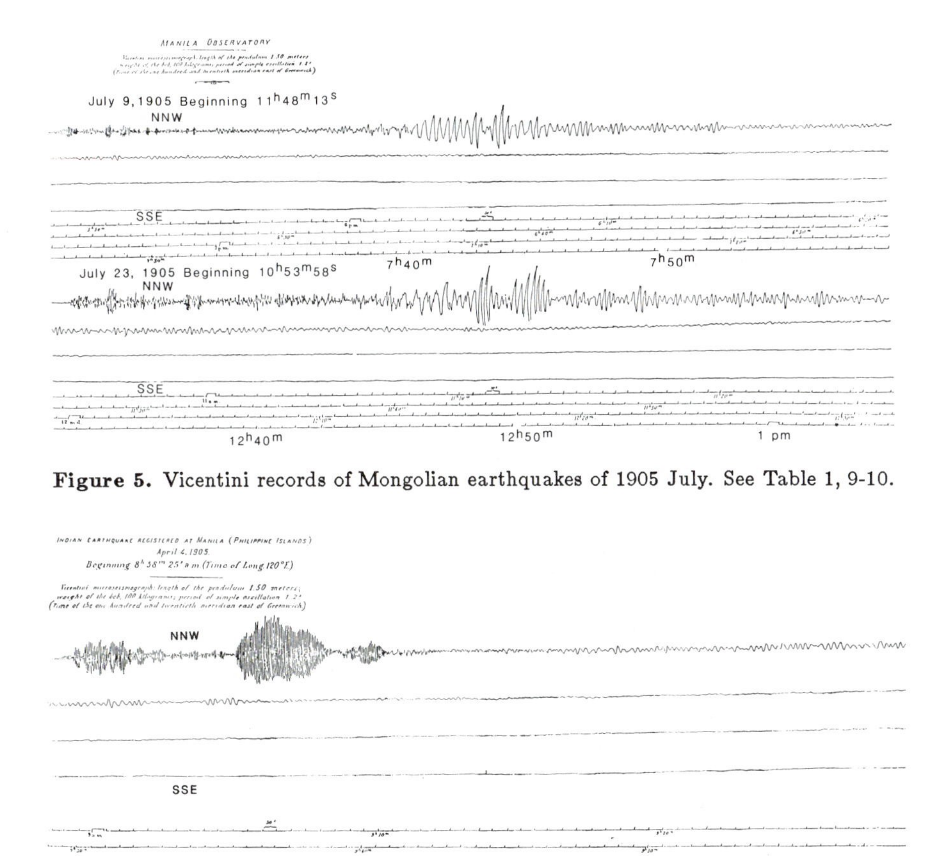

Figure 5. Vicentini records of Mongolian earthquakes of 1905 July. See Table 1, 9-10.

Figure 6. Vicentini record of Indian earthquake of 1905 April 4. See Table 1, 11.

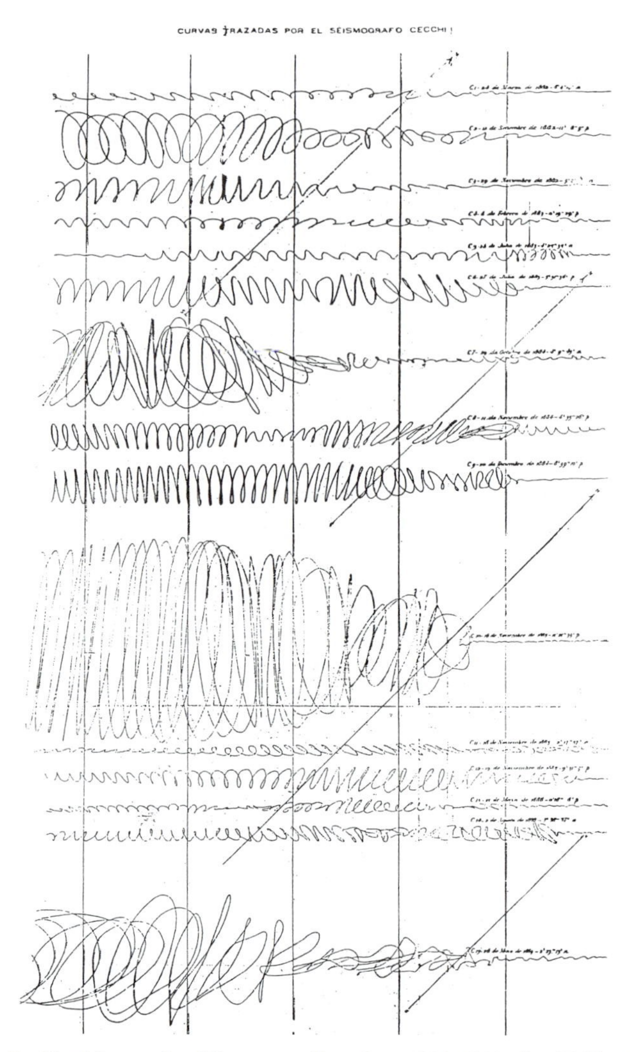

Figure 7. Cecchi records of Luzon earthquakes, 1882-1889. See Table 1, 12-26.

REFERENCES

The Philippine Archipelago (1900). Collection of Data, Volume II, Government press, Washington, 469 pp.

Maso, M. S. (1895). *Seismology in the Philippines*, Monograph of the Manila Observatory, 124 pp.

Manila Meteorological Observatory Verified Observations (1894). January to December.

Manila Observatory Monthly Bulletins (1897). January to December.

Philippine Weather Bureau Monthly Bulletins (1905). January to December.

Philippine Weather Bureau Monthly Bulletins (1907). January to December.

Repetti, W. C., S. J. (1946). Catalogue of Philippine Earthquakes, 1589-1899, *Bull. Seism. Soc. Am.*, **36**, 133-322.

Some Remarks on Historical Seismograms in Selected Developing Countries

K. L. Svendsen
CIRES, University of Colorado
Boulder, CO 80309, USA
and
NGDC/NOAA
Boulder, CO 80303, USA

1. Introduction

While making tours of South American and African geophysical institutions, 1979-1982, in connection with a project in geomagnetism, I took the opportunity to inquire about their seismology programs, especially as they relate to the Historical Seismogram Filming Project.

First, I did not worry about countries that have stations in the World-Wide Standard Seismograph Network (WWSSN). I assumed that the committee already had communicated with these countries and probably knew what was available. For those who are interested in further development of the world network, I have included a few comments on that subject also.

2. Countries Visited

The following are comments on the seismogram projects in each country visited.

Mexico. The Director of the Institute of Geophysics of the National University offered his complete cooperation in the project, but did not know what seismograms are in the archives.

Costa Rica. There are five stations operated by the university and eight stations operated by the electrical institute. All records will be made available to the project. One old station was operated by the National Museum from about 1910-1920. Unfortunately, no one knew where those records are stored.

Bolivia. The Observatorio San Calixto has seismograms dating from 1912. All records will be available to the project.

Chile. The records are at the University of Chile. Three stations operated from 1909 at Santiago, Copiapo, and Puntarenas. There were two horizontal components only, and there was a gap of about 10 years. All of these records have been edited and prepared for filming. The present network comprises Antofagasta, Peldehue, and Sombrero.

Argentina. The University of La Plata has many old records, even though many of them have disappeared. However, because many of the records were made with no damping, most probably were not good anyway. The University asked that all filming be done in Argentina with the assistance of their archivist.

Uruguay. Seismograms are available only for a few recent years; these were recorded by the University of the Republic.

Dominican Republic. For some years, one station with three short-period instruments has been operated by the Instituto Geografico Universitario of the University of Santo Domingo. Records are available through the local office of the Inter-American Geodetic Survey.

Morocco. There has been at least one seismograph in operation by the University of Rabat since about 1934; all seismograms have been microfilmed.

Cape Verde. The National Institute of Technical Investigation was given a seismograph by the U.S.S.R., but was not yet in operation at the time of my visit (1980).

Sierra Leone. The Geological Survey Department has a Chinese seismograph which records six channels, but is used only for engineering purposes. There is no other history of seismic work in Sierra Leone.

Ivory Coast. The Lamto station is well-known, though its equipment may not be the best. The Director said he would be happy to operate any better equipment if provided to them. All their records will be available for microfilming and could be handled by observatory personnel.

Togo. The French Office de la Recherche Scientifique et Technique Outre-mer (ORSTOM) organization used to operate a seismograph in Lome, but the site was noisy and recording was discontinued. The records were sent to the Paris office and presumably can be copied there.

Nigeria. The Physics Department of the University of Ibadan has only one seismometer for classroom work; similarly, the Physics Department of Ahmadu Bello University. The latter would, however, be interested in operating a WWSSN station.

Niger. The Bureau of Geological and Mining Research informed me that there had never been any seismographs operating in Niger.

Zaire. They have three seismographs located at Binza, Karavia, and (Bunia). They have a recurring problem getting photo paper and parts, so recording is intermittent. Copying of their records was not discussed, but they have technical staff who could probably handle it.

Zambia. The Geological Survey Department has a current cooperative program with the University of Finland, but said that any old records would have been obtained by the Goetz Observatory.

Malawi. The Geological Survey Department cooperates with the Goetz Observatory which manages the station at the Blantyre airport. Any old recordings would have been done by the Goetz Observatory.

Mozambique. Four stations are located in Maputo, Nampula, Tete, and Changelane, all operated by a visiting Soviet group at the Meteorological Service. Copying of the records was not discussed. No information was obtained on any earlier recordings.

Botswana. There have never been any seismic stations operated in Botswana. The Geological Survey of Botswana is interested in starting one, especially in the Okavango area. They have been discussing this matter with a Swedish group.

Lesotho. According to the Department of Physics of the University of Lesotho, no seismograph has ever operated in Lesotho.

Swaziland. According to the Department of Physics of University College, no seismograph has ever operated in Swaziland.

Mauritius. The geophysicists at Meteorological Services thought that there was no seismograph in operation, but that there might be some old records lying about. They promised to check, but I have had no further communication from them.

Madagascar. The Tananarive Observatory has had three seismographs in operation since 1970, five since 1981. All records can be copied at the University of Paris where they are sent for archiving.

Tanzania. The Department of Physics at the University of Dar es Salaam reported that seismographs have operated at several places, but none are operating at present. They did not know where the old records might be kept, but promised to find out. I have had no further communication from them.

Seychelles. There is no knowledge of any seismograph in operation in Seychelles.

Ethiopia. This is a well-known WWSSN station whose staff would like to cooperate fully with all international projects, but is presently having logistical problems. They have technical staff who could handle filming project, given the necessary equipment.

Djibouti and Reunion. Volcanological observatories in both of these countries operate several seismographs. All records are sent to the University of Paris, and presumably can be copied there.

3. Conclusion

Much of the information I have given here is sketchy, but as I stated in the beginning, I was not on a seismology mission. There was not enough time at the agency for further inquiry, or time to investigate at other agencies. However, some of the above remarks illustrate what is well known in regard to filming historical seismograms – that it is important to get the records copied as soon as possible before they are lost. Too often, when one inquires about historical seismograms, no one knows where the records are kept.

INDEX

C

D

E

H

I

J

P

Q

R

S

T

U

V

W

Y

Z